# Experimental Reversal of Acid Rain Effects

# Experimental Reversal of Acid Rain Effects

## THE GÅRDSJÖN ROOF PROJECT


*Edited by*

**HANS HULTBERG**

*Swedish Environmental Research Institute, Gothenburg, Sweden*

**RICHARD SKEFFINGTON**

*National Power PLC, UK*



JOHN WILEY & SONS

Chichester • New York • Weinheim • Brisbane • Singapore • Toronto

Baffins Lane, Chichester,
West Sussex PO19 1UD, England

National 01243 779777
International (+44) 1243 779777
e-mail (for orders and customer service enquiries):
cs-books@wiley.co.uk
Visit our Home Page on http://www.wiley.co.uk
or http://www.wiley.com



*Other Wiley Editorial Offices*

John Wiley & Sons, Inc., 605 Third Avenue,
New York, NY 10158-0012, USA

WILEY-VCH Verlag GmbH, Pappelallee 3,
D-69469 Weinheim, Germany

Jacaranda Wiley Ltd, 33 Park Road, Milton,
Queensland 4064, Australia

John Wiley & Sons (Asia) Pte Ltd, 2 Clementi Loop #02-01,
Jin Xing Distripark, Singapore 129809

John Wiley & Sons (Canada) Ltd, 22 Worcester Road,
Rexdale, Ontario M9W 1L1, Canada

***Library of Congress Cataloging-in-Publication Data***

A catalogue record for this book is available from the
Library of Congress

Experimental reversal of acid rain effects : the Gårdsjön Roof Project /
edited by Hans Hultberg and Richard Skeffington.
p. cm.
Includes bibliographical references and index.
ISBN 0-471-96151-5
1. Acid rain — Environmental aspects — Sweden — Gårdsjön Lake
Watershed. 2. Gårdsjön Roof Project. I. Hultberg, Hans.
II. Skeffington, R. A. III. Gårdsjön Roof Project.
QC926.57.S650373 1997
363.738′62′09485 — dc21 97-4585
CIP

***British Library Cataloguing in Publication Data***

A catalogue record for this book is available from the British Library

ISBN 0-471-96151-5

Typeset in 10/12pt Times from the author's disks by Dobbie Typesetting Ltd, Tavistock, Devon
Printed and bound in Great Britain by Bookcraft (Bath) Ltd, Midsomer Norton, Somerset
This book is printed on acid-free paper responsibly manufactured from sustainable forestation, for which at least two trees are planted for each one used for paper production.

# Contents

**BUCKINGHAM PALACE**

Acid rain is recognised as one of the greatest threats to the environment in industrialised countries, not only in Europe and North America but also increasingly in the developing world. It derives from the combustion of fossil fuels, the energy from which powered the Industrial Revolution and continues to underpin present day prosperity. However, the unintended consequence of emissions from fossil fuel burning has been damage to natural ecosystems, particularly in sensitive regions such as Scandinavia. Countries with relatively large acid gas emissions are ever more concerned about its effects and it is now accepted that acid emissions must be reduced. But how fast and how far is still under discussion and it is pleasing to see scientists from many countries working together in the jointly-funded Gårdsjön Roof Project in order to try and throw light on these issues.

I visited the Project in 1993 and was struck by the enthusiasm, commitment and huge effort from all those involved. I hope that the insights obtained at Gårdsjön will play their part in deciding how to meet the great environmental challenge of the next century: how to live in a suitable, yet prosperous, fashion on our small planet. As a contribution to this end I commend this book to you as a record of the results obtained from the Gårdsjön Roof Project in its first few years.

Andrew

HRH Prince Andrew, Duke of York

**VATTENFALL**

Sulphur and nitrogen oxides are spread by wind over large distances, whereby discharges in one country may imply loading on the environment in another. For Sweden, as an example, this is a major problem. In large areas the bedrock is granite and the soils often consist of a thin layer of till. In these areas the soil and the lakes are incapable of coping with more than minor amounts of acidic rain without becoming severely damaged in a long-term perspective. This was one of the reasons for the effects of sulphur emissions being demonstrated more than 25 years ago in Sweden.

Since then, research has provided us with profound knowledge of how complicated the acidification processes are, and how soil and water interact with living organisms. Parallel with the development of technology to restrict emissions, European countries have reached agreement on taking responsibility for protecting nature. During recent years we have noted with great pleasure that this has resulted in decreased sulphur deposition.

It was the hope of Vattenfall that research at Gårdsjön would contribute to explaining the mechanisms of acidification and the processes in the soil that lead to the long-term effect of acidification being reduced. These processes will start to dominate when acidic rain is restricted to below the critical load. The results achieved during the first five years of the project suggest that recovery in soil affected by acidic rain is a very complicated process. Today, we can only guess at what these processes are and how they interact. However, what we have already achieved is documentation on how natural processes respond to reduced acidification pressure and that recovery will take a long time — at present we cannot predict how long.

The results generated by the project demonstrate that the measures and economic inputs invested in the reduction of emissions were essential and that the investments have already started to pay for themselves in the form of recovery processes in acidified soil.

Vattenfall and the Swedish power industry wish to express their warmest thanks to all those who have participated in the research programme at Gårdsjön and to all those who, in different ways, have contributed to the accomplishment of the Covered Catchment Project.

Carl-Erik Nyquist
President and Chief Executive Officer

# Editors' Preface

The Gårdsjön Roof Project was conceived in November 1988. Earlier that year an agreement had been reached, after many years of difficult negotiations, to reduce emissions of acidifying substances from large combustion plants within the countries of the European Community by very substantial amounts. For the 12 countries in the European Community at the time, the reductions, from a base year of 1980, amounted to 23% by 1993, 42% by 1998 and 57% by 2003 for $SO_2$: and for $NO_x$, 10% by 1993 and 30% by 1998. A wider group of European nations had earlier committed themselves to smaller $SO_2$ and $NO_x$ reductions under protocols of the Convention on Long-range Transboundary Air Pollution of UNECE. But it was clear to us in late 1988 that even reductions of this magnitude would not put an end to the debates about acid rain. Would it be enough? Was it being done quickly enough? If not, how fast or far was it necessary to go? We knew these questions would be asked eventually, and that the research information which might provide the answers was accumulating very slowly. And that is why, in view of the long lead-time of research, we initiated the Gårdsjön Roof Project even while these agreements were being signed and ratified.

The Project was planned at a meeting between the Swedish Environmental Research Institute (IVL) and what was then the Central Electricity Research Laboratories of the English electricity generating industry. We were detached from the meeting and given 3 hours to produce a project plan, with costs. We came up with a 5-year programme costing £745 000. The Project has now been running for more than 5 years, and has certainly cost more than £745 000. But the objectives we set for the project have remained valid, and are quoted in Chapter 1. This book marks the completion of 5 years' work on the Gårdsjön Roof Project: it continues because it is clear that recovery from acidification is also going to take a long time in such environments, and because of the opportunities it gives for direct experimental testing of models and concepts about recovery from acidification.

We would like to thank our principal funders for their foresight in supporting this project: the Swedish State Power Board (Vattenfall) and subsequently ELFORSK; the Swedish Environmental Protection Agency; and the (English) Central Electricity Generating Board and subsequently the National Power–PowerGen Joint Environmental Programme. We would also like to thank the staff from IVL, especially Ingvar Andersson and Filip Moldan, who developed and ran the project; and all the other investigators who have contributed their time and energy to the research and to the present book. The clarity of individual chapters was improved by the comments of a number of referees. We hope the result will be of value in providing sound science on which to base rational policy decisions about acidification.

Hans Hultberg and Richard Skeffington
Gothenburg and Swindon, April 1997

# List of Contributors

**Dr B. Ingvar Andersson**, Swedish Environmental Research Institute (IVL), PO Box 47086, S-402 58, Gothenburg, Sweden

**Dr Claus Beier**, Risøe National Laboratory, Environmental Science and Technology Department, PO Box 49, DK-4000 Roskilde, Denmark

**Dr Kevin Bishop**, Swedish University of Agricultural Sciences, Department of Forest Ecology, S-901 83 Umeå, Sweden

**Dr Gunnar Borg**, Chalmers University of Technology (CTH), Department of Geology, S-412 96 Gothenburg, Sweden

**Dr Tor Erik Brandrud**, Norwegian Institute for Water Research (NIVA), PO Box 173, Kjelsås, N-0411 Oslo, Norway

**Dr Eva Brorström-Lundén**, Swedish Environmental Research Institute (IVL), PO Box 47086, S-402 58 Gothenburg, Sweden

**Dr R. Callenberg**, University of Bayreuth, Bayreuth Institute for Terrestrial Ecosystem Research (BITÖK), D-95440 Bayreuth, Germany

**Dr Ulf Carlsson**, Melica Miljökonsulter, Fjällgatan 3 E, S-413 17 Gothenburg, Sweden

**Dr Anna Clemensson-Lindell**, Swedish University of Agricultural Sciences, Department of Ecology and Environmental Research, PO Box 7072, S-750 07 Uppsala, Sweden

**Dr F. Courchesne**, Département de Géographie, Université de Montréal, CP 6128, Succursale Centre-Ville, Montréal (Québec), H3C 3J7, Canada

**Dr A. Dufresne**, Département de Géographie, Université de Montréal, CP 6128, Succursale Centre-Ville, Montréal (Québec), H3C 3J7, Canada

**Dr Martin Ferm**, Swedish Environmental Research Institute (IVL), PO Box 47086, S-402 58 Gothenburg, Sweden

**Dr Bob Ferrier**, The Macaulay Land Research Institute (MLURI), Craigiebuckler, Aberdeen AB9 2QJ, UK

**Dr Reiner Giesler**, Swedish University of Agricultural Sciences, Department of Forest Ecology, S-901 83 Umeå, Sweden

**Dr George Gobran**, Swedish University of Agricultural Sciences, Department of Ecology and Environmental Research, PO Box 7072, S-750 07 Uppsala, Sweden

**Professor Peringe Grennfelt**, Swedish Environmental Research Institute (IVL), PO Box 47086 S-402 58 Gothenburg, Sweden

**Dr Per Gundersen**, Danish Forest and Landscape Research Institute, Skovbrynet 16, DK-2800 Lyngby, Denmark

**Dr Conny Haraldsson**, Chalmers University of Technology and Gothenburg University, Department of Analytical and Marine Chemistry, S-412 96 Gothenburg, Sweden

**Professor Michael Hauhs**, University of Bayreuth, Bayreuth Institute for Terrestrial Ecosystem Research (BITÖK), D-95440 Bayreuth, Germany

**Dr R. Hoch**, University of Bayreuth, Bayreuth Institute for Terrestrial Ecosystem Research (BITÖK), D-95440 Bayreuth, Germany

**Dr René Hogervorst**, Free University, Amsterdam, Faculty of Biology, de Boelelaan 1087, NL-1081 HV Amsterdam, The Netherlands

**Professor Hans Hultberg**, Swedish Environmental Research Institute (IVL), PO Box 47086, S-402 58, Gothenburg, Sweden

**Dr Magne Huse**, Norwegian Forest Research Institute (NISK), PO Box 61, N-1432 Ås, Norway

**Dr Åke Iverfeldt**, Swedish Environmental Research Institute (IVL), PO Box 47086, S-402 58 Gothenburg, Sweden

**Dr Janne Kjønaas**, Norwegian Forest Research Institute (NISK), PO Box 61, N-1432 Ås, Norway

**Dr Frode Kroglund**, Norwegian Institute for Water Research (NIVA), Sörlandsavdelning Televeien 1, N-4890 Grimstad, Norway

**Dr Holger Lange**, University of Bayreuth, Bayreuth Institute for Terrestrial Ecosystem Research (BITÖK), D-95440 Bayreuth, Germany

**Dr Ying-Hua Lee**, Swedish Environmental Research Institute (IVL), PO Box 47086, S-402 58, Gothenburg, Sweden

**Dr Gunnar Lischeid**, University of Bayreuth, Bayreuth Institute for Terrestrial Ecosystem Research (BITÖK), D-95440 Bayreuth, Germany

**Dr C. Löfgren**, Swedish Environmental Research Institute (IVL), PO Box 47086, S-402 58, Gothenburg, Sweden

**Dr Ulla Lundström**, Department of Natural Science, University of Karlstad, S-651 88 Karlstad, Sweden

**Dr Benny Lyvén**, Department of Analytical and Marine Chemistry, Chalmers University of Technology and Gothenburg University, S-412 96 Gothenburg, Sweden

**Dr Filip Moldan**, Swedish Environmental Research Institute (IVL), PO Box 47086, S-402 58, Gothenburg, Sweden

**Dr Carl-Magnus Mörth**, University of Stockholm, Department of Geology and Geochemistry, S-106 91 Stockholm, Sweden

**Dr John Munthe**, Swedish Environmental Research Institute (IVL), PO Box 47086, S-402 58, Gothenburg, Sweden

**Dr Lars Nyberg**, Department of Hydrology, Uppsala University, Västra Ågatan 24, S-753 09 Uppsala, Sweden

**Dr Per Holm Nygaard**, Norwegian Forest Research Institute (NISK), PO Box 61, N-1432 Ås, Norway

**Dr Ulf Nyström**, University of Gothenburg, Department of Geology, Gulhedsgatan 5A, S-413 81 Gothenburg, Sweden

**Dr Per Östlund**, Swedish Environmental Research Institute (IVL), PO Box 21060, S-100 31 Stockholm, Sweden

**Dr Viveka Palm**, Swedish Environmental Research Institute (IVL), PO Box 21060, S-100 31 Stockholm, Sweden

**Dr Hans Persson**, Swedish University of Agricultural Sciences, Department of Ecology and Environmental Research, PO Box 7072, S-750 07 Uppsala, Sweden

**Dr Lennart Rasmussen**, Risøe National Laboratory, Environmental Science and Technology Department, PO Box 49, DK-4000 Roskilde, Denmark

**Dr Allan Rodhe**, Department of Hydrology, Uppsala University, Västra Ågatan 24, S-753 09 Uppsala, Sweden

**Dr Jan Seibert**, Uppsala University, Department of Hydrology, Västra Ågatan 24, S-753 09 Uppsala, Sweden

**Dr Richard Skeffington**, National Power Research and Engineering, Windmill Hill Business Park, Whitehill Way, Swindon, Wiltshire SN5 6PB, UK

**Professor Arne Stuanes**, Agricultural University of Norway, Department of Soil and Water Sciences, PO Box 5028 N-1432 Ås, Norway

**Professor Harald Sverdrup**, University of Lund, Department of Chemical Engineering II, PO Box 124, S-221 00 Lund, Sweden

**Dr Peter Torssander**, University of Stockholm, Department of Geology and Geochemistry, S-106 91 Stockholm, Sweden

**Dr H. A. Verhoef**, Free University, Amsterdam, Faculty of Biology, de Boelelaan 1087, NL-1081 HV Amsterdam, The Netherlands

**Dr Per Warfvinge**, University of Lund, Department of Chemical Engineering II, PO Box 124, S-221 00 Lund, Sweden

**Dr Tonie Wickman**, Department of Land and Water Resources, Civil and Environmental Engineering, Royal Institute of Technology, S-100 44 Stockholm, Sweden

**Dr R. F. Wright**, Norwegian Institute for Water Research (NIVA), PO Box 173, Kjelsås, N-0411 Oslo, Norway

**Dr H. R. Zoomer**, Free University, Amsterdam, Faculty of Biology, de Boelelaan 1087, NL-1081 HV Amsterdam, The Netherlands

# 1 Recovery from Acidification: the Background to the Covered Catchment Experiment at Gårdsjön

R. A. SKEFFINGTON, H. HULTBERG, B. I. ANDERSSON & K. H. BISHOP

This chapter places the Roof Experiment at Gårdsjön into context, and sets out its aims and objectives. Emissions of $SO_2$ are declining, but it is not clear how quickly they need to decline to meet ecological objectives. Increasing the rate of emission reduction is expensive, but may be necessary to promote recovery. The definition of recovery, and the appropriate biological and chemical criteria to use, are discussed. Current knowledge of recovery is briefly reviewed, with particular attention to the pioneering roof experiment: the RAIN Project. As a result of these considerations, the aim of the Gårdsjön Roof Project was defined as: "to assess the rate and direction of response of a forested catchment to a reduction in sulphate, nitrate and mercury deposition".

## INTRODUCTION: WHY IS RECOVERY AN IMPORTANT ISSUE?

### Trends in emissions

Acidification of the environment is a major environmental problem in Europe and North America, and it has the potential to become a major problem in Asia (Galloway, 1995). The complex syndrome of effects that can be attributed to acid deposition includes acidification and loss of fertility of soils; changes in soil fauna and flora leading to reductions in ecosystem functions; changes in species composition of natural communities; forest declines and yield reductions; and surface water acidification leading to substantial changes in populations of aquatic organisms, especially loss of fish populations. To ameliorate these problems, various political initiatives are under way aimed at reducing the emission of acidifying substances to the atmosphere. The most important of these is the Convention on Long-Range Transboundary Air Pollution (CLRTAP) of the United Nations Economic Commission for Europe (UNECE). This was signed by member states in 1979, and given more specific effect by various "Protocols", including the Sulphur (S) Protocol of 1985 and the $NO_x$ Protocol of 1988 (see e.g. UNECE, 1994; Hordijk,

*Experimental Reversal of Acid Rain Effects: The Gårdsjön Roof Project.*
Edited by Hans Hultberg and Richard Skeffington. 

1995). Sulphur emissions in Europe are currently controlled by the second S Protocol, signed in Oslo in 1994. This is expected to reduce S emissions relative to a 1980 base year by 51% by the year 2000, 56% by 2005 and 61% by 2010. A second $NO_x$ Protocol is being negotiated: this is expected to mandate a reduction in $NO_x$ emissions, though it is unlikely to be as large as that for S. The $NO_x$ Protocol may also include an attempt to reduce emissions of $NH_3$.

The protocols to the CLRTAP and related initiatives have already had some effect on emissions. Thus $SO_2$ emissions in Europe declined from 56.3 Mt in 1980 to 35.6 Mt in 1992, i.e. by 37% (Tuovinen *et al.*, 1994). Emissions of $NO_x$ and $NH_3$ hardly changed over the same period, being around 6.8 Mt of N in the case of $NO_x$ and 6.5 Mt of N for $NH_3$ (Tuovinen *et al.*, 1994), though the latter value is very uncertain. The lack of trend in $NO_x$ emissions disguises a reduction in emission from large boilers—for countries which were European Community members in 1988 this had to amount to 10% of 1980 emissions by 1993 and 30% by 1998 under the Large Combustion Plant Directive of the EC. This downward trend has however been offset by the growth of road transport and thus emissions from motor vehicles.

## The importance of the rate of recovery

Reductions in acid deposition should give biological benefits, but there are a number of scientific questions related to policy which need to be addressed. Among these are: (1) what is the end point?—in what state should ecosystems be when it can be said that emission reduction is sufficient; (2) what degree of emission reduction will be necessary to achieve this?; and (3) *how quickly* will recovery occur for given emission reduction scenarios? The critical loads approach (see below) has developed as a methodology for answering the second question, and in doing so has had to assume an answer to the first question. But the third question is also very important for emission control policy. This is because the cost of emission reduction increases with the rate of implementation. There are several reasons for this, but the major ones are:

1. The cost of fitting an end-of-pipe technology such as Flue Gas Desulfurization (FGD) to an existing installation is much higher than designing it into a new one. Retrofitting usually involves moving existing buildings and underground structures, longer duct lengths, changing ash handling systems, etc. For instance, the authoritative NAPAP study (South, 1990) estimated the capital cost in 1988 US\$ of fitting one type of FGD to a new 500 MW unit to be \$202 $kW^{-1}$, whereas retrofitting would cost \$222–\$389 $kW^{-1}$ dependent on site difficulty.
2. Research is in progress into cleaner technologies (such as gasification) which may have the potential in a few years time to deliver emission reductions at much lower cost than at present.
3. In standard forms of economic analysis, future expenditure is *discounted* (i.e. it has a lower value) relative to present expenditure. Thus a cost–benefit analysis of the control of acidifying emissions would show a lower cost if expenditure could be spread over a longer period.

If emission control can be delayed, the same abatement could be achieved at lower cost. Offsetting this, however, is the increased environmental damage which would probably ensue. This increased damage can take two forms. One is that acidified systems will recover more slowly than they otherwise would. The other is that emission reductions will not be sufficient to reverse the progress of acidification for some ecosystems, and thus the symptoms of acidification may intensify or spread to areas at present apparently unaffected. The Gårdsjön project addresses the first of these situations directly, though modelling should allow some of the results to be extended to the second. To provide a basis for political discussion and for balancing the costs against the benefits, it is necessary to know how quickly ecosystems will recover for a given degree of emission reduction. Many points of view about the right course of action are possible (see Chapter 24), but all discussion is groping in the dark unless this question can be answered. To answer it was one of the main aims of the Gårdsjön Roof Project.

**Recovery and critical loads**

As the second Sulphur Protocol and forthcoming $NO_x$ Protocol are based on the critical load concept, it is appropriate to consider how this relates to recovery issues. The critical load is defined as "a quantitative estimate of an exposure to one or more pollutants below which significant harmful effects on specified sensitive elements of the environment do not occur according to present knowledge" (Nilsson and Grennfelt, 1988). It is in other words a threshold deposition below which deleterious effects should not occur, and any ecosystem in which the critical load is not exceeded is said to be "protected". The ultimate aim is that deposition should be reduced to the extent that critical loads are not exceeded anywhere in Europe (see Bull, 1991). This concept of "protection" implies that actually *reversing* the damage already wrought by acidification is not an explicit goal of the UNECE negotiations. However, the extent to which reversibility is implicitly included depends on how the critical load is defined. A critical-load-based policy could potentially have several possible aims (Figure 1). Originally (Nilsson and Grennfelt, 1988), critical loads aimed at stabilizing the current situation in the case of soils (P-CL1 in Figure 1), and returning to a pre-industrial state in the case of waters (P-CL2). As the methodology has developed, however, the aim has changed so as to be to protect certain species of terrestrial plants or aquatic animals in the long term (see Bull, 1991; Hornung and Skeffington, 1993; Posch *et al.*, 1995). The calculated critical load then depends on the choice of species and the chemical criteria used to represent them in critical load calculations, and how appropriate these choices are for the areas of Europe to which they are applied. If a sensitive species is chosen (e.g. Species 1 in Figure 1) and if the soil or water has acidified to the extent that this species is not viable, then the calculated critical load will include an element of reversal of acidification (P-CL3). If, however, the criterion species is calculated to be unaffected at present, then the calculated critical load may allow further acidification (P-CL4). Irrespective of whether reversibility is implicitly included or not, the critical load approach as currently formulated does not give any indication about the appropriate rate of emission reduction (e.g. P-CL3 or P-CL5 in Figure 1 are both compatible with protecting Species 1; P-CL2 or P-CL6 are both compatible with returning to a

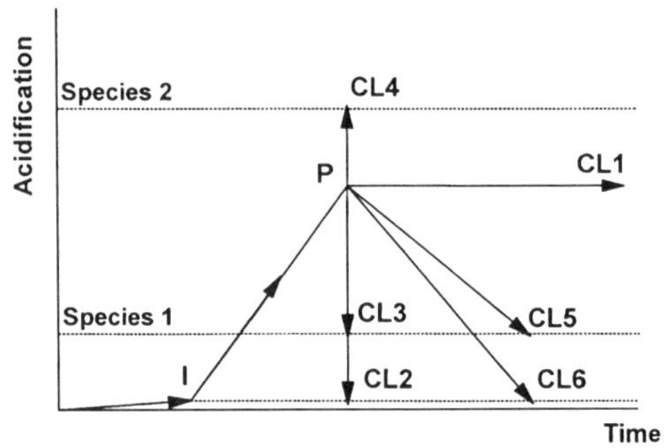


**Figure 1.1**. Schematic diagram of possible goals for the critical load approach, as applied to emission control policy. It is assumed that the slow rate of natural acidification was accelerated dramatically at the time of the Industrial Revolution (I), until the present state (P) was reached. The critical load could be designed to maintain the present state (CL1); return to a pre-industrial state (CL2); protect Species 1, which is relatively sensitive (CL3); or protect Species 2, which is less sensitive and would allow acidification to increase compared with the present state (CL4). At present, there are no criteria for judging how rapidly these critical loads are to be achieved: CL5 and CL6 give the same protection as CL3 and CL2 respectively, but would be less expensive to implement as they occur further into the future

pre-industrial state). In view of the usefulness of the critical load concept in policymaking so far (e.g. Hordijk, 1995), attempts are currently being made to incorporate a dynamic element into the methodology, which would enable such issues to be addressed. The results of the Gårdsjön Project should provide a vital input into this.

## WHAT IS MEANT BY RECOVERY?

### Biological recovery

Recovery must be defined before it can be quantified. Dise *et al.* (1994) discussed six possibilities in relation to biological recovery of aquatic ecosystems:

1. Return to the "pre-acidification" community structure as defined by species composition and abundance.
2. Return of "preacidification" species.
3. Return to a community showing normal functions.
4. Return of key "indicator species".
5. Presence of a healthy fish population.
6. A population of indicator species whose biological status may be predicted from empirical models without needing reference to the rate or intensity of acid deposition.

All these possibilities require some knowledge of what ecosystems ought to be like, or used to be like before they were acidified, at different levels of detail. A mixture of practical considerations and the desire to conserve whole ecosystems rather than just fish led Dise *et al.* (1994) to propose the return of healthy populations of sensitive key species, including fish where appropriate, as the criterion for recovery.

Criteria involving the return of key species have not been proposed for terrestrial systems (though lichen populations, for instance, could be used in this way). This may be because restoring a climax terrestrial community, such as a forest, may take much longer, or it may be because elimination of conspicuous populations of terrestrial organisms is not as common an event as elimination of fish populations in sensitive waters. Whatever the reason, recovery of terrestrial systems is most often discussed in the context of reduction in measures of forest damage rather than restoration of terrestrial communities.

The pragmatic definitions of recovery adopted by Dise *et al.* (1994) do not preclude some changes from the pre-acidification ecosystem. Hauhs (1994) suggested that re-establishment of a past "ecosystem state" is not possible merely by altering boundary conditions such as atmospheric deposition, because ecosystems are dynamic entities which can change irreversibly. Restoring a past state would require direct intervention, and perfect knowledge of what that state had been. Although ecosystems in their natural state fluctuate in composition, driven by random or cyclic changes in environmental conditions, the *presence* of key species is likely to be a rather more stable parameter and an appropriate biological criterion for recovery.

## Chemical criteria for recovery

The biological effects of acid deposition are mediated through essentially chemical processes such as reductions of pH or increases of inorganic aluminium concentrations in both soils and surface waters. Therefore it seems reasonable to calculate biological recovery from chemical parameters, which are in any case the entities which are directly manipulated by changing atmospheric deposition. It should not be assumed from this that biological recovery will follow chemical recovery automatically and immediately. Purely biological processes such as dispersal, reproductive behaviour, predator–prey relationships, or inter-specific competition are likely to have a profound effect on recovery of species. Direct intervention may be required if it is desired to speed the return to a pre-acidification ecosystem: fish may need to be stocked, for example.

The choice of a chemical criterion for recovery must be made carefully and critically. Acidification has apparently affected aquatic organisms by lowering surface water pH and increasing concentrations of aluminium and perhaps other toxic metals (Sayer *et al.*, 1993; Rosseland and Stuarnes, 1994). Calcium and perhaps other base cations can ameliorate this toxicity to some extent (Brown and Sadler, 1989). The reactions of these ions should therefore provide the best chemical criteria for recovery. These effects are interactive and complex (varying between different life stages of fish, for example) and thus hard to calculate as a single number to represent recovery. There is therefore a tendency to use acid neutralizing capacity (ANC) as the chemical criterion for recovery of acid waters. ANC can be defined as $(Na^+ + K^+ + Ca^{2+} + Mg^{2+}) - (SO_4{}^{2-} + NO_3{}^- + Cl^-)$ where the variables are ionic concentrations expressed in equivalence units. It is relatively easy to measure and calculate, and correlates well with fish status in synoptic surveys (e.g. Henriksen *et al.*, 1989). However, fish and other organisms do not respond *directly* to ANC, but to some of its constituent ions and other species with which it is correlated such as

$Al^{n+}$. ANC is merely correlated with the physiological ions, and these correlations may not hold for the recovery process. Thus ANC is positively correlated with pH and $Ca^{2+}$, but during recovery ANC and pH can be expected to rise in most circumstances, whereas $Ca^{2+}$ concentrations may fall (see below). There is some evidence that ANC should be used with caution. Cosby *et al.* (1994) used discriminant analysis of Norwegian lake survey data to show that ANC was slightly inferior to pH, $Ca^{2+}$ and Al as a fish status predictor, and similar results were obtained from the RAIN Project (see below). Sadler (1994) produced a discriminant function analysis of Norwegian lake data which can be adapted as a fishery status predictor, and this holds promise as a method for more accurate recovery predictions.

In terrestrial ecosystems, the chemical recovery criterion most often adopted has been the $Ca^{2+}:Al^{3+}$ ratio, or sometimes $(Ca^{2+} + Mg^{2+} + K^{+}):Al^{3+}$ ratio, in the soil solution (Sverdrup and De Vries, 1994). This is thought to be related to plant growth (Sverdrup and Warfvinge, 1993), though the link is not well established and the concept has been criticized (Högberg and Jensen, 1994). This criterion also suffers from the weakness that it is very difficult to make accurate predictions of inorganic Al concentrations.

## WHAT IS KNOWN ABOUT RECOVERY?

This section briefly summarizes current knowledge and hypotheses about recovery from acidification. A discussion of modern theoretical understanding of recovery is followed by a survey of empirical data (i.e. obtained from environmental monitoring) and of results from experimental manipulations. The theory is of course partly derived from the observations, but placing the discussion of theory first enables the data to be understood more readily. More detailed reviews are given by Wright and Hauhs (1991), Skeffington and Brown (1992), Steinberg and Wright (1994) and Bishop and Hultberg (1995) among others.

### The theory of recovery from acidification

The book by Reuss and Johnson (1986) provides a convenient summary of current biogeochemical theories in relation to acidification. In recovery, the same mechanisms operate in reverse, but some new issues are raised. As S deposition decreases, the flux of the mobile $SO_4^{2-}$ ion through catchments will begin to decline. To maintain charge balance, the flux of cations must decline as well. These cations may be "acidic" cations ($H^+$ or $Al^{n+}$) in which case water quality for most organisms will improve, or "base" cations ($Ca^{2+}$, $Mg^{2+}$, $Na^+$, $K^+$), in which case it may deteriorate. In practice, concentrations of both base and acidic cations will decrease, the ratio (change in non-marine $SO_4^{2-}$/change in non-marine base cations) being the "F-factor" of Henriksen (1984). The effect on fish status is a complex function of the relative and absolute concentrations of the acid and basic cations, but the reduction in $SO_4^{2-}$ flux may thus lead to a relatively rapid improvement in water quality. How rapid depends crucially on the extent and time-scale of release of catchment S (see below).

Skeffington and Brown (1992) also identify a slow phase of water quality improvement. There are several possible sources for the base cations being leached out of the soil, the major ones being atmospheric deposition; mineral weathering; release from biomass; and the soil cation exchange complex. If more base cations are leached than are produced by the first three of these sources, the soil cation exchange complex loses base cations—i.e it reduces in base saturation or acidifies. If base cation leaching can be reduced to a value which is less than the inputs, then base saturation will increase, improving soil quality which will in turn improve water quality. Unfortunately the cation exchange complex is normally large compared with weathering fluxes (see calculations in Bishop and Hultberg, 1995). This implies that recovery of soils will be slow, taking decades or centuries, and that if improvement in water quality has to depend on improvement in soil quality, that will be slow as well.

**Assumptions in acidification models**

Thus current theory, as embodied in hydrochemical models such as the MAGIC Model (Cosby *et al.*, 1985), implies that recovery will in most cases be slow unless there is a rapid response to decreased $SO_4^{2-}$ fluxes (e.g. Skeffington and Brown, 1992). However, there are considerable reservations about the predictions of these models. One is that they were developed to explain acidification, not recovery (Hauhs, 1994), and different processes may become more important in reversal. Bishop and Hultberg (1995) detail three categories of more specific reservations:

1. Unsatisfactory representation of certain processes which are modelled, such as aluminium solubility.
2. Absence of other processes from the models, most notably biological controls on soil chemistry.
3. Failure to account for spatial distribution of processes within the catchment.

These are detailed below. Empirical and experimental evidence on these points is described in the next two sections.

1. No model has yet been able to model changes in Al concentrations adequately. Current models assume equilibrium with $Al(OH)_3$, giving a cubic relationship between Al and $H^+$ in streamwater, which is not observed in the field (Neal *et al.*, 1989). This relationship implies that Al concentrations will decline rapidly with decreasing $SO_4^{2-}$, but there has been no systematic investigation of the available data aimed at testing this hypothesis. Such an investigation would be worth while, although data scattered through the literature suggest that Al concentrations do not decline rapidly with decreasing $SO_4^{2-}$. This was true in the RAIN Project (see below); in long-term data from calibrated Norwegian catchments (Kirchner and Lydersen, 1995), where $SO_4^{2-}$ in runoff declined significantly but inorganic Al did not decline at all; and in a survey of Galloway lakes (Wright *et al.*, 1994) which showed a 43% reduction in $SO_4^{2-}$ concentration, but only a 34% reduction in total Al. However, there are confounding factors such as concomitant soil acidification in many of these studies.

   Another important process is $SO_4^{2-}$ retention and release in soils, but models differ considerably in the way it is represented. Most soils contain quite a large store of S, enough if mobilized in the form of $SO_4^{2-}$ to maintain $SO_4^{2-}$ concentrations in drainage water for many years and thus to delay recovery. But it is not clear under what circumstances the rate of mobilization will be sufficiently great to affect recovery significantly. General principles suggest that soils will vary considerably in their reactions. Some soils, typically acid soils containing large amounts of Al and Fe hydrous oxides, are strong $SO_4^{2-}$ adsorbers (Chao *et al.*, 1964; Reuss and Johnson, 1986). This adsorbtion is at least partly reversible, so that if $SO_4^{2-}$ input decreases, the soil becomes a net producer of $SO_4^{2-}$. Harrison *et al.* (1989) in a study of 36 forest soils, found complete reversibility in less than 25%, and that on average, 36% of $SO_4^{2-}$ was irreversibly absorbed. Of commonly used hydrochemical models, some like MAGIC and SMART treat $SO_4^{2-}$ absorbtion/release as a completely reversible process, whereas others such as SAFE assume that $SO_4^{2-}$ is an inert "tracer" anion. Organically bound sulphur is normally a larger pool than adsorbed inorganic S in forest soils. Mitchell *et al.* (1992) suggested that only a small fraction of organic S in most soils is active in S cycling, and thus in control of $SO_4^{2-}$ concentrations in output waters. In wetland soils $SO_4^{2-}$ reduction is an important absorption mechanism, but this S tends to stay in the soil unless the wetland is allowed to dry out (Brown, 1985). None of the models treat the reactions of organic S. Chapters 6, 8 and 9 in the present book discuss this key issue in relation to the Gårdsjön catchments.

2. Biological processes will have a profound effect on chemical fluxes in catchments (e.g. Likens *et al.*, 1977) but are difficult to predict and hence tend to be neglected by models. Reduction in acid input might be expected to affect the activities of soil organisms, leading to changes in nutrient cycling and weathering rates, though there is so far little direct experimental evidence that this occurs. Boxman *et al.* (1995) found that reducing S and N input to a Scots pine stand by means of a roof reduced decomposition rates, though it increased the species diversity of microarthropods. Chapters 14–17 in the present book deal with changes in biological processes observed in the Gårdsjön roof catchment.
3. The spatial distribution of processes in real catchments may be an important issue in modelling recovery. Models have been developed to simulate homogeneous blocks of soil, treating them effectively as stirred tanks. However, the soils which control the acidity of runoff during spates, which may be what is significant biologically, are only a small fraction of the soils within the catchment. Perhaps then it is only these soils which need to recover to affect runoff water quality. Neither cation exchange capacity nor weathering rate are evenly distributed within the soil profile or horizontally over a catchment area. The hydrological transport system which links different areas has large spatial differentials in flux. The potential which these distributions have to alter the rate of response to changes in input chemistry is just beginning to be explored (see Chapters 5 and 21).

Current theories of acidification thus give some clues about what to expect as deposition is reduced, but leave a great deal of uncertainty about the outcome. Observational and experimental evidence is thus essential.

**Empirical data**

Emissions of $SO_2$ have now been generally declining in many parts of Europe and North America for 15–20 years, and hence deposition of $SO_4^{2-}$ has also declined.

The effects of this decline are observable in various long-term datasets (see Wright and Hauhs, 1991, for a review).

The most striking evidence of recovery, and the only evidence for recovery of fisheries after a reduction in S deposition, has been obtained round the Sudbury Smelter in Ontario, Canada (Gunn and Keller, 1990). Here a massive reduction in $SO_2$ emissions from a collection of large point sources has led to substantial water quality improvement in the surrounding lakes, accompanied by some recovery of fish populations. However, as argued by Skeffington and Brown (1992), the Sudbury region was acidified not because it is very sensitive, but because of the very large S deposition. The relatively high weathering rate of the Sudbury soils causes a rapid recovery once this deposition is reduced, and the situation thus cannot be taken as representative of more sensitive areas with lower weathering rates.

The largest dataset relating to recovery is the "1000-lake" survey of Norwegian lakes (Henriksen *et al.*, 1989). The chemical composition of these lakes was measured in 1974–75, and re-measured in 1986, and fish status was measured at a similar interval. Sulphate, $Ca^{2+}$ and $Mg^{2+}$ declined over the 10 year period, $H^+$ also declined but to a much smaller extent, but total Al increased. Complicating the picture was an increase in $NO_3^-$. Fish status declined markedly, with 48% of lakes which sustained a fish population in 1971–75 losing all fish by 1986, and a further 38% suffering a reduction in fish population (Henriksen *et al.*, 1989). As pH generally increased, this loss of fish was presumably due to the decline in $Ca^{2+}$ and increase in Al, although it may also be an effect of a lagged response of fish populations to earlier chemical conditions.

A more recent dataset showing chemical recovery was obtained by measurements of 50 lakes in Galloway, south-west Scotland, in 1979 and again in 1988 (Wright *et al.*, 1994). The data show a very large reduction in mean non-marine $SO_4^{2-}$ concentration, from 120 to 56 $\mu$eq $L^{-1}$, and a decrease in $NO_3^-$ concentration from 19 to 9 $\mu$eq $L^{-1}$. Balancing the reduction of 76 $\mu$eq $L^{-1}$ in strong acid anions is a reduction of 66 $\mu$eq $L^{-1}$ in non-marine base cations, an F-factor (Henriksen, 1984) of 0.87. There was no significant change in pH, though it is noticeable that the pH of all 11 lakes with a 1979 pH less than 4.8 had risen by 1988. Mean ANC rose by 10 $\mu$eq $L^{-1}$, and mean total Al fell from 168 to 110 $\mu$g $L^{-1}$. Fish status was not measured.

In contrast to these two surveys, which cover areas which are rather similar physiographically and geologically, Hauhs (1990) described a catchment study at Lange Bramke in the Harz Mountains of Germany on much deeper podzolic soils. Here $SO_4^{2-}$ in streamwater responded very slowly to changes in $SO_4^{2-}$ input, so that in spite of a 20–30% reduction in S deposition since 1986, there has been no change in $SO_4^{2-}$ output (Hauhs, 1994). This may be due to the large $SO_4^{2-}$ absorbtion capacity of the soil: clearly such catchments will respond much more slowly to changes in S deposition.

The conclusion from the Norwegian and Galloway studies appears to be that reductions in S deposition lead to a similar reduction in $SO_4^{2-}$ concentration in surface waters, but that much of this decline is balanced by reduced base cation concentrations. Increases in pH have been small. One difficulty with these studies is that the importance of identifying the aluminium species was not realized before

1980, so it is not possible to evaluate the extent to which Al toxicity has declined. Reduction of Al toxicity is likely to be a key to biological recovery. As a basis for prediction, the "1000-lake" and Galloway studies have the disadvantage that they are snapshots at two points in time, between which deposition and other environmental conditions have fluctuated considerably. More controlled studies are clearly necessary to provide a better basis for predictive modelling, but the synoptic surveys are essential to ground those results in the real world.

### Experimental evidence: early work

One early manipulation which reduced acid inputs to a whole catchment was carried out by Hultberg in 1984 (Hultberg and Grennfelt, 1992). The spruce forest on one of the micro-catchments near Lake Gårdsjön was clear-cut, effectively halving $SO_4^{2-}$ deposition by reducing the dry deposition component. In response, $SO_4^{2-}$ draining the clear-cut catchment was reduced by 50% within 2 years, and mean pH rose from 4.3 to 4.7. This suggests that rapid recovery of this type of catchment is indeed possible, but clear-cutting results in changes to processes other than S deposition which complicate the interpretation of the results.

### Experimental evidence: the RAIN Project

The idea of shielding a whole catchment from acid deposition and replacing it with clean rain derives from the RAIN (Reversing Acidification In Norway) Project (see Wright *et al.*, 1988, 1993). The Project began in 1983 with treatment starting in June 1984. Originally planned as a 5-year project, it finally terminated at the end of 1994. Three small catchments were instrumented at Risdalsheia in southern Norway in terrain typical of the surrounding upland area, with shallow, acidic, highly organic soils and sparse vegetation cover. Roofs were constructed over two, one (Kim) having the ambient acid deposition replaced by clean rain and the other (Egil) being a roofed control plot. The third plot (Rolf) was used as an unroofed control. In addition to the acid exclusion experiment at Risdalsheia, an acid addition experiment at Sogndal further north in Norway confirmed the effect of depositing acidity on a relatively pristine area. Because the RAIN Project was the pioneer roof experiment, and the only one to date apart from Gårdsjön roof to cover a whole catchment, it is worth considering the results in some detail. The results from the RAIN Project have also altered perceptions about acidification and recovery quite markedly as they have emerged.

Some of the major results from the RAIN Project are shown in Figure 1.2. This shows annual mean concentrations in runoff over the entire period of the experiment, from the roofed experimental and control catchments. Since treatment began on 13 June 1984, wet, non-marine $SO_4^{2-}$ deposition has been essentially completely eliminated on the treated catchment (Kim), though dry deposition has still continued to some extent. Using the RAIN Project team's own estimates of dry deposition to the roofed catchments, the overall reduction in non-marine $SO_4^{2-}$ deposition has varied year-on-year between 60 and 83% (Kim versus the roofed, acid catchment Egil). Dry deposition of $SO_2$ and particulate $SO_4^{2-}$ were estimated on the basis that the roofs did not affect it: it was therefore probably somewhat lower than estimated.

Non-marine $SO_4^{2-}$ took 8 years to reach a steady state after the start of treatment (Figure 1.2A). This presumably represents $SO_4^{2-}$ from stores in the catchment adjusting to the new input concentration. Examination of the flux data shows that inputs and outputs were approximately in balance from 1992 onwards (assuming dry deposition was calculated correctly). Between roof construction and 1992, the catchment produced approximately 180 meq $m^{-2}$ of $SO_4^{2-}$, which is approximately equivalent to the adsorbed and water-soluble S pool at the beginning of the experiment (Wright *et al.*, 1988). The *total* store of S before the experiment started was however about 1000 meq $m^{-2}$. If the response of $SO_4^{2-}$ at Risdalsheia is representative of catchments as a whole, these are important results, as they imply that significant $SO_4^{2-}$ leakage will last 5–10 years, even though there is enough stored S to increase output $SO_4^{2-}$ concentrations by 10 $\mu$eq $L^{-1}$ for about 100 years. The response of $NO_3^-$ is much quicker than $SO_4^{2-}$ (Figure 1.2B), falling well below the concentrations in the control catchment within a year. Though the catchment was previously N-saturated in the sense that it was probably leaking $NO_3^-$ before the roof was constructed, the store of organic N is apparently not released from the catchment following the decline in N deposition.

Non-marine strong base cations are reduced to concentrations close to zero very quickly (Wright, 1994), but of more relevance to water quality is the output of (total) $Ca^{2+}$, which declines from 13 $\mu$eq $L^{-1}$ in 1985 to 6 $\mu$eq $L^{-1}$ in 1991 (Figure 1.2C), in accordance with the mobile anion theory (see above). The decline in $Ca^{2+}$ runs in parallel with $SO_4^{2-}$, but at a much lower absolute value, the F-value ($\Delta SO_4^{2-}/\Delta Ca+Mg$) being between 0.1 and 0.2.

The response of runoff pH provided one of the major surprises of the RAIN Project (Figure 1.2C). The increase of pH in the "clean" treatment compared to the ambient treatment was very small until 1990, after 5-6 years of treatment. The major reason for this can be seen in Figure 1.2D. As $SO_4^{2-}$ concentrations declined, the concentration of dissociated organic anions (A-) increased, although the overall concentration of dissolved organic C (TOC) did not alter (Figure 1.2E). "Weak" organic acid anions were apparently dissociating to compensate for the reduction in strong acid anion concentrations. Although it was realized that this would happen, the extent to which organic acid dissociation has compensated for the decline in strong acid concentrations surprised the investigators. As an illustration of this, Wright *et al.* (1986) predicted after 1 year that "pH values may ultimately not go above 4.6 to 5 in runoff from the Kim catchment"—i.e. 10–25 $\mu$eq $L^{-1}$ $H^+$. In fact, after 10 years, the highest annual mean pH was 4.28 (53 $\mu$eq $L^{-1}$). Wright (1989) subsequently predicted the correct end-point pH in Kim runoff on the basis of 5 years' data, when the behaviour of organic acids had become clearer.

Another ion which has not behaved as expected is inorganic aluminium (Figure 1.2F). After a small decline in the first year, its concentration remained essentially unaffected for a further 5 years of treatment, though the conventional view would be that it should decline faster than $Ca^{2+}$ or $H^+$. Wright *et al.* (1990) used the MAGIC Model to predict a decline in $Al^{3+}$ from 10 to 5 $\mu$eq $L^{-1}$ between 1985 and 1987. The decline did take place, but not for a further 5 years: between 1990 and 1992. The

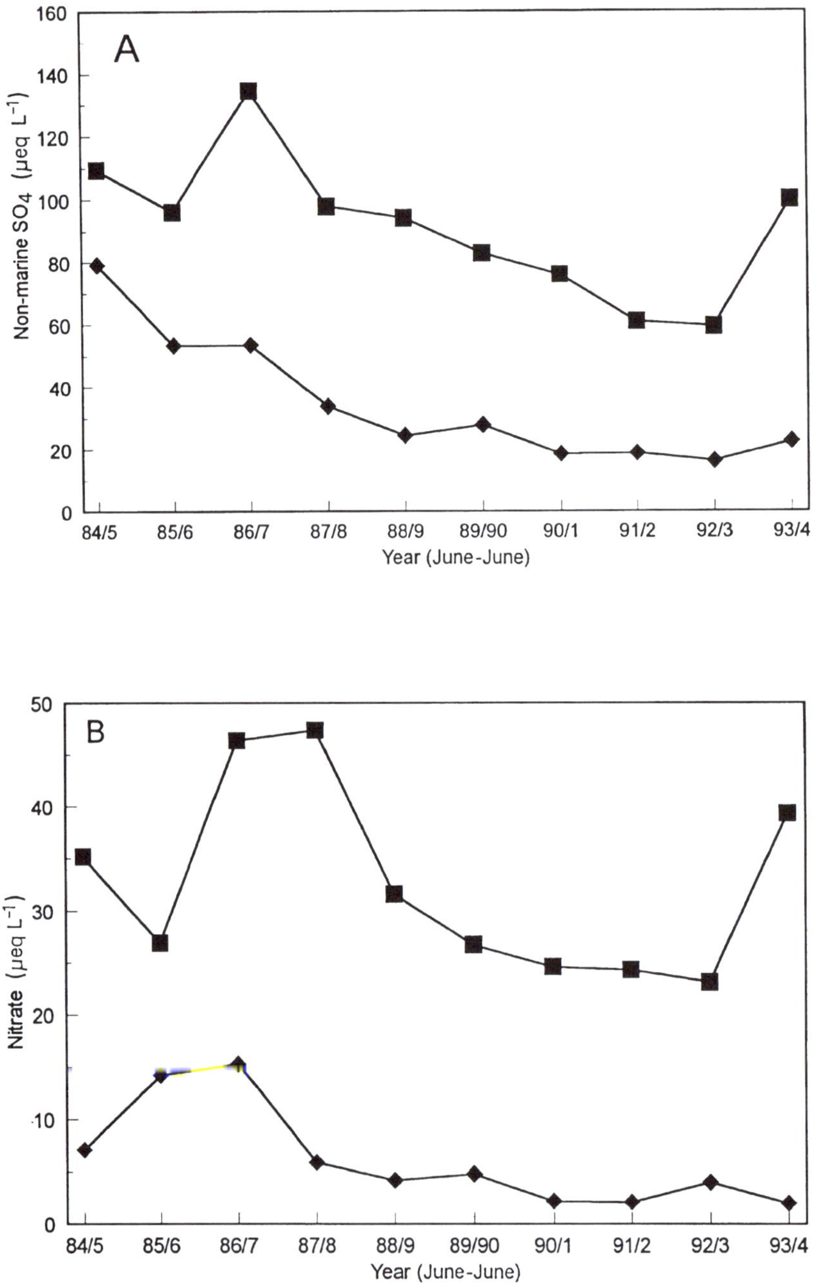


**Figure 1.2**. Volume-weighted annual mean concentrations of major species in runoff from two of the catchments at Risdalsheia (RAIN Project). Data are shown from the ambient roofed catchment "Egil" (squares) and the clean rain catchment "Kim" (diamonds). Note scales differ between graphs. Drawn from data kindly supplied by Dr R. F. Wright, and given in Wright (1994).

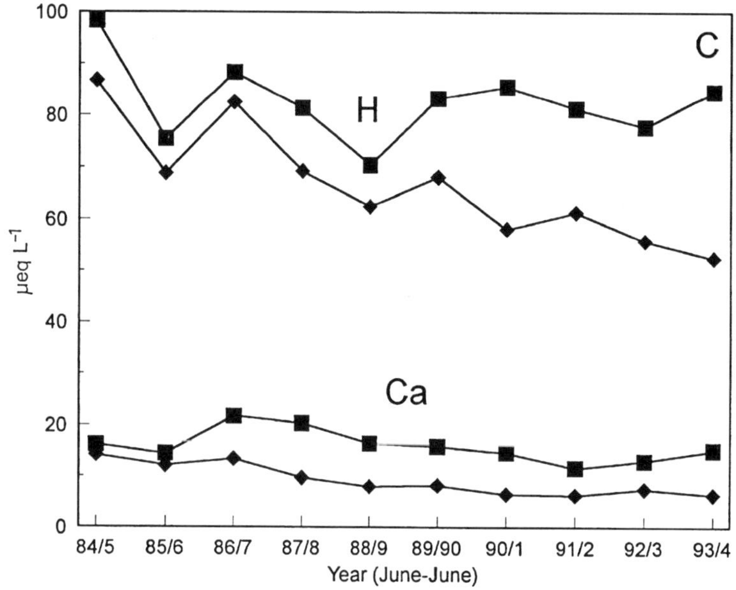


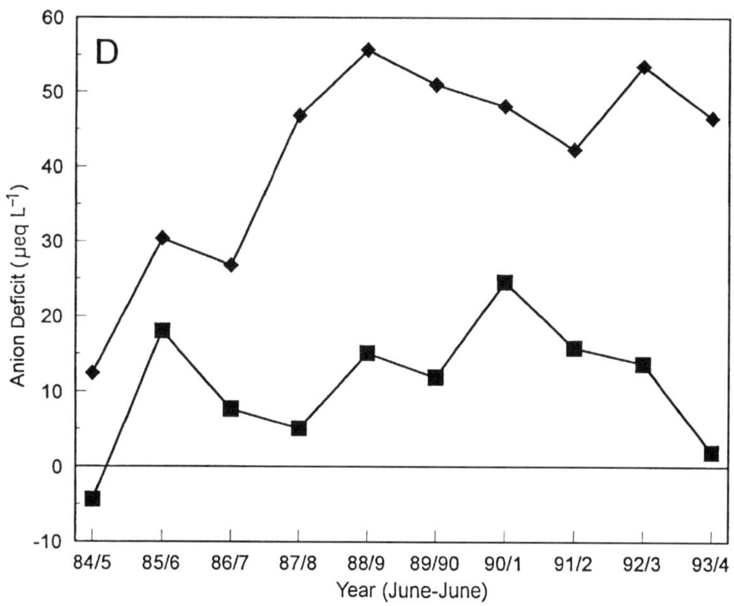


**Figure 1.2**. *continued*

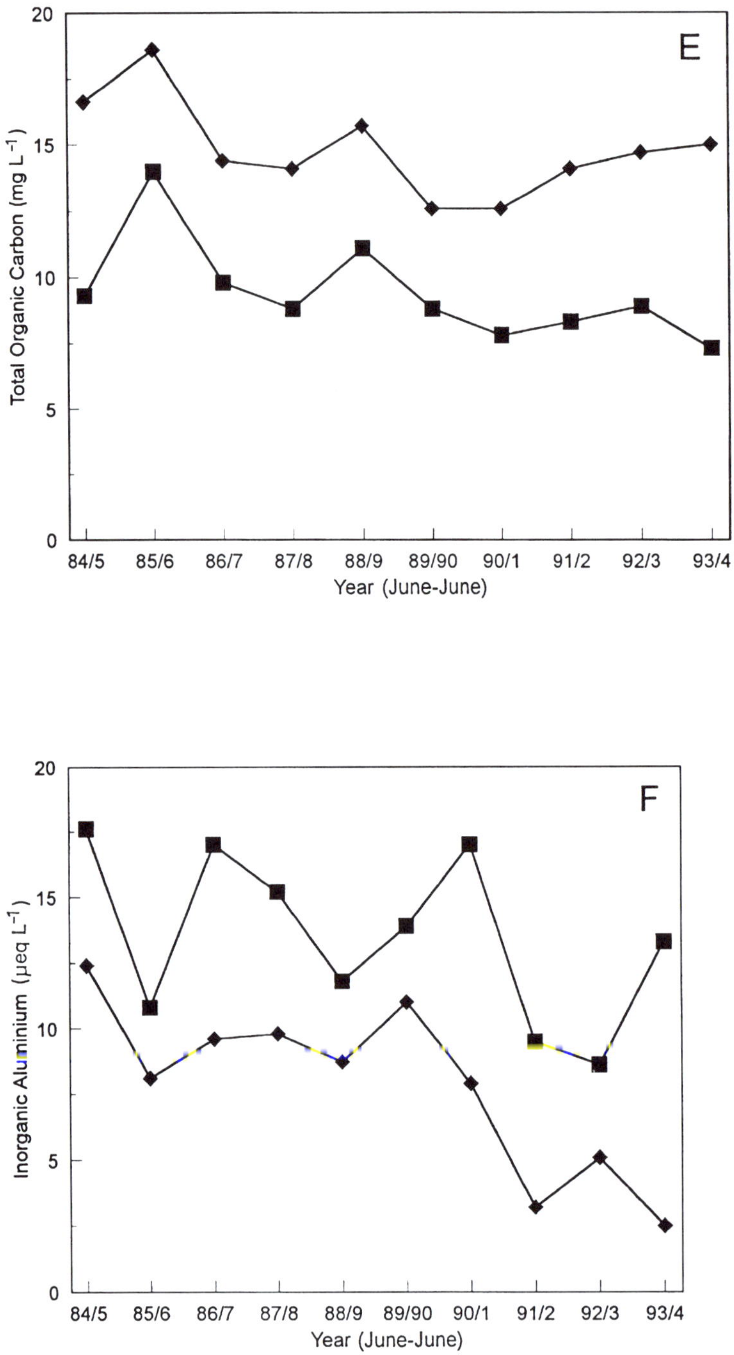


**Figure 1.2**. *continued*

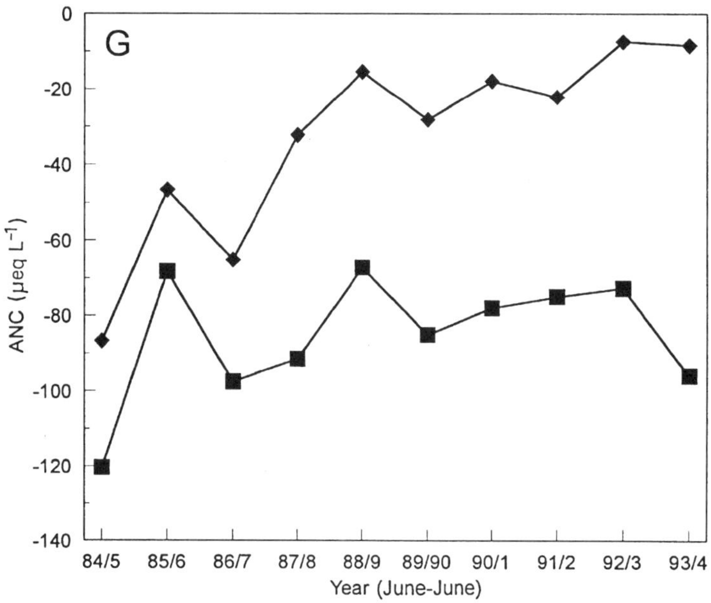


**Figure 1.2.** *continued*

responses of $Al^{3+}$ at Risdalsheia remain unexplained by any model, perhaps unsurprisingly as the models mostly derive their concepts from mineral soils, of which there is very little at Risdalsheia.

Finally, acid neutralizing capacity (ANC in Figure 1.2G) has responded to treatment, rapidly for the first 5 years and more slowly thereafter.

*Fish toxicity*

In Kim runoff, the pH has increased and the $Al^{3+}$ concentration (ultimately) declined, which should be regarded as an improvement in water quality from the point of view of aquatic organisms, but the $Ca^{2+}$ concentration has also declined, which is a deterioration. Is the improvement in pH and $Al^{3+}$ sufficient to compensate for the decline in $Ca^{2+}$? To answer this question an algorithm which calculates suitability for aquatic organisms from the water quality data is required. Sadler (1994) developed a set of discriminant functions from the Norwegian "1000-lake survey" dataset which allows lakes to be placed in various quality categories from a knowledge of water chemistry with a fair degree of accuracy. These functions have been used here in a rough assessment of the suitability for brown trout of the runoff from the Kim catchment, although it should be noted that this can only be indicative: Kim runoff is not fish habitat (see below) and the TOC concentrations are somewhat higher than the lakes in the original dataset. Sadler's functions are intended to place lakes into categories, but have been adapted to produce a "goodness of fish" function as follows. The discriminant functions for "good fish" and "decimated"—i.e. no fish—were calculated and the fish status parameter calculated as the difference (good – decimated). This function has the property that positive values denote placement in the "good" category, the more positive the better, and negative values in the "no fish" category, the more negative the worse. Its other properties are obscure. Its application to the RAIN Project data is shown in Figure 1.3. The lower lines in Figure 1.3 show the pH, Ca, Al discriminant functions for Kim and Egil—it can be seen that fish status at Kim does deteriorate slowly for the first 6 years of treatment, due to falling $Ca^{2+}$ concentrations. After 1990, the fall in Al concentration leads to a marked improvement. The values of the fish status function are well below zero: runoff at Kim and Egil is clearly unsuitable for fish according to this function.

Also shown on Figure 1.3 is a discriminant function (Sadler, 1994) based on ANC. Unsurprisingly, this resembles the ANC graph in Figure 1.2G, but in contrast to the pH, Ca, Al discriminant function, fish status at Kim shows a steady improvement, coming close to being suitable for fish at the end of the experiment. At Risdalsheia it was possible to make an experimental test between these models. In 1989, fisheries biologists from the Norwegian Institute for Water Research (NIVA) conducted some toxicity tests on brown trout in the runoff from the Risdalsheia catchments (Kroglund and Rosseland, 1992). All the fish died within 6 days, confirming the unsuitability of the waters for fish as predicted in Figure 1.3. However, the water from the treated catchment (Kim) was more rapidly toxic to all five strains of trout used than water from Egil, confirming the predictions of the pH, Ca, Al discriminant function rather than the ANC discriminant function, at Risdalsheia at least.

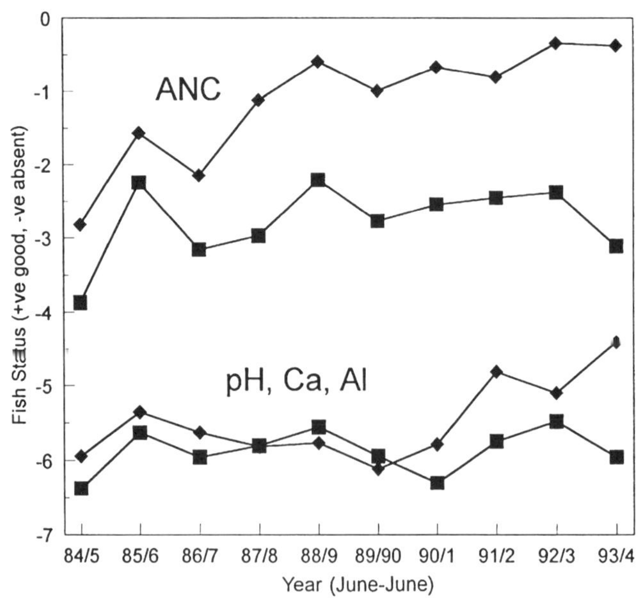


**Figure 1.3**. Calculated fish status in the Risdalsheia catchments, using discriminant functions described by Sadler (1994) based on ANC or pH, Al, Ca. All waters are unsuitable for fish, but the less negative the value of the fish status indicator, the better the water quality. All values are based on annual means

*Implications of the RAIN Project*

The results from the RAIN Project must be interpreted carefully in terms of the wider landscape. Fish clearly can never have lived in the runoff from the Kim catchment itself: the catchment is too small (860 $m^2$) to sustain water flow, but also produces an organic-rich leachate with pH and $Ca^{2+}$ both too low for trout, at least in the 1980s. Norwegian lakes which support fish typically have higher $Ca^{2+}$ concentrations and lower concentrations of organic acids (Henriksen *et al.*, 1989). This implies that they consist of a mixture of water derived from areas like Risdalsheia and from more mineral-rich areas, either as runoff from mineral soils or groundwater that has been in contact with weathering rock or superficial deposits. What the effect of deposition reduction on this mixture would be has not been examined, though it could probably be calculated given certain assumptions. The fact that at Risdalsheia the pH has been kept down by organic acids would be less important in this mixture as the organic acids would be diluted or oxidized, but the declining $Ca^{2+}$ concentrations would still be of relevance.

At Risdalsheia, S deposition was reduced by 80% for 10 years. Sulfate output took about 6 years to decline to close to a new steady state, but pH and inorganic Al only *started* to respond after about the same time. This is difficult to understand in terms of current models, but does imply a response time of about a decade for some improvement to appear. The decline in $Ca^{2+}$ did conform to expectations, as did the evidence for a very slow recovery in soil base saturation (1–3%), both of which imply long recovery times for soil and lake base status. The Risdalsheia results seem to confirm the two-stage recovery process as suggested by Skeffington and Brown (1992) with some recovery apparent in about a decade, but taking much longer where recovery of waters depends on soil recovery. This implies that some fraction of lakes (with particularly low $Ca^{2+}$) may not recover at all. The generality of these conclusions needs to be considered in view of the diversity of the soils and landscapes which have been affected by acid deposition.

## MERCURY

During recent years, both mercury (Hg) (Iverfeldt and Johansson, 1988) and methyl mercury (MeHg—Lee and Hultberg, 1990) in atmospheric deposition have been linked to a large-scale environmental problem, that of high MeHg content in fish in remote forested lakes and hydroelectric reservoirs in northern Europe and North America. Because of this Hg content, fish from more than 10 000 Swedish lakes are currently blacklisted for human consumption. An experiment in which deposition of S and N to a catchment was reduced by means of a roof would also provide an opportunity to investigate the effect of reduction of Hg and MeHg deposition. This section briefly considers the hypotheses which could be tested.

Atmospheric deposition and biogeochemical cycling of Hg and MeHg in forested catchments have been studied in the Lake Gårdsjön watershed for several years (Lee and Hultberg, 1990; Iverfeldt, 1991; Lee and Iverfeldt, 1991; Hultberg *et al.*, 1995a, b; Lee *et al.*, 1995). Of the Hg input to the soil, comprising wet plus dry deposition in throughfall and Hg in litterfall, 95% was retained in the organic soil horizons in the catchments (Driscoll *et al.*, 1994). Deposition of MeHg was about 2% of the total Hg deposition to the catchments, and 75% of the total MeHg input was retained in the catchment soils. The amount of MeHg retained in the soil corresponded to the input by litterfall, and the amount lost by runoff corresponded to the input by throughfall in the F1 catchment in Gårdsjön (Hultberg *et al.*, 1995a). The amounts of MeHg found in runoff from forests and wetlands, together with direct deposition of MeHg to the lake surface, are in most cases enough to explain the high MeHg concentrations found in fish in more than 10 000 remote, nutrient-poor and/or acidified lakes over large areas in Sweden (Hultberg *et al.*, 1995a).

As a result of the higher retention of Hg compared to MeHg in the soil, runoff had a higher proportion (4–9%) of MeHg than deposition (2%). Based on measurements of Hg in soil profiles (Lee *et al.*, 1995), a total pool of 15 to 20 kg $km^{-2}$ was found, while the pool of MeHg was only about 0.1 kg $km^{-2}$ or 0.5 to 0.7% of the total pool of Hg (Hultberg *et al.*, 1995b). On the basis of today's input of MeHg by throughfall and litterfall, rough estimates indicate that if this were the only source to the soil pool it would account for about 200 years of input. Based on present-day input of Hg by throughfall and litterfall, the pool of *total* Hg in the soil was accumulated during 400 to 500 years. Assuming that the natural Hg deposition was 5 to 10% of today's and that today's deposition had started by 1850, accumulation of Hg in the soils may have occurred over a period of 2000–3000 years (Hultberg *et al.*, 1995b).

The Forest Roof Catchment was designed to test the hypothesis that, as the Hg stored in the soils is bound strongly owing to low mineralization rates of the organic matter, losses by runoff of Hg and MeHg will depend on excess atmospheric deposition and catchment characteristics (Hultberg *et al.*, 1993b). The roof should decrease the input of Hg and MeHg in throughfall, and if this hypothesis is correct this should result in a decrease in MeHg and Hg losses in runoff. The results of the first 4 years' work are summarized in Chapter 12 of this book, and earlier accounts of the work have appeared in Hultberg *et al.* (1995b) and Munthe *et al.* (1995).

## AIMS OF THE GÅRDSJÖN PROJECT

Although a great deal was learned about recovery in the covered catchment experiment at Risdalsheia (see above), the diversity of ecosystems affected by acid deposition needs to be taken into account. Risdalsheia is typical of many areas of southern Norway, characterized by shallow, acidic organic soils and sparse vegetation cover. Many areas affected by acid deposition, for example in Sweden and Central Europe, differ substantially. They have deeper soils with a higher mineral content and are occupied by more productive ecosystems, typically Norway spruce (*Picea abies* L. Karst.) forest. Acidification is a serious problem in Sweden, with the fish population in 14 000 of its 83 000 lakes either eliminated or damaged by acidification (Bernes, 1991). It was clearly of benefit to undertake an acid exclusion experiment in typical Swedish terrain, so that the results could be compared and contrasted with those from the RAIN Project, and used for testing and validating models which would be more widely applicable. The Gårdsjön research area was selected as a suitable site not only because it had typical soils and vegetation but because of the large amount of data which had been collected over the last two decades (Andersson and Olsson, 1985) and the availability of expertise at IVL and infrastructure at the site itself.

### The Gårdsjön research site

The Gårdsjön area is reasonably typical of large areas of southern Sweden. Paleolimnological research (Renberg and Wallin, 1985) had demonstrated that an increased rate of acidification started within the Gårdsjön area during the 1950s. A few historical data from 1961/62 and measurements from 1969/70 illustrated how rapidly Lake Gårdsjön became chronically acidified (Hultberg, 1985a). Monitoring of lake water quality started in 1970 (Hultberg and Andersson, 1982; Hultberg, 1985a) and, with the initiation of the Lake Gårdsjön Project, monitoring of inputs and outputs started in catchment F1 in 1979 (Grennfelt *et al.*, 1985; Hultberg, 1985b). This was the first co-ordinated national research programme on acidification with the aim of quantifying the chemical fluxes between the atmosphere, the coniferous forest ecosystem and surface water, as well as the resulting chemical and biological effects in the terrestrial and aquatic ecosystems. The results from these studies were presented in Andersson and Olsen (1985).

Experimental research started in small catchments in the Lake Gårdsjön area in April 1984. During the years 1984 to 1995, chemical and biological effects on whole ecosystems have been studied after additions of dolomite limestone (Hultberg *et al.*, 1995c; Nyström *et al.*, 1995), clear-cutting of a whole catchment (Hultberg, 1992), increased acidification by elemental sulfur and additions of neutral sulfur (Hultberg et al., 1990), forest fertilization by nitrogen and low doses of limestone (Westling and Hultberg, 1990/91) as well as forest vitalization by additions of potassium and phosphorus. Two experiments started in April 1991 on the critical load of nitrogen through the NITREX Project (Hultberg *et al.*, 1993a) as well as the large-scale covered catchment experiment (Hultberg *et al.*, 1993b). The Gårdsjön Project has developed from being a purely Swedish research effort into an internationally known research area involving scientists from more than 10 countries in Europe, Asia and North America.

### Aims and objectives of the Roof Project

The Gårdsjön Roof Project was formulated at a meeting in 1988, and the aims and objectives were set out as follows:

*Aim:* to assess the rate and direction of response of a forested catchment to a reduction in sulphate, nitrate and mercury deposition.

*Specific objectives:* to answer the following major questions:

1. Does reducing acid input to this forest lead to an improvement in the quality of runoff water? Will the output of acidity and toxic ions such as aluminium be reduced, and if so by how much? Will there be an initial deterioration in water quality, as predicted by some models?

2. Will the runoff water be suitable for fish if acid input from the atmosphere is reduced? Or will the forest itself still produce toxic water?
3. How *fast* does any improvement in water quality take place? What implications does this have for the rate at which FGD and $NO_x$ reduction have to be fitted to power stations?
4. Do existing models of soil and water acidification and deacidification fit the response of the catchment? If not, how do they need to be modified?
5. Does a reduction in atmospheric mercury input lead to a reduction of mercury outputs, and can this be quantified?

These remain the aims of the Project.

The experiment has been funded by the Swedish Environmental Protection Agency, the Swedish Power industry (Vattenfall and now Elforsk), the English power industry (the Central Electricity Generating Board and then the Joint Environmental Programme of National Power and PowerGen), and IVL and outside investigators from their own funds. The cost after 6 years had reached 14 million SKr (£1.27m).

The present book aims to describe the design, construction and results of the Gårdsjön Roof Project at the time of writing (1995). The next section of the book puts the Gårdsjön area, and in particular the roofed catchment G1 ROOF, into context. The results of the experimental investigations and model development form the bulk of the volume. Finally, the results so far are considered in their wider context with chapters covering the problems of roof experimentation; the calculation of critical loads; and implications of the Gårdsjön Roof results for our knowledge of reversibility and for control of acidifying emissions.

## ACKNOWLEDGEMENTS

We thank Dr R. F. Wright for access to data from the RAIN Project, and permission to use it in this chapter.

## REFERENCES

Andersson, F. and Olsson, B. (eds) (1985) *Lake Gårdsjön, an acid forest lake and its catchment. Ecological Bulletins (Stockholm)*, **37**.

Bernes, C. (1991) Acidification and liming of Swedish freshwaters. *Monitor*, **12**, Statens Naturvardsverk, Stockholm.

Bishop, K. H. and Hultberg, H. (1995) Reversing acidification in a forest ecosystem: the Gardsjon Project. *Ambio*, **24**, 85–91.

Boxman, A. W., Van Dam, D., Van Dijk, H. F. G., Hogervorst, R. F. and Koopmans, C. J. (1995) Ecosystem responses to reduced nitrogen and sulphur inputs into two coniferous forest stands in The Netherlands. *Forest Ecology and Management*, **71**, 7–29.

Brown, D. J. A. and Sadler, K. (1989) Fish survival in acid waters. In: R. Morris, E. W. Taylor, D. J. A. Brown and J. A. Brown (eds), *Acid Toxicity and Aquatic Animals*. Cambridge University Press, Cambridge, pp. 31–44.

Brown, K. A. (1985) Sulphur distribution and metabolism in waterlogged peat. *Soil Biology and Biochemistry*, **17**, 39–45.

Bull, K. R. (1991) The critical loads/levels approach to gaseous pollutant emission control. *Environmental Pollution*, **69**, 105–123.

Chao, T. T., Harward, M. E. and Fang, S. C. (1964) Iron or aluminium coating in relation to sulfate adsorption characteristics of soils. *Soil Science Society of America Proceedings*, **28**, 632–635.

Cosby, B. J., Wright, R. F., Hornberger, G. M. and Galloway, J. N. (1985) Modelling the effects of acid deposition: assessment of a lumped parameter model of soil and stream water chemistry. *Water Resources Research*, **21**, 51–63.

Cosby, B. J., Bulger, A. J. and Wright, R. F. (1994) Predicting recovery of freshwater ecosystems: trout in Norwegian lakes. In: C. E. W. Steinberg and R. F. Wright (eds), *Acidification of Freshwater Ecosystems: Implications for the Future*. John Wiley & Sons Ltd, Chichester, pp. 355–373.

Dise, N., Ahlf, W., Brahmer, G., Cosby, B. J., Fott, J., Hauhs, M., Juttner, I., Kreutzer, K., Raddum, G. G. and Wright, R. F. (1994) Are chemical and biological changes reversible? In: C. E. W. Steinberg and R. F. Wright (eds), *Acidification of Freshwater Ecosystems: Implications for the Future*. John Wiley and Sons Ltd, Chichester, pp. 375–391.

Driscoll, C. T., Otton, J. K. and Iverfeldt, Å. (1994). Trace metals speciation and cycling. In: B. Moldan and J. Cerny (eds) *The Biogeochemistry of Small Catchments. SCOPE 51*, John Wiley & Sons Ltd, Chichester, pp. 299–322.

Galloway, J. N. (1995) Acid deposition: perspectives in time and space. *Water, Air and Soil Pollution*, **85**, 15–24.

Grennfelt, P., Larsson, S., Leyton, F. and Olsson, B. (1985) Atmospheric deposition in the Lake Gårdsjön area, SW Sweden. *Ecological Bulletins (Stockholm)*, **37**, 101–108.

Gunn, J. M. and Keller, W. (1990) Biological recovery of an acid lake after reductions in industrial emissions of sulphur. *Nature*, **345**, 431–433.

Harrison, R. B., Johnson, D. W. and Todd, D. E. (1989) Sulfate adsorption and desorption reversibility in a variety of forest soils. *Journal of Environmental Quality*, **18**, 419–426.

Hauhs, M. (1990) Lange Bramke: an ecosystem study of a forested watershed. In: D. C. Adriano and M. Havas (eds), *Acidic Precipitation: Volume 1, Case Studies*. Springer-Verlag, New York, pp. 275–305.

Hauhs, M. (1994) Does reduction in $SO_2$ and $NO_x$ emissions lead to recovery? In: C. E. W. Steinberg and R. F. Wright (eds), *Acidification of Freshwater Ecosystems: Implications for the Future*. John Wiley and Sons Ltd, Chichester, pp. 313–323.

Henriksen, A. (1984) Changes in base cation concentrations due to freshwater acidification. *Verh. Internat. Verein. Limnol.*, **22**, 692–698.

Henriksen, A., Lien, L., Rosseland, B. O., Traan, T. S. and Sevaldrud, I. S. (1989) Lake acidification in Norway: present and predicted fish status. *Ambio*, **18**, 314–321.

Högberg, P. and Jensen, P. (1994) Aluminium and uptake of base cations by tree roots—a critique of the model proposed by Sverdrup *et al. Water, Air and Soil Pollution*, **75**, 121–125.

Hordijk, L. (1995) Integrated assessment models as a basis for air pollution negotiations. *Water, Air, and Soil Pollution*, **85**, 249–260.

Hornung, M. and Skeffington, R. A. (eds) (1993) *Critical Loads: Concept and Applications*. ITE Report 28. HMSO, London.

Hultberg, H. (1985a) Changes in fish populations and water chemistry in Lake Gårdsjön and neighbouring lakes during the last century. *Ecological Bulletins (Stockholm)*, **37**, 64–72.

Hultberg, H. (1985b) Budgets of base cations, chloride, nitrogen and sulphur in the acid Lake Gårdsjön catchment, SW Sweden. *Ecological Bulletins (Stockholm)*, **37**, 133–157.

Hultberg, H. (1992) Biochemical cycling of sulphur and its effects on surface water chemistry in an acid catchment in SW Sweden. *Proceedings of International Symposium on Terrestrial Ecosystem Effects*, Commission of the European Communities, May 1991, Florence.

Hultberg, H. and Andersson, B. I. (1982) Liming of acidified lakes: Induced long-term changes. *Water, Air and Soil Pollution*, **18**, 311–331.

Hultberg, H. and Grennfelt, P. (1992) Sulphur and seasalt deposition as reflected by throughfall and runoff chemistry in forested catchments. *Environmental Pollution*, **75**, 215–222.

Hultberg, H., Lee, Y. H., Nyström, U. and Nilsson, S. I. (1990) Chemical effects on surface-, ground- and soil-water of adding acid and neutral sulphate to catchments in south-west Sweden. In: B. J. Mason (ed.), *The Surface Water Acidification Programme SWAP*. Cambridge University Press, Cambridge, pp. 167–182.

Hultberg, H., Dise, N. B., Wright, R. F., Andersson, B. I. and Nyström, U. (1993a) Nitrogen saturation induced during winter by experimental $NH_4NO_3$ addition to a forested catchment. *Environmental Pollution*, **84**, 145–147.

Hultberg, H., Andersson, B. I. and Moldan, F. (1993b) The covered catchment—An experimental approach to reversal of acidification in a forest ecosystem. In: L. Rassmusen, T. Brydges and P. Mathy (eds), *Experimental Manipulations of Biota and Biogeochemical Cycling in Ecosystems*. Ecosystems Research Report, **4**, CEC, Brussels, pp. 46–54.

Hultberg, H., Iverfeldt, Å. and Lee, Y. H. (1995a) Methylmercury input/output and accumulation in forested catchments and critical loads for lakes in southwestern Sweden. In: C. J. Watras and J. W. Huckabee (eds), *Mercury Pollution: Integration and Synthesis*. Lewis Publishers, Chelsea, MI, pp. 313–322.

Hultberg, H., Munthe, J. and Iverfeldt, Å. (1995b) Cycling of methyl mercury and mercury—Responses in the forest roof catchment to three years of decreased atmospheric deposition. *Water, Air and Soil Pollution*, **80**, 415–424.

Hultberg, H., Nilsson, S. I. and Nyström, U. (1995c) Effect on soils and leaching after application of dolomite to an acidified forested catchment in the Lake Gårdsjön watershed, south west Sweden. *Water, Air and Soil Pollution*, **85**, 1033–1038.

Iverfeldt, Å. (1991) Occurrence and turnover of atmospheric mercury in the Nordic countries. *Water, Air and Soil Pollution*, **56**, 251–265.

Iverfeldt, Å. and Johansson, K. (1988) Mercury in runoff from small watersheds. *Verh. Internat. Verein. Limnol.*, **23**, 1626.

Kirchner, J. W. and Lydersen, E. (1995) Base cation depletion and potential long-term acidification of Norwegian catchments. *Environmental Science and Technology*, **29**, 1953–1960.

Kroglund, F. and Rosseland, B. O. (1992) *Reversibility of Acidification: Fish Responses* in Experiments at Risdalsheia, Norway. Norwegian Institute for Water Research Report, Oslo.

Lee, Y.-H. and Hultberg, H. (1990) Methylmercury in some Swedish surface waters. *Environmental Toxicology and Chemistry*, **9**, 833–841.

Lee, Y.-H. and Iverfeldt, Å. (1991) Measurements of methyl mercury and mercury in runoff, lake and rainwater. *Water, Air and Soil Pollution*, **56**, 309–321.

Lee, Y.-H., Borg, G., Iverfeldt, Å. and Hultberg, H. (1995). Fluxes and turnover of methylmercury: Mercury pools in forest soils. In: C. J. Watras and J. W. Huckabee (eds), *Mercury Pollution: Integration and Synthesis*. Lewis Publishers, Chelsea, MI, pp. 329–341.

Likens, G. E., Bormann, F. H., Pierce, R. S., Eaton, J. S. and Johnson, N. M. (1977) *Biogeochemistry of a Forested Ecosystem*. Springer-Verlag, New York.

Mitchell, M. J., David, M. B. and Harrison, R. (1992) Sulphur dynamics of forest ecosystems. In: R. W. Howarth, J. W. B. Stewart and M. V. Ivanov (eds), *Sulphur Cycling on the Continents: Wetlands, Terrestrial Ecosystems and Associated Water Bodies. SCOPE 48*. John Wiley & Sons Ltd, Chichester, pp. 215–254.

Munthe, J., Hultberg, H., Lee, Y. H., Parkman, H., Iverfeldt, Å. and Renberg, I. (1995) Trends in mercury and methylmercury in deposition, runoff water and sediments in relation to experimental manipulations and acidification. *Water, Air and Soil Pollution*, **85**, 743–748.

Neal, C., Reynolds, B., Stevens, P. and Hornung, M. (1989) Hydrogeochemical controls for inorganic aluminium in acidic stream and soil waters at two upland catchments in Wales. *Journal of Hydrology*, **106**, 155–175.

Nilsson, J. and Grennfelt, P. (eds) (1988) *Critical Loads for Sulphur and Nitrogen*. NORD Report 1988:15. Nordic Council of Ministers, Copenhagen.

Nyström, U., Hultberg, H. and Lind, B. B. (1995) Can forest-soil liming mitigate acidification of surface waters in Sweden? *Water, Air and Soil Pollution*, **85**, 1855–1860.

Posch, M., de Smet, P. A. M., Hettelingh, J.-P. and Downing, R. J. (1995). *Calculation and Mapping of Critical Thresholds in Europe*. Report 259101004. RIVM, Bilthoven, The Netherlands.

Renberg, I. and Wallin, J.-E. (1985) The history of the acidification of Lake Gårdsjön as deduced from diatoms and *Sphagnum* leaves in the sediment. *Ecological Bulletins (Stockholm)*, **37**, 47–52.

Reuss, J. O. and Johnson, D. W. (1986) *Acid Deposition and the Acidification of Soils and Waters*. Springer-Verlag, New York.

Rosseland, B. O. and Stuarnes, M. (1994) Physiological mechanisms for toxic effects and resistance to acid water: an ecophysiological and ecotoxicological approach. In: C. E. W. Steinberg and R. F. Wright (eds), *Acidification of Freshwater Ecosystems: Implications for the Future*. John Wiley and Sons Ltd, Chichester, pp. 227–246.

Sadler, K. (1994) *Damage Functions for Predicting Fishery Status from Water Chemistry*. Report PT/94/330068/M. PowerGen, Ratcliffe-on-Soar, England.

Sayer, M. D. J., Reader, J. P. and Dalziel, T. R. K. (1993) Freshwater acidification — Effects on the early life stages of fish. *Reviews in Fish Biology and Fisheries*, **3**, 95–132.

Skeffington, R. A. and Brown, D. J. A. (1992) Timescales of recovery from acidification: implications of current knowledge for aquatic organisms. *Environmental Pollution*, **77**, 227–234.

South, D. W. (1990) Technologies and other methods for controlling emissions: performance, costs and applicability. In: P. M. Irving (Ed.), *Acidic Deposition: State of Science and Technology*, Report 25. National Acid Precipitation Assessment Program, Washington, DC.

Steinberg, C. E. W. and Wright, R. F. (eds) (1994) *Acidification of Freshwater Ecosystems: Implications for the Future*. John Wiley & Sons, Chichester.

Sverdrup, H. and De Vries, W. (1994) Calculating critical loads with the simple mass balance method. *Water, Air and Soil Pollution*, **72**, 143–162.

Sverdrup, H. and Warfvinge, P. (1993) *Soil Acidification Effect on Growth of Trees, Grasses and Herbs Expressed by the (Ca+Mg+K)/Al Ratio*. Technical Report, University of Lund, Lund, Sweden.

Sverdrup, H., Warfvinge, P. and Jonsson, C. (1993) Critical loads for acidity for forest soils, groundwater and first-order streams in Sweden. In: M. Hornung and R. A. Skeffington (eds), *Critical Loads: Concept and Applications*. ITE Symposium 28, HMSO, London, pp. 54–67.

Tuovinen, J.-P., Barrett, K. and Styve, H. (1994) *Transboundary Acidifying Pollution in Europe: Calculated Fields and Budgets 1985–93*. EMEP/MSC-W Report 1/94. Det Norske Meteorologiske Institutt, Oslo.

UNECE (1994) *Effects and Control of Long-range Transboundary Air Pollution*. ECE/EB.AIR/39. United Nations, Geneva.

Westling, O. and Hultberg, H. (1990/91) Liming and fertilisation of acid forest soil: short-term effects on runoff from small catchments. *Water, Air, and Soil Pollution*, **54**, 391–407.

Wright, R. F. (1989) RAIN Project: role of organic acids in moderating pH change following reduction in acid deposition. *Water, Air and Soil Pollution*, **46**, 251–259.

Wright, R. F. (1994) *RAIN Project: Risdalsheia Data Report for June 1990–May 1994*. Acid Rain Report, 36/1994. Norwegian Institute for Water Research, Oslo.

Wright, R. F. and Hauhs, M. (1991) Reversibility of acidification: soils and surface waters. *Proceedings of the Royal Society of Edinburgh*, **97B**, 169–191.

Wright, R. F. *et al.* (1986) Project RAIN: changing acid deposition to whole catchments. The first year of treatment. *Water, Air and Soil Pollution*, **30**, 47–63.

Wright, R. F., Lotse, E. and Semb, A. (1988) Reversibility of acidification shown by whole catchment experiments. *Nature*, **334**, 670–675.

Wright, R. F., Cosby, B. J., Flaten, M. B. and Reuss, J. O. (1990) Evaluation of an acidification model with data from manipulated catchments in Norway. *Nature*, **343**, 53–55.

Wright, R. F., Lotse, E. and Semb, A. (1993) RAIN Project: results after 8 years of experimentally reduced acid deposition to a whole catchment. *Canadian Journal of Fisheries and Aquatic Sciences*, **50**, 258–268.

Wright, R. F., Cosby, B. J., Ferrier, R. C., Jenkins, A., Bulger, A. J. and Harriman, R. (1994) Changes in acidification of lochs in Galloway, southwestern Scotland, 1979–1988 — the MAGIC model used to evaluate the role of afforestation, calculate critical loads, and predict fish status. *Journal of Hydrology*, **161**, 257–285.

# 2 The Covered Catchment Site: a Description of the Physiography, Climate and Vegetation of Three Small Coniferous Forest Catchments at Gårdsjön, South-west Sweden

**B. I. ANDERSSON, K. H. BISHOP, G. CHR. BORG, R. GIESLER, H. HULTBERG, M. HUSE, F. MOLDAN, L. NYBERG, P. H. NYGAARD & U. NYSTRÖM**

This chapter contains a description of the main features of the small catchments at Gårdsjön used for the Roof Project and related NITREX Project. Descriptions are given of the topography, geology, climate, hydrology, soils and vegetation. The elemental contents of various ecosystem compartments are tabulated for the three catchments and an assessment made of their relationships to other European and North American ecosystems.

The location of Lake Gårdsjön on igneous, rather acid bedrock with very thin soils of local origin and a mineral composition having slow to normal weathering rate renders this south-west Swedish forest land sensitive to acidification. Mature (80 to 100 years old) forests with Norway spruce (*Picea abies* (L.) Karst.) and some Scots pine (*Pinus sylvestris* L.) predominate while in the ground layer dwarf shrubs (mainly *Vaccinium myrtillus* L.) mixed or alternating with several species of mosses and the grass *Dechampsia flexuosa* constitute the most abundant vegetation elements in the catchment area surrounding Lake Gårdsjön. Generally podzolic soils of sandy to silty loam with only a small clay content occur interspersed with typical folisols on steep slopes or ridge areas and peaty soils in the valley bottoms of the larger catchments.

Annual precipitation amounts to *c.* 1150 mm and runoff is *c.* 580 mm, the climate being pronouncedly maritime with February as the coldest (−2.7°C) month and July the warmest (+16.0°C). Some precipitation occurs on almost 150 days of the year, though the typical rain intensity is fairly low (1.5 mm $h^{-1}$), usually lasting for several hours and caused by cyclonic weather systems.

The acidified soils have large amounts of exchangeable acidity and aluminium, while stores of exchangeable base cations are modest, resulting in a comparatively low base saturation. Acid deposition has resulted in substantial amounts of sulphur and nitrogen being stored in the soil profile, and the future response of the different fractions of these elements will be of importance for the recovery from acidification when loading decreases.

*Experimental Reversal of Acid Rain Effects: The Gårdsjön Roof Project.*
Edited by Hans Hultberg and Richard Skeffington. 

The presence of numerous small sub-catchments, relatively high rainfall, and good homogeneity between catchments make the Gårdsjön area suitable for manipulative catchment research.

## INTRODUCTION

The Lake Gårdsjön area is located approximately 15 km inland from the western coast of Sweden, on the west-facing slope of a hilly and gently raised region, Svartedalen (Figure 2.1). The geology and soils of the area, the climatic and hydrologic regimes including factors such as proximity to the coast, and general altitude (100 to 200 m a.s.l.) give this region some specific characteristics which are of importance for the response to the severe contemporary acidification stress to which it is subject.

Acidification research and surveys in the area started during the late 1960s when much of the area with numerous lakes already was chronically acidified (Avergård, 1971; Hultberg, 1977). Research focused on the effects of acidification, as well as liming which was initiated in 1973 (Grahn *et al.*, 1974; Hultberg and Andersson, 1982), and a reacidification study, part of which was prolonged until 1987 (Andersson and Hultberg, 1997). During 1978, much of the Swedish acidification research on lake ecosystems was concentrated in this area when the Lake Gårdsjön Project started as a major co-ordinated and collaborative effort (Andersson and Olsson, 1985). As part of that project, a number of papers were published describing the Lake Gårdsjön catchment (Olsson *et al.*, 1985), the more recent land-use history (Olsson, 1985), and the development of the lake and catchment since deglaciation more than 12 000 years ago (Wallin and Renberg, 1985). Following the intensive study period, approximately up to 1985 (cf. Dickson, 1988), there has been protracted monitoring, including experiments with lake biota and especially biogeochemical fluxes. Experimental manipulations in small catchments became possible after the liming of Lake Gårdsjön in 1982, since effects on the lake proper of the experiments were without significance in relation to the major perturbation of lake water chemistry caused by liming.

Monitoring of the small catchment F1 has been continuous since the start of the Lake Gårdsjön Project in 1978, generally as part of different catchment experiments. Currently it is one of many sites in the UNECE programme of Integrated Monitoring of Air Pollution Effects. The establishment of long-term data series combined with further experimental manipulations has appreciably increased the background information and understanding of biogeochemical cycling and structural components in both lake and terrestrial ecosystems (Hultberg, 1985; Nilsson, 1985; Hultberg and Grennfelt, 1986; Hultberg *et al.*, 1990; Hultberg and Grennfelt, 1992). This background knowledge, in combination with a number of favourable site conditions (cf. below) made it both natural and profitable to locate the acidification reversal study initiated in 1988 in catchments in the Lake Gårdsjön area (Chapter 1, this volume). In a parallel catchment, another large ecosystem manipulation with a threefold increase of current nitrogen deposition is being performed as part of the large multi-site NITREX Project (Dise and Wright, 1992).

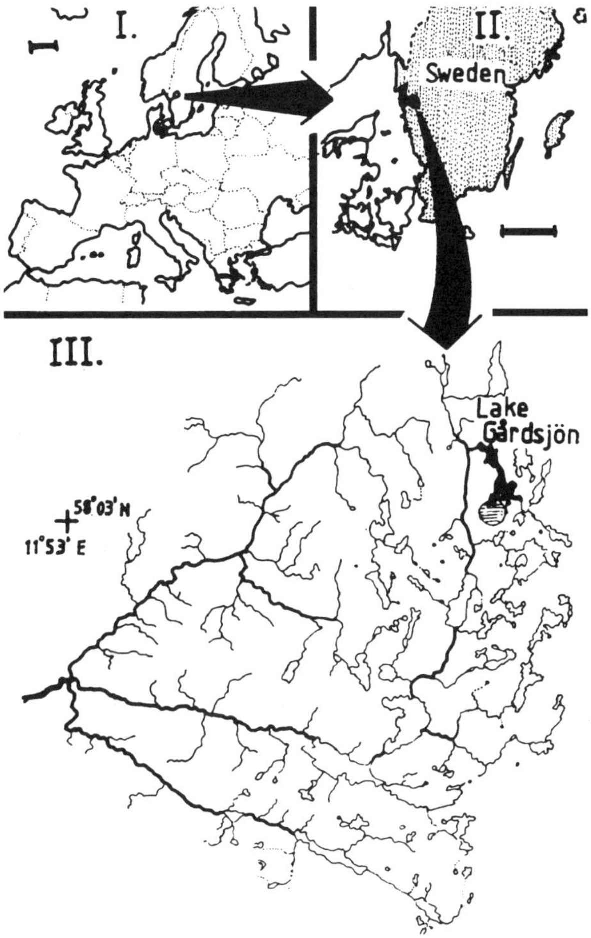


**Figure 2.1.** Location of the Lake Gårdsjön site near the Swedish west coast. Bars in I and II are 400 and 150 km respectively. The map in III shows the Anråseå river basin with Lake Gårdsjön and (circled) the experimental catchment areas. Scale is 1:100000 (1 cm on map equals 1 km in nature)

The intention of this chapter is to go beyond the earlier site descriptions (Olsson *et al.*, 1985; Johansson and Nilsson, 1988) which gave overviews for the Lake Gårdsjön area with some details of catchment F1. This description will especially focus on a more detailed outline of the experimental catchments; G1 (ROOF) and to a lesser extent G2 (NITREX) with some complementary recent information for catchment F1 (Control); and above all use newly measured data to present updated averages for climate and hydrology. The establishment of a large co-ordinated and partly collaborative research effort implies a more focused interest than ever before in these Gårdsjön catchments. The physical and chemical features with bearings on the biogeochemical cycles of importance for acidification recovery will be described together with other data pertinent for use in various catchment models. Finally, the acidified small catchments of Lake Gårdsjön will be placed in their context by comparison with other catchments of the temperate or boreal region in Europe and North America where acidification stress to ecosystems is observed.

## MORPHOLOGICAL FEATURES

### Geology

The central part of the Svartedalen region surrounding Lake Gårdsjön belongs to a major Swedish bedrock region, the south-west Swedish grey gneissic region. In the area, the gneissic region is intersected by alternating gneiss–granites creating an heterogeneous zone with bedrock "blocks" shifting between igneous and metamorphic rocks of different generations (cf. Samuelsson, 1980). Consequently, across the catchment of Lake Gårdsjön an alternating mixture of granite and gneissic granodiorite of older or younger origin is to be found. The complex tectonic pattern includes a system of prominent joints and fractures with either NNW–SSE or NNE–SSW direction, in general along the borders of the lithological units but not solely restricted to these areas. Thus small catchments in the area may be within either of the bedrock types, or even embrace two bedrock types if one of the major fracture zones traverses the valley bottom (Johansson *et al.*, 1988). The predominant mineral composition of the bedrock series classifies them as either intermediate or acidic with a dominance of the latter type, though minerals such as hornblende and biotite may sometimes occur within bedrock or till of these areas (Melkerud, 1983; cf. Olsson *et al.*, 1985).

The general large-scale appearance of the region is a gentle, SW-sloping sub-Cambrian peneplain. Only restricted amounts of glacial overburden material were deposited and, therefore, the Svartedalen area appears exposed and hilly with many barren bedrock outcrops which are an original feature and not caused by wave or glacial river erosion. The Quaternary deposits were formed when the ice sheet of the last glaciation melted away in the area, about 12 500 years ago. The predominating type of till is a sandy–fine sandy to silty till, typically thin and with a clay content of approximately 5%. The till is predominantly of a local origin (cf. Lindén, 1975; Melkerud, 1977), and its weathering rates relate to the mineral composition characteristic of the parental bedrock. Thicker deposits, reaching depths of several metres, are only found in some of the small valley bottoms or as restricted pockets (crevasse fillings). One kilometre north-east of Lake Gårdsjön is a restricted (currently exploited) glaciofluvial deposit located at the top of the area dividing runoff to the east (the River Göta) or to the west (the River Anråseå that carries the water from Lake Gårdsjön and another 50 lakes within the area). However, more prominent glaciofluvial or morainic structures are important landscape elements on both western and eastern margins of the Svartedalen area, e.g. the Dösebacka moraine on its south-east slopes.

The bedrock underlying the research catchments F1, G1 and G2 is a gneissic granodiorite of the younger generation. This younger formation is less foliated and usually not veined, unlike the older generation which dominates the western part of the Gårdsjön catchment. Fracture zones in NNW–SSE direction occur in catchment F1 (Fridh and Hayling, 1980) and with a NNE–SSW direction in G1. This does not seem to affect the more or less impervious nature of the catchment bedrock significantly, which facilitates hydrological measurements and biogeochemical monitoring.

**Topography and morphometric characteristics**

The outlet of Lake Gårdsjön is situated at 58° 04′ N and 12° 01′ E, while the covered catchment (G1 ROOF) and the G2 NITREX catchment are at the border of an adjacent catchment just south of the water divide at 58° 03′ N and 12° 01′ E respectively. A few hundred metres north-east from these manipulation catchments is the control catchment F1, which is situated within the Lake Gårdsjön catchment proper (Figure 2.2). One characteristic feature of this part of the Svartedalen region is a pronounced small-scale morphological variation in the landscape which is manifested in a mosaic of lakes, wetlands and forested areas, with only moderate altitudinal differences. The small-scale topography is especially pronounced in the area south of Lake Gårdsjön where a large number of adjacent small valleys are located, many of which are very similar and have been used for manipulation experiments during 10 years of small catchment research (cf. Hultberg and Grennfelt, 1992; Andersson, 1993).

Within the catchment of Lake Gårdsjön, the highest sea level following the deglaciation appears at about 130 m a.s.l. As land upheaval was initially rapid, however, this is not very obvious on the ground and only occasionally can land terraces and other formations created as coastline elements be clearly defined (Melkerud, 1983). Catchments F1 Control and G1 ROOF, with altitudes of 113 to 141 and 123 to 143 m a.s.l. respectively, include this zone around the highest marine shore level without any obvious effects on morphology or soil characteristics (Melkerud, 1983; Borg, 1995). All of the third experimental catchment, G2 (NITREX), is probably located above the highest marine limit as the altitude spans 133 to 146 m a.s.l.

All three catchments have the same appearance of a small valley with a flat, or gently sloping valley bottom, with steep sides. In the larger catchment, F1, the valley bottom harbours an ephemeral brooklet, while both the G1 and G2 catchments are too small for runoff to form brooklets. However, there were small discharge areas at the bottom ends of these catchments where weir dams were constructed. Catchment boundaries are usually well-defined water divides on ridges, and only in the G1 catchment is there an large upper flat saddle area. The aspect of both G1 and G2 is towards S–SW and the longer F1 valley faces N to NW in its lower end while its upper part is rather flat and more exposed to winds from a sector E to SW. This exposure is enhanced because the areas surrounding the uppermost part of the F1 catchment to the east and south were clear-cut in 1976, including also a minor part of F1 proper.

As the F1 Control catchment is larger than the manipulated catchments G1 and G2 (Table 2.1), the exact mapping which was performed for the latter two catchments was not feasible. This mapping was a prerequisite for building of the roof

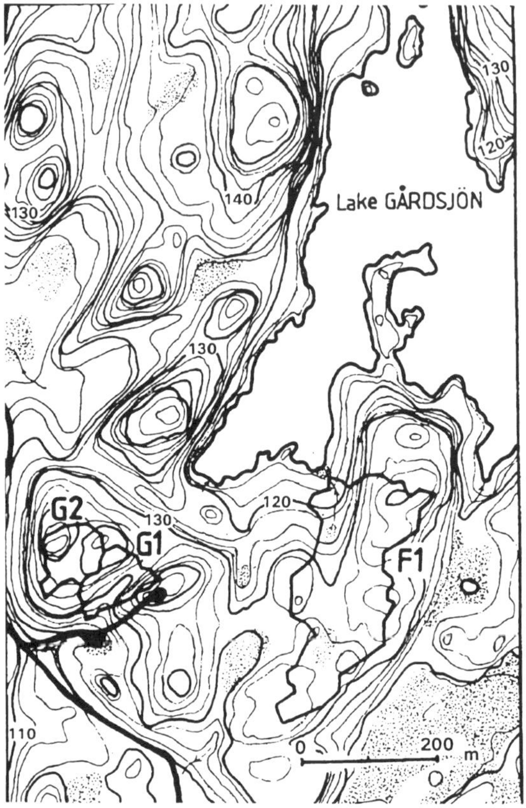


**Figure 2.2.** Detailed topographic map over the experimental area at the southern end of Lake Gårdsjön. F1 is the Control, G1 and G2 are the ROOF and NITREX manipulation catchments

**Table 2.1.** Some characteristics of the experimental and the control catchment at Gårdsjön

| Catchment | Size ($m^2$) | Altitude highest (m a.s.l.) | Weir Altitude (m a.s.l.) | Type | Brooklet | Manipulation |
|---|---|---|---|---|---|---|
| F1 | 37 000 | 141 | 115 | Valley | Yes | None |
| G1 | 6 300 | 143 | 123 | Valley | No | Deacidification |
| G2 | 5 200 | 146 | 133 | Open Valley | No | Nitrogen increase |

in G1 but also served the purpose of enabling an exact three-dimensional positioning of scientific equipment, which became important especially for the detailed hydrological research. In F1, only a restricted area close to the water weir was mapped in detail, but this was where most of the equipment for intensive studies was located. A 5×5 m gridnet is the foundation for the map, embracing the catchment areas in a local co-ordinate system starting from the south-westernmost point. The system is referenced into the ordinary topographic system by a fix point south-west of the G1 weir, therefore altitudes (z-axis) are real. Figure 2.3 gives an overview of many installations and specific protected areas for intensive research in catchment G1 ROOF. This detailed morphometric representation would have been impossible to construct without the local mapping, as the spatial resolution, especially the altitudinal morphometry, on ordinary topographic maps is not accurate enough.

## CLIMATE

### Methods

Climatic measurements starting in winter 1990 have been made on a 15-m-high mast placed on a hilltop site at 150 m a.s.l., just north of Gårdsjön. Climatic measurements in the forest at catchment F1 and beneath the roof in catchment G1 began later, in 1992. Wind speed and direction are measured with a Young wind monitor, model 05103, mounted close to the top of the mast. Relative humidity is measured by Rotronics MP100 sensors, installed both at the mast (where it also measures temperature) and in catchments F1 and G1. For measurements of temperatures at two levels above ground surface in the forested catchments ventilated small tubes with a Pt 100 sensor are used. Similar sensors are also installed to give soil temperatures at depths of 5, 15 and 30 cm. On the climate mast, global radiation is measured by a Li 200SA pyranometer sensor (Lamda Instruments).

### Wind

Wind measurements at the ground surface give a pronounced dominance of winds from the south-west (21.4%), about double the frequency of any other direction (Figure 2.4). This SW-centred wind component was more pronounced than given by Olsson *et al.* (1985) with data from the meteorological station on the island of Vinga some 50 km SSW of Gårdsjön and is also different from the geostrophic winds (cf. Chapter 3, this volume). The frequency of daily (24 hour) averages for winds from the 90° south to west sector was >50% with the opposite direction (north-east) being the second most important. Furthermore, only winds from this south-west sector reach high speeds (0.8% > 10 m $s^{-1}$) as a daily average and, notably, only one single day (24 hour period) was totally calm during 4 years of measurement. Forests of the region are well adapted to high winds which are particularly common in the autumn, and wind damage is usually minor. However, a few trees are blown down each winter season in catchment F1, though probably more as a result of root rots than as an effect of exceptionally strong wind.

### Temperature

No temperature measurements existed for the area before measurements started in 1990. At the climate mast, temperature is registered at 2 m and in the forest at 0.25 m and 1.75 m above the ground respectively. As 4 years is a rather short observation period, an estimate of the long-term normal temperature value of about 6.3°C was

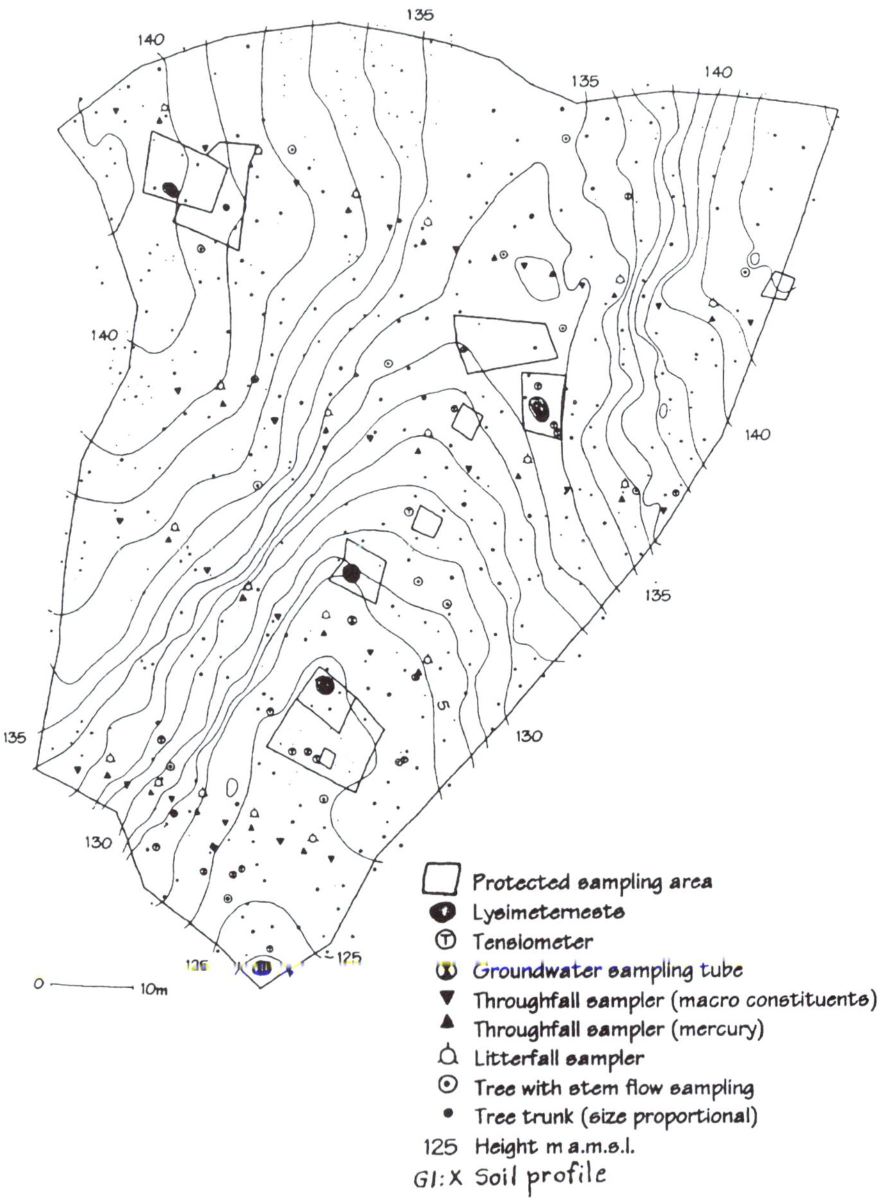


**Figure 2.3.** The original sampling installations in catchment G1 before the start of the experiment by roofing in April 1991. Similar installations still run in catchment G2 and part of the control catchment F1

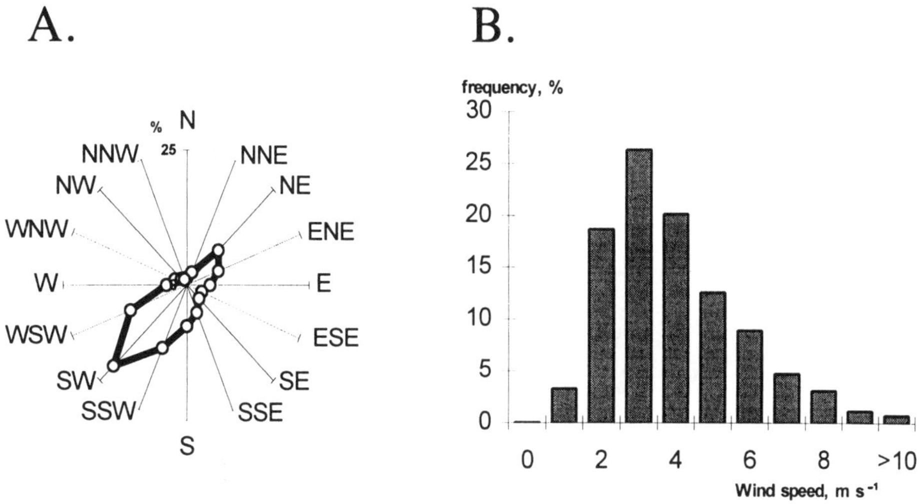


**Figure 2.4.** (A) Frequencies of different wind directions measured at the Gårdsjön climate station, based on hourly means. (B) Frequency distribution of the wind speeds (m $s^{-1}$), gas 24 h means, measured at Gårdsjön between 1991 and 1994

reached by regression calculations based on surrounding weather stations with long-term records (cf. values from Hultberg and Andersson, 1982, and Dise and Wright, 1992, of 6.6 and 6.4°C respectively). The reference period for these stations is generally 1961–90 (cf. Alexandersson *et al.*, 1991).

The coldest months are January–February with temperatures of about −2.6 to −2.7°C (Figure 2.5). However, weather during the winter is usually quite variable, principally due to occasional shifts in routing of the passages of the dominating cyclonic depressions, and snow cover seldom stays throughout the season. Mild humid winters are succeeded by a comparatively dry period during late spring–early summer. The warmest month is July with an estimated long-term normal average temperature of 16.0°C.

The annual mean of 7.3°C, observed during the measurement period, is notably above the estimated long-term normal value for the reference period 1961 to 1990. The difference is due to the fact that five out of six winters in the observation series have been unusually mild. It is currently impossible to judge whether this recent string of mild winters is a coincidence due to natural weather pattern oscillations or an emerging manifestation of a pattern related to anticipated global changes. The few years of data from measurements in the forest in catchment F1 and G1 show only minor differences (Chapter 4, this volume) in spite of the different aspects of the two catchments. A rather typical pattern of annual temperature variation is given by the measurements in catchment F1 from April 1993 to March 1994, which is the third year of the manipulation experiment, though the summer period was rather warmer and drier than usual (Figure 2.6).

Continuous monitoring of the temperature and humidity is one of several means whereby the influence of the roof built in catchment G1 is assessed (Chapters 4 and

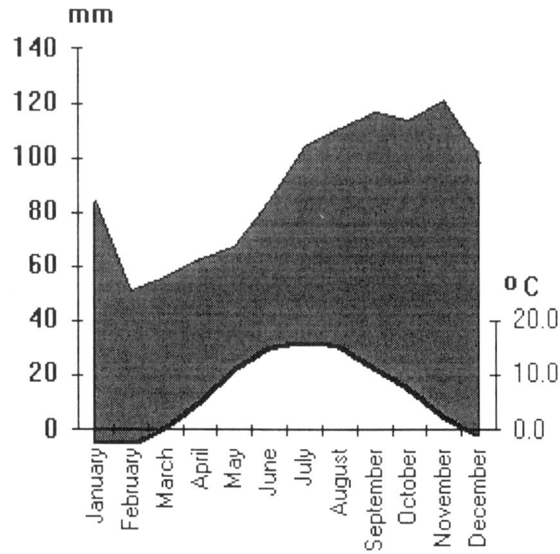


**Figure 2.5.** Climate diagram. The upper line shows long-term monthly reference values for precipitation (1961 to 1991 data) and the lower line the same for temperature. Estimates based on regressions with nearby regular weather stations

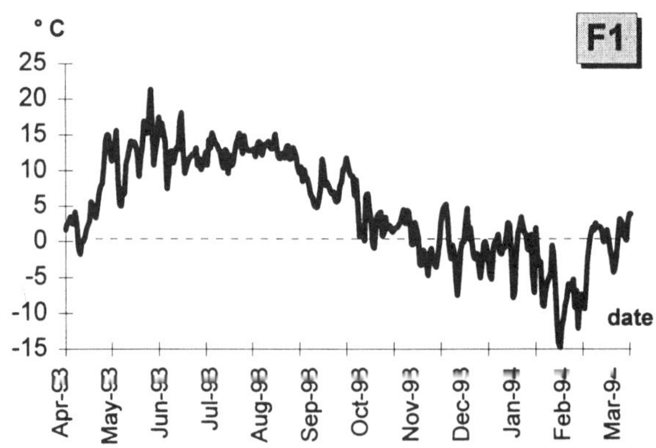


**Figure 2.6.** Air temperature measured during a single year. Sensor placed 1.75 m above ground in the forest at catchment F1 (Control)

22, this volume). Due to the variable winters and the snow cover, soil freezing probably also shows substantial inter-annual variation. Even during "normal" winters, stable frost periods sometimes with much snow may suddenly be interrupted by prolonged thaw periods followed by frost on bare ground. The bare soil below the roof will be more exposed to these variable temperature regimes as a protective snow cover does not exist and this may increase the incidence of frozen soil in the G1 catchment (Chapter 4, this volume).

## Precipitation and humidity

Autumn rains dominate, with October and November being the months with the largest amounts of precipitation (Figure 2.5). The markedly drier period in May is clearly reflected in much lower values of relative humidity (RH) during the early summer season (Figure 2.7). Only during protracted drought events (1992 summer, 1993 spring, and 1994 summer) are there observations of low RH values which otherwise are rather high even in summers. During the late autumn or winter the relative humidity is close to or equal to 100% for long periods.

Because of the relative proximity to the coast, even the gently raised Svartedalen forest region, with an altitudinal difference of at most 200 m, contributes a strong orographic component to the precipitation. Low-altitude weather stations at the coast only receive annual precipitation amounts in the range of about 700 to 900 mm, while the raised areas further inland, including Lake Gårdsjön, receive about 1000 to 1100 mm. During the measurement period (1978 to 1995) at Lake Gårdsjön, precipitation has decreased during the summer and increased during the winter (Johansson and Nilsson, 1988) compared with the period 1931–60, when July and August had the highest precipitation amounts. Also annual sums of precipitation during the observation period have been higher, an average of 1137 mm compared with 1074 mm for the reference period (Johansson and Nilsson, 1988).

## Radiation

As part of the weather/climate monitoring, global radiation is measured with a sensor located at 2 m height on the weather mast just north of Gårdsjön. These measurements indicate very low daily averages during November through January (minimum often being $<1\,\mathrm{W\,m^{-2}}$) steadily increasing to reach the highest daily averages of approximately $300\,\mathrm{W\,m^{-2}}$ from late May to late July, though even in this summer period radiation may be as little as $35\,\mathrm{W\,m^{-2}}$ during cloudy and rainy days. The hourly maximum is reached around noon during bright sunny summer days at midsummer, and amounts to approximately $850\,\mathrm{W\,m^{-2}}$.

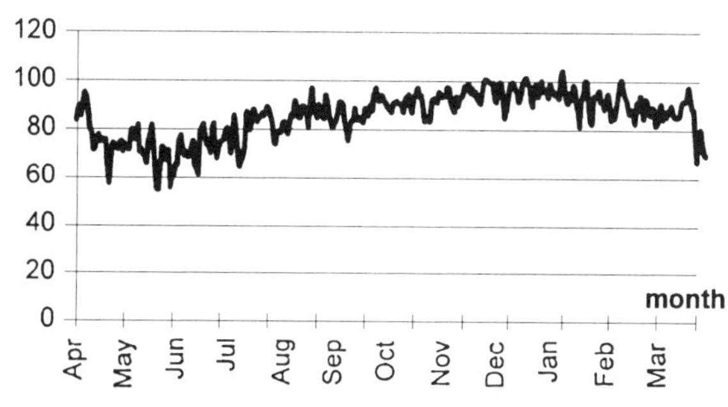


**Figure 2.7.** Daily mean relative humidity in open air at the Gårdsjön climate station (measured data during 4 years)

### Growing season

The climatic factors influencing the growing season can be defined in somewhat different ways. Olsson *et al.* (1985) gives a range of the growing season in the Lake Gårdsjön area from 210 to 240 days (average daily air temperature $>3°C$), while Borg (1995) estimates the period to be only 180 days based on the assumption that the first and last four consecutive days with an average temperature $>5°C$ (Odin *et al.*, 1983) initiate, and respectively terminate the growing season. Hence, this latter estimate gives a period spanning approximately from late April to mid-October which is consistent with other estimates from southern Sweden. During the unusually mild winter seasons in the latest period, the response of vegetation in accordance with the above definitions has been variable with species. Thus, the predominating Norway spruce (*Picea abies* (L.) Karst.) has not "responded" while birches (*Betula pendula* Roth. and *B. pubescens* Ehrh.) and many of the plant and moss species in the field and ground layers have started to grow very early in the spring or even been photosynthetically active throughout the winter season with radiation being the limiting factor. Different species, and in the case of Norway spruce, different clones, have different strategies for winter dormancy. These species-specific reaction patterns may be crucial for future changes when anticipated scenarios of global warming, might profoundly decrease the abundance of Norway spruce in natural forests in south and middle Scandinavia (Holten and Carey, 1992).

## HYDROLOGICAL CHARACTERISTICS

### General

The hydrology of forest soils is dependent on a pattern created from climatic factors, morphometric factors and intrinsic soil factors. This complex of factors regulating water and solute transport through forest catchment soils is important for both the development of acidification and its reversal (cf. Nyberg, 1995a; Chapters 5 and 21, this volume). A good understanding of the hydrology of the experimental catchments is thus imperative for the understanding of biogeochemical cycling. Products of the elemental transformations and reactions, in the form of different ions, will be transported out of catchments by water either as discharge to surface waters or as leachate to deeper groundwater aquifers. Apart from some specific in-lake changes, lakes also largely reflect the sum of all reactions taking place in the catchment soils as typically 80 to 90% of Swedish lake water has passed through soil. Due to varying soil thickness and other soil characteristics, flowpaths and transit times for the passage of water may be very different in different parts of the catchment, which may thus differ considerably in their influence on water chemistry.

Typically for the Gårdsjön region, there is a snowmelt period, usually coinciding with somewhat higher spring precipitation. However, this is not as prominent as in more northern areas with larger snow accumulation and low midwinter runoff (cf. Table 2.2, based partially on Johansson and Nilsson, 1988). On the contrary, runoff amounts are fairly high at Gårdsjön throughout the winter period, reflecting the balance between snow accumulation and intermittent melting periods with only

**Table 2.2.** Estimated monthly average runoff in catchment F1 based on regression calculations for the period 1951 to 1980 and observations during the period 1979 to 1993. Also shown are seasonal averages for the summer (June through August), an approximate growing season, the winter season, and an annual average for calendar years

| Month | Jan. | Feb. | Mar. | Apr. | May | Jun. | Jul. | Aug. | Sep. | Oct. | Nov. | Dec. |
|---|---|---|---|---|---|---|---|---|---|---|---|---|
| Runoff (mm) | 62 | 47 | 53 | 59 | 30 | 19 | 14 | 23 | 29 | 54 | 76 | 77 |
| Periods | Jun.–Aug. | | Apr.–Sep. | | Oct.–Mar. | | | | YEAR | | | |
| | 56 | | 173 | | 371 | | | | 544 | | | |

occasional impacts from soils freezing. During late spring–early summer the lower precipitation and the concomitant onset of forest transpiration decreases runoff considerably. Evapotranspiration outweighs a rather high average monthly precipitation resulting in low or very low runoff throughout the summer up to and including September. The combined effect of increased monthly precipitation amounts and the decrease in evapotranspiration as the dormancy season starts results in high runoff values during the autumn. This pattern can clearly be seen in the 15-year (1979 to 1994) records of precipitation and runoff from catchment F1 (Figure 2.8).

The hydrological years during the recorded period include a few very wet and a few very dry years, and variation around the average is greater for runoff compared with precipitation. This is due to the "wet years" in general being characterized by unusually high summer precipitation and sometimes cooler summers, reducing evapotranspiration and thus increasing summer run-off, even though the annual precipitation sum may not be unusually high. As is apparent from Tables 2.3 and 2.4, extreme months occur every now and then during the observation period. Hence, due to single events of precipitation, high monthly sums occurred during June 1980, 1987 and 1991, with the latter two also resulting in high runoff whereas

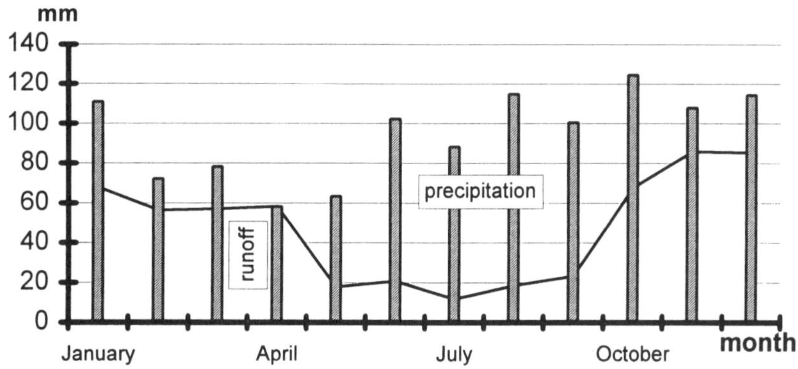


**Figure 2.8.** Average monthly precipitation (columns) and runoff (line) as observed at Lake Gårdsjön during a 15-year period, 1979/80 up to and including 1993/94. Note large difference between precipitation and runoff during the summer months

**Table 2.3.** Measured precipitation (mm) by month at Lake Gårdsjön, corrected values from station T2. Annual sums and average are for hydrological years

| | 1978/79 | 79/80 | 80/81 | 8 /82 | 82/83 | 83/84 | 84/85 | 85/86 | 86/87 | 87/88 | 88/89 | 89/90 | 90/91 | 91/92 | 92/93 | 93/94 | Average | 95% |
|---|---|---|---|---|---|---|---|---|---|---|---|---|---|---|---|---|---|---|
| October | | 67 | 122 | 153 | 114 | 137 | 207 | 99 | 150 | 111 | 132 | 135 | 85 | 99 | 81 | 121 | 120.8 | 17.5 |
| November | | 234 | 130 | 198 | 227 | 47 | 79 | 79 | 196 | 97 | 48 | 37 | 41 | 148 | 152 | 32 | 116.5 | 36.7 |
| December | | 100 | 177 | 38 | 199 | 79 | 75 | 246 | 134 | 79 | 101 | 84 | 127 | 53 | 60 | 151 | 113.5 | 29.7 |
| January | | 36 | 98 | 05 | 190 | 184 | 59 | 75 | 38 | 188 | 50 | 210 | 119 | 77 | 126 | 168 | 114.9 | 30.5 |
| February | | 36 | 58 | 33 | 57 | 30 | 15 | 05 | 77 | 171 | 157 | 202 | 79 | 53 | 40 | 05 | 67.9 | 31.0 |
| March | | 19 | 135 | 79 | 90 | 39 | 72 | 123 | 111 | 81 | 143 | 68 | 57 | 67 | 16 | 126 | 81.7 | 20.2 |
| April | | 10 | 10 | 45 | 92 | 39 | 119 | 46 | 34 | 67 | 57 | 84 | 37 | 134 | 43 | 63 | 58.6 | 18.2 |
| May | | 33 | 104 | 101 | 84 | 74 | 26 | 77 | 83 | 55 | 69 | 46 | 31 | 45 | 61 | 23 | 60.8 | 13.5 |
| June | 52 | 189 | 114 | 92 | 77 | 83 | 135 | 24 | 233 | 54 | 34 | 115 | 214 | 15 | 52 | 88 | 98.3 | 32.3 |
| July | 102 | 119 | 85 | 47 | 39 | 50 | 61 | 82 | 96 | 182 | 78 | 92 | 57 | 72 | 180 | 10 | 84.3 | 22.6 |
| August | 123 | 193 | 43 | 121 | 19 | 80 | 135 | 154 | 163 | 116 | 92 | 95 | 90 | 159 | 148 | 110 | 115.1 | 22.2 |
| September | 129 | 117 | 73 | 126 | 180 | 109 | 118 | 57 | 119 | 116 | 61 | 109 | 92 | 78 | 55 | 254 | 112.0 | 24.5 |
| Annual sum | 407 | 1153 | 1148 | 1136 | 1369 | 952 | 1099 | 1068 | 1432 | 1318 | 1022 | 1278 | 1029 | 1001 | 1014 | 1150 | 1144.6 | 73.1 |

**Table 2.4.** Observed monthly runoff values (mm) at the weir in catchment F1 for hydrological years

| | 1979/80 | 80/81 | 81/82 | 82/83 | 83/84 | 84/85 | 85/86 | 86/87 | 87/88 | 88/89 | 89/90 | 90/91 | 91/92 | 92/93 | Average | 95% |
|---|---|---|---|---|---|---|---|---|---|---|---|---|---|---|---|---|
| October | 51 | 95 | 100 | 75 | 78 | 127 | 49 | 62 | 44 | 79 | 51 | 54 | 42 | 23 | 66.4 | 14.2 |
| November | 194 | 120 | 183 | 215 | 24 | 55 | 52 | 169 | 47 | 28 | 20 | 24 | 121 | 135 | 99.1 | 36.4 |
| December | 52 | 190 | 12 | 148 | 69 | 72 | 172 | 67 | 34 | 79 | 61 | 92 | 44 | 68 | 82.9 | 26.7 |
| January | 22 | 42 | 3 | 126 | 75 | 6 | 65 | 10 | 144 | 40 | 121 | 116 | 70 | 113 | 68.0 | 25.9 |
| February | 3 | 64 | 77 | 14 | 26 | 2 | 14 | 36 | 165 | 114 | 158 | 42 | 47 | 28 | 56.4 | 27.8 |
| March | 5 | 117 | 149 | 50 | 9 | 26 | 100 | 40 | 35 | 106 | 47 | 44 | 50 | 19 | 57.0 | 23.1 |
| April | 67 | 22 | 27 | 63 | 56 | 140 | 59 | 111 | 92 | 20 | 43 | 30 | 76 | 10 | 58.4 | 19.5 |
| May | 6 | 16 | 40 | 37 | 3 | 16 | 37 | 35 | 6 | 10 | 3 | 3 | 33 | 2 | 17.8 | 8.0 |
| June | 26 | 21 | 16 | 5 | 6 | 16 | 5 | 118 | 2 | 2 | 13 | 61 | 0 | 0 | 20.8 | 16.9 |
| July | 51 | 18 | 2 | 2 | 1 | 1 | 1 | 19 | 4 | 0 | 17 | 15 | 0 | 32 | 11.7 | 7.9 |
| August | 63 | 1 | 0 | 0 | 0 | 18 | 18 | 84 | 5 | 0 | 0 | 1 | 4 | 61 | 18.4 | 15.2 |
| September | 63 | 3 | 22 | 23 | 3 | 63 | 28 | 62 | 31 | 0 | 6 | 1 | 18 | 4 | 23.3 | 12.4 |
| Annual sum | 604 | 710 | 630 | 758 | 350 | 542 | 601 | 813 | 608 | 477 | 540 | 484 | 506 | 497 | 580.0 | 87.7 |

runoff generally is low or even absent in June. Apart from these few unusually wet years, summer seasons (June to August) generally had low runoff, in particular during the summers 1982 to 1984, 1988, 1989, 1992 and 1994. Among these summers, 1983, 1984, 1989, 1992 and 1994 had comparatively low summer precipitation sums.

Examples of a dry and a wet year are shown in Figure 2.9 for the control catchment F1. More frequent appearance of heavy rainfall during 1986/87 resulted in a pronounced increase in runoff during this year, especially during the summer season when the precipitation amount was almost 300 mm larger during 1986/87 compared with the reference period. This difference in summer precipitation was the major fraction of the interannual difference (1432 mm 1986/87 compared with 952 mm 1983/84). The precipitation difference caused a much larger proportional difference in runoff (Table 2.4): 813 mm in 1986/87 compared with only 350 mm in 1983/84 when there was virtually no runoff between mid-June and early October.

### The water balance

For long-term data series there is practically no change in storage of water either in the soils or in lakes such as Lake Gårdsjön. Johansson and Nilsson (1988) estimated the long-term reference values for the water balance 'Precipitation ($P$)=Evapotranspiration ($E$)+Runoff ($Q$)' based on datasets from nearby weather and hydrological stations to be:

$$1071 = 554 + 517$$
$$P = E + Q$$

During 1979/80 to 1984/85, Johansson and Nilsson (1988) measured the following values for Lake Gårdsjön: precipitation=1143 mm, evapotranspiration=538 mm and runoff=605 mm; while the values for catchment F1 were: 1143 mm, 527 mm and 616 mm respectively. In both cases, $P$ is measured precipitation with corrections according to the formulas given by Eriksson (1980), $Q$ is measured runoff and $E$ is the evapotranspiration estimated as the difference $P-Q$. The longer data series that is now available for catchment F1 (1979/80 to 1993/94) gives the following values for the annual water balance:

$$1145 = 558 + 587$$

Most of the increase in precipitation results in a disproportional increase of runoff, which may be caused by a shift towards higher winter season precipitation (Johansson and Nilsson, 1988). Currently, there seems to be a weak trend towards lower runoff values since the start of the experimental manipulation (Figure 2.10). Observations from other national Swedish network stations (SMHI) located at the west coast suggest this may be a regional phenomenon. The implications of these hydrological changes for the assessment of the experimental deacidification is not clear but one effect may be a delayed "discovery" of responses as the control may be in a parallel state of change caused by a decreased acid load.

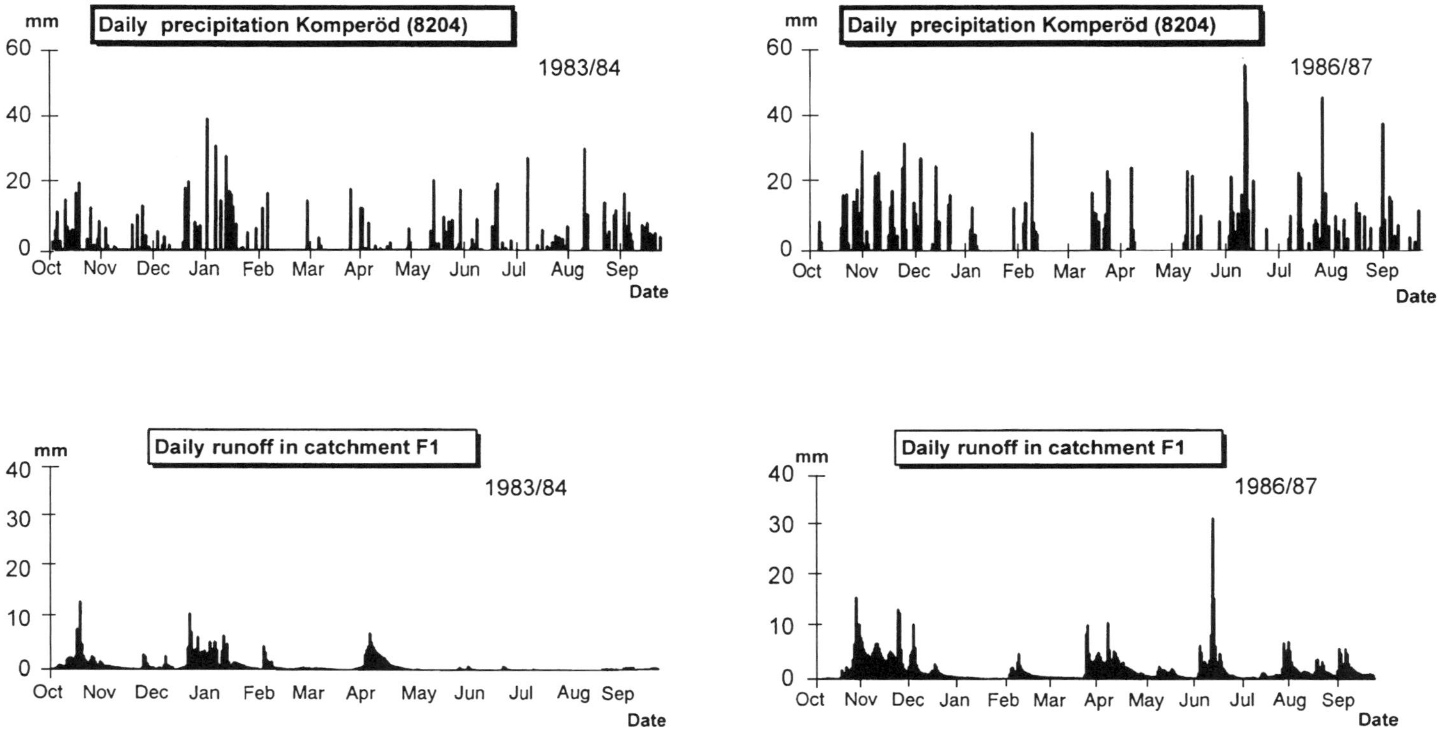


**Figure 2.9.** Examples of a dry (1983/84) and a wet hydrological year (1986/87) at Lake Gårdsjön, with daily values of corrected precipitation and runoff (measured in catchment F1) given in millimetres

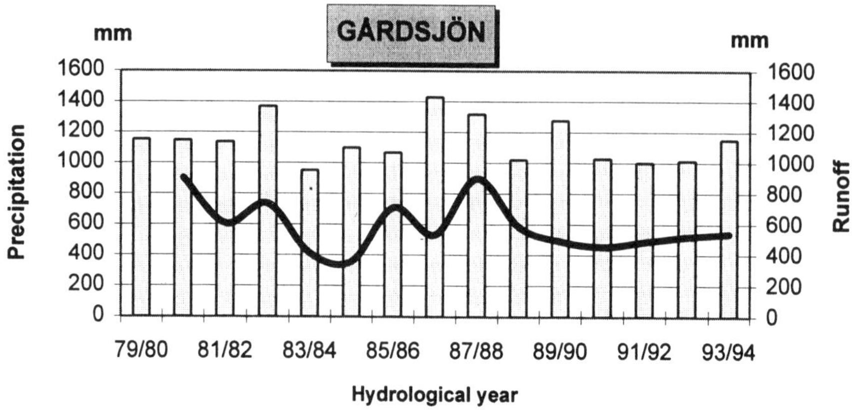


**Figure 2.10.** Observed annual sums of precipitation (columns) and runoff (line) at Lake Gårdsjön for hydrological years

**Soil hydrology**

The hydrological characteristics of the G1 catchment soils are explored in detail in Chapter 5 of this volume. One marked characteristic of the forest soils of the Gårdsjön area is a very high saturated hydraulic conductivity in the uppermost layers, reaching values of $10^{-3}$ m $s^{-1}$. While B horizon values in the G1 catchment are in contrast much lower at $5 \times 10^{-5}$ m $s^{-1}$ to $5 \times 10^{-4}$ m $s^{-1}$ (Nyström, 1985; Borg, 1995; Nyberg, 1995b). Similar conductivity values ($10^{-4}$ to $10^{-3}$ m $s^{-1}$) were given by Johansson and Nilsson (1985) for catchment F1. The important implication of this is a large increase in superficial flow as groundwater reaches the more conductive upper soils during heavy precipitation events (Nyberg, 1995a; Chapter 5, this volume). These, for till soils, high values of saturated hydraulic conductivity are apparently connected to the very large porosity and high concentrations of organic carbon, even in the Bh horizon. Nyberg (1995b) assumes that these features are related to the specific morphometry of these small valleys, resulting in a strongly concentrated flow routing down along the valley bottoms.

## SOILS

The bedrock of catchments in the area surrounding Lake Gårdsjön is overlain by generally thin podzolic and peaty soils with the latter type occurring in valley bottoms with little or virtually no slope. These soils are developed from a glacial till deposited at the decay of the ice sheet and originate predominantly from local bedrock, and are therefore of the same or very similar lithology. The soil depths in catchments G1 and G2 were measured in 1990 (Stuanes *et al.*, 1992) by pushing a steel rod down to the bedrock at each node in the 5 by 5 m gridnet constructed for the purpose of detailed mapping. The distribution of soil depths in G1 and G2 is shown in Figure 2.11. In catchment F1 a seismic survey in the central lower valley portion was performed in 1980 (cf. Fridh and Hayling, 1980; Melkerud, 1983). Detailed maps are not available for catchment F1 but the seismic measurements from one transect along the main axis of the catchment and a couple of oblique transects give an overview. The general picture is that the soils are as thin as in G1 and G2 on the steeper sides and towards the water divide, but soil depths in the valley bottom are much greater (mean 2.5 m; maximum >3.4 m) compared with minimum depths of 1.2 m for G1 and 1.4 m for G2.

### Texture and soil types

Mapping of soil texture, soil types and chemical composition was also performed during 1990 before the experimental manipulations started. Samples of humus layer soil were taken on a 5×10 m gridnet and mineral soil was sampled in a 10×10 m gridnet across the whole G1 and G2 catchments but only within a restricted sampling area in the lower part of catchment F1 (Kjønaas and Stuanes, 1991). Earlier analyses exist for catchment F1, however, which cover a wider area (Olsson *et al.*, 1985; Cronan and Goldstein, 1989; Nilsson, unpublished in Hultberg, 1994). In addition, samples were taken from three soil pits dug to represent the major ground vegetation–moisture regimes from each of G1 and G2 (Kjønaas *et al.*, 1992) and one profile in F1. One soil profile each from F1 and G1 was also sampled by Borg (1995) during 1989.

Melkerud (1983) summarises measurements from 26 soil samples distributed around the lake, though concentrating on catchments F1 and another experimental

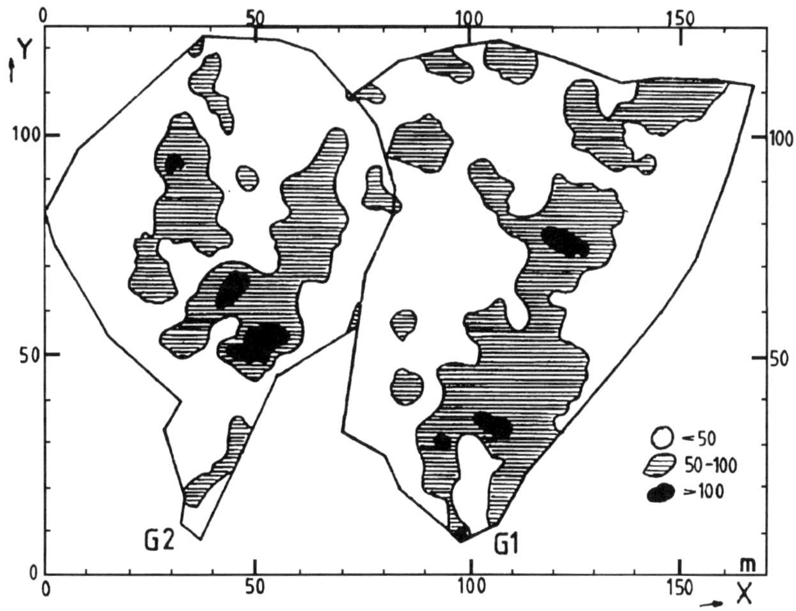


**Figure 2.11.** Soil depths in the manipulated catchments G1 and G2. Values for given depth classes are in centimetres

catchment (F3 in the north-east). The average soil texture was fine sandy to silty till usually with a clay content of approximately 5% and with insignificant amounts of gravel. A typical range of the grain size distributions in the podzols of the F1 area is shown in Figure 2.12, where silt is more prominent in one and the clay fraction more notable in the other, with the sand fraction predominating in both. Samples from the soil pits in catchments G1 and G2 have a similar grain size distribution (Figure 2.13) with the soils generally classified as sandy–fine sandy to silty loam (Kjønaas *et al.*, 1992). The sample G1 (Borg) is from an upper slope area of the valley where the organic content even in the mineral B horizons of the soil is very high, presumably as an effect of the convergent flowpaths occurring within this area (Nyberg, 1995b). The other two samples from G1 and G2 respectively are more typical for the catchment soils as a whole. A prominent mor layer occurs universally at the surface although the thickness may be rather variable and, especially in wetter (*Sphagnum* dominated) areas, there is a tendency to develop a thicker peaty or organic surface soil. The importance of this organic top layer is even more pronounced in the very thin soils on the sides of the catchments. These would generally be classified as bedrock outcrops in Swedish surveys where the limit value is <0.3 m soil depth (cf. map in Olsson *et al.*, 1985). The organic top layer is here usually underlain by a very thin mineral soil portion with sand or fine sand as the dominant size fraction (Figure 2.13).

The soils in the G1 ROOF and G2 NITREX catchments are classified according to the Canadian system as orthic humic podzols, orthic ferro-humic podzols, and with typical folisols at the shallow outcrops (Kjønaas *et al.*, 1992). Soils of the F1 area were similarly classified as orthic ferro-humic or humic podzols by Olsson *et al.* (1985) who also gives a more detailed description of mineralogical and elemental composition of clays and till samples based on a set of typical orthic podzol profiles.

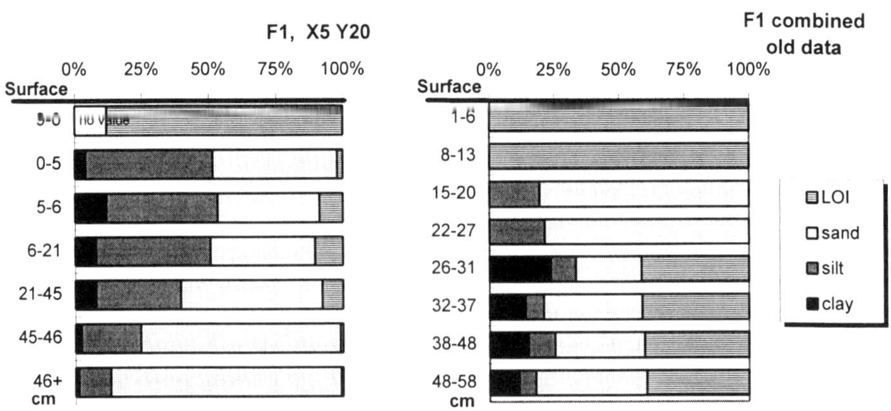


**Figure 2.12.** Soil texture and loss on ignition (LOI) (relative distribution) from older (1980) measurements and from one profile sampled during the current study

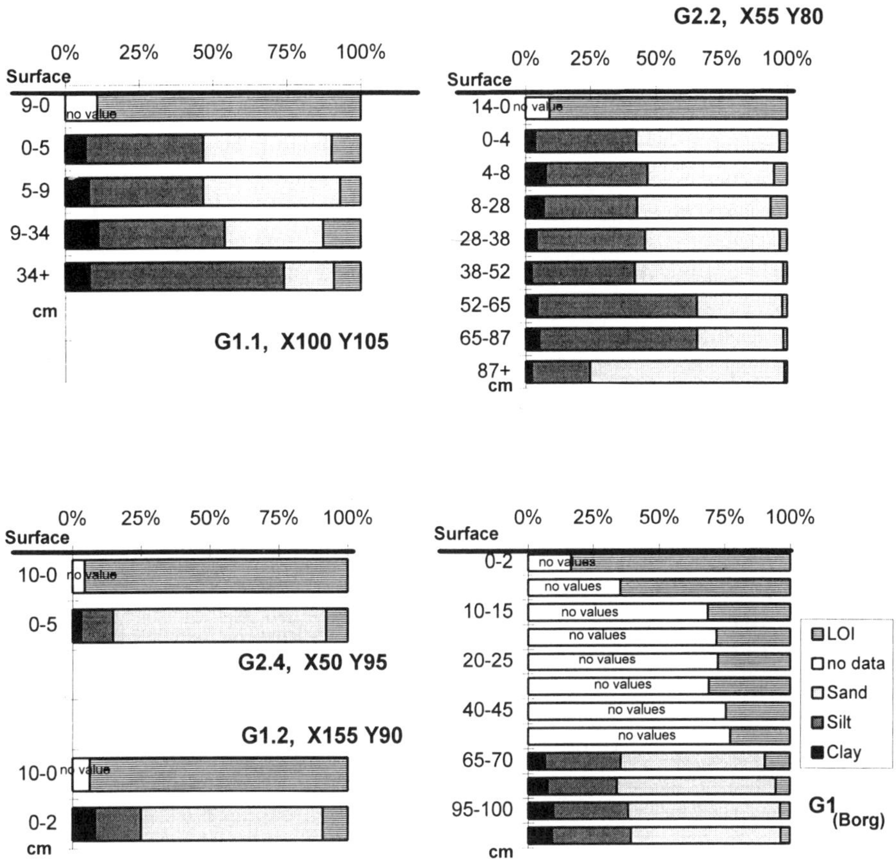


**Figure 2.13.** Soil texture and loss on ignition (LOI) (relative distribution) from a sloping valley bottom profile ($G1_{(Borg)}$) and deeper slope areas (G2.2 and G1.1), with podzol compared with two folisol profiles (G2.4 and G1.2) typical for the outcrop areas close to water divides. Based on Kjønaas *et al.* (1992) and unpublished data from Borg (1989)

## Mineral composition and chemical characteristics

The following general description is based on Melkerud (1983) (also cf. Olsson *et al.*, 1985) with samples mainly from and around the F1 Control area of which some points are very close to the catchments G1 and G2.

The soil mineral components derive from morainic material that was influenced by water during the post-glacial isostatic rebound, and there are some differences in the amount of various minerals within soil fractions and slight decreases or enrichments occur compared with the parental bedrock/till material. The clay fraction comprises predominantly (>75%) primary minerals with a distribution that is typical for a podzol profile. The bleached layer ($A_2$) is dominated by smectite. Here mica and illite are totally missing, but they increase deeper in the horizon as the weathering effect diminishes. Vermiculite, with different inter-layer hydroxy-aluminium polymers, constitutes the dominant clay mineral in the B horizon while easily weathered chlorite is essentially absent from these soils. In the silt fraction, quartz and feldspar minerals are predominant with fractions of 50–70% for quartz and about 15–20% for each of plagioclase and potassium feldspar. Other constituents, like amphiboles, epidote and different clay minerals, amount to less than 5%. An apparently even distribution within this fraction also reveals a clear increase of quartz and decrease of plagioclase and biotite compared with the underlying bedrock (also cf. Table 20.2, this volume, derived from the same source). Distribution of major elements in the indigenous till matrix is strikingly homogeneous throughout soils of the Lake Gårdsjön area (Olsson *et al.*, 1985) and several of the elements show signs of redistribution or leaching due to preferential weathering in the surface horizons. For further details see Melkerud (1983). For catchments G1 and G2, Sverdrup *et al.* (Chapter 20, this volume) have derived a mineralogical composition as the basis for their weathering estimates.

Samples, both from soil profiles and the grid sampling, were chemically analysed for base cations and nitrogen (some samples also for organic carbon and sulphur) using the methods described by Ogner *et al.* (1991). Exchangeable cations and other extractable elements (Al, Ca, K, Mg, Na, P and S) were determined in 1 M $NH_4NO_3$ extracts by ICP (inductively coupled plasma emission spectroscopy). The cation exchange capacity (CEC) is defined as $CEC_{NH_4NO_3}$=exchangeable acidity$+Na^++K^++Mg^{2+}+Ca^{2+}+Mn^{2+}+NH_4^+$. The base saturation (BS) is defined as $BS_{NH_4NO_3}=(Na^++K^++Mg^{2+}+Ca^{2+}+NH_4^+)/\ CEC_{NH_4NO_3}\times100\%$. The $NH_4^+$ value is included in the calculations only if it is available from a separate extraction with 1 M KCl (Kjønaas *et al.*, 1992).

Some selected physical and chemical properties of the soil from profiles sampled in the pits are given in Table 2.5. As is to be expected, the organic top layer, with values of ignition loss (LOI, %) around or slightly above 90%, has generally very high amounts of nitrogen, exchangeable acidity and cations. The content of total nitrogen (organically bound) is about an order of magnitude larger in the organic top layers, usually well above 1000 mmol $kg^{-1}$ as compared with 100 to 200 mmol $kg^{-1}$ in the eluvial and B horizons. Sample G1.3 deviates from this general pattern with very high carbon and nitrogen contents extending to a substantial depth in the profile. This profile is from the sloping valley bottom area mentioned above (Figure 2.13) where flows from upslope areas converge, but it is located close to prominent peaty soil on the flat bottom area and is therefore comparable to the G2.3 profile. The amounts of exchangeable aluminium are unusually high in parallel with a high organic content whereby the base saturation (BS) becomes comparatively low even in the mor layer. The "flowpath" location of the two profiles is probably much the same, with a large lateral component of water flow through the soil.

The C/N ratios are rather high in the soils of G1 and G2 based on the profile samples. Averages for the organic and the eluvial layers were around 40, and the ratio was 29.5 for the uppermost B horizon. These values are higher than the F1 profile, and also higher than the earlier reported ratios of 25 to 32 for the humus layer and about 20 to 25 for the uppermost mineral soil (Ae and upper B or Bh horizon) given by Olsson *et al.* (1985) for catchment F1. In spite of fewer samples this indicates that there may be some differences between G1 and G2 compared with F1.

Soil chemistry statistics based on the grid-samples for G1 and F1 are given in Table 2.6. The overall picture is similar to the profile samples. The organic top layers are very similar between the catchments, while the eluvial horizons tend to differ, with higher amounts of exchangeable acidity, cation exchange capacity (CEC), BS and single exchangeable cations in the soils of catchment G1. This is probably due to a larger content of organic material in the Ae horizon of the G1 catchment. While the highest values of CEC, exchangeable Al and acidity occur in the upper B horizon of catchment F1, exchangeable cations and BS were highest in the G1 Bh soil (Figure 2.14). This pattern could be due to original differences between the F1 and G1 soils, minor differences in elemental loadings resulting from the different slope aspects, and differences in water flowpaths between these catchments. However, these soil differences may above all reflect the differences in the size and heterogeneity of the two sampled areas, with F1 grid sampling covering a much smaller, flatter and relatively homogeneous area close to the weir.

## VEGETATION

### General and historic description

The Svartedalen region where Lake Gårdsjön is located comprises the most western part of the boreo-nemoral region bordering the Nemoral zone at the coast. The latter is characterised by the occurrence of several deciduous species which abound in more southern regions. The typical mixed conifer–deciduous appearance of the old boreo-nemoral forest stands is different from the present-day total dominance of coniferous trees due to human management over extensive areas as in Svartedalen. The Gårdsjön site is situated just east of the border for naturally occurring Norway spruce (*Picea abies* (L.) Karst.) which, as a floral element from the north-east, colonized this region rather late when it reached northern areas of the provinces Bohuslän and Västergötland a few hundred years AD (cf. Wallin and Renberg, 1985). Based on the pollen records from Lake Gårdsjön sediments, the same authors indicate a restricted abundance (*c*. 5%) in the region from around AD 1400 and with little change up to about AD 1850. During this period, Scots pine (*Pinus sylvestris* L.) was the most abundant species together with deciduous tree species such as birch (*Betula* spp.), oak (*Quercus* spp.) and alder (*Alnus glutinosa* [L.] Gaertn.), while typically more demanding nemoral deciduous species were rare.

The pollen stratigraphic record indicates that from AD 1850 and onwards the Norway spruce increased at the expense of the deciduous trees and especially the oak (*Quercus robur* L.). This change would mainly be an effect of human management, including large-scale planting of Norway spruce during the most recent period as the established forestry practice, initiated already during the late 19th century (cf. Wallin and Renberg, 1985). Such large-scale recent changes were not, however, verified by Olsson (1985) for the area located around Gårdsjön. Based on archives of written historical records and verbal information from elderly people in the area, he described local forestry and land-use history during the last centuries. The vicinity of Lake Gårdsjön seems to have experienced rather small forestry changes with conifers and especially Norway spruce being of major importance for a period embracing

**Table 2.5.** Some characteristics of representative soil profiles from the different vegetation–moisture regimes in the three catchment areas (Kjønaas *et al.*, 1992; Kjønaas and Stuanes 1996). Bulk density was determined on sieved soil samples

| Horizon | Depth (cm) | Sand (%) | Silt (%) | Clay (%) | pH ($H_2O$) | LOI (%) | Total N ($mmol_c\,kg^{-1}$) | C/N | Exch. acid. ($mmol_c\,kg^{-1}$) | CEC ($mmol\,kg^{-1}$) | BS (%) | Exch. Al ($mmol\,kg^{-1}$) | Bulk den. ($kg\,m^{-3}$) |
|---|---|---|---|---|---|---|---|---|---|---|---|---|---|
| **Profile G1.1, Coordinates X100Y105, Loam over silty loam, Orthic Ferro-Humic Podzol** | | | | | | | | | | | | | |
| LFH | 9–0 | | | | 3.53 | 89.2 | 951 | 44.3 | 204.7 | 387.6 | 46.9 | 10.5 | 252 |
| Aeh | 0–5 | 48.0 | 44.5 | 7.5 | 3.61 | 11.0 | 97 | 53.3 | 57.3 | 69.4 | 17.4 | 11.6 | 747 |
| Bf | 5–9 | 49.5 | 41.8 | 8.6 | 3.56 | 7.5 | 78 | 43.8 | 66.7 | 71.3 | 6.4 | 19.9 | 921 |
| Bhf | 9–34 | 38.1 | 49.1 | 12.8 | 4.06 | 14.7 | 169 | 34.0 | 77.1 | 81.8 | 5.7 | 28.8 | 774 |
| Bf | 34+ | 18.5 | 72.5 | 9.0 | 4.36 | 10.2 | 156 | 25.1 | 35.3 | 38.2 | 7.6 | 13.7 | 882 |
| **Profile G1.2, Coordinates X155Y90, Loamy sand, Typical Folisol** | | | | | | | | | | | | | |
| LFH | 10–0 | | | | 3.66 | 93.8 | 1174 | 37.5 | 176.8 | 378.7 | 52.7 | 8.9 | 245 |
| Aeh | 0–2 | 79.1 | 16.3 | 4.6 | 3.85 | 10.0 | 131 | 36.9 | 45.6 | 59.1 | 22.7 | 8.9 | 745 |
| **Profile G1.3, Coordinates X115Y70, Silty loam, Orthic Ferro-Humic Podzol** | | | | | | | | | | | | | |
| LFH | 8.5–0 | | | | 4.00 | 90.7 | 1391 | | 308.1 | 359.3 | 14.0 | 97.5 | 215 |
| Bhf1 | 0–6 | 31.3 | 50.5 | 18.2 | 3.74 | 21.4 | 430 | 23.5 | 132.2 | 144.7 | 7.8 | 40.9 | 643 |
| Bhf2 | 6–32 | 37.8 | 55.7 | 6.5 | 4.20 | 30.3 | 515 | 24.3 | 86.7 | 95.2 | 8.4 | 32.2 | 657 |
| Bhf3 | 32–5 | 43.3 | 52.8 | 3.9 | 4.31 | 18.2 | 275 | 26.4 | 51.6 | 56.3 | 8.2 | 19.4 | 789 |
| Bf | 54–7 | 23.5 | 64.9 | 11.5 | 4.28 | 5.9 | 97 | 23.2 | 29.3 | 32.4 | 9.5 | 11.4 | 1080 |
| 2BC | 79–9 | 74.3 | 22.9 | 2.8 | 4.27 | 2.1 | 19 | 35.1 | 14.8 | 16.8 | 11.8 | 5.1 | 1549 |
| **Profile F1.1, Coordinates X5Y20, Sandy loam, Orthic Ferro-Humic Podzol** | | | | | | | | | | | | | |
| LFH | 9–0 | | | | 3.59 | 88.3 | 1271 | 32.3 | 183.1 | 370.7 | 50.1 | 10.3 | 263 |
| Ae | 0–5 | 47.2 | 48.8 | 4.0 | 3.52 | 2.2 | 39 | 27.8 | 19.3 | 22.1 | 12.2 | 3.9 | 1205 |
| Bhf1 | 5–6 | 41.4 | 45.8 | 12.7 | 3.55 | 9.7 | 147 | 28.9 | 98.6 | 104.5 | 5.6 | 27.6 | 923 |
| Bhf2 | 6–21 | 43.4 | 47.9 | 8.7 | 3.96 | 11.8 | 150 | 31.7 | 67.1 | 72.4 | 7.2 | 23.2 | 835 |
| Bf | 21–4 | 57.1 | 34.5 | 8.4 | 4.19 | 8.6 | 103 | 30.7 | 44.3 | 52.1 | 14.9 | 16.4 | 1024 |
| BC | 45–4 | 75.2 | 22.5 | 2.3 | 4.32 | 1.1 | 13 | 19.2 | 11.6 | 13.4 | 13.2 | 4.3 | 1515 |
| C | 46+ | 86.8 | 12.0 | 1.2 | 4.59 | 0.8 | 7 | | 10.1 | 11.2 | 9.4 | 3.5 | 1395 |

| | | | | | | | | | | | | | |
|---|---|---|---|---|---|---|---|---|---|---|---|---|---|
| **Profile G2.1, Coordinates X75Y105, Sandy loam, Orthic Humic Podzol** | | | | | | | | | | | | | |
| LF | 14–5 | | | | 3.74 | 93.8 | 1128 | 39.4 | 150.1 | 373.6 | 59.0 | 6.3 | 203 |
| FH | 5–0 | | | | 3.45 | 93.1 | 857 | 52.1 | 240.6 | 475.3 | 49.1 | 10.3 | 228 |
| Aeh | 0–4 | 50.5 | 44.3 | 5.2 | 3.58 | 10.1 | 115 | 46.4 | 57.5 | 70.3 | 18.1 | 12.7 | 825 |
| Bf | 4–8 | 50.5 | 42.9 | 6.6 | 3.63 | 7.3 | 74 | 48.4 | 64.3 | 72.3 | 11.0 | 19.0 | 899 |
| Bhf | 8–18 | 48.2 | 46.9 | 4.9 | 4.18 | 11.7 | 134 | 34.8 | 50.5 | 57.4 | 11.9 | 19.3 | 803 |
| Bh | 18–3 | 52.1 | 44.8 | 3.1 | 4.26 | 6.9 | 91 | 29.3 | 29.6 | 32.9 | 10.0 | 11.3 | 1003 |
| BC | 30+ | 53.4 | 42.9 | 3.7 | 4.13 | 3.5 | 44 | 28.4 | 17.8 | 19.1 | 7.0 | 6.5 | 1259 |
| **Profile G2.2, Coordinates X55Y80, Sandy loam, Gleyed Humo-Ferric Podzol** | | | | | | | | | | | | | |
| LFH | 14–0 | | | | 3.66 | 90.7 | 945 | 45.8 | 178.8 | 374.0 | 51.8 | 7.9 | 232 |
| Ae | 0–4 | 56.1 | 40.2 | 3.8 | 3.74 | 3.1 | 36 | 44.0 | 34.7 | 37.7 | 8.0 | 8.7 | 1082 |
| Bf1 | 4–8 | 50.7 | 41.1 | 8.2 | 3.85 | 5.3 | 62 | 37.6 | 58.0 | 61.4 | 5.5 | 21.6 | 1014 |
| Bf2 | 8–28 | 54.2 | 38.3 | 7.5 | 4.22 | 6.7 | 82 | 32.5 | 24.4 | 26.2 | 6.8 | 10.0 | 991 |
| BC | 28–3 | 52.8 | 42.8 | 4.4 | 4.24 | 3.0 | 36 | 30.1 | 13.1 | 14.2 | 7.8 | 5.4 | 1199 |
| C | 38–5 | 57.4 | 40.3 | 2.3 | 4.32 | 1.5 | 16 | | 7.6 | 8.3 | 8.8 | 3.1 | 1326 |
| 2Cg1 | 52–6 | 33.3 | 62.1 | 4.7 | 4.45 | 1.9 | 25 | | 9.8 | 10.8 | 9.0 | 4.2 | 1213 |
| 2Cg2 | 65–8 | 33.4 | 61.3 | 5.3 | 4.31 | 1.5 | 16 | | 8.5 | 9.9 | 14.1 | 3.5 | 1242 |
| 3Cg | 87+ | 75.0 | 22.8 | 2.3 | 4.33 | 1.1 | 9 | | 5.8 | 6.5 | 10.3 | 2.2 | 1560 |
| **Profile G2.3, Coordinates X35Y90, Sandy loam over silt, Orthic Humic Podzol** | | | | | | | | | | | | | |
| LF | 33–21 | | | | 3.98 | 93.2 | 1205 | 35.1 | 423.7 | 428.8 | 11.4 | 129.7 | 183 |
| F | 21–17 | | | | 4.00 | 88.2 | 1019 | 37.7 | 534.8 | 573.4 | 6.7 | 175.7 | 208 |
| Oh1 | 17–0 | | | | 4.13 | 66.5 | 1240 | 25.5 | 204.6 | 223.1 | 8.3 | 71.0 | 538 |
| Oh2* | 17–0 | | | | 4.41 | 25.2 | 476 | 22.7 | 66.3 | 72.2 | 8.2 | 24.8 | 452 |
| Bh | 0–11 | 41.8 | 50.8 | 7.4 | 4.40 | 9.4 | 177 | 24.5 | 38.8 | 41.3 | 6.2 | 13.9 | 842 |
| CB | 11–2 | 58.9 | 37.8 | 3.4 | 4.42 | 4.7 | 74 | 17.3 | 25.1 | 26.5 | 5.4 | 9.1 | 1064 |
| 2Cg | 26+ | 15.2 | 79.0 | 5.9 | 4.42 | 3.9 | 66 | | 15.8 | 16.9 | 6.2 | 5.1 | 1074 |
| **Profile G2.4, Coordinates X50Y95 Loamy sand, Typical Folisol** | | | | | | | | | | | | | |
| LF | 10–0 | | | | 3.57 | 95.6 | 1002 | | 210.1 | 385.1 | 45.1 | 7.7 | 235 |
| Aeh | 0–5 | 83.9 | 12.7 | 3.4 | 3.76 | 8.6 | 104 | 39.3 | 41.4 | 45.4 | 8.8 | 9.3 | 883 |

* Oh2 consists of large pockets within Oh1, filled with oxides.

**Table 2.6.** Arithmetic average values of pH, LOI (%), CEC ($mmol_c\,kg^{-1}$), BS (%), total nitrogen ($mmol\,kg^{-1}$) and exchangeable base cations ($mmol_c\,kg^{-1}$), Al, Fe, Mn ($mmol\,kg^{-1}$), and acidity ($mmol_c\,kg^{-1}$) in the different soil layers of the ROOF and control catchment. Regular sampling performed based on the gridnet (Kjønaas and Stuanes, 1996)

| | pH ($H_2O$) | LOI | CEC | BS | Ex. acid | Ex. Na | Ex. K | Ex. Mg | Ex. Ca | Ex. Al | Ex. Fe | Ex. Mn | Total N |
|---|---|---|---|---|---|---|---|---|---|---|---|---|---|
| **G1(ROOF)** | | | | | | | | | | | | | |
| *Humus Layer (132 samples)* | | | | | | | | | | | | | |
| Maximum | 5.70 | 96.2 | 481.5 | 68.8 | 391.1 | 37.2 | 31.8 | 35.3 | 99.1 | 126.3 | 8.0 | 3.6 | 1838 |
| Minimum | 3.50 | 21.6 | 118.1 | 7.6 | 96.3 | 2.0 | 2.1 | 1.4 | 1.6 | 3.8 | 0.2 | 0.0 | 356 |
| Mean | 3.79 | 87.0 | 352.5 | 43.7 | 193.6 | 11.7 | 17.0 | 20.1 | 43.6 | 27.1 | 1.7 | 1.4 | 1175 |
| Standard deviation | 0.25 | 10.0 | 57.1 | 14.5 | 55.8 | 5.3 | 5.7 | 7.4 | 20.6 | 30.8 | 1.3 | 0.8 | 206 |
| *Bleached Horizon (35 samples)* | | | | | | | | | | | | | |
| Maximum | 4.16 | 28.1 | 159.7 | 36.5 | 152.4 | 3.9 | 6.5 | 5.3 | 15.0 | 51.8 | 4.0 | 0.5 | 361 |
| Minimum | 3.58 | 1.6 | 21.6 | 4.5 | 20.0 | 0.5 | 0.1 | 0.2 | 0.3 | 3.4 | 0.1 | 0.0 | 31 |
| Mean | 3.79 | 11.1 | 65.0 | 19.2 | 52.0 | 1.6 | 2.3 | 1.6 | 2.9 | 12.2 | 0.9 | 0.1 | 147 |
| Standard deviation | 0.13 | 6.6 | 29.2 | 8.5 | 24.2 | 0.9 | 1.3 | 1.2 | 3.0 | 8.6 | 0.8 | 0.1 | 81 |
| *Upper B Horizon (Mainly Bh, 18 samples)* | | | | | | | | | | | | | |
| Maximum | 4.51 | 30.8 | 159.7 | 36.5 | 152.4 | 3.9 | 6.9 | 5.3 | 15.0 | 51.8 | 5.9 | 0.5 | 463 |
| Minimum | 0.13 | 1.6 | 21.6 | 4.5 | 20.0 | 0.5 | 0.1 | 0.2 | 0.3 | 3.4 | 0.0 | 0.0 | 31 |
| Mean | 3.79 | 11.5 | 67.6 | 14.5 | 58.0 | 1.5 | 2.1 | 1.4 | 2.3 | 16.8 | 1.4 | 0.1 | 170 |
| Standard deviation | 0.69 | 7.5 | 29.9 | 8.0 | 27.0 | 0.8 | 1.7 | 1.3 | 2.9 | 9.5 | 1.6 | 0.1 | 110 |
| *Lower B Horizon (44 samples)* | | | | | | | | | | | | | |
| Maximum | 4.53 | 21.5 | 141.3 | 14.4 | 124.7 | 4.5 | 2.5 | 2.8 | 3.5 | 40.3 | 4.3 | 1.1 | 276 |
| Minimum | 3.60 | 3.4 | 13.8 | 5.1 | 12.1 | 0.5 | 0.3 | 0.2 | 0.3 | 4.7 | 0.0 | 0.0 | 36 |
| Mean | 4.11 | 9.9 | 72.7 | 8.8 | 66.0 | 1.4 | 1.2 | 0.8 | 1.1 | 22.6 | 1.2 | 0.1 | 135 |
| Standard deviation | 0.24 | 3.9 | 27.9 | 2.3 | 24.9 | 0.8 | 0.6 | 0.6 | 0.8 | 8.1 | 1.1 | 0.2 | 56 |

| | | | | | | | | | | | | | |
|---|---|---|---|---|---|---|---|---|---|---|---|---|---|
| **F1(Control)** | | | | | | | | | | | | | |
| *Humus Horizon (33 samples)* | | | | | | | | | | | | | |
| Maximum | 4.18 | 96.1 | 541.5 | 58.9 | 391.6 | 18.5 | 35.5 | 34.4 | 82.5 | 110.1 | 6.1 | 2.2 | 1417 |
| Minimum | 3.50 | 45.0 | 156.0 | 9.4 | 122.1 | 2.7 | 3.0 | 2.9 | 2.7 | 5.7 | 0.4 | 0.1 | 780 |
| Mean | 3.69 | 82.7 | 362.5 | 42.1 | 204.0 | 10.4 | 16.4 | 21.1 | 43.9 | 23.7 | 1.8 | 0.9 | 1049 |
| Standard deviation | 0.16 | 13.1 | 76.1 | 13.6 | 56.7 | 3.7 | 6.7 | 8.1 | 20.9 | 22.1 | 1.3 | 0.6 | 144 |
| *Bleached Horizon (21 samples)* | | | | | | | | | | | | | |
| Maximum | 4.39 | 19.7 | 109.8 | 30.7 | 90.8 | 2.1 | 3.1 | 3.8 | 3.5 | 26.3 | 1.2 | 0.1 | 396 |
| Minimum | 3.55 | 0.0 | 29.6 | 6.7 | 24.3 | 0.4 | 0.2 | 0.3 | 0.4 | 3.4 | 0.1 | 0.0 | 43 |
| Mean | 3.64 | 6.2 | 45.4 | 15.9 | 38.3 | 1.0 | 1.2 | 0.9 | 1.5 | 8.7 | 0.5 | 0.0 | 123 |
| Standard deviation | 0.83 | 4.8 | 19.7 | 6.0 | 17.3 | 0.5 | 0.7 | 0.7 | 0.9 | 5.9 | 0.3 | 0.0 | 90 |
| *Upper B Horizon (Mainly Bh, 9 samples)* | | | | | | | | | | | | | |
| Maximum | 4.51 | 16.2 | 151.4 | 10.7 | 138.4 | 1.5 | 2.0 | 1.4 | 3.6 | 44.8 | 4.5 | 0.0 | 259 |
| Minimum | 3.92 | 6.2 | 49.8 | 7.3 | 45.2 | 0.8 | 0.4 | 0.2 | 0.9 | 15.7 | 0.2 | 0.0 | 104 |
| Mean | 4.25 | 13.5 | 80.6 | 9.0 | 73.3 | 1.2 | 1.2 | 0.8 | 1.7 | 24.8 | 1.2 | 0.0 | 200 |
| Standard deviation | 0.23 | 2.9 | 28.0 | 1.0 | 25.6 | 0.3 | 0.5 | 0.4 | 0.8 | 7.9 | 1.3 | 0.0 | 43 |
| *Lower B-Horizon (16 samples)* | | | | | | | | | | | | | |
| Maximum | 4.50 | 14.3 | 95.4 | 13.4 | 86.4 | 1.6 | 1.9 | 0.9 | 2.6 | 29.7 | 1.7 | 0.0 | 210 |
| Minimum | 4.05 | 6.9 | 29.7 | 7.0 | 26.5 | 0.4 | 0.8 | 0.2 | 0.5 | 9.1 | 0.1 | 0.0 | 16 |
| Mean | 4.31 | 11.0 | 64.0 | 9.2 | 58.2 | 1.0 | 1.4 | 0.5 | 1.2 | 20.3 | 0.7 | 0.0 | 149 |
| Standard deviation | 0.13 | 2.1 | 18.4 | 1.4 | 17.0 | 0.3 | 0.3 | 0.2 | 0.5 | 5.6 | 0.5 | 0.0 | 46 |
| *BC and C Horizon (10 samples)* | | | | | | | | | | | | | |
| Maximum | 4.58 | 5.2 | 34.2 | 12.7 | 32.2 | 0.6 | 0.6 | 0.4 | 0.5 | 11.1 | 0.2 | 0.0 | 77 |
| Minimum | 4.27 | 2.3 | 13.4 | 5.8 | 11.7 | 0.3 | 0.1 | 0.0 | 0.2 | 4.3 | 0.0 | 0.0 | 23 |
| Mean | 4.45 | 3.7 | 23.9 | 8.0 | 22.1 | 0.4 | 0.4 | 0.2 | 0.3 | 7.9 | 0.1 | 0.0 | 48 |
| Standard deviation | 0.08 | 1.1 | 7.1 | 1.8 | 6.9 | 0.1 | 0.2 | 0.1 | 0.1 | 2.4 | 0.0 | 0.0 | 19 |

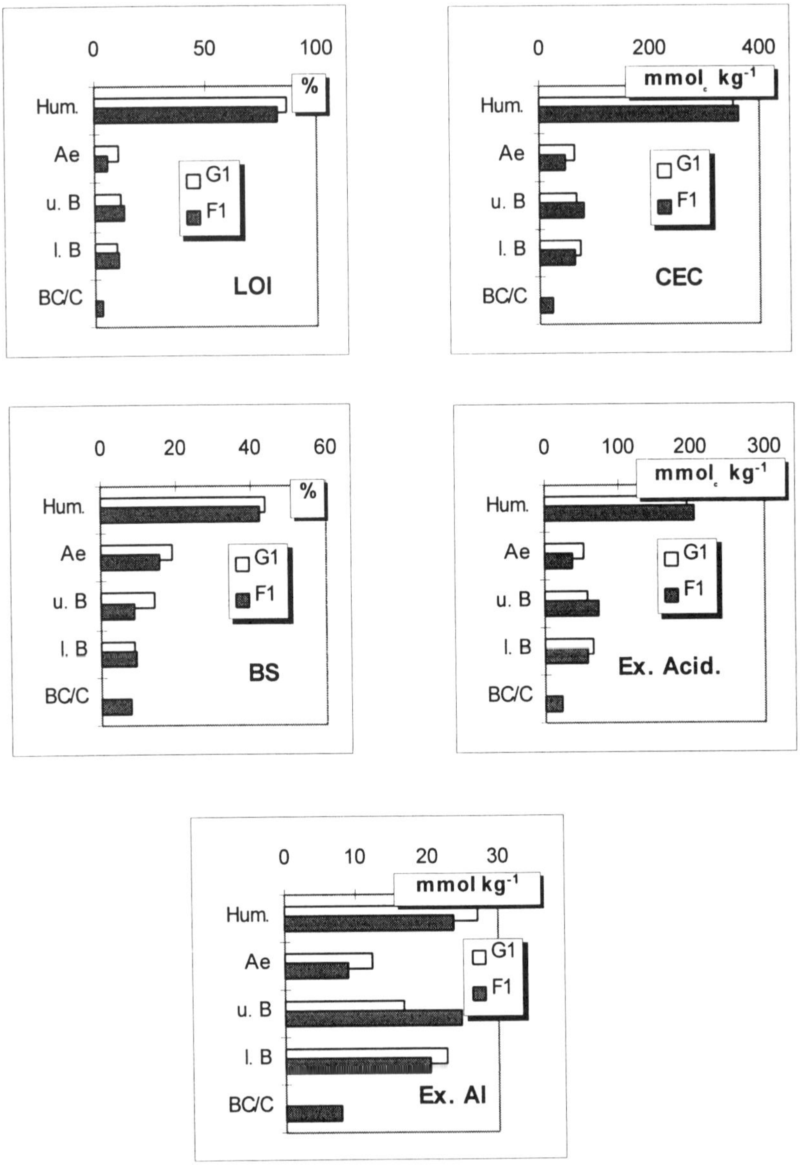


**Figure 2.14.** Average concentrations of organic material (LOI), cation exchange capacity (CEC), base saturation (BS), and exchangeable acidity and aluminium in the soil layers of catchments F1 (Control) and G1 (ROOF)

more than 200 years. However, land-use changes include cessation of a former, but not very intensive, practice of cattle grazing on the outfield forest grounds. This habit, which decreased slowly during the period and finally totally ceased around 1950, was of some importance for forest regeneration and generally created a more open forest with little understorey.

At the turn of the 19th century and during the first decade of the 20th century, an extensive exploitation cutting was performed in the region as in many other surrounding regions. Olsson (1985) does not assume this to have been an ordinary clear-felling, in the modern sense, but rather a selective dimension cutting taking away most of the larger pine and spruce timber. The residual included at least a number of suppressed small trees that comprise the giants of the current forest generation which can be identified based on their girdle of very densely spaced twigs at about 2 m height, much the same appearance as displayed by the currently occurring suppressed Norway spruce understorey. To what extent this old cutting should be regarded as clear-felling depends on the density and closure of the old stand of which little is known, although heavily grazed forests tended to be rather open as regrowth was effectively hampered. The abundance and "small size" of these former suppressed trees indicates that grazing could have been slight, and consequently the old stand rather dense. Hence, in the area south of Lake Gårdsjön where the experimental and control catchments are located, this old cutting may well be classified as approaching a modern clear-felling, though without the excessive "cleaning up" of the understorey which has been recent practice.

**Present forest structure**

Today, the forest in the Lake Gårdsjön area is dominated by mature, 80–100-year-old Norway spruce. On some ridges and hilltops, a drier and more open Scots pine forest dominates and on outcrop areas only sparse pine forest occurs. For some catchment areas like G1 ROOF there is virtually not a single deciduous tree. Occurrence of understorey is rather variable and in some more intensively or lately managed areas almost absent, otherwise it comprises mostly young spruce and more rarely young pine or in openings regrowth of birch. Single specimens of alder and birches (both *Betula pendula* Roth. and *B verrucosa* Ehrh.) or small thickets of alder buckthorn (*Rhamnus frangula* L.) or willows (*Salix caprea* L., *Salix* spp.) occur with a preference for wetter parts of the catchments. Single specimens of mountain-ash (*Sorbus aucuparia* L.), aspen (*Populus tremula* L.), and in F1 also hazel (*Corylus avellana* L.) are dispersed throughout the catchments. In the more open areas on hilltops, junipers (*Juniperus communis* L.) are a sparse but regularly occurring element of the flora.

In catchment G1 ROOF, there is a small windblown "clearing" at the upper saddle area with some regrowth of Norway spruce and a few oaks, but apart from this the stand is mature and comparatively dense (Stuanes *et al.*, 1992). In catchment G2 NITREX, the western ridge area and also the central part are rather open due to exposed bedrock or thin soils while the narrow lower end naturally had sparse vegetation further thinned by late management with only a few large pine trees as canopy builders. As in catchment G1, the relative openness of the outlet area just above the runoff dam has resulted in a regrowth of birch and tiny pine. In Figure 2.15 the canopy closure is shown as a ground projection of major tree canopies. The larger size of catchment F1 Control means that variation is more pronounced within this catchment, e.g. a pine regrowth occurs in the upper part (following clear-felling in 1976), a bog/fen area is located in the southern part and a rare hurricane in 1969

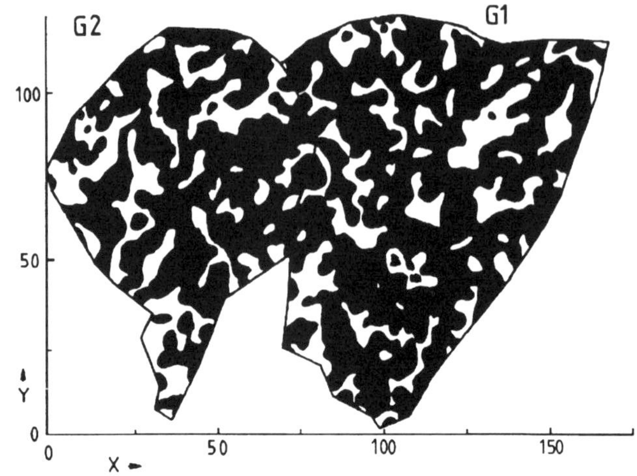


**Figure 2.15.** Catchment areas in G1 (ROOF) and G2 (NITREX) covered by canopy projections of larger trees (black)

blew down an area at the eastern border, now occupied by a very dense thicket of birch and young Norway spruce. A few larger trees, usually Norway spruce suffering from root rots, are blown down during most winter seasons due to strong winds and thus generate small openings in this already fairly open stand.

### Vegetation associations and dominant floral elements

Ground and field layer vegetation varies with density of the forest and the humidity of the soil, being better developed in areas with most light penetration. The acidic mor layer, poor nutrient and light conditions in this managed mature forest result in an ordinary and sparse flora with herbs being especially scarce. There are five main vegetation associations in the area:

1. A forest dominated by Norway spruce or mixed conifers with dwarf shrubs, especially *Vaccinium myrtillus* L., as ground cover.
2. A Norway spruce forest with the grass *Deschampsia flexuosa* (L.) Trin as prominent ground vegetation.
3. A Norway spruce forest with a ground flora usually dominated by the moss *Dicranum majus* Sm. though occasionally other moss species occur as field layer components.
4. Small patches of heather *Calluna vulgaris* (L.) Salisb. occur in gaps in the forest canopy on the drier outcrops in combination with several species of the lichen genus *Cladonia* as the most abundant species.
5. Mosses of the genus *Sphagnum*, especially *S. girgensohnii* Russ., are very dominant in discharge areas, along the brooklets as in F1, or in marginal areas of fens or the small bogs. As with (4), these occur largely under gaps in the forest canopy.

Sub-dominant ground and field layer species include the dwarf shrub *Vaccinium vitis-idea* L., usually in drier more open ridge/hilltop areas, and the mosses *Pleurozium schreberi* (Brid.) Mitt. and *Hylocomium splendens* (Hedw.) Schimp. in fresh but fairly open *Vaccinium* dominated parts. Locally abundant in F1 is the large fern *Pteridium aquilinum* (L.) (bracken), and in some of the more humid areas *Phegopteris polypodioides* Feé. In the middle fairly open area on the eastern slope of F1, several species of herbs occur, including *Maianthemum bifolium* (L.) F. W. Schm., *Melampyrum pratense* L., *Oxalis acetosella* L., *Potentilla erecta* (L.) Räusch. and the tall grass *Calamagrostis arundinacea* (L.) Roth. In the wetland areas are sub-oceanic components, like *Erica tetralix* L. and *Narthecium ossifragum* (L.) Huds, which have a distinct western Swedish distribution, as well as dominant tussock-building species such as *Trichophorum caespitosum* (L.) Hartm. and *Eriophorum vaginatum* L. Another western sub-oceanic plant, the small fern *Blechnum spicant* (L.) Roth., is also observed sparsely within the Gårdsjön area. Typical features of the relatively undisturbed forests at Lake Gårdsjön (not managed for several decades) include a rich occurrence of tree-living lichens of the genus *Usnea,* and the fallen trunks in different states of degradation which are seldom seen in managed stands. However, the most abundant arbusculoid lichen on the trees of the area is *Hypogymnia physodes* (L.) Nyl., which is a species rather insensitive to acidification and polluted air.

A combination of vegetation associations and soil moisture regimes for each of the catchments are shown in the vegetation maps (Figures 2.16, 2.17, and 2.18) modified from originals produced by NISK (G1 and G2) and the Swedish Environmental Protection Agency (F1). Details and also definition for the different units identified are somewhat variable between the G1–G2 catchments and the F1 catchment, but basically they give a good picture of the intra- and inter-catchment variation that occurs around Lake Gårdsjön.

## ELEMENTAL STORES

### Soils

Soils at Gårdsjön were sampled on a number of occasions: 1981 (Olsson *et al.*, 1985), 1989 (Borg, unpublished); 1991 (Grodzinska *et al.*, 1991; Nilsson, 1992), and finally the current samplings performed as parts of the ROOF and NITREX projects (Kjønaas *et al.*, 1992; Kjønaas and Stuanes, 1996). The different datasets that can be compiled vary in their resolution from data obtained from a few soil pits to stratified random or regular sampling covering the whole catchments. Data also vary with respect to identification and confinement to the dominant soil layers and analytical procedures have also varied between the samplings, although generally well-known standard methods have been applied.

Average concentration values for exchangeable elements in the F1 (Control) and G1 (ROOF) catchments are given in Table 2.6. As the sampling area in F1 was areally restricted compared with the G1 catchment, data from F1 were checked against independent samplings (Olsson *et al.*, 1985; Grodzinska *et al.*, 1991; Nilsson, 1992). The different sampling intensities and areal distributions, coupled with a risk

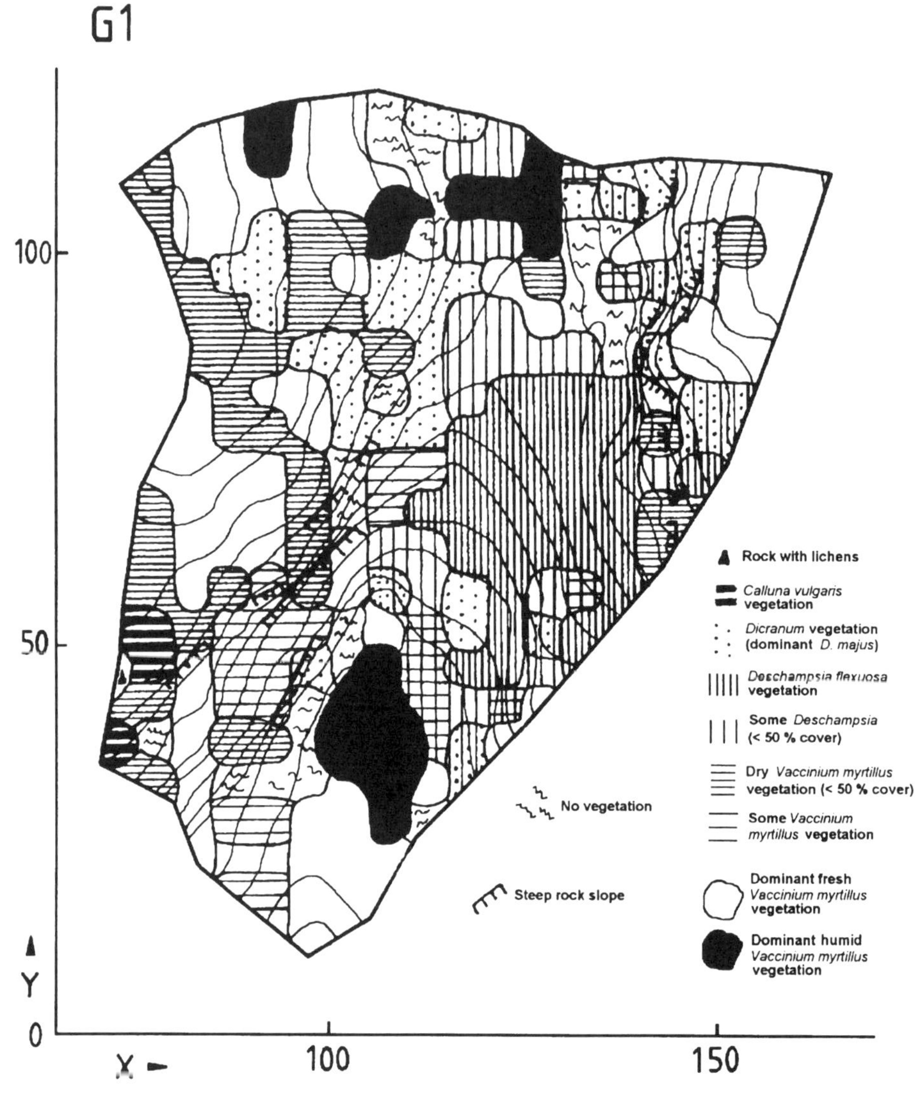


**Figure 2.16.** Map showing vegetation in the ROOF (G1) catchment. Modified from map made by Ødegaard, NISK (Nygaard, *et al.*, 1993). X : Y scale in metres: contour interval 1 m

of significant change during the period between samplings, makes it impossible to judge which dataset should be regarded as more representative for catchment F1. Therefore, calculations are based on the last sampling to minimise differences compared to G1 and, in a few cases, very different values indicated by use of other datasets are commented on. This assessment of pools for the different elements may be premature as there are large gaps in the data, but it was essential to produce as good an estimate as possible for modelling purposes.

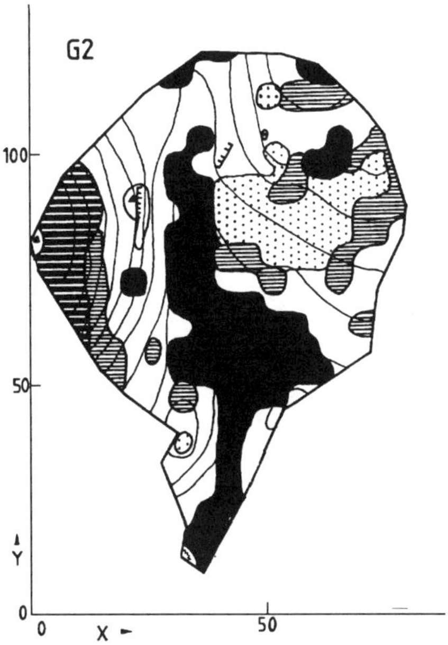


**Figure 2.17.** Vegetation map of the G2 (NITREX) catchment. Modified from map made by Ødegaard, NISK (Nygaard, *et al.*, 1993). Equidistance between lines is 1 m. Key as Figure 2.16

The total sizes of catchment pools for exchangeable ions, different sulphur fractions, total nitrogen, and carbon were determined based on the average concentrations in the respective soil layer (Tables 2.5 and 2.6) and the estimated mass of the respective soil layers as given in Table 2.7 and finally summed. For the F1 Control catchment where no detailed assessment of soil layers is available, average layer depths are related to G1 with the B and C horizon increased to simulate the larger average depth in F1.

The results of the pool assessments (Table 2.8) indicate some small differences between the catchments, with G1 being somewhat higher in carbon and nitrogen. Carbon in the soil is estimated by assuming a fraction of 50% of the loss on ignition values to be carbon. The total area-based amounts of Mg, K and especially Ca tend to be larger in F1 as average soil thickness is greater. Roughly normalising data to a standard soil depth of 0.5 m decreases this difference to a large extent, and the overall picture is a large similarity between catchments. Differences occurring may also be attributable to the higher total amount of organic material (peat) as a result of more wetland areas occurring in F1.

Adsorbed sulphur pools in the soils of F1 and G1 seem to be similar and amount to approximately 6 kmol $ha^{-1}$ (lower part of Table 2.8) despite the uncertainties regarding mean concentration and representativeness of only two soil profiles from G1. Values for total amount of sulphur within the catchment soils are as given in Mörth and Torssander (1995) for G1, about 46 kmol $ha^{-1}$, and as calculated for F1

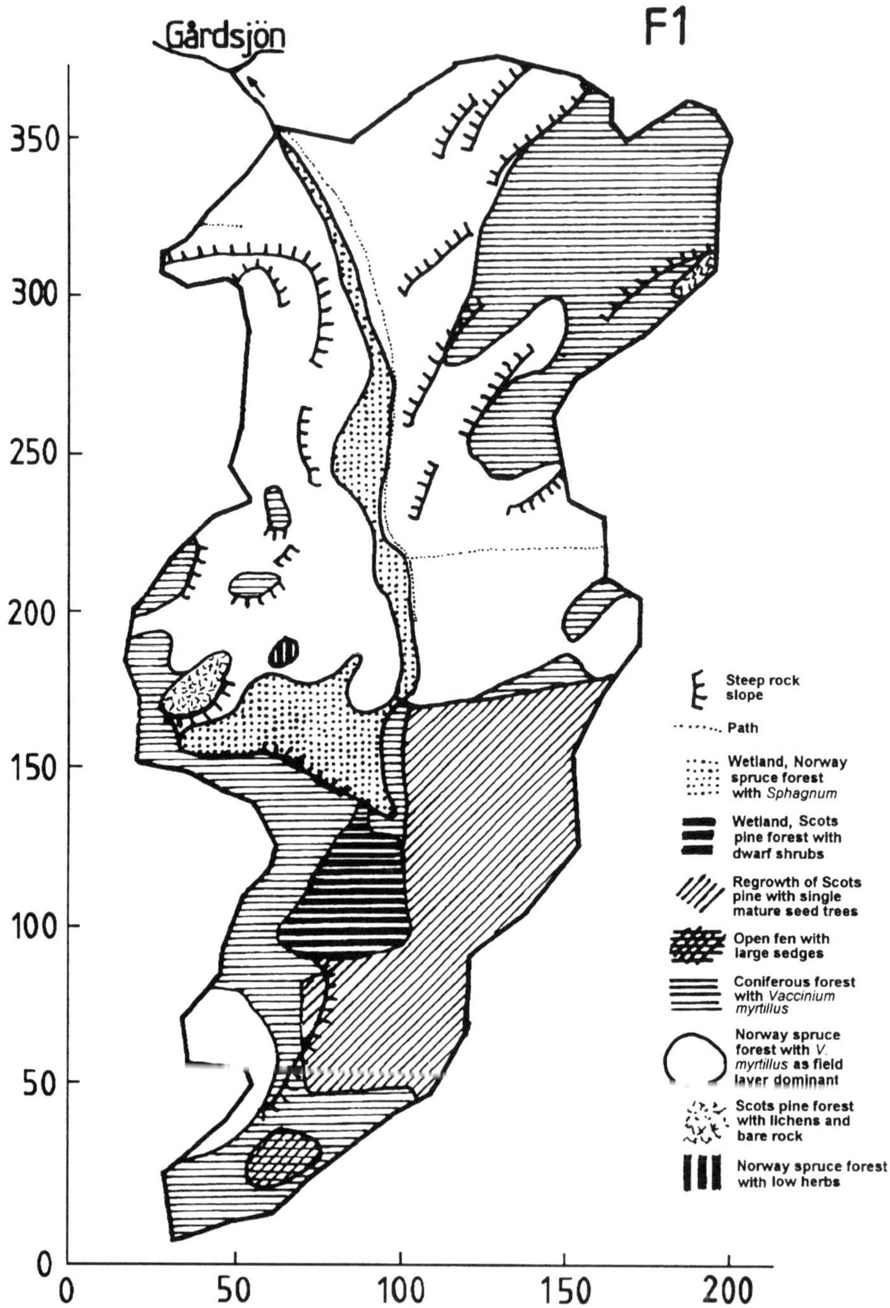


**Figure 2.18.** Dominant tree associations in the control catchment (F1). Map modified from S. Bråkenhielm (Department of Environmental Assessment, Swedish University of Agricultural Sciences, 1992). X : Y scale in metres

**Table 2.7.** Some soil layer parameters calculated for the three catchments. Bulk density estimates and soil layer thickness for G1 are based on Nyberg (1995b) and Nisk (unpublished)

| Horizon | Average soil depth (cm) | | | Bulk density in layers ($t\,m^{-3}$) | | | Total soil mass in layer ($t\,ha^{-1}$) | | | Volume fraction in layer (%) | | |
|---|---|---|---|---|---|---|---|---|---|---|---|---|
| | F1* | G1 | G2* | F1 | G1 | G2 | F1 | G1 | G2 | F1 | G1 | G2 |
| LFH | 8 | 8.8 | 8 | † | 0.18 | † | 140 | 160 | 140 | 12 | 20 | 21 |
| A2+AB (Ae) | 5 | 5.2 | 4 | 1.05 | 0.88 | 1.15 | 530 | 460 | 460 | 8 | 12 | 11 |
| B | 21 | 20.6 | 20 | 0.72 | 0.61 | 0.98 | 1500 | 1300 | 2000 | 32 | 48 | 53 |
| BC+C | 30 | 7.1 | 4 | 1.06 | 1.25 | 1.46 | 2900 | 890 | 580 | 42 | 17 | 11 |
| Peat soil whole area | 4 | 1.3 | 2 | ‡ | ‡ | 0.495 | 200 | 60 | 100 | 6 | 3 | 5 |
| Average | | | | 0.78 | 0.64 | 0.83 | | | | | | |
| Sum | 68 | 43 | 38 | | | | 5200 | 2800 | 3200 | 100 | 100 | 100 |

*Layers assumed proportional with G1.
†Assumed similar to G1.
‡Assumed similar to G2.

based on Olsson *et al.* (1985), approximately 95 kmol ha$^{-1}$. This is a large difference, although if the same rough standardisation to equal soil depth as above is used, they end up rather similar at *c.* 65 kmol ha$^{-1}$. If this is used for adsorbed sulphur, the $SO_4$ amounts increase to around 7 kmol ha$^{-1}$ in G1 and decrease to about 4 kmol ha$^{-1}$ in F1, indicating a higher concentration of adsorbed $SO_4$ present in the roofed catchment, especially in the B horizon. However, the uncertainty in the two estimates may imply that calculated differences should not be exaggerated. As is indicated from other areas of southern Sweden subject to high acid sulphur loading (Eriksson *et al.*, 1992), the depth of the acidification "horizon" (including raised sulphur content) in the soil profile may reach 1–2 m. This implies that there may be, as indicated by Table 2.8, a significant amount of sulphur even in the deeper till layer of the lower valley bottom in catchment F1.

There are some significant differences between the present sampling and some earlier ones. Thus data collected by Nilsson (1992) indicate a substantially higher amount of exchangeable calcium in F1, as the humus layer concentration given was about double the concentration used here. This higher value may be more representative for the catchment as such considering the more restricted sampling area used by Kjønaas and Stuanes (1996). Also the total CEC based on Nilsson (1992) should be higher as values are higher in the uppermost soil layers, while an estimate of base saturation would be lower as especially the humus pool observed by Nilsson (1992) was substantially lower. The analyses performed by Grodzinska *et al.* (1991) in F1, to estimate the total elemental pools, were reasonably similar for sulphur to the earlier estimate based on Olsson *et al.* (1985). Hence, if the approximately 65 kmol ha$^{-1}$ is increased by another 20 kmol ha$^{-1}$ to compensate for the humus layer amount, the total would sum to about 85 based on Grodzinska *et al.* (1991) compared with about 95 kmol ha$^{-1}$ as calculated by us. For nitrogen the two estimates are almost the same value, approximately 800 kmol ha$^{-1}$, indicating that these numbers may be rather accurate. The total base cation stores based on

**Table 2.8.** Estimated storages of carbon (0.5*LOI), sulphur, nitrogen and exchangeable cations in the soils of catchment G1 (ROOF) and F1 (control). Units are kmol ha$^{-1}$ except where stated: totals through the soil profile are given in both kmol ha$^{-1}$ and kg ha$^{-1}$. Principally based on Kjønaas *et al.* (1992) and Kjønaas and Stuanes (1996)

| | Weight of layer (t ha$^{-1}$) | Carbon | Na | K | Mg | Ca | Al | Fe | Mn | Total N | *Adsorbed $SO_4^{2-}$ | | | Total S |
|---|---|---|---|---|---|---|---|---|---|---|---|---|---|---|
| | | | | | | | | | | | $H_2O$ | $N_aHCO_3$ | Total | |
| **F1** | | | | | | | | | | | | | | |
| | | | | | | | | | | | | | | 8.5 |
| LFH | 140 | 5000 | 1.5 | 2.4 | 3.0 | 6.3 | 3.4 | 0.3 | 0.1 | 150 | | | 0.2 | 6.2 |
| A2+AB (Ae) | 530 | 790 | 0.5 | 0.6 | 0.5 | 0.8 | 4.6 | 0.3 | 0.0 | 65 | | | 2.0 | 19 |
| B | 1500 | 7200 | 1.5 | 1.9 | 0.9 | 2.0 | 32 | 1.2 | 0.0 | 240 | | | 4.0 | 49 |
| BC+C | 2900 | 5400 | 1.1 | 1.1 | 0.6 | 0.9 | 22 | 0.3 | 0.0 | 140 | | | | 12 |
| Peat[†] | 200 | 6800 | 2.1 | 3.2 | 4.2 | 8.7 | 4.7 | 0.4 | 0.2 | 210 | | | 6.1 | 95 |
| SUM | 5200 | 25100 | 6.7 | 9.3 | 9.1 | 19 | 67 | 2.4 | 0.4 | 800 | | | 190 | 3000 |
| SUM (C and N in tonnes, else kg) | | 300 | 150 | 360 | 220 | 750 | 1800 | 130 | 21 | 11 | | | | |
| **G1** | | | | | | | | | | | | | | |
| LFH | 160 | 5700 | 1.8 | 2.7 | 3.2 | 6.8 | 4.3 | 0.3 | 0.2 | 180 | | | | 7.4 |
| A2+AB (Ae) | 460 | 2100 | 0.7 | 1.1 | 0.7 | 1.3 | 5.6 | 0.4 | 0.0 | 68 | 0.3 | 0.2 | 0.5 | |
| B | 1300 | 5300 | 1.7 | 1.8 | 1.2 | 1.7 | 26 | 1.5 | 0.1 | 180 | 0.7 | 3.5 | 4.2 | |
| BC+C[‡] | 890 | 1500 | 0.5 | 0.4 | 0.3 | 0.4 | 8.1 | 0.4 | 0.0 | 48 | 0.3 | 1.1 | 1.4 | |
| Peat[†] | 63 | 2300 | 0.7 | 1.1 | 1.3 | 2.7 | 1.7 | 0.1 | 0.1 | 74 | | | | |
| SUM | 2800 | 16900 | 5.5 | 7.0 | 6.6 | 13 | 45 | 2.7 | 0.5 | 550 | 1.2 | 4.8 | 6.1 | 45 |
| SUM (C and N in tonnes, else kg) | | 200 | 130 | 270 | 160 | 520 | 1200 | 150 | 27 | 8 | 40 | 150 | 190 | 1500 |

*Calculations based on Olsson *et al.* (1985) for F1; Mörth and Torssander (1995) and Torssander and Mörth (Chapter 8) for G1.
[†]On soils only analysed from a G2 profile, the concentrations assumed equal to LFH concentrations.
[‡]Concentrations assumed equal with 0.4×lower B horizon.

Grodzinska *et al.* (1991) were as follows: Ca and Mg *c.* 600 kmol $ha^{-1}$, K *c.* 100, and Na only 35 kmol $ha^{-1}$. For potassium and sodium, the estimates should probably be increased substantially (10 and 20% respectively) as the exchangeable humus fraction was not included in the analyses.

From these assessments it becomes obvious that there are large pools of carbon and nitrogen in the soils at Gårdsjön, generally with a C/N ratio between 30 and 40 in the humus layer, and decreasing in the mineral soil layers (also cf. Olsson *et al.*, 1985). Dependent on average soil depth, the total sulphur pool estimates vary between 50 and 100 kmol $ha^{-1}$ and a substantial part of this stored sulphur derives from the deposition during more than 100 years of input from anthropogenic sources (Chapter 8, this volume). The directly water-leachable fraction may be only 1-1.5 kmol $ha^{-1}$ while the totally adsorbed and probably leachable $SO_4$ fraction are somewhere around 6 kmol $ha^{-1}$. Reactions involving the large pool of organically bound sulphur may then be crucial for the rate and total time that will elapse until recovery has occurred.

### Forest stands and vegetation

All three catchments were surveyed for vegetation composition, and biomass estimates were made during 1990 (Nygaard *et al.*, 1992; Kjønass and Stuanes, 1997 and unpublished). This survey included measurement of diameter at breast height and total height of all trees in the catchments G1 and G2, but only a restricted number of trees in the lower part of F1. For the F1 catchment an earlier biomass estimate, including chemical analyses of different components, was performed during the 1980s (Hallbäcken, 1980). However, during 1991 Grodzinska *et al.* (1991) surveyed the whole F1 area, whereby assessment of the standing biomass of boles can be made for all three catchments (Table 2.9). The 1980 assessment, besides biomass estimates, also included analyses of elemental concentrations in major vegetation units and an estimate of the annual increment. Assuming that these concentrations are still valid, estimates of the elemental pools in vegetation can be performed for G1 and G2 by proportioning in accordance with their biomass.

The density and total biomass of the forest stands in the catchments are significantly different with G1 having the highest total biomass of trees mainly due to a greater number of larger trees if compared with F1 and a greater total number of stems if compared with G2. The differences are especially pronounced for the Norway spruce stands of the respective catchments (Figure 2.19). Due to the regrowth area in the upper part of catchment F1, there is a higher total number of trees per hectare in this catchment. However, both this pine regrowth and a larger number of small spruce in the more open forest of the F1 catchment contribute only a small amount of biomass (Figure 2.20). The most notable difference among the catchments is that younger birch, and several other deciduous tree species take a significant share (almost half of pine) in the F1 area. This is in contrast to the manipulated catchments where the fraction of deciduous trees is virtually zero.

For the calculation of total biomass and elemental stores, total volumes of the boles have been used and a density of 0.5 g $cm^{-3}$ assumed equally valid for all wood. Biomass of other vegetation and the litter layer as estimated during the 1980

**Table 2.9.** Forest stand characteristics of the experimental and control catchments at Gårdsjön. Based on Nygaard *et al.* (1992) and Kjønaas and Stuanes (1996)

| | | G1 (ROOF) | G2 (NITREX) | F1 (Control) |
|---|---|---|---|---|
| Total no. of trees | (No. $ha^{-1}$) | 587 | 534 | 653 |
| % Norway spruce | | 83.0 | 81.0 | 63.6 |
| % Scots pine | | 17.0 | 19.0 | 13.7 |
| % deciduous trees | | | | 22.7 |
| Average height, spruce | (m) | 18.1 | 18.0 | 13.4 |
| Average height, pine | (m) | 17.6 | 15.8 | 13.4 |
| Average height, birch | (m) | | | 10.6 |
| Average D.B.H., spruce | (cm) | 24.7 | 23.8 | 19.7 |
| Average D.B.H., pine | (cm) | 25.4 | 27.0 | 24.9 |
| Average D.B.H., birch | (cm) | | | 13.7 |
| Total basal area | ($m^2$) | 31.4 | 26.5 | 23.3 |
| Basal area, spruce | ($m^2$) | 26.1 | 20.1 | 15.7 |
| Basal area, pine | ($m^2$) | 5.3 | 6.4 | 5.0 |
| Basal area, birch | ($m^2$) | | | 2.1 |
| Total bole volume | $m^3 ha^{-1}$ | 283 | 216 | 179 |
| Bole volume spruce | $m^3 ha^{-1}$ | 238 | 155 | 131 |
| Bole volume pine | $m^3 ha^{-1}$ | 45.4 | 45.5 | 32.8 |
| Bole volume deciduous | $m^3 ha^{-1}$ | | | 15.3 |
| Bole volume birch | $m^3 ha^{-1}$ | | | 11.9 |

D.B.H.=diameter at breast height.

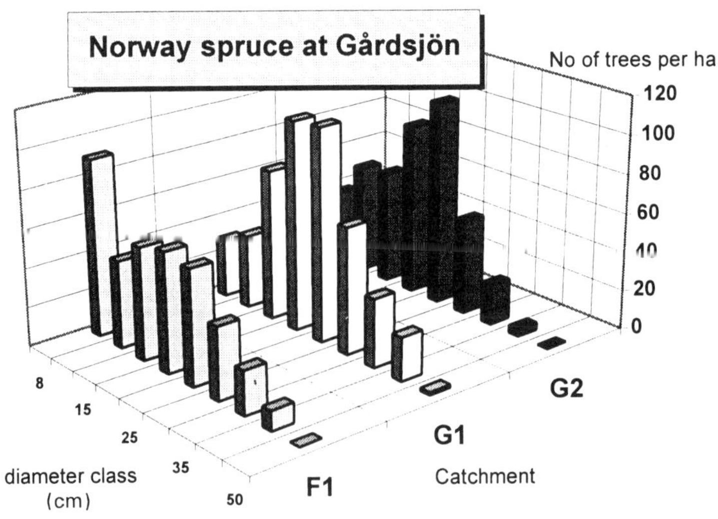


**Figure 2.19.** Structure of the Norway spruce (*Picea abies* (L.) Karst.) stands in the control and manipulated catchments based on abundance of diameter classes

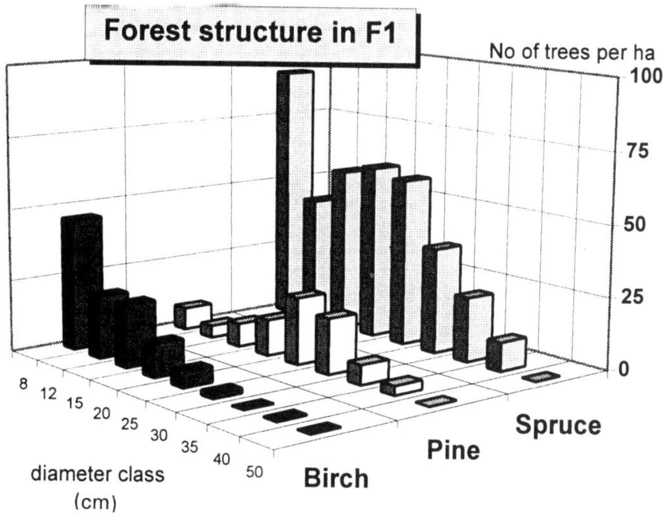


**Figure 2.20.** Composition and size distributions of the forest stand in catchment F1 (Control)

**Table 2.10.** Estimated storages of major elements in the live and dead vegetation biomass from small catchments at Lake Gårdsjön. Total storage is given in both kmol $ha^{-1}$ and kg $ha^{-1}$. Elemental composition data from Hallbäcken (1980, unpublished), Grodzinska *et al.* (1991) and Nygaard *et al.* (1992)

| | Dry weight (t $ha^{-1}$) | Storages (kmol $ha^{-1}$) | | | | | |
|---|---|---|---|---|---|---|---|
| | | Na | K | Mg | Ca | S | N |
| **F1 (Control)** | | | | | | | |
| Tree above-ground | 110 | 0.6 | 3.3 | 7.0 | 25 | 2.8 | |
| Total live tree | 140 | 0.6 | 4.0 | 8.1 | 29 | 3.2 | 39 |
| Ground and field layer | 2.0 | 0.0 | 0.3 | 0.5 | 0.5 | 0.1 | 1.9 |
| Dead tree incl. mosses and litter | 14 | 0.1 | 0.3 | 2.2 | 3.1 | 1.9 | 9.0 |
| Total sum | 160 | 0.8 | 4.5 | 11 | 32 | 5.2 | 50 |
| Total sum (kg) | | 18 | 180 | 260 | 1300 | 170 | 700 |
| **G1 (ROOF)** | | | | | | | |
| Tree above-ground | 180 | 1.0 | 5.7 | 12 | 43 | 4.8 | |
| Total live tree | 240 | 1.1 | 6.8 | 14 | 50 | 5.6 | 55 |
| Ground and field layer | 2.4 | 0.0 | 0.3 | 0.5 | 0.6 | 0.1 | 2.2 |
| Dead tree incl. mosses and litter | 20 | 0.2 | 0.4 | 2.8 | 4.1 | 2.4 | 11 |
| Total sum | 270 | 1.3 | 7.6 | 17 | 54 | 8.2 | 69 |
| Total sum (kg) | | 30 | 300 | 430 | 2200 | 260 | 960 |

**Table 2.11**. Comparison of the Gårdsjön site characteristics with other sites in Europe and North America

| Site | Location | Altitude (m a.s.l.) | Climate | Geology | Soils | Forest dominant tree | Precipitation: Amount (mm $yr^{-1}$) | Precipitation: Conc. ($\mu mol_c L^{-1}$) $SO_4$-S | $NO_3$-N | $NH_4$-N | pH in runoff (or deep soilwater[#]) | Source | Remarks |
|---|---|---|---|---|---|---|---|---|---|---|---|---|---|
| Gårdsjön | SW Sweden | 100–150 | Maritime temperate | Gneiss, granite, sandy-loamy till | Orthic ferro-humic podzols (Folisoils) | Mixed coniferous Norway spruce, 80–100 years old | *c.* 1150 | 69 | 40 | 33 | 3.9–4.2 | | Deposition currently decreasing |
| Svartberget | N Sweden | 235–310 | Continental cold | Gneiss, thick glacial till | Iron podzols and some peat | Scots pine and Norway spruce, incl. small mire | *c.* 700 | 35 | 16 | 14 | 4.5–5.5 | Bishop, 1991 | Notable acid events at snowmelt |
| Risdalsheia | S Norway | *c.* 300 | Mountainous maritime | Granite 30–50% outcrops | Organic-rich podzol (loam) | Sparce natural Scots pine and birch | *c.* 1450 | *c.* 77 | *c.* 50 | *c.* 40 | 4.1 | Wright *et al.*, 1993 | |
| Klosterhede | W Denmark | 27 | Maritime temperate | Very thick sandy deposits | Podzol Haplorthod | Planted Norway spruce 75 years old | *c.* 1000 | 100 | 50 | 35 | 4.1[#] | Beier and Rasmussen, 1993 | |
| Solling | Central Germany | above 500 | Semi-continential temperate | Triassic sandstone covered with loess | Dystric Cambisols podzolized | Planted Norway spruce 60 years old | *c.* 1100 | 72 | 48 | 69 | 4.2[#] | Beier and Rasmussen, 1993 | |

| | | | | | | | | | | | | | |
|---|---|---|---|---|---|---|---|---|---|---|---|---|---|
| Höglwald | S Germany | *c.* 540 | Semi-continental temperate | Miocene sediments, upper layer with loess | Haplodult | Planted Norway spruce 85 years old | *c.* 800 | 75 | 41 | 55 | 5.2[#] | Beier and Rasmussen, 1993 | |
| Llyn Brianne | Central Wales | above 275 | Sub-oceanic temperate | Paleozoic shales are dominant | Peat, stagnopodzol stagnogley and podzolized brown soils | Heath-moorland, partially planted with Sitka spruce | *c.* 1800 | *c.* 130 | *c.* 20 | *c.* 25 | 4.9 | Hornung *et al.*, 1990 | |
| Loch Fleet | SW Scotland | 340–470 | Maritime temperate | Granite, notable outcrops | Peat, minor areas with minerogenic soils | *c.* 10% Sitka spruce rest moorland with *Calluna* and *Molinia* | *c.* 2150 | *c.* 45 | *c.* 15 | *c.* 20 | 4.7 | Howells and Dalziel, 1988 | |
| Woods Lake | Adirondack, New York, US | 600–725 | Continental? temperate | Granite | Spodozols | Deciduous species (red spruce) | *c.* 1250 | *c.* 60 | *c.* 35 | *c.* 20 | 5.9 | Johannes *et al.*, 1985; Cronan, 1985; Schofield *et al.*, 1985 | Deposition values from around 1980 |
| Bear Brook | W Maine, US | 250–475 | Maritime temperate | Metamorphic pellites with incl. of gneiss and granite | Spodozols | Deciduous species, conifers occur at top- and lowland areas | *c.* 1000 | 28 | 16 | 10 | 6.0–6.7 | Norton *et al.*, 1992 | |
| Turkey Lake | S Ontario, Canada | 340–630 | Continental temperate | Precambrian greenstone till influenced by granite | Podzols | Deciduous species, conifers most in lowland areas | 1200–1400 | *c.* 24 | *c.* 37 | | 4.1–5.0 | Jeffries *et al.*, 1988; Vet *et al.*, 1988 | |

assessment in catchment F1 is proportioned in the other two catchments. However, the analyses during 1980 did not include the nitrogen concentration of tree units and therefore data on needles from 1991/92 and an estimate of net annual uptake rate, approximately 0.4 kmol $ha^{-1}$ (40 meq $m^{-2}$ $yr^{-1}$), for F1, is used to calculate the store in tree stands.

Estimates of the total amounts of major elements such as base cations, sulphur and nitrogen stored in the trees and other vegetation units are presented in Table 2.10. For potassium and sulphur the amounts stored in the vegetation approach what is stored as the exchangeable or total leachable soil fraction (Table 2.8) and for magnesium and especially for calcium the vegetation stores are substantially larger than the exchangeable soil store. It is also notable that concentrations of K, Mg, Ca, N and, to a lesser degree S, are much enriched in the vegetation of the ground and field layers as compared with the total tree biomass. This is due to universally low concentrations in the wood while leaves and needles have high concentrations of most elements. Also the dead moss–litter fraction shows an enrichment above tree wood concentration and thus constitutes the major part of the combined "dead" compartment.

The differences observed between the catchments reflect the importance of how forests are managed, as the much higher biomass in G1 would be rather typical for production forests and, as shown by Andersson *et al.* (Chapter 23, this volume), this has implications for the critical load estimates. The situation has important implications for sustainable management practice as the amount of base cations removed by either conventional timber–pulp wood or whole-tree harvesting differ significantly in relation to the exchangeable pools in the forest soil, which are currently leached at an increased rate due to acidification (cf. Hallbäcken, 1992; Warfvinge *et al.*, 1993). Thus, the future forestry management of such "poor" and consequently very sensitive systems like the Gårdsjön catchments may be crucial for the development during the next rotation period. Despite the emission reductions achieved by international agreements, the impact of forestry may be very significant for both the rate of recovery and the ultimate final soil situation as suggested by several studies (cf. Nilsson *et al.*, 1982; Rosén, 1991; Warfvinge *et al.*, 1992).

## THE GÅRDSJÖN CATCHMENTS IN PERSPECTIVE

It was shown above that the experimental catchments G1, G2 and F1 are reasonably typical of the Gårdsjön area, but how representative is Gårdsjön in a wider regional context? How do the Gårdsjön catchments compare with other intensively studied catchments around the northern hemisphere? Among these catchments substantial variation occurs in the soil characteristics which, in combination with a variation in annual air pollution load and composition of acidifying substances, may hamper both interpretation and especially the possibility of generalisation of results from the Gårdsjön reversibility study. On the European scale, Gårdsjön is subject to a rather modest deposition of both sulphur and nitrogen (Chapter 3, this volume), as are many other well-known research areas. However, chemical and physical soil characteristics, climate and hydrology make this site among the most vulnerable in

Europe. This is recognised by the low critical load calculated for the site (Chapter 23, this volume).

As is obvious from Table 2.11, Gårdsjön shares some specific characteristics like the strong sea-salt impact with several other intensively studied areas in Europe. The podzolic soils and the dominance of coniferous forests make Gårdsjön representative of a rather common Scandinavian situation. In contrast to this, the North American areas generally subject to acidification stress are located within a temperate deciduous biome and their soils may be different and more similar to Central European soils. Furthermore, for Gårdsjön like most of the other European sites, there is a significant dry deposition of both nitrogen and sulphur, generally amounting to somewhere between 100 and up to 200% respectively of the wet deposition. The Bear Brook area receives a substantial dry deposition while this fraction contributes far less at more remote areas such as Turkey Lake, Woods Lake and especially at Svartberget in northern Sweden. In view of this, what is learnt about acidification reversibility and rates of recovery from this covered catchment study at Gårdsjön should apply to large areas of Scandinavia and, by use of process-orientated biogeochemical models (Chapters 10 and 19, this volume), find more general applicability. As Gårdsjön is among the most sensitive areas in Europe, results from the project should provide guidance for the international agreements aimed at improving the situation over vast areas.

## ACKNOWLEDGEMENTS

Financial support for the different studies contributing data to this site description came from the Swedish national power industry (Vattenfall and Elforsk), Great Britain's National Power and PowerGen, IVL's research programme, the Swedish Environmental Protection Agency, the Commission of European Communities via the European Network of Catchments Organised for Research on Ecosystems (ENCORE) and the Nitrogen Saturation Experiments (NITREX).

## REFERENCES

Alexandersson, H., Karlström, C. and Larsson-McCann, S. (1991) *Temperature and Precipitation in Sweden 1961–90. Reference Values.* SMHI Meteorologi, Norrköping, no. 81, 87 pp. (In Swedish, *Temperaturen och Nederbörden i Sverige 1961–90. Referensnormaler*).

Andersson, B. I. (ed.) (1993) *Project Catalogue for the Gårdsjön Covered Catchment Experiment.* IVL Report B 1077, Gothenburg, November 1993, 83 pp.

Andersson, B. I. and Hultberg, H. (1997) Lake Lysevatten—a study of acidification and liming effects in a forest lake ecosystem in southwestern Sweden. IVL Report **B 1250**, 132 pp.

Andersson, F. and Olsson, B. (eds) (1985) Lake Gårdsjön—an Acid Forest Lake and its Catchment. *Ecological Bulletins (Stockholm)*, **37**.

Avergård, I. (1971) *pH Conditions in Lakes of Western Sweden During 1970–71.* Länsstyrelserna i O, N, P, F, G län, Fiskeristyrelsen och Naturvårdsverket, 88 pp. (In Swedish, *pH-förhållandena i västsvenska sjöar 1970–71.*)

Beier, C. and Rasmussen, L. (eds) (1993) *The EXMAN-Project. Experimental Manipulation of Forest Ecosystems in Europe.* Ecosystems Research Report 7, Commission of the European Communities, Brussels, 124 pp.

Bishop, K. H. (1991) Episodic increases in stream acidity, catchment flow pathways and hydrograph separation. Doctoral thesis, Jesus College, Cambridge, 246 pp.

Borg, G. Ch. (1989) Tables of characteristic soil properties from G1 and F1. (Unpublished data partially included in the present paper.)

Borg, G. Ch. (1995) Soil physics and hydrology—calibration of a simulation model to small catchments at Lake Gårdsjön. (Manuscript to be published as IVL Report.)

Cronan, S. C. (1985) Biogeochemical influence of vegetation and soils in the ILWAS watersheds. *Water, Air and Soil Pollution*, **26**, 355–371.

Cronan, C. S. and Goldstein, R. A. (1989) ALBIOS: a comparison of aluminum biogeochemistry in forested watersheds exposed to acidic deposition. In: Adriano, C. and Havas, M. (eds), *Advances in Environmental Science, Acidic Precipitation*, Volume I. Springer-Verlag, New York, pp. 113–135.

Dickson, W. (ed.) (1988) *Liming of Lake Gårdsjön—An Acidified Lake in SW Sweden*. The National Swedish Environmental Protection Agency, Report 3426, Solna, Sweden. 327 pp.

Dise, N. B. and Wright, R. F. (eds) (1992) *The NITREX Project (Nitrogen Saturation Experiments)*. Ecosystems Research Report No. 2, Commission of the European Communities, Brussels.

Eriksson, B. (1980) *The Water Balance of Sweden. Annual Mean Values (1931–1960) of Precipitation, Evaporation and Runoff*. SMHI Report no. 18. Norrköping, Sweden, 34 pp.

Eriksson, E., Karltun, E. and Lundmark J.-E. (1992) Acidification of forest soils in Sweden. *Ambio*, 21, 150–154.

Fridh, B. and Hayling, K. (1980) *Seismic Survey at Gårdsjön during 1980*. Mimeographed. Department of Geology, GU and CTH. (In Swedish, *Seismisk undersökning, Gårdsjön, 1980*).

Grahn, O., Hultberg, H. and Landner, L. (1974) Oligotrophication—a self-accelerating process in lakes subjected to excessive supply of acid substances. *Ambio*, **3**, 93–94.

Grodzinska, K., Braniewski, S., Godzik, B. and Kiszka, J. (1991) *Studies in Lake Gårdsjön Catchment, June 1991*. Mimeographed. Report for the joint Polish–Swedish Program within the field of effects of air pollution on forests, 116 pp.

Hallbäcken, L. (1980) Tabulated data on forest biomass, forest growth and elemental composition from catchment F1. (Unpublished but partially included by Hultberg, 1985, and partially used in the present paper as basis for calculation.)

Hallbäcken, L. (1992) Long-term changes of base cation pools in soil and biomass in a beech and a spruce forest of southern Sweden. *Zeitschrift für Planzenernährung und Bodenkunde*, **155**, 51–60.

Holten, J. I. and Carey, P. D. (1992) *Responses of Climate Change on Natural Terrestrial Ecosystems in Norway*. NINA, Forskningsrapport 029, January 1992, 59 pp.

Hornung, M., Roda, F. and Langan, S. J. (eds) (1990) *A Review of Small Catchment Studies in Western Europe Producing Hydrochemical Budgets*. Air Pollution Research Report 28, Commission of the European Communities, Brussels, 186 pp.

Howells, G. and Dalziel, T. R. K. (eds) (1988) *Loch Fleet Project. A Report of the Intervention Phase (2) 1986–1987*. Life Sciences Section, CERL, Leatherhead, UK, 102 pp. with 5 annexes.

Hultberg, H. (1977) Thermally stratified acid water in late winter—a key factor inducing self-accelerating processes which increase acidification. *Water, Air and Soil Pollution*, **7**, 279–294.

Hultberg, H. (1985) Budgets of base cations, chloride, nitrogen and sulphur in the acid Lake Gårdsjön catchment SW Sweden. *Ecological Bulletins (Stockholm)*, **37**, 133–158.

Hultberg, H. (1994) *Response Following Soil Liming to a Small Catchment at Gårdsjön. Subproject Chemistry and Hydrology*. Mimeographed. Slutrapport till SNV. (In Swedish, *Effekter av markkalkning i Gårdsjöns avrinningsområde. Delprojekt: Kemi och Hydrologi*.)

Hultberg, H. and Andersson, B. I. (1982) Liming of acidified lakes: Induced long-term changes. *Water, Air and Soil Pollution*, **18**, 311–331.

Hultberg, H. and Grennfelt, P. (1986) Gårdsjön Project: lake acidification, chemistry in catchment runoff, lake liming and microcatchment manipulations. *Water, Air and Soil Pollution*, **30**, 31–46.

Hultberg, H. and Grennfelt, P. (1992) Sulphur and seasalt deposition as reflected by throughfall and runoff chemistry in forested catchments. *Environmental Pollution*, **75**, 215–222.

Hultberg, H., Lee, Y. H., Nyström, U. and Nilsson, S. I. (1990) Chemical effects on surface-, ground- and soil-water of adding acid and neutral sulphate to catchments in southwest Sweden. In: B. J. Mason (ed.), *SWAP, The Surface Water Acidification Programme*. Cambridge University Press, Cambridge, pp. 167–182.

Jeffries, D. S., Kelso, J. R. M. and Morrison, I. K. (1988) Physical, chemical, and biological characteristics of the Turkey Lakes watershed, central Ontario, Canada. *Canadian Journal of Fisheries and Aquatic Sciences*, **45** (suppl. 1), 3–13.

Johannes, A. H., Altwicker, E. R. and Clesceri, N. L. (1985) The Integrated Lake–Watershed Acidification Study: atmospheric inputs. *Water, Air and Soil Pollution*, **26**, 339–353.

Johansson, S. and Nilsson, T. (1985) Hydrology of the Lake Gårdsjön area. *Ecological Bulletins (Stockholm)*, **37**, 86–97.

Johansson, S. and Nilsson, T. (1988) Lake Gårdsjön—The hydrological balance of an acidified forested catchment area 1979–1986. In: Dickson, W. (ed.), *Liming of Lake Gårdsjön—An Acidified Lake in SW Sweden*. The National Swedish Environmental Protection Agency, Report 3426, Solna, Sweden, pp. 9–37.

Johansson, S., Eriksson, E. and Wolf, A. (1988) Lake Gårdsjön—The chemistry of the ground water in an acidified forested catchment area 1979–1986. In: Dickson, W. (ed.), *Liming of Lake Gårdsjön—An Acidified Lake in SW Sweden*, The National Swedish Environmental Protection Agency, Report 3426, Solna, Sweden, pp. 37–95.

Kjønaas, O. J. and Stuanes, A. O. (1991) Nitrex Project—Gårdsjön. Status Report for 1988–1990. NIVA report o-87208, pp. 22–24.

Kjønaas, O. J. and Stuanes, A. O. (1996) Data on soil characteristics. (Partially unpublished but included in the present paper.)

Kjønaas, O. J., Stuanes, A. O. and Huse, M. (1992) Soil chemical and physical characteristics. In: Dise, N. B. and Wright, R. F. (eds), *NITREX Gårdsjön*. Report no. 2/92. Norwegian Institute for Water Research, Norway, pp. 23–26.

Kjønaas, O. J., Stuanes, A. O. and Huse, M. (1997) Effects of weekly nitrogen additions on N cycling in a coniferous forest catchment, Gårdsjön, Sweden. *Forest Ecology and Management* (in press).

Lindén, A. (1975) *Till Petrographical Studies in an Archæan Bedrock Area in Southern Central Sweden*. Striæ 1. Societas Upsaliensis Pro Geologia Quartenaria, pp. 1–57.

Melkerud, P.-A. (1977) Connections between till and bedrock geology. A transect across the Cambrio-Siluric region of the province of Östergötland. PhD thesis, Department of Quaternary Research, University of Stockholm, 163 pp. (In Swedish with English summary, *Samband mellan morän och berggrund. En travers över Östergötlands kambrosiluromrảde.*)

Melkerud, P.-A. (1983) *Quaternary Deposits and Bedrock Outcrops in an Area Around Gårdsjön, South-west Sweden, with Physical, Mineralogical and Geochemical Investigations*. Reports on Forest Ecology and Forest Soils No. 44. Swedish University of Agricultural Sciences, Uppsala, Sweden, 87 pp.

Mörth, C.-M. and Torssander, P. (1995) Sulfur and oxygen isotope ratios in sulfate during an acidification reversal study at Lake Gårdsjön, western Sweden. *Water, Air and Soil Pollution*, **79**, 261–278.

Nilsson, S. I. (1985) Budgets of aluminum species, iron and manganese in the Lake Gårdsjön catchment, in SW Sweden. *Ecological Bulletins (Stockholm)*, **37**, 120–133.

Nilsson, S. I. (1992) Data on exchangeable fractions in soil from F1. (Unpublished but included in Hultberg, 1994.)

Nilsson, S. I., Miller, H. G. and Miller, J. D. (1982) Forest growth as a possible cause of soil and water acidification: an examination of the concepts. *Oikos*, **39**, 40–49.

Norton, S. A., Wright, R. F., Kahl, J. S. and Scofield, J. P. (1992) The MAGIC simulation of surface water acidification at, and first year results from, the Bear Brook Watershed Manipulation, Maine, USA. *Environmental Pollution*, **77**, 279–286.

Nyberg, L. (1995a) Soil- and groundwater distribution, flowpaths, and transit times in a small till catchment. Doctoral dissertation, Institute of Earth Sciences, Department of Hydrology, Uppsala University, 147 pp.

Nyberg, L. (1995b) Water flow path interactions with soil hydraulic properties in the soil at Gårdsjön, Sweden. *Journal of Hydrology*, **170**, 255–275.

Nygaard, P. H., Kjønaas, O. J., Stuanes, A. O. and Huse, M. (1992) Vegetation. In: Dise, N. and Wright, R.F. (eds), *Nitrex Gårdsjön*. Report no. 2/92. Norwegian Institute for Water Research, Norway, pp. 39–44.

Nygaard, P. H., Huse, M. and Stuanes, A. O. (1993) Vegetation. In: *Nitrex Gårdsjön. Status Report for 1991–1992; the First Year of Treatment*. Report no. 3/93. NIVA report 2865, pp. 33–38.

Nyström, U. (1985) Transit time distributions of water in two small forested catchments. *Ecological Bulletins (Stockholm)*, **37**, 97–101.

Odin, H., Eriksson, B. and Perttu, K. (1983) *Temperature Climate Maps for Swedish Forestry*. Reports on Forest Ecology and Forest Soils, **45**, 57 pp.

Ogner, G., Opem, M., Remedios, G., Sjøtveit, G. and Sørlie, B. (1991) *The Chemical Analysis Program of the Norwegian Forest Research Institute, 1991*. Norwegian Forest Research Institute, Ås, Norway.

Olsson, B. (1985) Land-use history of the Lake Gårdsjön area, SW Sweden. *Ecological Bulletins (Stockholm)*, **37**, 35–46.

Olsson, B., Hallbäcken, L., Johansson, S., Melkerud, P.-A., Nilsson, S.I. and Nilsson, T. (1985) The Lake Gårdsjön area—physiographical and biological features. *Ecological Bulletins (Stockholm)*, **37**, 10–28.

Rosén, K. (1991) *Harvest of Fire-wood by Clear Felling and Thinning—Ecological Effects*. Meddelande No. 5. Department of Forest Soil, Swedish Agricultural University, Uppsala, 59 pp.+appendices. (In Swedish, *Skörd av Skogsbränslen i Slutavverkning och Gallring—Ekologiska Effekter*.)

Samuelsson, L. (1980) Major features of the Proterozoic of the Göteborg region, Sweden. *Geologiska Föreningen Stockholm, Förhandlingar*, **102**, 141–144.

Schofield, C. L., Galloway, J. N. and Hendry, G. R. (1985) Surface water chemistry in the ILWAS basins. *Water, Air and Soil Pollution*, **26**, 403–423.

Stuanes A. O., Andersson, I., Dise, N. B., Hultberg, H., Kjönaas, J., Nygaard, P. H. and Nyström, U. (1992) Gårdsjön, Sweden. In: Dise, N. B. and Wright, R. F. (eds), *The NITREX Project*. Ecosystems Research Report No. 2. Commission of the European Comunities, pp. 24–34.

Vet, R. J., Sirois, A., Jeffries, D. S., Semkin, R. G., Foster, N. W., Hazlett, P. and Chan, C. H. (1988) Comparison of bulk, wet-only, and wet-plus-dry deposition measurements at the Turkey Lakes watershed. *Canadian Journal of Fisheries and Aquatic Sciences*, **45** (suppl. 1), 26–37.

Wallin, J.-E. and Renberg, I. (1985) Vegetation history of the Lake Gårdsjön area, Bohuslän, SW Sweden. *Ecological Bulletins (Stockholm)*, **37**, 31–35.

Warfvinge, P., Sverdrup, H., Ågren, H. and Rosén, K. (1992) Effects of air pollution on future forest growth. In: Lorentzon, S.E. (ed.), *Skogspolitiken inför 2000-talet*. Statens Offentliga Utredningar, SOU 1992:76. Vol. III, Bilaga II till huvudbetänkande av 1990 års skogspolitiska komitté. Jordbruksdepartementet, Stockholm. Allmänna Förlaget, pp. 267-412. (Effekter av luftföroreningar på framtida skogstillväxt) (In Swedish).

Warfvinge, P., Falkengren-Grerup, U., Sverdrup, H. and Andersen, B. (1993) Modelling long-term cation supply in acidified forest stands. *Environmental Pollution*, **80**, 209–221.

Wright, R. F., Lotse, E. and Semb, A. (1993) RAIN Project: Results after 8 years of experimentally reduced acid deposition to a whole catchment. *Canadian Journal of Fisheries and Aquatic Sciences*, **50**, 258–268.

# 3 Atmospheric Deposition to the Gårdsjön Research Area

**MARTIN FERM & HANS HULTBERG**

Atmospheric inputs to open fields and forests at Gårdsjön have been quantified by modelling and measurements. Deposition to the Gårdsjön research area has been monitored for the time period 1980/81 to 1993/94. Modelled data on total sulphur deposition cover the period from 1900 to 1994 and gave input similar to the measured wet deposition, while throughfall deposition was two times higher. The deposition of acidifying and eutrophying species to the Gårdsjön catchment is not the highest in Sweden, but is higher than average. The sulphur deposition reached a maximum during the 1980s and is now declining. Since 1989, both wet and dry deposition have been reduced by about 50%. No decline of the nitrogen deposition has been observed. Both sulphur and nitrogen deposition, as well as that of many heavy metals, are to a major extent caused by long-range transport of pollutants from the south or south-west or south-east. The dry deposition process is very important in this hilly and forested area and results in a doubling of the sulphur deposition and three to four times higher deposition of natural sea salt in the forest, while the fog and cloud deposition processes are negligible.

## INTRODUCTION

Deposition is usually divided into three categories: dry, wet and "occult" or cloud water deposition. Dry deposition of gases and particles is not only dependent on the reactivity of the gas and the size distribution of the particles, but to a great extent on the surface where dry deposition occurs. A forest, especially a coniferous forest, acts as an efficient trap for both gases and particles (Hultberg and Grennfelt, 1992). Deposition will further increase if the forest is not homogeneous, with frequent edges, and the ground is hilly (Hasselroth and Grennfelt, 1987). The Gårdsjön research area is hilly and forested mainly with mature Norway spruce, which results in high levels of dry deposition (Grennfelt *et al.*, 1985; Hultberg, 1985).

Wet deposition is independent of the surface. The amount of precipitation at Gårdsjön is relatively high, about 1100 mm per year, and relatively evenly distributed through the year, but with a maximum in the autumn. "Occult" or cloud deposition partly resembles dry- and partly wet deposition. It is independent of chemical reactivity, but depends on the structure of the receptor. Fog or cloud deposition is

*Experimental Reversal of Acid Rain Effects: The Gårdsjön Roof Project.*
Edited by Hans Hultberg and Richard Skeffington. 

probably not significant at Gårdsjön, but may still have an effect due to the high ion concentration in the droplets.

Deposition of air-borne nitrogen compounds can lead to soil and streamwater acidification as well as eutrophication. The majority of this acidification is caused by sulphur and its secondary pollutants, but the end products of the atmospheric transformations of nitrogen oxides (nitric acid and particulate nitrate) also contribute. Ammonia (plus particulate ammonium) can also acidify if nitrified in the soil to nitrate and if this is leached out.

Acidification of soil will increase the role of weathering of certain minerals and loss of base cations from the soil. Base cations are essential for trees so atmospheric inputs of base cations therefore have a positive effect in acidified forests and surface waters.

## METHODS

At Rörvik (a station about 70 km south of Gårdsjön), air concentrations of $SO_2$ and particulate $SO_4^{2-}$ were sampled using a cellulose prefilter and a gas wash bottle with hydrogen peroxide. The $SO_4^{2-}$ content in the $H_2O_2$ solution was analysed using ion chromatography (IC). The sulphur content of the prefilter was first analysed using X-ray fluorescence and, after October 1986, IC was used. Impregnated filters directly exposed to the air were used to measure the total ammonium and total nitrate concentrations in air (Ferm *et al.*, 1988). After 1993 a filter pack consisting of a Teflon prefilter, a sodium carbonate impregnated filter and an oxalic acid impregnated filter in series were used to determine the $SO_2$, particulate $SO_4^{2-}$ and the total $NH_4^+$ and total $NO_3^-$ concentrations.

Throughfall was collected in open buckets with polyethylene bags inside in the winter, and with funnels attached to bottles with polyethylene bags during the other seasons. The throughfall deposition was determined by pooling the throughfall from 28 randomly placed collectors. Wet deposition was also collected using bulk collectors. The base cations were analysed using atomic absorption spectroscopy. The surrogate surfaces for estimating the dry and fog depositions are described in the text.

## RESULTS AND DISCUSSION

### Concentrations in air

Air pollutant concentrations are measured in the Gårdsjön area with techniques specially developed for estimating deposition on a monthly basis. A filter pack, intermittently exposed for 2 minutes per hour during one month is used to obtain particulate concentrations. Diffusive samplers are used to measure monthly averages of $SO_2$, $NO_2$, $NH_3$ and $HNO_3$ (Ferm and Svanberg, 1997; Ferm and Rodhe, 1997). At Rörvik, the concentrations of a large number of pollutants are measured on a daily or shorter basis to study long-range transport. Several comparisons of the concentrations at the two sites have shown that the data are very similar (Grennfelt *et al.*, 1985). Data from Rörvik will be used to represent long-term concentrations in air at Gårdsjön in this paper, because longer time series are available together with a better time resolution. Continuous measurements of sulphur dioxide and particulate sulphate started in 1978. In 1986 measurements of the sum of nitric acid and particulate nitrate (total nitrate) in the air started at Rörvik, together with measurement of total ammonium.

As can be seen from Figure 3.1, the concentration of $SO_2$ at Rörvik has decreased from 1978 to 1994. The decrease started in 1981–83, but increased again in 1984 and 1985. A strong and steady decrease again occurred after 1989 and by 1994 the concentration was reduced by more than 50% compared with the 1980s and about 75% compared with the late 1970s. Particulate $SO_4^{2-}$ showed a similar pattern but decreased more slowly: less than 50% compared with the late 1970s. Total $NO_3^-$ decreased a little during 1988, 1989 and 1991 and the same pattern is true for total $NH_4^+$, but the decrease was more pronounced for $NH_4^+$ and totalled some 30% between 1987 and 1994.

### Deposition

Table 3.1 shows all the mean deposition data measured at various intervals since the start of the Gårdsjön project. Data for 1980/81 for heavy metals were obtained from Grahn and Rosén (1983), phosphorus, DOC and organic nitrogen were from Persson and Broberg (1985) and the rest from Hultberg (1985) except for water fluxes and $H^+$ in precipitation. The 1983–86 data were all obtained from Westling and Hultberg (1990/91). Data for 1991/93 have not been published earlier except for the mercury and methyl mercury data (Hultberg *et al.*, 1994; Munthe *et al.*, 1995). The measurements in Table 3.1 show that $H^+$ and $SO_4^{2-}$ deposition has declined in both rain and throughfall. Sulphur deposition is about two times higher in throughfall compared with input by precipitation. Sea salts have increased over time in both precipitation and throughfall. The sea salts and DOC are three to four times higher in throughfall compared to wet deposition (Potts, 1978; Hultberg and Grennfelt, 1992). K and Mn have increased in throughfall but not in precipitation, which indicates that these species have increased as a result of increased leaching of

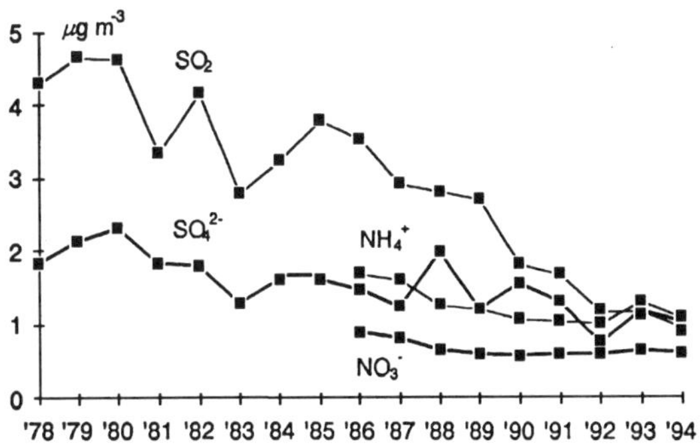


**Figure 3.1**. Annual average air concentrations of particulate $SO_4^{2-}$; $SO_2$ (both starting in 1978); total $NO_3^-$ and total $NH_4^+$ (both starting in 1986). The concentrations are expressed in $\mu$g S or N $m^{-3}$

**Table 1.1**. Deposition (kg $ha^{-1}$ $yr^{-1}$) of different compounds by precipitation, throughfall and cloud/fog during different time periods. Fog deposition was calculated from active and passive fog samples, see text

| | Bulk precipitation | | | Throughfall | | | Fog/cloud |
|---|---|---|---|---|---|---|---|
| | 80/81 | 83/86 | 91/93 | 80/81 | 83/86 | 91/93 | 1994 |
| mm | 1087 | 1038 | 1033 | 728 | 708 | 661 | 6.1 |
| $H^+$ | 0.54 | 0.48 | 0.43 | 0.78 | 0.78 | 0.45 | 0.004 |
| $SO_4$-S | 11.4 | 10.4 | 8.5 | 25.1 | 25.2 | 18.5 | 0.23 |
| Na | 8.5 | 11.6 | 15.7 | 42.9 | 42.5 | 56.5 | 0.58 |
| Cl | 16.5 | 22.3 | 26.1 | 81.6 | 83 | 102.1 | 0.82 |
| $NH_4$-N | 6.3 | 5.0 | 4.9 | 4.7 | 6.0 | 3.6 | 0.15 |
| $NO_3$-N | 4.6 | 4.9 | 5.4 | 5.7 | 6.8 | 7.0 | 0.16 |
| Org-N | 0.5 | 2.2 | 0.3 | 2.5 | 2.5 | 3.0 | — |
| Total-N | 11.4 | 12.1 | 10.7 | 12.9 | 15.3 | 13.6 | — |
| K | 0.6 | 2.1 | 1.8 | 12.9 | 15.2 | 19.9 | 0.05 |
| Ca | 1.7 | 2.4 | 1.9 | 10.2 | 10.8 | 9.2 | 0.06 |
| Mg | 1.2 | 1.8 | 2.0 | 6.6 | 7.0 | 8.7 | 0.07 |
| Mn | 0.02 | 0.14 | 0.09 | 0.90 | 0.98 | 1.24 | $210^{-3}$ |
| DOC | 13 | — | 11.9 | 59 | — | 68.2 | — |
| P | 0.06 | 0.23 | 0.07 | 0.25 | — | 0.08 | — |
| Fe | 0.19 | — | — | 0.25 | — | — | — |
| Al | 0.20 | — | — | 0.46 | — | — | — |
| Cu | 0.007 | — | | 0.013 | — | — | — |
| Zn | 0.32 | — | — | 0.28 | — | — | — |
| Cd | 0.002 | — | — | 0.002 | — | — | — |
| Pb | 0.06 | — | — | 0.05 | — | — | — |
| V | 0.003 | — | — | 0.008 | — | — | — |
| Co | $210^{-4}$ | — | — | $4 \times 10^{-4}$ | — | — | — |
| Cr | 0.002 | — | — | 0.006 | — | — | — |
| Ni | 0.004 | — | — | 0.008 | — | — | — |
| Hg | — | — | $1.2 \times 10^{-4}$ | — | — | $1.6 \times 10^{-4}$ | — |
| MeHg | — | — | $4.0 \times 10^{-6}$ | — | — | $1.8 \times 10^{-6}$ | — |

internally circulated K and Mn in the tree crowns (Hultberg and Ferm, 1995). The importance of internally circulated K and Mn is also obvious from the difference between input by precipitation and throughfall, with about 10 times higher input by throughfall. No trends are obvious in $NH_4^+$, $NO_3^-$, organic-N or total-N, but throughfall deposition of $NH_4^+$ is lower than bulk deposition, which shows that canopy retention of $NH_4^+$ is higher than for $NO_3^-$ The measurements of heavy metal deposition show that Fe, Zn, Cd, Pb and total Hg are similar in precipitation and throughfall input while Al, Cu, V, Co, Cr and Ni are about two times higher in throughfall (Grahn and Rosén, 1983). Methyl mercury is retained in the tree canopies and throughfall is about half of the wet deposition (Hultberg *et al.*, 1994).

Wet deposition of $H^+$, $NH_4$-N, $NO_3$-N and non-marine $SO_4$-S are shown in Figure 3.2. The trend of non-marine wet deposition of $SO_4^{2-}$ resembles the atmospheric $SO_2$ concentration more than the particulate $SO_4^{2-}$ concentration. Ammonium and $NO_3^-$ concentrations are very well correlated with one another in both precipitation and air, while concentrations in air are not so well correlated with concentrations in precipitation. Figure 3.2 further shows that the non-marine sulphate and the hydrogen ion deposition decreased in recent years, but the deposition of nitrogen compounds did not.

Dry deposition of sulphur to the forest has been estimated from throughfall measurements (Hultberg and Grennfelt, 1992). The throughfall technique can also be used to estimate roughly the minimum dry deposition of $NO_3^-$ in Gårdsjön, but not $NH_4^+$ because there is a strong retention in the canopy by uptake and transformation of this compound. This effect is very obvious at low depositions, but probably negligible at other sites with very high N loads. The throughfall technique (Table 3.1) gave a dry deposition of 1.8 kg N $ha^{-1}$ $yr^{-1}$ for $NO_3^-$ which seems low, and a negative value for $NH_4^+$ (−0.6). Dry deposition of nitrogen compounds was therefore also estimated using air concentrations at Gårdsjön together with relationships between air concentrations and deposition at a forest 60 km distant. The measured total S and N deposition at Gårdsjön is also generally much higher (about 400%) than the critical loads of sulphur and at the same level as the critical loads for total nitrogen (Chapter 22, this volume).

An attempt to estimate the deposition of sulphur during the 20th century to the Gårdsjön area has also been performed (Ferm *et al.*, 1989). An average ratio between the calculated deposition according to the EMEP model originating from a certain country and that country's emission during 1979–87 was therefore calculated for the 14 most important countries. The sulphur emission from each of those countries before 1979 was estimated using consumption figures (UN data) for oil and coal together with their sulphur content. The result is presented in Figure 3.3, which shows that modelled total S deposition in 1993 and 1994 was slightly less than at the turn of the century. The measured wet deposition of sulphur was similar to the modelled total sulphur deposition. "Real" S deposition measured as throughfall in the forested catchments at Gårdsjön is two times higher than modelled (Table 3.2). This is partly due to the fact that the total deposition to a forest is higher than to other surfaces in the grid. Sweden, like many other countries in northern Europe, is to a large extent forested, so in reality total deposition in these countries may be higher than total deposition as estimated by the EMEP model.

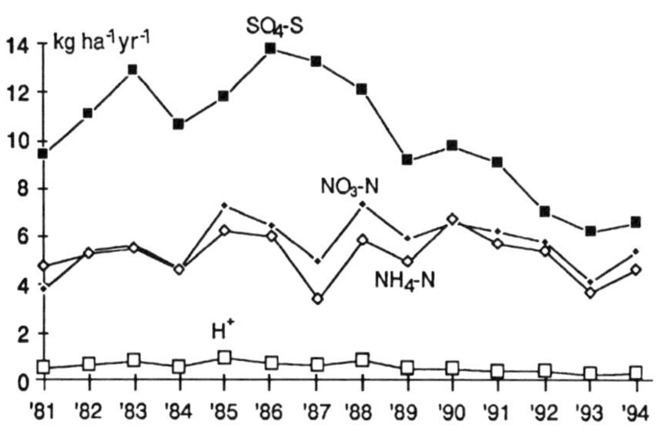


**Figure 3.2.** Wet deposition in kg $ha^{-1}$ $yr^{-1}$ of $H^+$, $NH_4$-N, $NO_3$-N and non-marine $SO_4$-S at Gårdsjön

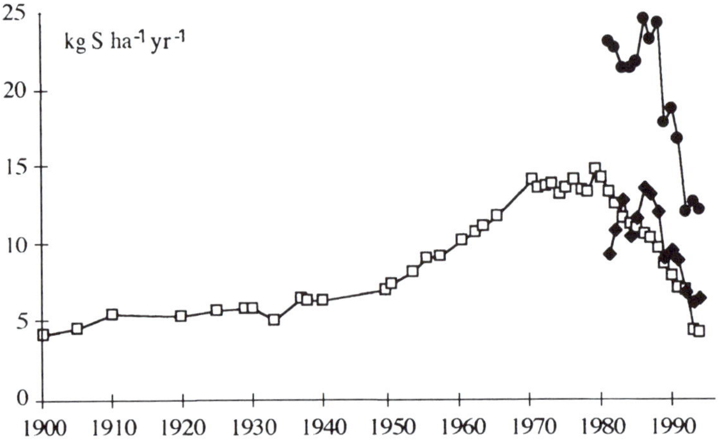


**Figure 3.3**. Estimated sulphur deposition (circles) together with measured wet deposition (diamonds) and throughfall deposition (unfilled squares) to the Gårdsjön research area between 1900 and 1994

**Table 3.2**. Measured and modelled annual sulphur and inorganic nitrogen deposition (all in kg $ha^{-1}$ $yr^{-1}$) in forested catchments at Gårdsjön

| Measured | $SO_4$-S | $NO_3$-N | $NH_4$–N |
|---|---|---|---|
| Wet deposition | 10.2 | 6.1 | 5.3 |
| Dry deposition | 9.0 | ≈3 | ≈2 |
| Σ | 19.2 | ≈9 | ≈7 |
| Modelled | 8.6 | 5.4 | 3.5 |

$NH_4^+$ and perhaps also $NO_3^-$ can be transformed to organic nitrogen in throughfall. This transformation can be interrupted by protecting the throughfall collector from light or by adding a preservative such as iodine (Ferm, 1993). The organic nitrogen formed has a tendency to stick to the walls of the throughfall collector. This tendency ceases if iodine is added. Table 3.3 shows some data from Gårdsjön. The measurements are unfortunately not continuous, but sporadic, and do not allow a comparison with the estimated dry deposition.

## Input of base cations

Throughfall deposition of base cations can be divided into marine and non-marine wet deposition and marine dry deposition (assuming Na is conservative as it passes through the canopy; Hultberg, 1985; Hultberg and Grennfelt, 1992). The rest originates from non-marine dry deposition and recirculation in the forest stand (Hultberg and Ferm, 1995). Results from 1981–94 are shown for $Ca^{2+}$ in Figure 3.4 and $K^+$ in Figure 3.5. A decline mainly in internal circulation is obvious for $Ca^{2+}$ after 1989, which is similar to the decrease in dry deposition of sulphur measured in throughfall (Hultberg and Ferm, 1995) as well as in precipitation in Gårdsjön

**Table 3.3.** Nitrogen deposition (kg N $ha^{-1}$) in throughfall (TF) and bulk precipitation (Bulk) in dark unpreserved collectors and collectors preserved with iodine ($I_2$)

| | Days | $NH_4^+$ | $NO_3^-$ | Org-N | Tot-N |
|---|---|---|---|---|---|
| TF | 385 | 3.8 | 5.9 | 4.9 | 14.5 |
| TF $I_2$ | 385 | 3.6 | 6.3 | 7.1 | 16.9 |
| Bulk | 168 | 2.5 | 2.9 | 0.2 | 5.6 |
| Bulk $I_2$ | 168 | 2.8 | 3.0 | 0.1 | 5.9 |

(Lövblad *et al.*, 1995). Recirculation of $K^+$ first increased with time over the period 1981–91 but decreased again during 1992 and 1994. Non-marine dry deposition of $Ca^{2+}$ and $K^+$ is included in the recirculation values.

It is difficult to obtain independent estimates of non-marine dry deposition of base cations. An attempt has been made by using a surrogate surface that resembles the needles. The sampler consists of thin Teflon strings mounted under a small roof to protect them from precipitation. The strings are washed monthly and the leachate is analysed. The dry deposition of base cations is calculated from the base cation to sodium ratio on the strings multiplied by the net throughfall deposition of sodium to the forest (Ferm and Hultberg, 1995). Results from a one-year study are shown in Table 3.4. The technique can also be applied to particulate sulphate deposition. The particulate $NH_4^+$ deposition estimated in the same way as the base cations was 1.4 kg N $ha^{-1}$ $yr^{-1}$. It can be seen that most of the throughfall flux of $Ca^{2+}$ (52%) and $K^+$ (74%) is derived from recirculation, while only 12 and 4% respectively is non-marine dry deposition. The input of base cations can also be obtained through model calculations (Westling *et al.*, 1995). Good agreement between model

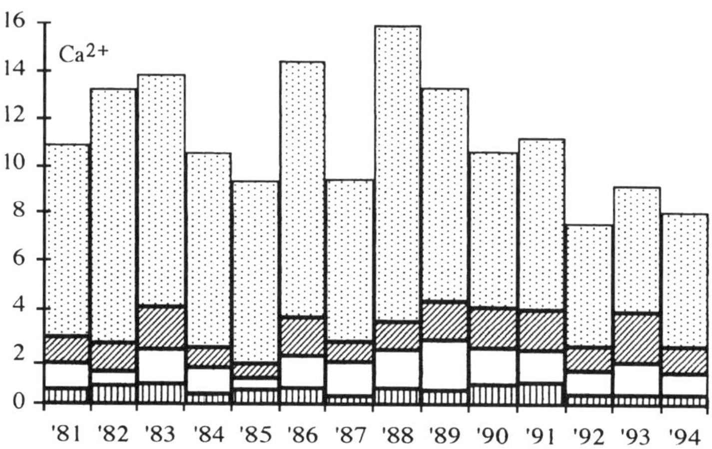


**Figure 3.4.** Throughfall deposition of $Ca^{2+}$ (kg $ha^{-1}$ $yr^{-1}$) as a function of year. The bars from bottom represent marine wet deposition, non-marine wet, marine dry and the largest on top is internal circulation plus some (about 12%) non-marine dry deposition

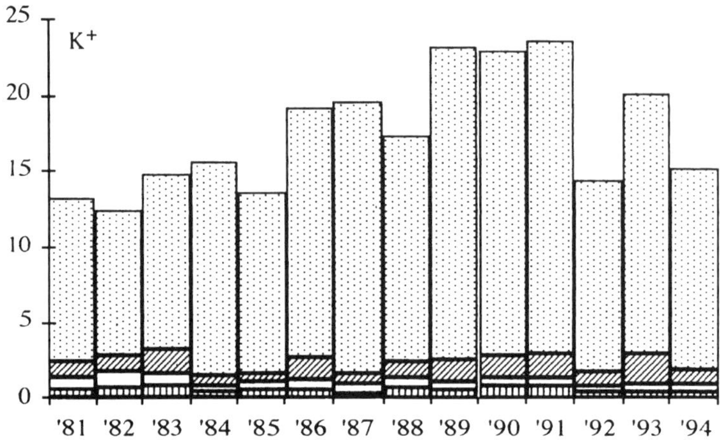


**Figure 3.5**. Throughfall deposition of $K^+$ (kg ha$^{-1}$ yr$^{-1}$) as a function of year. The bars from bottom represent marine wet deposition, non-marine wet, marine dry deposition and the largest on top is internal circulation plus a small amount (about 4%) of non-marine dry deposition

calculations and the surrogate surface approach has been obtained for $Ca^{2+}$ (Ferm and Hultberg, 1995).

## Fog and cloud deposition

The deposition of cloud and fog (Table 3.1) was measured using the same approach as for the base cations. The only difference was that the Teflon strings were only exposed during fog. Fog was registered using a laser beam. When the backscattered light gave a constant signal, there was fog without precipitation. The signal was connected to a logger that also triggered the engine that lifted the Teflon strings out of a cylinder. The advantage of a passive fog collector is that the deposition of fog can be calculated. Evaporation of the collected water does, however, occur. During

**Table 3.4**. Estimated origin of base cations and sulphate in throughfall (Catchment G2, 1994). The deposition is given in kg ha$^{-1}$ yr$^{-1}$. The "recirculation" figure for $SO_4^{2-}$ (kg S ha$^{-1}$ yr$^{-1}$) represents $SO_2$ uptake, and is thus an input to the system, in contrast to the other ions. WD—wet deposition; DD—dry deposition

| | $Ca^{2+}$ | $K^+$ | $Mg^{2+}$ | $SO_4^{2-}$ |
|---|---|---|---|---|
| WD, marine | 0.5 | 0.5 | 1.7 | 1.1 |
| WD, non-marine | 0.9 | 1.3 | −0.1 | 6.6 |
| DD, marine | 0.8 | 0.8 | 2.7 | 1.8 |
| DD, non-marine | 0.7 | 0.5 | 0.0 | 1.4 |
| Recirculation | +3.1 | +8.7 | +1.0 | +3.1 |
| Throughfall | 5.9 | 11.7 | 5.3 | 13.9 |

radiation fog there is an inversion layer near the ground implying that the wind speed and the fog deposition are very low. When the cloud base is so low that parts of the forest are in the clouds, the deposition will be very high.

An active fog collector was also used. Evaporation of the collected water does not occur with this technique. The ion concentrations in the fog water were measured on a monthly basis with the active collector. The ion concentrations were on the average six times higher in fog than in precipitation. The deposition of fog was, however, only around 3% of the wet deposition and 2% of the total deposition on an annual basis. During 1994 there was fog during 0.9% of the time. The highest fog deposition took place in March. The deposition of fog water (6.1 mm) was estimated from the deposition of an ion via fog and the concentration of the same ion in the fog water. The figures were surprisingly similar for different ions. Even though deposition via fog is negligible on the Swedish west coast (see also Table 3.1), fog may have an effect on trees because the concentrations of all ions are so high.

## Origin of the air pollutants

The concentrations of pollutants in air depend very much on the wind direction. Measurements were made on a 24-hour basis and were divided into wind trajectory

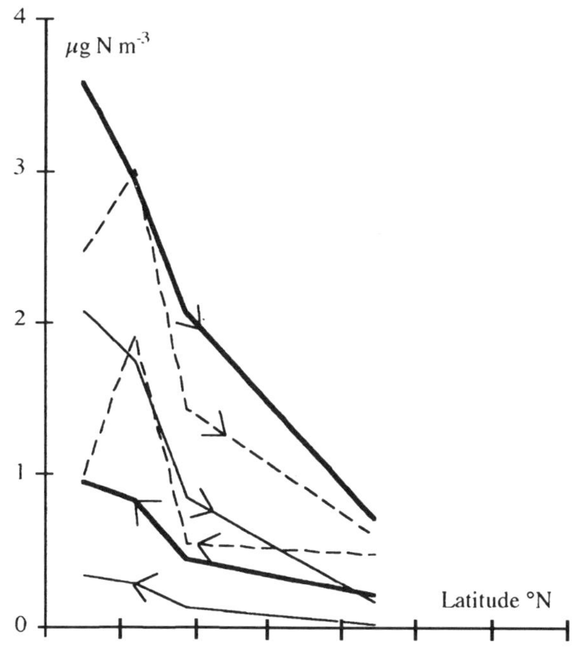


**Figure 3.6**. Air concentrations of $NH_3$ + particulate $NH_4^+$ (bold lines), $HNO_3$ + particulate $NO_3^-$ (thin lines) and $NO_2$ (dotted lines) as a function of the latitude of the sampling stations. The upper line of all species represents wind trajectory from the south and the lower from the north

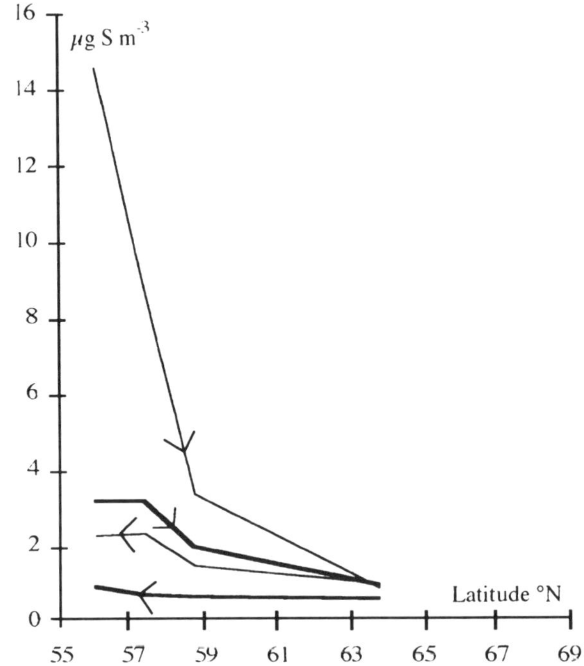


**Figure 3.7.** Air concentrations of $SO_2$ (thin lines) and particulate $SO_4^{2-}$ (bold lines) as a function of the latitude of the sampling stations. The upper line of both species represents wind trajectory from the south and the lower from the north

sectors. Sweden is long and narrow almost in a north–south direction. The average concentration of nitrogen and sulphur compounds when the wind blows from the south at a height corresponding to 800 mbar is shown in Figures 3.6 and 3.7 respectively as a function of latitude at four stations (Vavihill, Rörvik, Velen and Bredkälen in the middle of Sweden). The average concentrations when the wind comes from the north are shown in the same figures. As can be seen the concentrations are about five times higher for $SO_2$ during southern winds than during northern. $NO_2$ shows a maximum concentration at Rörvik, which may be due to local sources, especially during northern winds, when the air passes Gothenburg. This illustrates in a qualitative way that the major part of these species in the air over Sweden are imported from emission sources in Europe outside Sweden. Gårdsjön has a latitude of 58° 04′ N and Rörvik 57° 25′ N.

Import and export of air pollutants takes place in all wind directions. In Figure 3.8 the average concentrations of nitrogen and sulphur compounds are shown for eight 3-day trajectory sectors at Rörvik. The data represent 60% of the time between February 1986 and December 1988. During the rest of the time (40%), the trajectories could not be allocated to a certain sector. As can be seen from the figure, the highest concentrations occur when the wind trajectory comes from south or south-east. $NO_2$ is an exception with a rather uniform wind rose. The most prevailing trajectory sector is south-west.

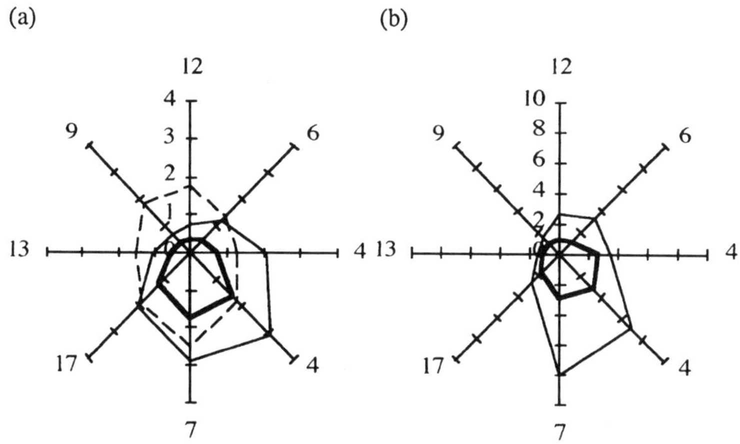


**Figure 3.8**. (a) average concentrations of total $NH_4^+$ (thin line), total $NO_3^-$ (bold line) and $NO_2$ (dotted line) as a function of wind trajectory sector. (b) average concentrations of $SO_2$ (thin line) and particulate $SO_4^{2-}$ (bold line). Concentrations are expressed in $\mu g\ m^{-3}$ and were taken from Rörvik (1986–88). The frequencies as percentages are shown at the ends of the axes

The origin of the pollutants can be estimated in a quantitative way by using computer models containing emission inventories, transformation and deposition rates together with meteorological information. In Table 3.5, emissions and depositions for Sweden and Europe are presented. In Table 3.6 the origin of the sulphur and nitrogen deposition to the Gårdsjön area has been estimated. The original calculations were made by Tuovinen *et al.* (1994). The grid size for the calculation is 150 × 150 km and a large part of the grid where Gårdsjön is situated also comprises Norway, including Oslo. The calculations in Table 3.6 used an average of four grids in west Sweden to minimize the effects of local anomalies.

**Table 3.5**. Emissions and deposition in Sweden and Europe. The figures are an average for 1985–93 (Tuovinen *et al.*, 1994)

| | Emission ($kt\ yr^{-1}$) | Deposition from own sources ($kt\ yr^{-1}$) | Deposition total ($kt\ yr^{-1}$) | Areal deposition total ($kg\ ha^{-1}\ yr^{-1}$) |
|---|---|---|---|---|
| **Sweden** | | | | |
| $SO_2$-S | 66 | 25 | 230 | 5.1 |
| $NO_X$-N | 78 | 16 | 129 | 2.9 |
| $NH_4$-N | 39 | 23 | 82 | 1.8 |
| **Europe** | | | | |
| $SO_2$-S | 16 525 | 6 555 | 12 426 | 25.2 |
| $NO_X$-N | 4 951 | 994 | 3 442 | 7.0 |
| $NH_4$-N | 5 757 | 3 425 | 4 883 | 9.9 |

**Table 3.6.** Modelled average origin of oxidized sulphur, nitrogen and reduced nitrogen compounds to four grids near the Gårdsjön research area during 1985–93. In the table the deposition is given as a percentage of the calculated total deposition

| Country of origin | $SO_4^{2-}$ (%) | $NO_3^-$ (%) | $NH_4^+$ (%) |
|---|---|---|---|
| Sweden | 10.8 | 9.2 | 27.9 |
| United Kingdom | 13.1 | 17.6 | 3.4 |
| GDR | 18.3 | 4.1 | 5.1 |
| FRG | 6.1 | 18.1 | 8.8 |
| Denmark | 6.1 | 7.3 | 14.3 |
| Poland | 10.5 | 6.3 | 7.4 |
| France | 2.0 | 4.7 | 3.4 |
| The Netherlands | 1.2 | 4.8 | 4.6 |
| USSR (European) | 3.5 | 2.6 | 4.0 |
| Czechoslovakia | 4.9 | 3.6 | 1.1 |
| Norway | 2.9 | 3.2 | 5.4 |
| Belgium | 1.7 | 2.2 | 1.1 |
| Others | 6.4 | 7.2 | 3.3 |
| Σ inattributable | 12.5 | 9.6 | 9.9 |

**Comparison of deposition with other areas**

Today the highest deposition of acidifying compounds is found in the northern hemisphere (Galloway and Rodhe, 1991). Somewhat higher deposition is found in Europe than in North America (Hultberg and Likens, 1991). Relatively few data have been published from north-east Asia (Dianwu and Jiling, 1988). Shen *et al.*, (1993) studied a Masson pine forest in the suburb of Chongqing in China and found a sulphur deposition of 200 kg S $ha^{-1}$ $yr^{-1}$.

The EMEP calculations (Tuovinen *et al.*, 1994) give maximal depositions to grids ($150 \times 150$ km) in Europe of 75 kg S $ha^{-1}$ $yr^{-1}$ (Germany and Czech Republic), 12.5 kg N $ha^{-1}$ $yr^{-1}$ of $NO_3^-$ (Germany) and 28 kg N $ha^{-1}$ $yr^{-1}$ of $NH_4^+$ (The Netherlands). Those depositions are only half of those observed in forests using the throughfall technique (Hauhs *et al.*, 1989), as was the case in the Gårdsjön area (Hultberg, 1985; Hultberg and Grennfelt, 1992). The maximum deposition in Europe is consequently about eight times higher for $SO_4^{2-}$, three for $NO_3^-$ and six times higher for $NH_4^+$ than in the Gårdsjön area. The deposition to Gårdsjön is, however, high compared with other sites in Sweden. Lövblad *et al.*, (1995) have estimated the range of deposition to spruce stands in Sweden to be about 1.5 to 25 kg S $ha^{-1}$ $yr^{-1}$, 1.5 to 13 kg $NO_3$-N $ha^{-1}$ $yr^{-1}$ and 2 to 13 kg $NH_4$-N $ha^{-1}$ $yr^{-1}$.

## REFERENCES


Dianwu, Z. and Jiling, X. (1988) Acidification in South-western China. In: H. Rodhe and R. Herra (eds), *Acidification in Tropical Countries. SCOPE 36*. J. Wiley & Sons Ltd, Chichester, pp. 317–346.

Ferm M. (1993) Throughfall measurements of nitrogen and sulphur compounds. *International. Journal of Analytical Chemistry*, **50**, 29–43.

Ferm, M. and Hultberg, H. (1995) Deposition of base cations to a forested catchment area. *Water, Air and Soil Pollution*, **85**, 2229–2234.

Ferm, M. and Rodhe, H. (1997) Measurements of air concentrations of $SO_2$, $NO_2$ and $NH_3$ at rural and remote sites in Asia. *Journal of Atmospheric Chemistry*, **27**, 17–29.

Ferm, M. and Svanberg, P.-A. (1997) Cost-efficient techniques for urban- and background measurements of $SO_2$ and $NO_2$. *Atmospheric Environment* (in press).

Ferm, M., Areskoug, H., Hanssen, J.-E., Hilbert, G. and Lättilä, H. (1988) Field intercomparison of measurements techniques for total $NH_4^+$ and total $NO_3^-$ in ambient air. *Atmospheric Environment*, **22**, 2275–2281.

Ferm, M., Borg, G. Ch., Gould, C. and Sjöberg, K. (1989) Estimation of the accumulated acid load in some different regions in Sweden. IVL Report 1989-10-16 L89/338 (in Swedish).

Galloway, J. N. and Rodhe, H. (1991) Regional atmospheric budgets of S and N fluxes: how well can they be quantified. *Proceedings of the Royal Society of Edinburgh*, **97B**, 61–80.

Grahn, O. and Rosén, K. (1983) Deposition and transport of metals in some acidic catchments in south-west, middle and northern Sweden. Swedish EPA, *SNV PM* 1686–1687 (in Swedish).

Grennfelt, P., Larsson, S., Leyton, P. and Olsson, B. (1985) Atmospheric deposition in the Lake Gårdsjön area SW Sweden. *Ecological Bulletins (Stockholm)*, **37**, 101–108.

Hasselrot, B. and Grennfelt, P. (1987) Deposition of air pollutants in a wind exposed forest edge. *Water, Air and Soil Pollution*, **34**, 153–143.

Hauhs, M., Rost-Siebert, K., Raben, G. Paces, T. and Vigerust, B. (1989) Summary of European data. In: J. L. Malanchuk and J. Nilsson (eds). *The Role of Nitrogen in the Acidification of Soils and Surface Waters*. Nordic Council of Ministers, Miljörapport: 1989:10, 5, 1–37.

Hultberg, H. (1985) Budgets of base cations, chloride, nitrogen and sulphur in the acid Lake Gårdsjön catchment, SW Sweden. *Ecological Bulletins (Stockholm)*, **37**, 133–157.

Hultberg, H. and Ferm, M. (1995) Measurements of atmospheric deposition and internal circulation of base cations to a forested catchment area. *Water, Air and Soil Pollution*, **85**, 2235–2240.

Hultberg, H. and Grennfelt, P. (1992) Sulphur and sea salt deposition as reflected by throughfall and runoff chemistry in forested catchments. *Environmental Pollution*, **75**, 215–222.

Hultberg, H. and Likens, G. E. (1991) Sulphur deposition to forested catchments in Northern Europe and North America — Large-scale variations and long-term dynamics. In: S. E. Schwarts and W. G. N. Slinn (eds), *Proceedings of 5th International Conference on Precipitation Scavenging and Atmosphere–Surface Exchange*, Vol. 3, Hemisphere Publ. Corp., USA, pp. 1343–1365.

Hultberg, H., Iverfeldt, Å. and Lee, Y.-H. (1994) Methyl mercury input/output and accumulation in forested catchments and critical loads for lakes in south-western Sweden. In: C. J. Watras and J. W. Huckabee (eds), *Mercury Pollution — Integration and Synthesis*. Lewis Publishers, Chelsea, MI, pp. 313–322.

Lövblad, G., Kindbom, K., Grennfelt, P., Hultberg, H. and Westling, O. (1995) Deposition of acidifying substances in Sweden. In: H. Staaf and G. Tyler (eds), *Effects of Acid Deposition and Tropospheric Ozone on Forest Ecosystems in Sweden. Ecological Bulletins (Stockholm)*, **44**, 17–34.

Munthe, J., Hultberg, H. and Iverfeldt, Å. (1995) Mechanisms of deposition of methylmercury and mercury to coniferous forests. *Water, Air and Soil Pollution*, **80**, 363–371.

Persson, P. and Broberg, O. (1985) Nutrient concentrations in the acidified Lake Gårdsjön: The role of transport and retention of phosphorus, nitrogen and DOC in watershed and lake. *Ecological Bulletins (Stockholm)*, **37**, 158–175.

Potts, M. J. (1978) The pattern of deposition of air-borne salt of marine origin under a forest canopy. *Plant and Soil*, **50**, 233–236.

Shen, J., Qianxue, Z., Guiru, J., Hongshou, T., Fuzhu, Z., Zongwei, F., Okita, O. and Ogura, N. (1993) Pollution and deposition of $SO_2$, $SO_4^{2-}$ and $HNO_3$ in Chongqing suburb forest area. In: K. Satake (Ed.), Proceedings of the International Workshop on *Development and Application of Biogeochemical Methods in Acid Rain Research*, Tskuba and Kusatsu, Japan, 8–12 March, 1993, pp. 54–70.

Tuovinen, J.-P., Barrett, K. and Styve, H. (1994) *Transboundary Acidifying Pollution in Europe: Calculated Fields and Budgets 1985–93*. The Norwegian Meteorological Institute, EMEP/MSC-W Report 1/94.

Westling, O. and Hultberg, H. (1990/91) Liming and fertilisation of acid forest soil: Short-term effects on runoff from small catchments. *Water, Air and Soil Pollution*, **54**, 391–407.

Westling, O., Hultberg, H. and Malm, G. (1995) Total deposition and tree canopy internal circulation of nutrients in a strong acid gradient in Sweden, as reflected by throughfall fluxes. In: L. O. Nilsson and R. F. Hüttl (eds), Proceedings from *IUFRO Symposium*, Halmstad, Sweden, June 1992, 639–647.

# 4 Catchment-Scale Acidification Reversal Experiment at Gårdsjön, South-west Sweden — Assessment of the Experimental Design

**FILIP MOLDAN, INGVAR ANDERSSON, KEVIN BISHOP & HANS HULTBERG**

At the Gårdsjön Covered Catchment Experiment, atmospheric deposition is almost entirely excluded from the forested headwater catchment G1 ROOF by means of a plastic roof that prevents precipitation from reaching the ground. In the first 4 years of the treatment, 2960 mm of artificial precipitation was applied (18 600 $m^3$) over the course of 339 sprinklings. The throughfall input of sulphate to the catchment was reduced by the treatment from 128 meq/$m^2$/yr to 25 meq/$m^2$/yr. The inputs of nitrate declined from 63 meq/$m^2$/yr to 4 meq/$m^2$/yr and ammonium (31 meq/$m^2$/yr before the roof) was excluded entirely. This deposition regime is designed to simulate a major reduction in acid deposition. There are many advantages to such catchment manipulations, but given the technical difficulties of controlling as many experimental variables as this project entails, it is important to evaluate the extent to which the catchment manipulation simulates the deposition reduction intended. In this paper we assess the extent to which this artificial regime simulates the reduction of acid deposition that would occur if atmospheric emissions were reduced to pre-industrial levels.

The change of atmospheric input was accomplished as planned without causing major changes in microclimate or hydrological regime. Several compromises, however, were necessary. Sprinkling solution does not represent pre-industrial throughfall ideally and several other aspects were altered as well, such as the spatial–temporal dynamics of water input and internal cycling of forest nutrients. We believe these inevitable compromises generally tend to slow the rate of recovery which would have been induced by removal of all anthropogenic deposition to this gravely acidified forest ecosystem. None of these deviations from an ideal manipulation, however, significantly diminishes the capacity of the experimental results to contribute to the understanding of acidification recovery. Nevertheless, careful consideration must be given to these compromises, particularly with regard to the alteration in the internal cycling of nutrients when evaluating the covered catchment experiment.

## INTRODUCTION

Transparent roofs are used in many different plot- or catchment-scale manipulation studies (Wright *et al.*, 1988a; Jansson 1990; van Dijk *et al.*, 1992; Beier and Rasmussen,

*Experimental Reversal of Acid Rain Effects: The Gårdsjön Roof Project.*
Edited by Hans Hultberg and Richard Skeffington. 

1993) because a roof provides a large degree of control over water and elemental inputs into a study area.

The Gårdsjön Covered Catchment Experiment started with pre-treatment data collection at the end of 1988. The roof became operational in April 1991, and was planned to continue for a minimum of 5 years. The purpose of the roof was to entirely cover a catchment that was large enough to include the range of hydrobiogeochemical interactions that may influence the dynamics of recovery from acidification. Interpretation of plot-scale experiments is most commonly based on soil water sampling by means of lysimeters. Large uncertainties, however, are involved in the extrapolation of soil solution chemistry to runoff chemistry, and hence the recovery of the aquatic ecosystem. The catchment scale of the experiment makes it possible to measure the response of runoff directly.

Despite the advantages of conducting manipulations on the catchment scale, there are also drawbacks. High building costs, labour-intensive execution of the treatment and maintenance are among the most obvious. Of more importance for the interpretation of the results of the catchment manipulation are the difficulties of controlling the manifold aspects of the catchment environment required to simulate a reduction in just part of the atmospheric deposition without inadvertently introducing other changes. The temporal and spatial distribution of rainfall, temperature and relative humidity or light conditions are examples of just some of the catchment parameters which are disturbed by the manipulation (see Chapter 22, this volume). Placing the roof below the crown canopy at Gårdsjön adds a range of other parameters which must be controlled since the roof interrupts the natural cycling of nutrients from crown to forest floor. The placement of the roof below the canopy has some advantages as well. Heating of the ecosystem due to a literal "greenhouse effect" is reduced. Dry deposition and fog deposition are decreased, reducing the uncertainty associated with estimating these inputs. The possibility of manipulating the throughfall instead of the above-canopy rain could also be an advantage.

To achieve the pre-industrial deposition, the $SO_4^{2-}$, $NO_3^-$, $NH_4^+$ and $H^+$ in current deposition has to be reduced substantially. It is assumed that deposition of sea salts and organic anions has not been affected by human activities and should not, therefore, be changed. It could be argued that the deposition of the base cations from the atmosphere which are not of marine origin should also be reduced to some extent under the roof, since it is believed that a substantial part of contemporary, non-marine base cation deposition is due to fuel combustion, industrial processes, traffic and agriculture (Hedin *et al.*, 1994). These alterations in deposition should be accomplished without altering other factors such as microclimate, water regime or biogeochemical cycling between the crown canopy and forest floor via throughfall and litterfall.

Mercury deposition has also been reduced as part of the experimental treatment. While Hg is not directly responsible for acidification, it is a pollutant which potentially compromises the functioning of the ecosystem. A supplementary objective of the Gårdsjön Covered Catchment project was therefore to study the effects of a reduction in atmospheric mercury deposition.

This paper presents the implementation of the roof experiment, the design of the pre-industrial atmospheric deposition which the experiment seeks to reproduce, and then identifies discrepancies between the manipulation that was intended and what has actually occurred.

## METHODS

### Site description

Three catchments are involved in the experiment, the covered G1 ROOF catchment of 0.63 ha and two reference catchments with throughfall and runoff monitoring. One of the references is the nearby 3.7 ha F1 CONTROL which has no treatment, and the other is the adjacent G2 NITREX (Dise and Wright, 1992) of 0.52 ha where the current manipulation — increased nitrogen deposition — does not affect throughfall or runoff volume measurements. The G2 NITREX deposition data have been used to complement the data from the untreated F1 CONTROL because G1 and G2 share the same south-west aspect while F1 faces north.

The G1 ROOF catchment is located near Lake Gårdsjön, where acidification research has been conducted for over two decades (Andersson and Olsson, 1985; Dickson, 1988). The watershed of Lake Gårdsjön is approximately 12 km from the sea coast, and situated on a gentle ridge where the watershed altitude ranges between 113 and 170 m a.s.l. The climate is maritime with a mean annual temperature of 6.3°C and a small month to month temperature variation over the course of a year (Chapter 2, this volume). Snow cover does not usually persist during the winter. The mean annual precipitation of 1100 mm is relatively high for the area, since it is enhanced by an orographic effect. A detailed description of the Gårdsjön Lake site is given by Andersson and Olsson (1985).

The G1 ROOF catchment has a closed stand of 80 to 100-year-old Norway spruce (*Picea abies* (L.) Karst.) forest, with some Scots pine (*Pinus sylvestris*) present. The bedrock is gneissic with intermediate to acid chemical composition. The soils are generally thin podzols, and the mean soil depth for the whole catchment is 43 cm. The F1 and G2 catchments have approximately the same type of soils and forest cover as G1 ROOF. A detailed description of the experimental catchments is given by Andersson *et al.* (Chapter 2, this volume).

The deposition in the Gårdsjön area is dominated by sea salt which forms about two-thirds of the elements in ambient throughfall on an equivalent basis. The annual throughfall input of Cl is about 100 kg/ha. The S and N deposition in the area is intermediate relative to that in other areas of Europe. During the 6 year period from 1989 to 1994, F1 CONTROL received an average of 19 kg/ha/yr of $SO_4$-S in throughfall, of which about 5 kg is of marine origin (calculated from $Na^+/SO_4^{2-}$ sea-salt ratio, assuming all $Na^+$ is of marine origin), as well as 7 kg of $NO_3$–N and 4 kg of $NH_4$–N. The atmospheric input to the area is discussed in Ferm and Hultberg (Chapter 3, this volume).

### Roof design

The plastic roof covers the 6300 $m^2$ catchment G1 ROOF in its entirety. The roof itself is somewhat larger, about 7000 $m^2$, because it overlaps the catchment at the edges (Figure 4.1a). The supporting structure of the roof is constructed of untreated wood (Figure 4.1b). The roof itself consists of polycarbonate corrugated plastic sheets placed on top of the wooden frame 2 to 4 m above the ground (Figure 4.1c). The 0.8 mm thick roof sheathing is transparent, and transmits 90% of the light at wavelengths above 400 nm. Each tree (385 in total) penetrates the roof through an opening large enough to allow the trees to move with the wind (Figure 4.1d). The roof is divided into eight irregular sections.

Precipitation intercepted by the roof is drained outside the catchment area through a system of gutters. Water from the major part (70% of the roof area) flows to a central gutter (Figure 4.1e) which leads to a V-notch weir where the throughfall amount is registered and sampled. The slope of each roof section approximately follows the topography of the catchment. Overlaps between roof sections are left unsealed to improve ventilation. The construction is strong enough to bear snow or walk on. Litterfall is swept from the roof and stored in a dry area until it is redistributed manually underneath the roof. The roof is scrubbed and cleaned with pressurised water semi-annually to maintain its transparency.

During the construction period of the roof (November 1990 to March 1991), particular attention was paid to minimising site disturbance. A series of wooden walkways were constructed, and in so far as possible the movement of material and personnel was restricted to these walkways. The roof construction is supported by some 440 anchor points which were designed to minimise soil disturbance. In most places, polyester-coated concrete blocks were used as anchor points for the roof structure. The soil dug up to position these blocks was removed from the catchment. In steep areas, or where the bedrock was close to the surface, anchor holes were drilled directly into the bedrock. The rock dust generated by the drilling was vacuumed and removed from the site.

### Sprinkling system design

The solution is sprinkled under the roof by 275 centrifugal irrigating sprinklers placed 0.3 to 1 m above the ground and spaced in a regular 5×5 m grid across the catchment. Each sprinkler has a pressure-reducing feedback device which provides an even water distribution regardless of the individual sprinkler's position. The maximum sprinkling intensity is 3 mm per hour. The frequency and duration of sprinkling events is dependent on natural precipitation. Watering is typically done once or twice a week. The watering system installed at G1 ROOF is operational only when the air temperature is above the freezing point. When the precipitation occurred as snow, the equivalent of water was sprinkled during the subsequent snow melt. Water for irrigation is pumped 1 km from Lake Gårdsjön, mechanically filtered, and deionised by ion exchanger columns. Lake water has the advantage that it is continuously available, whereas intercepted water is irregular and limited in supply. To ensure the quality of deionised water, ion exchanger columns are automatically regenerated if the conductivity of purified water reaches 0.2 mS/m. Control analysis of the deionised water with conductivity close to this limit did not reveal a concentration of any analysed ions above detection limit at the IVL laboratory except occasionally for the appearance of chloride in concentrations of less than 3 $\mu$eq/l. The $Cl^-$ which passed the anion exchanger (associated with hydrogen) was estimated to represent less than 1% of an ambient chloride input to the area.

The simulated throughfall (Figure 4.2) is created from deionised lake water by addition of sea water (which simulates the sea salts), as well as potassium and calcium. Several hundred litres of sea water are needed every month. They are taken from the sea at the far edge of the nearby archipelago to minimise variability in the salinity of the sea water. The proportion in which the sea water has been added was calculated to reproduce sea-salt deposition measured in contemporary throughfall under the assumption that all sodium in throughfall is of marine origin. The actual composition of the solution applied each year is subject to adjustments to cope with the high variability of sea-salt deposition as measured at the reference sites over different time scales.

To simulate the internal forest cycling of K and Ca from the canopy to the soil via throughfall, these elements were added to the artificial throughfall. The $K^+$ was added in form of KCl, or $KNO_3$ after December 1994, to approximately the same concentration as in ambient throughfall. Calcium was added in the form of $CaCl_2$ hydrate, but not in amounts that yield concentrations equal to that found in ambient throughfall. These salts of K and Ca were used because they are pH neutral.

## Throughfall/stemflow measurements

Deposition data for catchments F1 CONTROL, G2 NITREX and G1 ROOF prior to treatment are based on throughfall measurements. Throughfall is collected in 15 collectors on F1 CONTROL, 28 on G2 NITREX and 26 on G1 ROOF. The samples are collected monthly or fortnightly. The sampling on the G1 ROOF catchment, however, was terminated after the emplacement of the roof sheathing began in January 1991. One volume-weighted composite sample is blended from all collectors at each site for each sampling occasion. Samples are routinely analysed for pH, $SO_4^{2-}$, $Cl^-$, $NO_3^-$, Kjeldahl-N, $NH_4^+$, $Ca^{2+}$, $Mg^{2+}$, $Mn^{2+}$, Fe, $Na^+$, $K^+$, DOC, total P, electrical conductivity and colour.

The openings in the roof for each tree trunk are not sealed. Therefore the stemflow is not prevented from reaching the ground. The stemflow was monitored on 17 trees in catchment G1 ROOF for 18 months in 1991 and 1992. The monitoring was terminated in October 1992 because of the relative unimportance of stemflow contribution to both water and element budgets (Table 4.1). Stemflow volume was less than 1% of the throughfall volume and stemflow deposition of all major ions was between 1 and 4% of throughfall deposition.

Bulk deposition is monitored by two collectors located in an open area near the F1 CONTROL catchment. Samples are collected and analysed in the same manner as throughfall.

## Runoff and soil water measurement

The G1 ROOF catchment is too small to form a natural stream. To capture the water leaving the catchment, a dam was created by inserting a plastic panel vertically into the soil at the catchment's lowest point, and sealing it to the bedrock 2.5 m below the soil surface. The intercepted groundwater, soil water and occasional

(a)
(b)
(c)

**Figure 4.1.** Configuration of the roof. (a) From outside the G1 catchment on the east side. Two overlapping sections can be seen. (b) From underneath showing the method of construction. (c) At the lower end of the catchment. Part of a walkway can be seen in the lower left corner. (d) Tree penetrations from above. The baffle plate on the upslope side prevents water and litter falling down the hole left to allow for tree movement. (e) Looking up the central gully of the roof

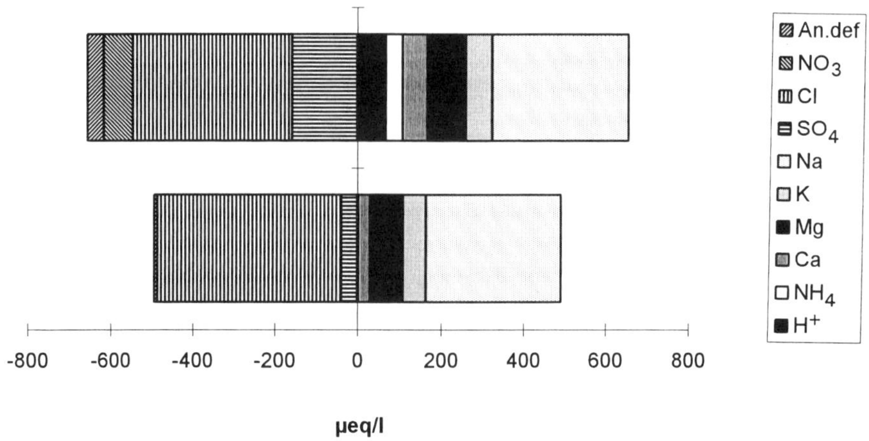


**Figure 4.2.** Mean composition of ambient throughfall at Gårdsjön (top bar) and applied sprinkling water (lower bar)

**Table 4.1.** Eighteen months' stemflow flux measured at 17 trees at G1 ROOF expressed as a percentage of throughfall flux measured at adjacent G2 NITREX catchment

| $H_2O$ | $H^+$ | Ca | Mg | Na | K | $NH_4$ | $NO_3$ | $SO_4$ | Cl |
|---|---|---|---|---|---|---|---|---|---|
| 0.5 | 2.0 | 4.4 | 2.6 | 2.2 | 2.9 | 0.9 | 1.4 | 2.4 | 2.4 |

overland flow water (which is all regarded as runoff) is drained by a tube at the level of the soil surface and gauged continuously. The runoff is sampled automatically at a frequency proportional to the flow to provide representative samples for accurate flux calculations. Sub-samples from each millimetre of runoff are mixed together and analysed as a 2 week, volume-weighted, composite sample. The chemical parameters which are routinely analysed are pH, $SO_4^{2-}$, $Cl^-$, $NO_3^-$, Kjeldahl-N, $NH_4^+$, $Ca^{2+}$, $Mg^{2+}$, $Mn^{2+}$, Fe, $Na^+$, $K^+$, total Al, organic Al, DOC, total P, electric conductivity and colour.

Suction lysimeters are installed at three sites on the G1 ROOF catchment with several depths measured at each site. Samples are collected fortnightly or monthly. The samples are analysed for all the elements in runoff with the addition of B, Cu, Si and Zn.

Similar runoff and soil solution monitoring is performed at the F1 CONTROL and G2 NITREX catchments.

## Supporting data

For background climatic data, on-line monitoring of wind direction and wind speed, air temperature, relative humidity, global radiation and open field precipitation has been established in several steps since 1989. Since 1992 there has also been monitoring of soil temperature at depths of 30, 15 and 5 cm, and air temperature at heights of 25 and 175 cm as well as relative humidity both underneath the roof on the

G1 ROOF catchment and on the F1 CONTROL site. Data are stored with a 15 minute time resolution. Soil water suction is monitored in several profiles on the G1 ROOF, G2 NITREX and F1 CONTROL catchments by an automatic tensiometer system. These data are also stored with 15 minute time resolution.

# RESULTS

## Change of solute inputs

*Sulphur, nitrogen, hydrogen ion*

Due to the treatment, all nitrate, ammonium and non-marine sulphate was removed from the artificial throughfall (Table 4.2, Figure 4.2). The input of $SO_4^{2-}$ was therefore reduced by 80% from *c.* 128 meq/m$^2$/yr (mean of two years before treatment) to 25 meq/m$^2$/yr, which is the sulphate from sea water, during the four treatment years. The throughfall input of 63 meq/m$^2$/yr of nitrate was excluded entirely during the first three years of the treatment. In the fourth year it was decided to use $KNO_3$ instead of KCl for potassium additions. Therefore 17 meq/m$^2$/yr of $NO_3^-$ was added in 1994 and about 40 meq/m$^2$/yr of $NO_3^-$ will be added in the future years of the experiment. The ammonium input of 31 meq/m$^2$/yr was reduced by the roof to zero. Hydrogen ion input was reduced from 61 to 2 meq/m$^2$/yr.

*Sea salt*

Large temporal variation of sea-salt input is a common feature at Gårdsjön. On an annual basis, the 10 year mean $Cl^-$ deposition at F1 CONTROL was 271 meq/m$^2$/yr (Hultberg and Grennfelt, 1992) with a standard deviation of 70 meq/m$^2$/yr or 26% of the mean. The ambient sea-salt deposition could not be reproduced in real time by sprinkling without devoting considerable resources specifically to that aspect of the experimental treatment. Given the time needed to analyse and evaluate throughfall and sprinkling samples, the periods of high and low sea-salt deposition to the control were reproduced by sprinkling only with a certain time lag. Therefore the sea-salt input to G1 ROOF is not the same as that to the control sites when looking at individual seasons or years (Figure 4.3).

During the first treatment year, more sea water than appropriate was added to the artificial throughfall as the proportion was calculated from the first two analyses of sea water, which happened to have the lowest salinity ever recorded in the coastal area where sea water was collected. This yielded *c.* 50% too many sea-salt ions (Table 4.2). Higher sea-salt deposition during the first treatment year may have temporarily increased both the runoff acidity by means of the salt-effect mechanism, as well as the soil base saturation (Wright *et al.*, 1988b). Sea-salt input was reduced in the second treatment year to compensate for the excess in the first year (Figure 4.3). The fact that since the onset of the treatment, years with high, intermediate or low sea-salt deposition are not occurring simultaneously at the G1 ROOF and control catchments must be kept in mind when comparing the year to year difference between G1 ROOF and F1 CONTROL, but should not pose insurmountable difficulties in the interpretation of the experimental results.

**Table 4.2.** Fluxes of major ions in throughfall at F1 CONTROL (April 1989–March 1995), G2 NITREX (April 1989–March 1995) and at G1 ROOF (April 1989–March 1991), and flux in sprinkling at G1 ROOF after the roof was finished (April 1991–March 1995). Numbers in parentheses are marine part of the flux, calculated from ratio to Na in sea water, assuming all Na of marine origin. SBC, sum of base cations (including $NH_4$); SAA, sum of acid anions; ANC, acid neutralising capacity (SBC–SAA). Units are mm/yr (water) and meq/m$^2$/yr

| | | Pre-treatment | | Treatment | | | |
|---|---|---|---|---|---|---|---|
| | | Apr. 89–Mar. 90 | Apr. 90–Mar. 91 | Apr. 91–Mar. 92 | Apr. 92–Mar. 93 | Apr. 93–Mar. 94 | Apr. 94–Mar. 95 |
| G1 ROOF | $H_2O$ | 692 | 627 | 707 | 623 | 618 | 1012 |
| F1 CONTROL | $H_2O$ | 691 | 657 | 646 | 647 | 755 | 852 |
| G2 NITREX | $H_2O$ | 678 | 638 | 631 | 649 | 710 | 814 |
| G1 ROOF | $H^+$ | 70 | 51 | 2 | 2 | 2 | 3 |
| F1 CONTROL | $H^+$ | 60 | 50 | 51 | 45 | 35 | 44 |
| G2 NITREX | $H^+$ | 63 | 52 | 51 | 49 | 43 | 53 |
| G1 ROOF | Ca | 76 (12) | 35 (7) | 14 (14) | 8 (7) | 32 (7) | 20 (9) |
| F1 CONTROL | Ca | 71 (12) | 38 (8) | 50 (12) | 52 (14) | 35 (6) | 41 (8) |
| G2 NITREX | Ca | 46 (11) | 31 (7) | 40 (9) | 40 (12) | 26 (5) | 30 (7) |
| G1 ROOF | Mg | 82 (61) | 47 (35) | 72 (72) | 38 (38) | 35 (35) | 45 (45) |
| F1 CONTROL | Mg | 86 (63) | 59 (41) | 73 (60) | 90 (73) | 47 (33) | 59 (44) |
| G2 NITREX | Mg | 68 (59) | 46 (35) | 58 (48) | 72 (62) | 37 (26) | 46 (36) |
| G1 ROOF | Na | 267 | 154 | 315 | 168 | 156 | 199 |
| F1 CONTROL | Na | 276 | 182 | 266 | 322 | 146 | 192 |
| G2 NITREX | Na | 261 | 155 | 211 | 274 | 116 | 157 |
| G1 ROOF | K | 53 (6) | 38 (3) | 41 (7) | 31 (4) | 35 (3) | 52 (4) |

| | | | | | | | |
|---|---|---|---|---|---|---|---|
| F1 CONTROL | K | 66 (6) | 57 (4) | 45 (6) | 43 (7) | 56 (3) | 39 (4) |
| G2 NITREX | K | 47 (6) | 31 (3) | 34 (5) | 35 (6) | 44 (3) | 28 (3) |
| G1 ROOF | $NH_4$ | 40 | 22 | 0 | 0 | 0 | 0 |
| F1 CONTROL | $NH_4$ | 37 | 26 | 32 | 22 | 17 | 28 |
| G2 NITREX | $NH_4$ | 33 | 21 | 36 | 26 | 30 | 31 |
| G1 ROOF | $NO_3$ | 80 | 46 | 0 | 0 | 0 | 17 |
| F1 CONTROL | $NO_3$ | 66 | 51 | 56 | 51 | 36 | 51 |
| G2 NITREX | $NO_3$ | 63 | 42 | 54 | 52 | 40 | 55 |
| G1 ROOF | $SO_4$ | 147 (32) | 108 (19) | 38 (38) | 20 (20) | 19 (19) | 24 (24) |
| F1 CONTROL | $SO_4$ | 150 (33) | 122 (22) | 127 (32) | 118 (39) | 92 (18) | 101 (23) |
| G2 NITREX | $SO_4$ | 130 (32) | 103 (19) | 110 (26) | 106 (33) | 84 (14) | 88 (19) |
| G1 ROOF | Cl | 328 | 184 | 401 | 225 | 240 | 276 |
| F1 CONTROL | Cl | 358 | 223 | 300 | 394 | 170 | 236 |
| G2 NITREX | Cl | 298 | 178 | 238 | 323 | 130 | 188 |
| G1 ROOF | SBC | 518 | 296 | 442 | 245 | 258 | 317 |
| F1 CONTROL | SBC | 536 | 362 | 466 | 529 | 301 | 359 |
| G2 NITREX | SBC | 455 | 284 | 379 | 447 | 252 | 293 |
| G1 ROOF | SAA | 555 | 338 | 439 | 245 | 259 | 318 |
| F1 CONTROL | SAA | 574 | 396 | 483 | 563 | 299 | 389 |
| G2 NITREX | SAA | 491 | 323 | 402 | 481 | 254 | 331 |
| G1 ROOF | ANC | −37 | −42 | 3 | 0 | −1 | 0 |
| F1 CONTROL | ANC | −38 | −34 | −17 | −34 | 3 | −29 |
| G2 NITREX | ANC | −36 | −39 | −23 | −34 | −2 | −38 |

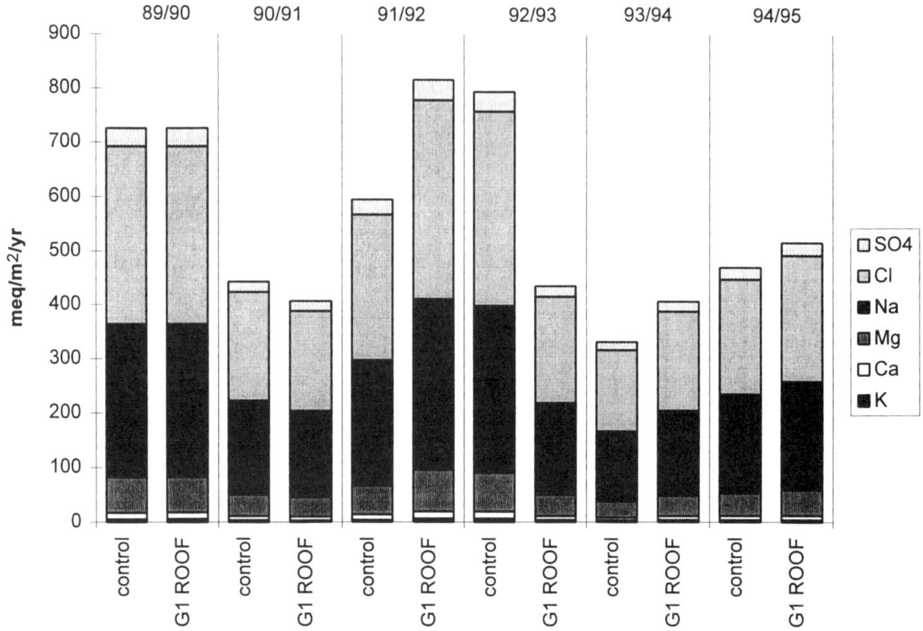


**Figure 4.3.** Annual sea salt deposition to the roofed catchment between 1989 and 1995

It should also be noted that there is a possibility that the design target for $Cl^-$ in the clean throughfall is several per cent too low. If we regard all $Na^+$ in throughfall at Gårdsjön as marine, then there is, on average, about 9% of the $Cl^-$ in throughfall deposition (*c.* 70 $\mu$eq/l) which is of "non-marine" origin (Hultberg and Grennfelt, 1992), calculated from the known ratio between $Na^+$ and $Cl^-$ in sea water. Since it is not known whether this "non-marine" $Cl^-$ has a natural or anthropogenic origin, it may be an underestimate of the natural $Cl^-$ level to set it according to the marine $Na^+$ concentration.

*Calcium*

The non-marine calcium in the artificial throughfall was 70% less than in the throughfall at the control sites over the four years of the experiment. The throughfall $Ca^{2+}$ flux has three major origins: marine deposition, non-marine deposition and canopy exchange. Some of this, especially the dry deposition of non-marine calcium, is anthropogenic (Hedin *et al.*, 1994). The canopy leaching of $Ca^{2+}$ may be also enhanced anthropogenically, by elevated $SO_2$ concentrations in the air and by precipitation acidity (Skeffington and Sutherland, 1995). Therefore it could be argued that only a certain portion of contemporary non-marine $Ca^{2+}$ from dry deposition and canopy leaching should be re-introduced to the clean throughfall, if the experiment seeks to irrigate with a "pre-industrial throughfall". Ferm and Hultberg (Chapter 3, this volume) estimated the origin of the 30 meq/m$^2$/yr $Ca^{2+}$ in throughfall at G2 NITREX in 1994 as consisting of 22% marine deposition, 27% non-marine deposition and 51% canopy exchange.

The marine part of deposition at G1 ROOF was restored by sea water addition. However, the amount of non-marine $Ca^{2+}$ added to the sprinkling solution in the form of $CaCl_2$ hydrate, amounted only to about 10 meq/m$^2$/yr compared with 46 meq/m$^2$/yr of non-marine $Ca^{2+}$ in G1 ROOF throughfall before the treatment. Therefore the roof causes a net calcium loss of about 36 meq/m$^2$/yr which was previously either deposited to the canopies from the atmosphere, or taken up from the soil by the trees and then leached from the canopy with precipitation.

This annual loss can be compared with the estimated exchangeable soil $Ca^{2+}$ pool of *c.* 2600 meq/m$^2$, of which more than half of is in the non soil layer (Chapter 2, this volume). The main pathway for the $Ca^{2+}$ circulation in the spruce forest is litterfall, and that $Ca^{2+}$ flux is restored at G1 ROOF by spreading the litter beneath the roof. About 65 meq/m$^2$/yr is estimated to reach the soil in litterfall (Chapter 17, this volume). Therefore, it was hypothesised when designing the experiment, that any net export of $Ca^{2+}$ induced by the roof should not have a negative effect upon the tree health or runoff chemistry over the planned five years of the experiment. The $Ca^{2+}$ losses (and associated ANC losses), however, are similar in size to the estimated annual alkalinity generation by weathering of *c.* 40 meq/m$^2$/yr (Chapter 10, this volume). To avoid the potential impact of an inadvertent $Ca^{2+}$ depletion on the recovery process, the addition of more $Ca^{2+}$ than the present 20% of ambient non-marine $Ca^{2+}$ is being considered if the experiment is to proceed for several more years.

### *Potassium*

In contrast to $Ca^{2+}$, the amount of $K^+$ added to the clean sprinkling water brought the $K^+$ input up to a level that was similar to that found in ambient throughfall. Before the treatment, the annual throughfall flux of $K^+$ at G1 ROOF was *c.* 46 meq/m$^2$/yr. Ferm and Hultberg (Chapter 3, this volume) estimated that 74% of throughfall $K^+$ at G2 NITREX was from a canopy exchange, 15% was from non-marine deposition and 11% was of marine origin. Litterfall flux was estimated to 23 meq/m$^2$/yr at G1 ROOF by Stuanes *et al.* (Chapter 17, this volume). Thus, throughfall is the principal medium for internal forest circulation of $K^+$. The estimated exchangeable $K^+$ pool in the soil is 700 meq/m$^2$ (Chapter 2, this volume). Therefore, the roof would cause a relatively rapid depletion of exchangeable $K^+$ from the soil, if $K^+$ was not kept at pre-treatment levels in the artificial throughfall during the duration of the experiment.

### *Magnesium and sodium*

No further addition of magnesium or sodium was done as about 80% of magnesium and 100% of sodium found in throughfall at Gårdsjön F1 CONTROL is reproduced by the sea water addition.

*DOC, DON*

Dissolved organic carbon (DOC) input from throughfall has been greatly reduced since the onset of the experiment. The pre-treatment throughfall DOC concentration was 0.65 mmol DOC/l, but the concentration is only *c.* 0.1–0.2 mmol DOC/l in the sprinkling water. The Gårdsjön lake water used for irrigation has a contemporary DOC concentration around 0.3 mmol DOC/l (lake outlet mean 3/91–9/93). The origin of DOC in the sprinkling water is most likely a fraction of the lake water DOC which passed the filtering and ion exchange treatments, combined with a small amount of DOC introduced by sea water addition. The input of dissolved organic N (DON) was also greatly reduced from 73 mmol/l before treatment to 10–15 mmol/l.

The absence of this organic carbon posed a dilemma for the design of the artificial throughfall. The missing DOC has a number of properties of significance for the natural ecosystem, among others as a source of energy for soil micro-organisms. One property of the DOC in throughfall which is of particular significance for the roof experiment is the dissociation of organic anions. In the pre-treatment throughfall, the organic anions amounted to *c.* 5% of the negative charge input, or 30 $\mu$eq/l. (This concentration of organic anions was inferred from the difference between measured major cations and anions in throughfall.)

Organic carbon was not added to the irrigation water in any form to compensate for this shortfall. The main reason was that little is known about the exact nature of DOC in throughfall at Gårdsjön. The addition of some readily available form of DOC (acetates, oxalates or glucose) was considered, but decided against because these forms are not satisfactory substitutes for natural DOC. Addition of DOC would also increase the risk of biological growth within the irrigation system which could lead to malfunction.

*ANC*

The ANC input (calculated as the sum of $Ca^{2+}$, $Mg^{2+}$, $Na^+$, $K^+$ and $NH_4^+$ minus the sum of $SO_4^{2-}$, $NO_3^-$ and $Cl^-$) increased from −40 meq/m²/yr in throughfall to +3 meq/m²/yr in the sprinkling water. Addition of organic anions associated with, for example, $Ca^{2+}$ would have increased the ANC input for the sprinkling water by an additional *c.* 20 meq/m²/yr. However, several compromises were necessary, largely due to the inability to reproduce the 30 $\mu$eq/l (*c.* 20 meq/m²/yr) of dissociated organic anions in throughfall. (N.B. Since there is very little $H^+$ or $HCO_3^-$ in the design target for the clean throughfall these organic anions are the source of the ANC, as defined above.)

**Changes in water regime**

During the four years of treatment, 2852 mm of throughfall recorded at the control sites were replaced with 2960 mm of sprinkling under the roof. The capacity of the watering system was sufficient to induce runoff with dynamics similar to that at the reference sites. Sprinkling was not synchronised automatically with ambient rain events. The time shift between natural rain events and sprinkling was usually a few days. As a consequence, there was a similar time shift in the G1 ROOF runoff peaks, which typically came after those observed at the reference sites (Figure 4.4).

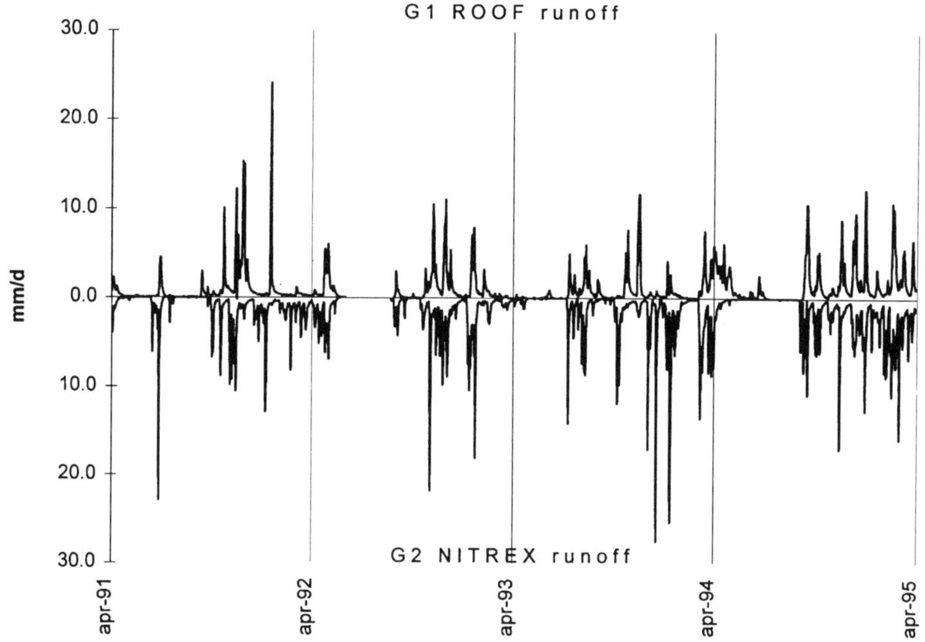


**Figure 4.4.** Four year hydrograph from the 0.63 ha catchment G1 ROOF (upwards) and adjacent 0.52 ha catchment G2 NITREX (downwards). The occasional dissimilarities in the runoff peaks are due to the time and duration shifts between ambient precipitation and sprinkling under the roof

Larger differences between ambient and artificial throughfall volumes occurred occasionally due to the inability to operate the watering system at temperatures below freezing point. This was most apparent in winter 93/94 (Figure 4.5), when ambient rain and snow precipitation from December 1993 to February 1994 was replaced with sprinkling during the March to April 1994 snowmelt, as well as later on during May and June 1994. The change of soil moisture regime was, however, presumably not as large as suggested from comparison of monthly natural throughfall and sprinkling volumes (Figure 4.5), because a certain portion of winter precipitation does not enter the soil until spring snowmelt which was simulated by enhanced spring watering (cf. Figure 4.4). The study of soil water tension revealed that the chosen way of simulating the natural precipitation induced soil moisture dynamics comparable to those observed at G1 ROOF catchment before the treatment (Chapter 5, this volume). One area of potential concern, however, is the moisture regime of the mor layer. Neither soil moisture nor tension were measured in that soil horizon. It is conceivable that the absence of many small rain events may have made the mor layer drier. While this would not lead to noticeable changes in runoff, it may be of importance for fine roots, mycorrhizae and the flora of the field layer.

The precipitation intensity at Gårdsjön is variable (Figure 4.6a, b). The frequency distribution of the intensity of open field rain and throughfall are similar to each

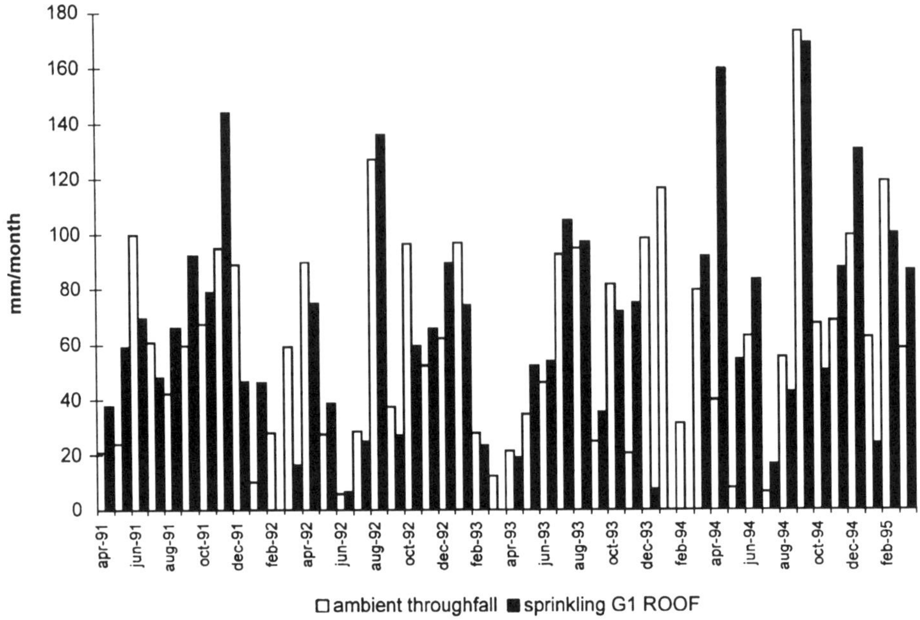


**Figure 4.5.** A comparison between ambient throughfall and monthly sprinkling water input to the catchment G1 ROOF

other. The most frequent precipitation intensity is 0.2 mm/h (Figure 4.6a), which is the sensitivity of the tipping bucket rain gauges which were used. However, these very low intensity rains of less than 0.5 mm/h contribute less than 20% of rain or throughfall volume (Figure 4.6b). The intensity of sprinkling was less variable, with a median value of 2.6 mm/h and 90% of sprinkling events between 1.9 and 2.8 mm/h. This variation was unintentional, although welcomed, and it resulted from a variety of technical reasons such as the amount of water in storage before sprinkling or increased resistance of sprinklers at temperatures around the freezing point.

With the sprinkling system used, the highest watering intensity is about 3 mm/h, precluding the simulation of rains with a higher intensity. If the sprinkling intensity is decreased by reducing the pressure in the watering system, the spatial distribution gets uneven and only soil close to sprinklers gets watered. Therefore low intensity rains were also not simulated. About 55% of throughfall volume comes with an intensity lower then 2.6 mm/h and 45% with a higher intensity. Therefore the median sprinkling intensity of 2.6 mm/h seems to be an appropriate compromise. Substitution of the range of throughfall intensities occurring naturally at Gårdsjön with a relatively uniform, and relatively high, sprinkling intensity is a simplification demanded mainly by technical constraints.

The spatial distribution of precipitation was changed due to the treatment. In the spruce forest, the throughfall volume generally increases with the distance from the stem to the edge of the crown canopy (Beier *et al.*, 1993). There are no direct measurements of the spatial distribution of sprinkling water at the G1 ROOF

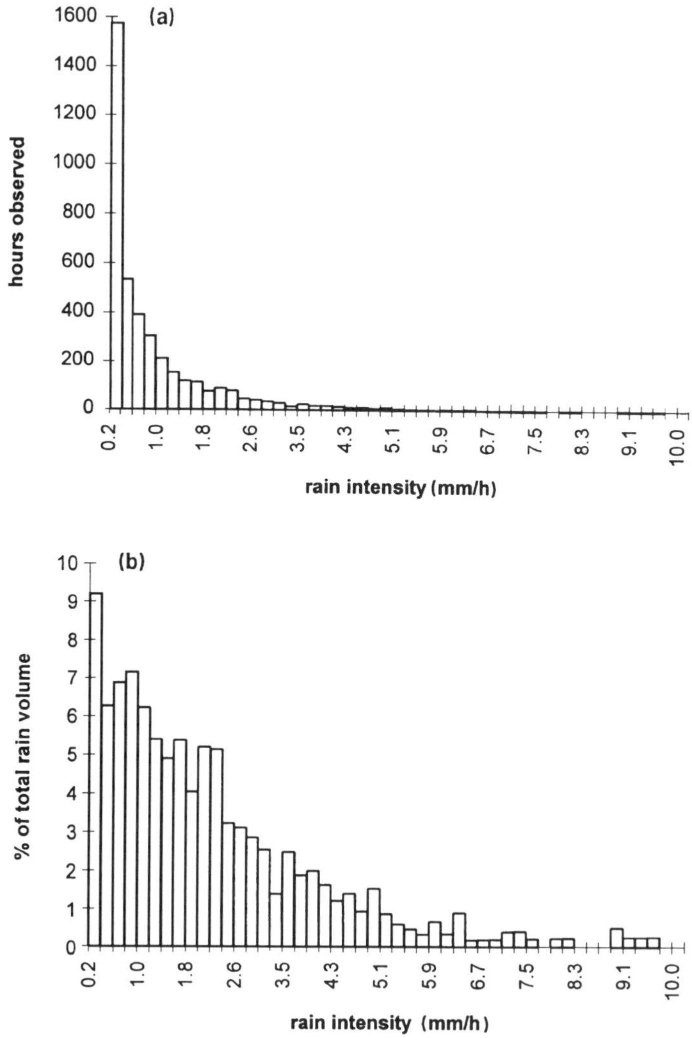


**Figure 4.6.** (a) The number of hours when rain with different intensities was observed at Gårdsjön and (b) the relative contribution of rainfall volume caused by rains with different intensities

catchment. From the design of the sprinkling system it could be assumed that water distribution is relatively even in the space in between the stems. In contrast to natural precipitation, the water droplets are sprinkled parallel to the soil surface. Therefore the water input near the stems is enhanced by a "stem catch" effect, especially where the regularly distributed individual sprinklers are close to the tree trunks (Gundersen *et al.*, 1995). This phenomenon may be of importance when interpreting lysimeter data, which could be biased by changes in the spatial distribution of water fluxes when comparing pre-treatment and treatment data. It is of lesser importance when interpreting the runoff data which are available at Gårdsjön due to the catchment-scale integration of the experiment.

The substitution of throughfall with sprinkling did not affect the ratio between runoff and evapotranspiration. The G1 ROOF catchment water yield was consistent through the six measured years, two before and four after the roof was constructed. On average, 51% (46–60%) of the throughfall water or sprinkling water input left the G1 ROOF catchment as runoff each year. It should be noted that the runoff water yield from throughfall on G1 ROOF (51%) differed consistently from that observed for F1 CONTROL (75%) and G2 NITREX (71%). This lower runoff yield from G1 ROOF could result either from a higher evapotranspiration rate from the G1 ROOF catchment, incorrectly determined size of the catchment or from runoff bypassing the catchment outlet (e.g. via flow in fractured bedrock).

One way of testing the water budget is to determine the catchment $Cl^-$ input/output budget. Assuming that there is no source of $Cl^-$ in the G1 ROOF catchment, then any flow of water from the catchment which is bypassing the outlet weir generates a deficit in the chloride input/output budget. After six years, there is a 4% excess of $Cl^-$ input compared with output from the G1 ROOF catchment (Figure 4.7), while the F1 CONTROL $Cl^-$ input is balanced with the output within less than 1%. (The $Cl^-$ flux from G2 NITREX could not be used for the comparison, because the nitrogen addition there causes significant changes in the G2 NITREX runoff chemistry.) The Cl deficit from the G1 ROOF catchment may not be significant because it is within the estimated accuracy of the runoff flux calculation.

However, some degree of runoff bypass has been inferred from tracer experiment results (Lange *et al.*, 1996). The mean annual evapotranspiration calculated as a difference between throughfall or sprinkling and runoff is 348 mm at G1 ROOF, *c.* 177 mm at F1 CONTROL, and 205 mm at G2 NITREX. Even if the 4% deficit in $Cl^-$ output does represent water leaving the catchment in some way other than through the catchment outlet or evapotranspiration though, it accounts for only *c.* 15 mm of runoff annually assuming that the missing runoff has the average measured runoff $Cl^-$. This does not explain the large difference in the runoff yield between G1 ROOF and the two reference catchments, G2 NITREX and F1 CONTROL.

A biomass inventory on the G1 ROOF, F1 CONTROL and G2 NITREX catchments revealed that there was significantly more standing biomass on G1 ROOF. The total bole volume was 283, 201 and 179 $m^3$/ha at G1, G2 and F1 respectively (Chapter 2, this volume). It is not certain that the large dissimilarity in runoff volume could be explained by the difference in forest cover, but with only a 4% deficit in the $Cl^-$ after five years, we have no alternative explanation for the bulk of the difference in runoff yield. Therefore, it was concluded that the most likely reason for a lower water yield from G1 ROOF is higher evapotranspiration from that catchment associated with the higher biomass.

### Changes in climate under the roof

A profile of soil temperature (5, 15 and 30 cm depth) and air temperature (25 and 175 cm height) as well as relative humidity (175 cm height) was measured at 15 minute intervals at G1 ROOF and F1 CONTROL starting in April 1992 after the

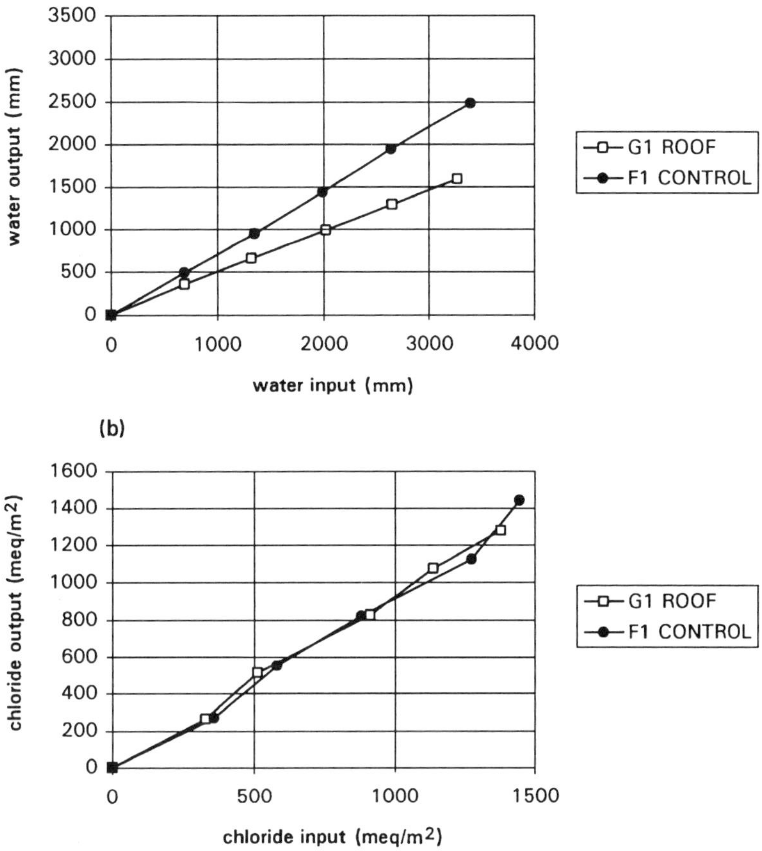


**Figure 4.7.** Accumulated inputs and outputs of water (a) and Cl (b) at G1 ROOF and F1 CONTROL between April 1989 and March 1995

roof was constructed. At both sites, sensors were placed under the canopy. Daily mean values from three years of measurement (April 1992–March 1995) were used for comparison. With the large number of daily observations, even very small differences in 3 year mean seasonal temperatures and relative humidity were statistically significant at a $p \leqslant 0.05$ level using a paired t-test.

In general, there were only relatively small differences between G1 ROOF and F1 CONTROL soil temperature, air temperature or relative humidity during the three years that microclimate was measured after the roof was built. The average soil temperature of the whole measured profile was 6.5°C, compared with a mean air temperature of 6.4°C. All measured temperatures followed similar seasonal patterns with a maximum in July–August and a minimum in February–March (Figure 4.8). The differences between air and soil temperature underneath the roof and outside the roof vary seasonally as well (Figure 4.8). The air temperature under the roof was

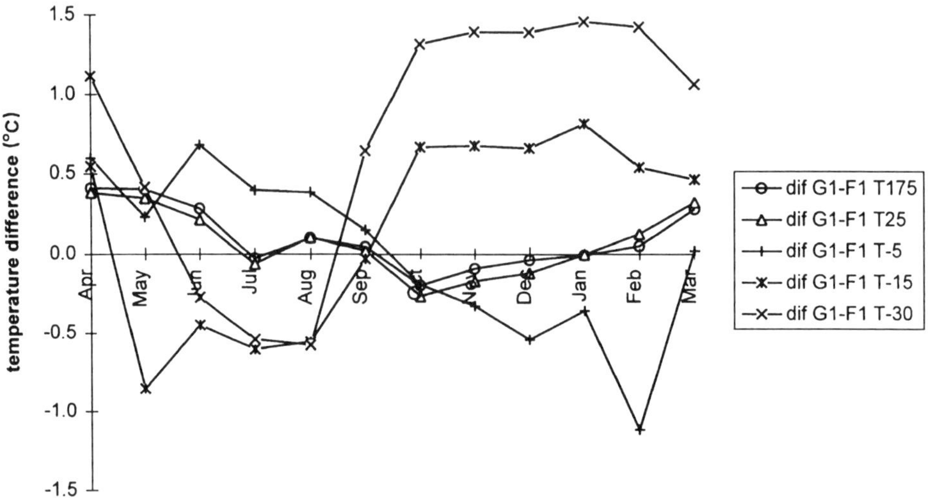


**Figure 4.8.** The difference in seasonal soil and air temperatures between G1 ROOF and F1 CONTROL

warmer on average during the late winter, spring and summer, and approximately the same as the control for the rest of the year. The soil at 5 cm followed the seasonal pattern of the air temperatures while the difference in the soil temperature measured at 15 and 30 cm depth followed an inverse pattern, with the roof being colder during the spring and summer and warmer during the rest of the year. The importance of these differences should not be overestimated, because they are often within the ±0.3°C uncertainty range of the temperature measurements. The large degree of small-scale spatial variation in soil temperature also reduces the ability to draw conclusions about differences between the G1 ROOF and F1 CONTROL catchments on the basis of a single profile of soil temperature measurements from each catchment.

Gunderssen et al. (1995) have hypothesised that colder summers and warmer winters under a roof could be expected due to partial reflection of radiation, thus reducing the heat input to the soil during the summers, and reducing radiation from the soil during the winter. Several other factors could contribute to such differences, e.g. reduced ventilation or differences in the temperature of irrigating water.

On average, there was no difference in relative humidity measured in the forest below and outside the roof. Daily means occasionally differed considerably due to the different timing of rain and irrigation events.

The difference in aspect between G1 ROOF and F1 CONTROL where the G1 catchment faces south-west and the F1 faces to the north is an important factor presumably causing dissimilarities in temperature and humidity between the two sites. Since the microclimate measurements began only after the roof was installed, the extent to which the differences result from the roof cannot be ascertained.

**Light regime**

Construction of the roof caused a reduction in the irradiation of the forest floor. The high transparency of the polycarbonate roofing material at wavelengths greater than 400 nm is offset to some extent by the wooden roof framework, litter deposited on the roof between the periodic cleanings, and occasional algae growth, especially at difficult-to-clean overlaps on the sheet edges. Gunderssen et al. (1995) reported 50% light reduction under a roof of similar construction that resulted in a 20% decrease of moss cover after several years which was related mainly to lack of light. Similar light reduction could be expected at Gårdsjön.

**Litterfall input**

Litterfall is a major component in the internal circulation of several forest nutrients. Thus it is essential to maintain this component of the cycle as well as that of the throughfall component. Litterfall which is deposited on the roof is first swept and transported from the roof. Then it is spread by hand under the roof as evenly as possible. However, this procedure involves technical and logistical difficulties which entail compromises. Ideally, clean pre-industrial litterfall would be simulated by using the same amount of litterfall from the same type of forest from a less polluted site. That would solve the problem of introducing part of the unwanted anthropogenic component of deposition associated with placing local litterfall under the roof. The anthropogenic pollution input is especially important for mercury where it represents about 50% of total deposition (Munthe *et al.*, 1995). To collect about 2.5 t of litter annually (Chapter 17, this volume) from some clean site and to spread it below the roof at Gårdsjön was not feasible. Therefore, the litterfall swept from the roof is used. There is, however, often a delay of several months between the litterfall reaching the roof and its being redistributed under the roof. The litter which lays on the roof is also leached by precipitation falling on the roof and fine material could be transported by that water which is diverted from the catchment. Quantification of this mass flux with water drained from the roof is problematic due to the flashiness of runoff from the roof. Some indication of what can be expected, however, comes from a similar roof experiment in Solling, Germany, where all the water from the roof is collected in storage tanks and re-used for sprinkling. The important conclusion from the chemical analysis of fine sediment deposited at the bottom of these storage tanks was, that up to 1 kg/ha/yr of phosphorus is lost from the catchment if this material is discarded (Lamersdorf and Blanck, 1995). It is likely that a similar loss of phosphorus is caused by the roof at Gårdsjön.

## CONCLUSIONS

The Gårdsjön roof was designed to simulate pre-industrial deposition conditions on a forested catchment. The roof worked as planned to remove all nitrate, ammonium and non-marine sulphate entering the catchment in throughfall. The concentration of $H^+$ in the artificial precipitation sprinkled under the roof was also reduced considerably compared to the levels in contemporary precipitation at Gårdsjön. The temporal fluctuation in sea-salt deposition was not reproduced in real time, but the magnitude of the year to year variation and the total sea-salt input after three years were reproduced reasonably well. The replacement of the natural precipitation with irrigation supplied through a sprinkling system replaced the water flux through the G1 ROOF catchment and generated soil moisture conditions similar to those in the catchment prior to the roof. From the measurements of temperature and air humidity which commenced after the roof was built, no considerable effect attributable to the roof was found.

A crucial feature of the experiment was the need to maintain the internal forest circulation of nutrients which was interdicted by the roof. The input of Na and most Mg was reproduced satisfactorily by addition of sea water to the sprinkling solution. Both K and Ca are leached from the crown canopy in significant quantities. As the throughfall is intercepted by the roof, their cycle is disturbed. All of the K and *c.* 30% of Ca needed to compensate for the leaching of these elements from the canopy was added to the sprinkling solution. It was hypothesised that reduced input of Ca should not be critical for the soil base status or for the vegetation given the relative magnitude of soil exchangeable Ca, Ca flux in litterfall and planned duration of the experiment.

The use of $Cl^-$ and $NO_3^-$ as counter ions for $K^+$ and $Ca^{2+}$ results in a noticeable reduction of ANC inputs in the artificial throughfall, because the $Cl^-$ and $NO_3^-$ are strong acid anions, as opposed to the weak organic acid anions which they replaced in natural throughfall. Furthermore, the total input of base cations in throughfall was reduced by 10–15% due to the reduced input of calcium (down by *c.* 36 meq/m$^2$/yr and magnesium (down by *c.* 12 meq/m$^2$/yr). Both the lack of base cations as well as additional, strong acid, chloride ions will reduce the atmospheric ANC input in the experiment. The average annual deficit in the ANC input relative to the experimental design target is comparable to the estimated production of ANC by primary weathering in acidification-sensitive areas of Sweden.

The litterfall cycle is restored only after considerable delays during which time litter was deposited on the roof, washed with several precipitation events, swept and stored prior to re-distribution underneath the roof. The associated loss of base cations, N, P or C has yet to be fully evaluated, as well as the amount of pollution introduced to the catchment by distributing litter from the canopies exposed to the ambient atmosphere. Examination of the intercepted throughfall collected from the roof, however, suggested that leaching of litterfall while on the roof was not significant for any of the major ions, since the intercepted throughfall did not differ consistently from throughfall on the control sites.

## Effects on rate of recovery

When evaluating the results of the experiment, it is necessary to bear in mind the potential for artefacts created by the experimental method. In many cases discussed here, compromises that could lead to artefacts were inevitable due to technical, financial or logistical constraints. The full implications of the material presented in this paper will need to be considered by those who evaluate the results of the experiment, but we wish to conclude this paper by assessing whether the deviations of the experiment from a genuine reduction in acid deposition will tend to accelerate or retard the rate of recovery from acidification on the G1 ROOF catchment.

There are three principal categories in which the verisimilitude of the experiment has been assessed: climate, throughfall and litterfall. The climate parameters were generally well reproduced, and it is not clear whether any of these climate factors will hasten or delay the recovery. The extent to which the amount of elements in litterfall deviated from the experimental design targets is not yet known, but it is likely that if there are deviations, they would be losses of base cations and nutrients (including P) leached from the litter prior to distribution under the roof as well as unintended inclusion of acid pollutants. These, and possibly even the slowing down of the internal circulation by delays in spreading the litter on the soil, would tend to delay recovery.

The tendency for the experiment to retard recovery is even more evident in the simulation of the other component of the forest's internal nutrient circulation, namely throughfall where the deficit in ANC input by the sprinkler solution relative to the design target is similar in size to the ANC estimated to be generated by primary weathering.

Therefore, the rate of recovery in the Gårdsjön roof experiment will probably be less than the rate at which recovery would occur if acid deposition were actually reduced to the same extent as the experiment seeks to simulate. At this point, however, it is not possible to quantify that underestimate of the potential recovery rate.

## ACKNOWLEDGEMENTS

Much of this project was funded by the Swedish Environmental Protection Agency and Great Britain's National Power and PowerGen, as well as the Swedish national power industry (Vattenfall and Elforsk). We thank all our colleagues who contributed with their dedicated work in the field and laboratory, but especially Jan Tobisson, Inger Torbrink, Pia Carlsson and Ulf Carlsson. Jake Peters is acknowledged for discussions about catchment water budget. Furthermore, we would like to acknowledge our colleagues involved in this or other roof experiments for many invaluable discussions and their openness in discussing all that could go wrong. These are Dick Wright, Per Gunderssen, Claus Beier, Lennart Rasmussen, Arne Stuanes, Michael Hauhs, Mike Bredemeier, Richard Skeffington and many, many more.

## REFERENCES

Andersson, F. and Olsson, B. (eds) (1985) *Lake Gårdsjön, an acid forest lake and its catchment. Ecological Bulletins (Stockholm)*, **37**.

Beier, C. and Rasmussen, L. (eds) (1993) *The EXMAN project*. Ecosystem Research Reports 7, European Union, Brussels, 124 pp.

Beier, C., Hansen, K. and Gundersen, P. (1993) Spatial variability of throughfall fluxes in a spruce forest. *Environmental Pollution*, **81**, 257–267.

Dickson, W. (ed.) (1988) *Liming of Lake Gårdsjön*. SNV report 3426, Solna, Sweden, 327 pp.

Dise, N. B. and Wright, R. F. (eds) (1992) *The NITREX Project (Nitrogen Saturation Experiments)*. Ecosystem Research Reports, 2, European Union, Brussels, 101 pp.

Gundersen, P., Andersen, B. R., Beier, C. and Rasmussen, L. (1995). Experimental manipulations of water and nutrient input to a Norway spruce plantation at Klosterhede, Denmark. 1) Unintended physical and chemical changes by roof experiments. *Plant and Soil*, **168–169**, 601–611.

Hedin, L. O., Granat, L., Likens, G. E., Buishand, T. A., Galloway, J. N., Butler, T. J. and Rodhe, H. (1994) Steep declines in atmospheric base cations in regions of Europe and North America. *Nature*, **367**, 351–354.

Hultberg, H. and Grennfelt, P. (1992) Sulphur and seasalt deposition as reflected by throughfall and runoff chemistry in forested catchments. *Environmental Polllution*, **75**, 215–222.

Jansson, P. E. (ed.) (1990) *The Skogaby Project*. Swedish University of Agricricultural Sciences, Uppsala, and University of Lund, 90:2

Lamersdorf, N. P. and Blanck, K. (1995) Evaluation of fine material input with throughfall for a spruce forest in Solling, FRG, by means of a roof construction. In: A. Jenkins, R. C. Ferrier and C. Kirby (eds), *Ecosystem Manipulation Experiments: Scientific Approaches, Experimental Design and Relevant Results*. Ecosystem Research Reports, 20, European Union, Brussels, pp. 168–170.

Lange, H., Lischeid, G., Hoch, R. and Hauhs, M. (1996) Water flowpaths and residence times in a small headwater catchment at Gårdsjön (Sweden) during steady state stormflow conditions. *Water Resources Research* **32**, 1689–98.

Munthe, J., Hultberg, H. and Iverfeldt, Å. (1995) Mechanisms of deposition of methylmercury and mercury to coniferous forests. *Water, Air, and Soil Pollution*, **80**, 363–371.

Skeffington, R. A. and Sutherland, P. M. (1995) The effects of $SO_2$ and $O_3$ fumigation on acid deposition and foliar leaching in the Liphook forest fumigation experiment. *Plant, Cell and Environment*, **18**, 247–261.

van Dijk, H. F. G., Boxman, A. W. and Roelofs, J. G. M. (1992) Effects of a decrease in atmospheric deposition of nitrogen and sulphur on the mineral balance and vitality of a Scotch pine stand in the Netherlands. *Forest Ecology and Management*, **51**, 207–215.

Wright, R. F., Lotse, E. and Semb, A. (1988a) Reversibility of acidification shown by whole-catchment experiments. *Nature*, **334**, 670–675.

Wright, R. F., Norton, S. A., Brakke, D. F. and Frogner, T. (1988b) Experimental verification of episodic acidification of freshwaters by sea salts. *Nature*, **334**, 422–424.

# 5 The Hydrology of the Covered Catchment: Water Storage, Flowpaths and Residence Times

K. H. BISHOP, M. HAUHS, L. NYBERG, J. SEIBERT, F. MOLDAN, A. RODHE, H. LANGE & G. LISCHEID

The hydrology of the G1 ROOF catchment was analysed using a variety of hydrometric and tracer techniques that sought to identify flowpaths and residence times. These studies found that runoff response to precipitation was rapid once a water storage threshold is passed (200 mm). Water storage within the catchment varied from 125–245 mm, with a 3 year average of 191 mm. The proportion of this storage which was held in the unsaturated zone above the water table went from over 90% at low flow to a minimum of 40% at peak flow. Much of the runoff was transmitted in the upper three decimetres of the soil. Some tracer studies found that the residence time of water within the catchment can be described by an exponential distribution. The mean residence time varies from 5 days at the highest observed flow rates to 175 days at the lowest observed flow rates. A small portion of the water can move at rates of several metres per hour during periods of high flow. The spatial pattern of residence times within the catchment, as observed from a step-shift of the $\delta^{18}O$ in precipitation after the roof was built, was consistent with the general picture of flowpaths arrived at from hydrometric measurements.

While these investigations have identified many hydrological features typical of Fenno-Scandian till catchments, the shallow soils, high organic matter content in the B horizon, and strong lateral concentration of flow across the concave topography of the catchment are important to bear in mind when generalizing the hydrochemical response of the Covered Catchment Experiment to other catchments.

## INTRODUCTION

The Gårdsjön Covered Catchment Experiment (CCE) is providing empirical evidence about how quickly runoff hydrochemistry can respond to large reductions in acid deposition. There are two principal reasons to expect that catchment hydrology is of importance for the rate of recovery. One is water's role as a reagent

*Experimental Reversal of Acid Rain Effects: The Gårdsjön Roof Project.*
Edited by Hans Hultberg and Richard Skeffington. 

in the reactions which affect acid deposition after it enters the catchment, and the second is water's role as a transport medium that moves the products of soil hydrochemical reactions from one reaction site to another within the catchment, and ultimately to the aquatic ecosystem as runoff.

There is a strong spatial differentiation within the catchment of both water pathways and the chemistry of the soil environment traversed by those flowpaths. Furthermore, the length of time water is in contact with any specific soil environment along its flowpath affects the degree to which reactions can progress. Therefore, both the flowpaths and residence times of water may be of importance for the response of an ecosystem to the chemistry of atmospheric inputs. Unfortunately, the complex spatial and temporal patterns of catchment hydrology have required great simplifications in the representation of catchment hydrology used in hydrochemical models. There is a possibility that these hydrological simplifications are a source of error when modelling the recovery of ecosystems from acid deposition (Nyberg *et al.*, 1993). Therefore, an investigation of catchment hydrology is a central component of the CCE.

Progress has been made in the last two decades on some aspects of the hydrological basis for ecosystem models, most notably in new observation technologies that allow an increased spatial and temporal resolution in the study of water transport and chemistry. One result of such studies in catchments similar to those at Gårdsjön was isotopic identification of the large amount of water displaced from the catchment in the stormflow hydrograph, relative to that of precipitation from the event which generated the hydrograph (Rodhe, 1987). Useful as this recognition was in dimensioning the amount of water stored in a catchment model, that advance did little to elucidate the pathways and residence times of the water flowing through the catchment (Bishop, 1991).

The difficulty of quantifying the spatial and temporal patterns of water flow has kept catchment hydrology a source of uncertainty in hydrochemical modelling. The complete control of precipitation inputs to the entire covered catchment and intensive monitoring, however, provide a unique opportunity to define better the hydrochemical dynamics, as well as to explore their significance for the chemical and biological response of the ecosystem (Bishop and Hultberg, 1995). This paper seeks to synthesize the results of several different studies which have employed a variety of physical, hydrometric and tracer methods into a description of catchment hydrology that will be of use to those interested in the CCE and the interpretations of its results. A subsequent and related chapter (Hauhs *et al.*, Chapter 21, this volume) deals with the various hydrological models that have been applied to this CCE data.

From a hydrological standpoint, the CCE is distinguished by the concerted effort to combine data on two central, but distinct hydrological concepts (i.e. water flowpaths and residence times). Each of these concepts alone includes a set of instrumentational, experimental and modelling issues that do not necessarily address the other concept. The two concepts have rarely been applied in parallel and, to our knowledge, have not been integrated successfully enough to provide a satisfactory basis for hydrochemical modelling. We will review some of the key assumptions underlying the concepts of water flowpaths and residence times.

## MECHANISMS CONTRIBUTING TO WATER TRANSPORT

Many Fenno-Scandian catchments on till soils exhibit a characteristic pattern of runoff generation which has been termed the "transmissivity feedback mechanism" (Bishop, 1991). The shallow till soils of these catchments have a large potential to rapidly increase lateral flows in response to infiltration. This occurs because the saturated conductivity of the soil often increases dramatically towards the soil surface. Increased water inputs raise the water table into progressively more conductive layers which transmit an increasing proportion of the water input at faster rates. This observation does not imply or exclude a non-isotropic hydraulic conductivity.

The water table rises until the transmissivity of the soil profile is sufficient to allow the water coming into the profile from both upslope and the soil surface to flow laterally downslope without additional accumulation of water within the soil profile, and hence without further rise in the water table. When water inputs to the profile decrease, the high level of water storage in the profile (and the high water table) cannot be sustained. As the water table falls, the largest, most conductive pores in the temporarily saturated superficial soil layers are drained, reducing the transmissivity, and hence the downslope flow of water. The long-term chemical consequences of such flow regimes may lead to chemical stratification in the horizontal plane perpendicular to the vertical flow through the superficial soil horizons.

This macroscopic view of flow transport processes has a parallel microscopic aspect. Flow through a porous medium is distributed over a large range of pore size classes and fracture networks. Depending on the effective size of these pores, the networks can be characterized by different transport regimes, e.g. typical transport times. Furthermore, the diffusive and dispersive exchange of water among these networks may depend on the differences in flow rates, and the frequency with which they become filled and active. Water held in some classes of pores can be considered immobile, relative to the situation in other classes of pores. Chemical gradients may also develop between the different pore classes in a soil, complicating the macroscopic differences in soil chemistry mentioned above.

Viewed in their full range of possible complications, these macroscopic and microscopic processes are intimately interrelated, e.g. providing the physical explanations for relations between the local water content and water potential or the mobile and immobile fractions of pore water.

Such a full picture of all potential processes encompassing all potential scales, however, is beyond the scope of hydrological experiments in the CCE where operational models are needed for the specific task of supporting the interpretation of the recovery from acidification. Simplifications are inevitably required to arrive at these operational models, but the simplifications should be guided by rational choices between two complementary hydrological concepts, flowpaths and residence times. These lines of abstraction are assumed to encompass the key processes for understanding and predicting the hydrology of the CCE.

### Flowpaths

The flowpath concept seeks to first define the trajectories of water from entry points at different locations on the catchment to the outlet under varying flow conditions.

Similar trajectories are then aggregated into flowpaths. This aggregation is informed by several features of catchment hydrology. The first is that in soils near saturation, large and rapid changes in the amount and direction of flow can occur. Under saturated conditions, changes in water potential can be transmitted almost instantaneously. This may lead to spatially correlated responses in groundwater potentials across a catchment. Such synchronized changes in hydraulic potential occur over macroscopic scales that are large compared with local heterogeneities of the soil. The rapid signal propagation of changes in hydraulic potential, however, is not necessarily coupled to a corresponding set of hydrochemical changes.

A common goal of hydrological analysis is to relate macroscopic flowpath structures to local hydrological variables such as water table or soil water potentials. This might be accomplished by dividing the catchment vertically into zones of saturated versus unsaturated transport, or horizontally into zones of diverging or converging streamlines. These flowpath categories can be further resolved in the temporal dimension to try to characterize average flow conditions or different types of episodes. Definition of flowpaths can also involve determination of how much of the spatial heterogeneity in catchment topography or hydrological parameter functions (pF, conductivity) is necessary to account for the local variability of these parameters, and how much of the spatial synchronization of hydrological responses can be explained by the spatial variability.

**Residence times**

The residence time concept seeks to derive a complementary view of catchment hydrology in terms of the distribution of water ages at a given point of observation (with age defined relative to when water entered the catchment). Regardless of the large spatial variability of the pore and fracture systems that contribute to runoff, any fixed observation point can be influenced in a characteristic and reproducible way. Therefore, it may be possible to determine a characteristic age distribution of water reaching any sampling point inside the catchment over time intervals that are large compared with the small-scale variability of natural rainfall events. Experiments with both natural and/or artificial hydrological tracers are used to derive breakthrough curves from which these age distributions can be estimated for different flow situations.

A typical goal is to see whether simple classifications of age distributions can be abstracted from the variability in the transient flow conditions, i.e. characteristic ages of soil water with respect to a given sampling point and boundary conditions. The ages observed at a given location inside the catchment are related to the area draining towards that point and the transient flow conditions within that area. The spatial aggregation of these local distributions into an estimate of the age distribution at the outlet (i.e. the transit time distribution for water in the catchment as a whole) is another facet of the residence time concept.

So far, there are few studies such as that for the Maimai catchment in New Zealand (McDonnell *et al.*, 1990) which have attempted to synthesize the flowpath and residence times concepts into one consistent picture of catchment hydrology. The hydrological studies in the CCE seek to make a contribution to that synthesis

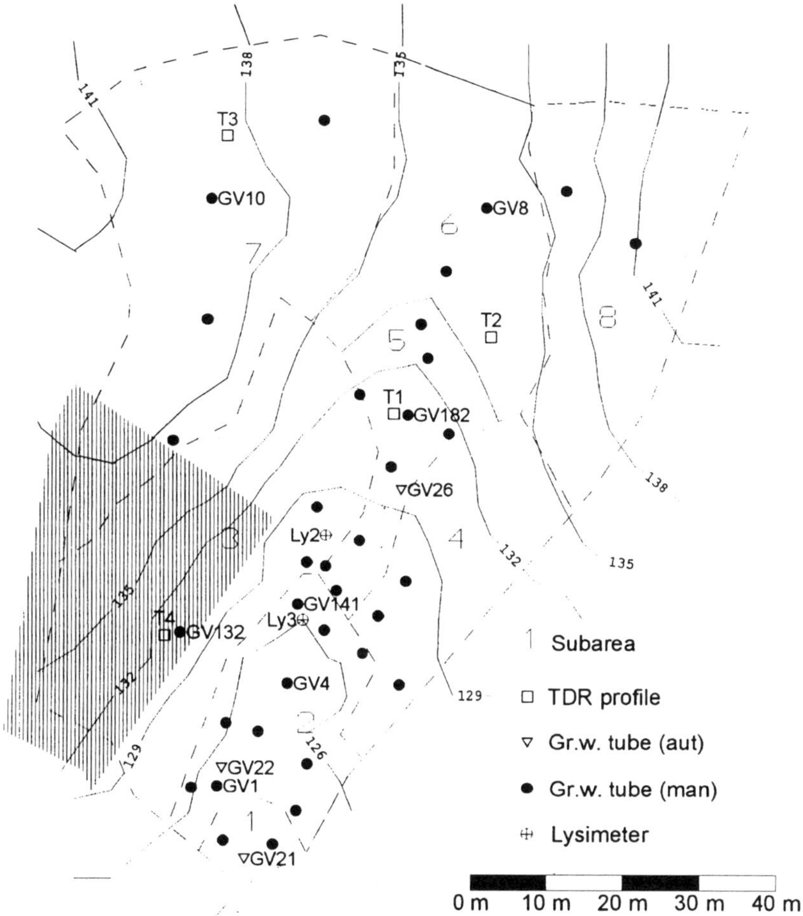


**Figure 5.1.** The G1 ROOF catchment showing the location of groundwater wells (manually monitored are closed circles, automatically monitored are open triangles) and TDR observation profiles. The boundaries of hydrologically similar subareas are also shown, together with topographic contours. The shaded area was irrigated in the steady-state, LiBr tracer experiments

which can be used in the task of understanding and predicting the observed hydrochemical response.

## STUDY SITE

The roof experiment is being conducted on the 6300 $m^2$ G1 ROOF catchment (Figure 5.1). This catchment is one of a number of small valleys in the vicinity of Lake Gårdsjön that define fairly similar mini-catchments of 0.5 to 5 ha. Two other such mini-catchments, the adjacent G2 NITREX catchment, and the nearby F1 CONTROL catchment, serve as references for the G1 ROOF catchment.

The topography of G1 ROOF features a 100 m long valley with steep side slopes. The bedrock on the site is gneissic granodiorite covered by a shallow glacial till upon which a podzol soil has developed. In the valley bottom, the podzols which are on the slopes give way to local areas of peat.

The overstorey vegetation on the catchment is dominated by 80–100 year old Norway spruce. Yearly annual precipitation is about 1100 mm, and the average annual temperate is 6.4°C. The maritime influence on the climate moderates the winters, so a persistent snow cover is unusual in the area.

The catchment did not have a natural water course, so a trench was dug at the lower end of the central valley in 1989 where the subsurface flow from the catchment converges. A dam was made to collect the water from this trench so that runoff rate and water chemistry could be measured. A fuller description of the catchment may be found in Andersson *et al.* (Chapter 2, this volume).

## METHODS

### Hydrological inputs and outputs

Since the roof was built, the acidified precipitation has been replaced with sprinkler irrigation distributed by 275 nozzles spread across the catchment. The chemistry of this irrigation water simulates unacidified throughfall on this site (see Moldan *et al.*, Chapter 4, this volume). The irrigation system has a maximum sprinkling rate of 2.8 mm $h^{-1}$, and the median rate of irrigation was 2.6 mm $h^{-1}$.

Precipitation was measured with tipping bucket rain gauges in the open field, and in bulk collectors under the tree canopy prior to construction of the roof. Runoff was collected in a 0.7 $m^3$ tank. The number of times this tank filled and emptied was recorded until April 1991, and thereafter was supplemented by 15 minute measurements of the rate at which the tank filled. Interception has been estimated from the difference between open field precipitation and throughfall, while evapotranspiration has been estimated from the difference between throughfall and runoff. On some occasions transpiration has been determined from sap flow measurements.

In the attempt to define the flowpaths and residence times of water, a number of different methods were used, including determination of soil hydraulic properties (Nyberg, 1995a), time domain reflectometry (TDR) surveys of the spatial distribution of water (Nyberg, 1996), modelling of water storage from the dynamics of the water table (Bishop *et al.*, 1995a) and tracer experiments (Nyberg, 1995b; Rodhe *et al.*, 1997).

### Soil properties

The depth of the soil was gauged with a steel probe at 240 regularly spaced points across the catchment (NISK, 1990). The thickness of the humus layer, O, Ae, B and C horizons was measured at another 60 regularly spaced points.

Sampling for soil physical properties on G1 ROOF was conducted on four profiles. One profile was collected and analysed by Borg (1993), and the others by Nyberg (1995a) with some of the measurements conducted by NISK (1990). Soil water retention in these profiles was measured using the pressure plate method, and porosity was estimated from the water content at saturation. Organic content was measured as loss on ignition. Since the porosity was very high in some of the mineral soils, the organic content in the soil was also expressed as the bulk organic density, which is the organic contribution to the soil bulk density [$kg\,m^{-3}$], multiplied by the loss on ignition [$kg\,kg^{-1}$].

Saturated hydraulic conductivity ($K_s$) was measured in the laboratory on undisturbed cores (5 cm thick and 7.5 cm in diameter) using a constant head permeameter. *In situ* estimates of $K_s$ variation with depth in the soil profile were made from the relationship between catchment runoff ($Q\,[m^3\,d^{-1}]$) and water table ($z_g$ [m]; Eq. 5.1) at locations where water level measurements were available (e.g. near GV21 and GV182 in Figure 5.1). This calculation was based on the assumptions that lateral flow occurred only in the saturated soil, that the flow occurred in accordance with Darcy's law, that the flow traversing a topographic contour was a time-invariant proportion of the catchment runoff, and that the hydraulic gradient ($\Delta h/\Delta l$) was either equal to the local topographic gradient or defined by nearby groundwater measurements. In Eq. 5.1, $Q_w[m^2\,d^{-1}]$ is the proportion of the catchment runoff traversing a topographic contour per metre of contour length (Seibert, 1993).

$$K_s(z_g) = \frac{dQ_w/dz_g}{\Delta h/\Delta l} \tag{5.1}$$

### Hydraulic potential

The transient response of the water table was monitored with pressure transducers at 15 minute intervals in three groundwater wells (Figure 5.1), and soil water potential was monitored with tensiometers at 12 locations at 30 minute intervals. To increase the spatial resolution of the water table observations, manual measurements were made in 40 groundwater wells every other week between April 1990 and November 1993.

### Water storage

To aid in the characterization of the catchment hydrology, a set of eight subareas were demarcated on the basis of similarities in water table dynamics, slope and soil depths (Table 5.1; Figure 5.1; Seibert, 1993). Due to the good correlation between water level measurements at different points within the catchment (Nyberg, 1995b), mean daily water tables in the subareas were calculated from the relationship between the manual water table measurements in each subarea and those in one of the groundwater wells with a 15 minute record of water table levels.

From the water table, soil porosity and distribution of soil depths in the subarea, a groundwater storage was calculated as a function of the automatically recorded

**Table 5.1.** Classification of hydrologically similar catchment subareas in Figure 5.1

| | Slope | |
|---|---|---|
| Groundwater | Steep | Moderate |
| Almost always | 1 | 2/6 |
| Usually after rain events | 5 | 7 |
| Only when very wet | 3/4/8 | — |

(From Seibert, 1993).

water level (Seibert, 1993; Bishop *et al.*, 1995a). Storage in the unsaturated zone was calculated from the average soil moisture retention properties calculated by Nyberg (1995b), and an assumption that the vertical profile of hydraulic potential in the unsaturated zone was in a state of equilibrium (i.e. the total potential gradients were constant throughout the soil profile down to the water table). Vertical infiltration to groundwater and equilibration of the unsaturated soil profile was assumed to occur within the daily time-step of this water-storage model. During dry spells, this assumption will underestimate the tension in the unsaturated zone created by evapotranspiration, and hence overestimate water storage in the unsaturated zone. This problem was greatest when the water table reached the bottom of key measurement wells. Water flow from the catchment outlet ceased at roughly the same time as the water table reached the bottom of these groundwater wells, and hence the bottom of the soil profile in much of the catchment's central valley.

To provide an independent measure of soil water content, time domain reflectometry (TDR) was used. Four soil profiles were monitored at 10 cm depth intervals between 1991 and 1993. On two occasions during spring 1993, the lateral distribution of soil water content in the upper 30 cm of the soil was mapped using TDR with 15 cm and 30 cm probes inserted vertically into the soil (Nyberg, 1996). In 1995, six TDR probes were temporarily installed along a transect during the second steady-state flow experiment.

A simple turnover time ($T$ [d], Eq. 5.2), defined as the long-term storage ($S$ [mm]) divided by mean output from the catchment ($O$ [mm d$^{-1}$]) was also calculated from the water storage data

$$T = \frac{S}{O} \tag{5.2}$$

## Tracers

Hydrological tracers were used to complement the hydrometric data with information more indicative of water residence time. One tracer was provided by the switch from natural precipitation to irrigation water. The source of the irrigation water was Lake Gårdsjön which has a $\delta^{18}O$ that is several per mille $\delta$-units higher than the precipitation in the area as a result of evaporative isotopic fractionation.

Thus, when irrigation began, there was a step-shift in the $\delta^{18}O$ of the water input, from which a transit-time distribution was calculated assuming time-invariant flowpaths (Rodhe *et al.*, 1997). The response to this shift was monitored in soil water with suction lysimeters, groundwater wells and runoff.

To trace the flow through specific portions of the catchment, simultaneous injections of $^{3}H$ and $^{36}Cl$ were made on two occasions at 20 points along rows perpendicular to flowpaths at a specific depth. The movement of the tracer plume was monitored in the soil and runoff.

To avoid the difficulty of interpreting tracer experiments conducted under changing flow conditions, the sprinkling system was temporarily rearranged twice to establish steady-state flow in a subcatchment of about 1950 $m^2$ (Lange *et al.*, 1996). To this end, a 1000 $m^2$ area on the west flank of the catchment (shaded area in Figure 5.1) was constantly watered at 3.8 mm $h^{-1}$ in November 1993, and 3.1 mm $h^{-1}$ in June 1995. In both experiments, a LiBr amendment was made to the irrigation water for 15 minutes after steady-state flow had been achieved. The same level of irrigation, and hence steady-state flow, was then maintained after the Br breakthrough in the runoff occurred and Br concentrations had fallen back to at most 25% of the peak Br concentrations.

During the second steady-state flow experiment conducted during the vegetation period, 16 sapflow measurement devices were installed on trees of this subcatchment. The total transpiration loss from this subcatchment could thus be derived from an extrapolation of these trees to the whole stand (96 trees) and directly from the input–output difference in instantaneous flow.

## RESULTS

The mean depth of the soil measurements was 41 cm, but there was a large variation in depth, even at adjacent measurement points. The texture of the soil was dominated by sand and silt, with 40–43% each of those components, and a clay content of *c.* 6% in the four profiles investigated. The porosity of the soil was high, compared both with other catchments in Gårdsjön, and in Scandinavia. About half of the mineral soil samples from the four profiles had a porosity between 75 and 85%, with the highest porosity in the B horizon (Table 5.2).

The loss on ignition was also extremely high in many of the soils, with values as high as 30% in the B horizon, but considerably lower values in the A and E horizon.

**Table 5.2.** Soil horizon properties (adapted from Nyberg, 1995b)

| Horizon | Porosity (%) | Mean depth (cm) | Volumetric water content, $\Theta_\Psi$ | Mean water storage (mm) April 1991–May 1992 |
|---|---|---|---|---|
| O/LFH | 93 | 10 | $\Theta_\Psi = 0.29\Psi - 0.261$ | 30 |
| $A_e$ | 63 | 5 | $\Theta_\Psi = 0.40\Psi - 0.102$ | 20 |
| B | 76 | 23 | $\Theta_\Psi = 0.53\Psi - 0.082$ | 90 |
| C | 52 | 5 | $\Theta_\Psi = 0.37\Psi - 0.078$ | 51 |

$\Psi$ is the pressure potential in metres of water.

The bulk organic density (a measure of organic content designed not to be biased by the high porosity in a soil) was also extremely high in the B horizon compared with what has been reported from other Nordic soils (Nyberg, 1995a). Soil water retention at all tensions was higher in the B horizon than in other horizons. The largest capacity for drainable water was found in the humic layer, and the largest capacity for plant available water was in the B horizon (Figure 5.2).

The geometric mean value of $K_s$ in more than one hundred soil samples analysed in the laboratory was $3 \times 10^{-5}\,m\,s^{-1}$, with a standard deviation of 50% (Nyberg, 1995a). The value of $K_s$, however, decreased strongly with depth. The mean $K_s$ for the Ae horizon was $7 \times 10^{-5}\,m\,s^{-1}$, which was twice that in the B horizon. The average $K_s$ in the uppermost 10 cm of the mineral soil was *c.* 13 times greater than in the lowermost 10 cm layer samples (80–90 cm). Two samples of peat soil from a depth of between 10 and 30 cm in the central valley both had very low $K_s$ values ($2 \times 10^{-6}\,m\,s^{-1}$) and a high porosity (over 85%).

The artificial irrigation regime with a uniform watering intensity rate of *c.* $2.6\,mm\,h^{-1}$ did not simulate the frequent light rains which produce a small amount of annual runoff or short episodes of high intensity rainfall. None the less, the irrigation regime has succeeded in generating runoff hydrographs that are generally similar, though offset in time, to those on the reference catchment or before the roof was built (see Moldan *et al.*, Chapter 4, this volume). The relationship

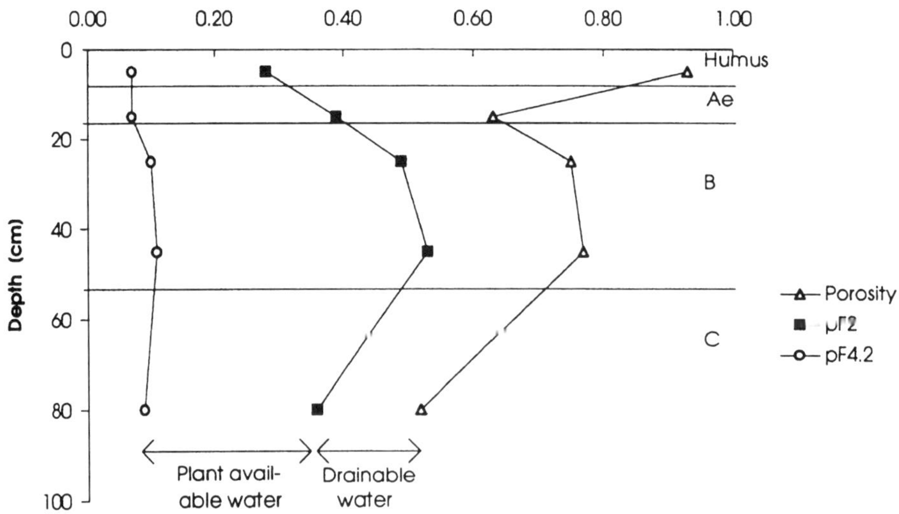


**Figure 5.2.** Mean soil water retention at different depths in three soil profiles from G1 ROOF which were investigated in detail. Drainable water is that which the soil loses when going from a soil water tension in equilibrium with the atmosphere, and a tension of pF 2 (the nominal field capacity). Water held with a tension between pF 2 and pF 4.2 (the nominal wilting point) is regarded as plant-available (from Nyberg, 1995a)

between water table and runoff, as well as the response in soil water suction to precipitation, did not change noticeably either after the roof became operational.

The annual runoff yield of throughfall remained consistently *c.* 51% (46–60%) in the 2 years before the roof and 4 years after the roof. Three years after the roof was built, the average sprinkler irrigation of 650 nm yr$^{-1}$ (i.e. simulated throughfall) had resulted in 312 mm yr$^{-1}$ of runoff. During this period, the interception loss was 370 mm yr$^{-1}$, so the water yield, as a percentage of precipitation, was *c.* 30% from G1 ROOF. This is lower than that observed on neighbouring catchments G2 NITREX and F1 CONTROL where runoff yields from precipitation were *c.* 45%.

**Hydraulic potential**

There was a strong and characteristic relationship between water table and catchment runoff in all of the groundwater wells when they were not empty (Figure 5.3). There was also a characteristic "threshold" water table, above which there was a much greater increase in catchment runoff for each centimetre of increase in water table than below the "threshold". The slope of the flow/water table relationship was 10 to 1000 times greater above the threshold, indicating a similar range of increase in $K_s$ towards the ground surface above the threshold. On average, the rate of increase in flow (and $K_s$ as calculated using Equation 5.1) was about 100 times greater above the threshold.

The depth of the threshold varied from well to well, but showed a tendency to occur at *c.* 25 cm in the valley bottom and deeper (*c.* 60 cm) on the slopes and upper part of the catchment. The frequency of higher water tables was also greater in the valley than on the side slopes where the groundwater wells were dry during much of the 3 year study period (Figure 5.4). When the water table did rise on those slopes, it

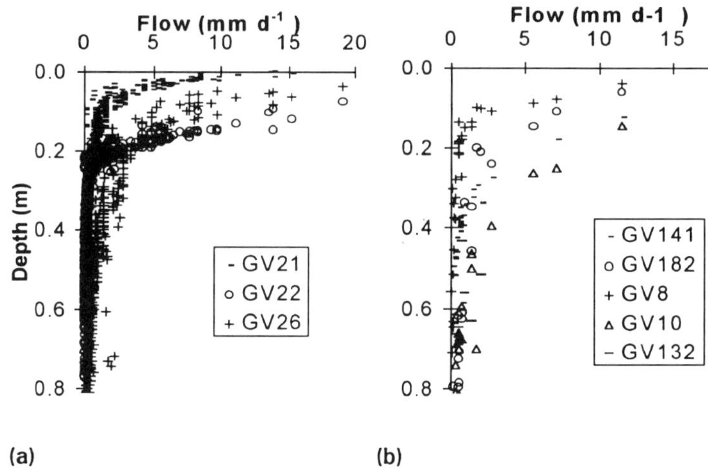


**Figure 5.3.** The relationship between water table and daily runoff in (a) the automatically monitored groundwater wells and (b) several of the manually monitored groundwater wells representative of those where groundwater was present regularly (from Bishop *et al.*, 1995a)

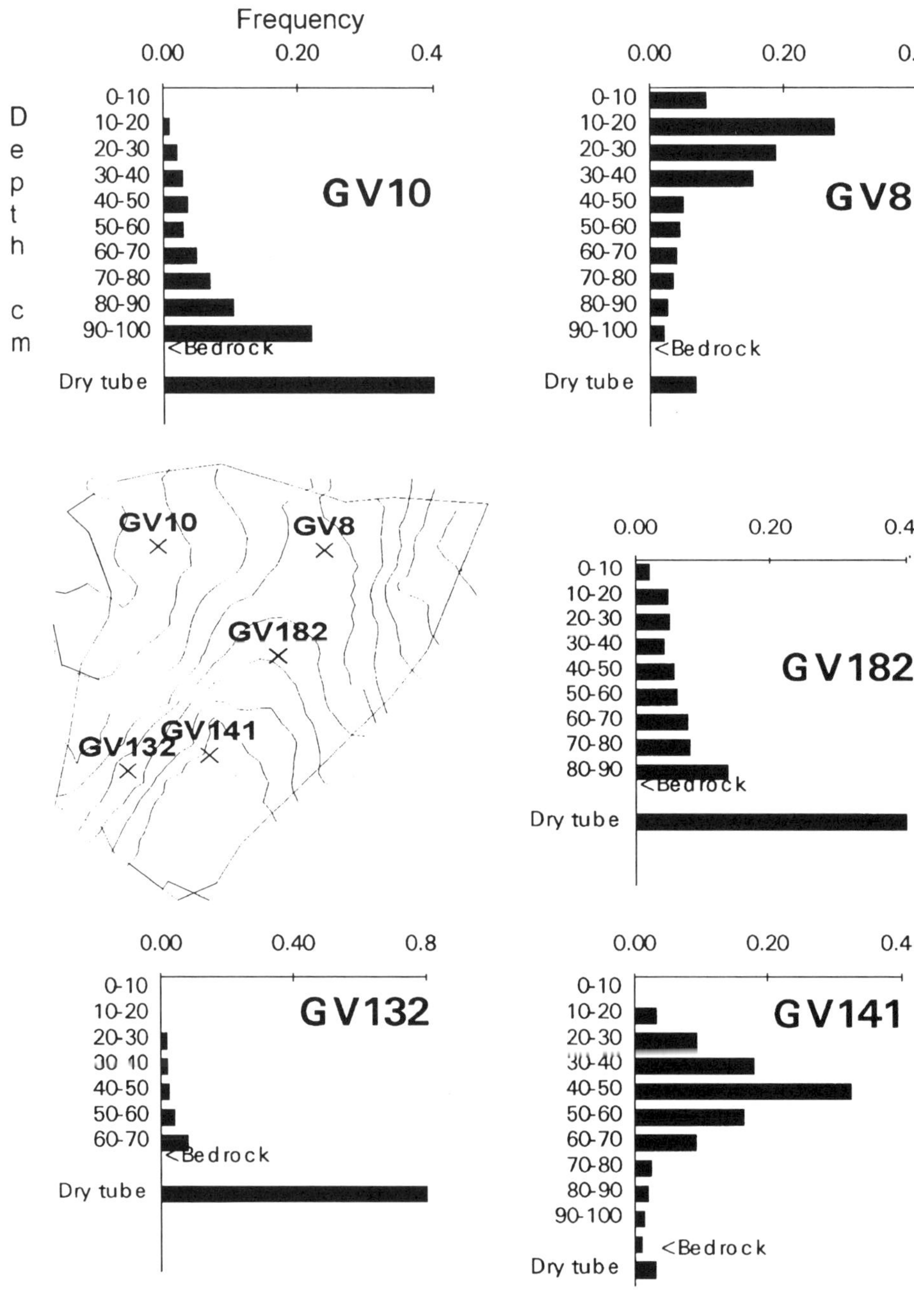


**Figure 5.4.** The relative frequency of water table levels at different locations and depths in the catchment based on fortnightly manual observations between April 1991 and November 1993 (from Nyberg, 1995b). N.B. The diagram for GV132 has a different $x$-axis scale

did not rise above the B horizon. In some of the wells located in the central valley bottom, superficial water tables were common, with groundwater reaching up into the humus layer and Ae horizon during periods of peak flow. During the steady-state flow conditions, a large part of the boggy central valley bottom above the outlet weir became saturated to the surface even though it did not directly receive any sprinkled water (Lange *et al.*, 1996).

The tensiometers in the catchment responded quickly to precipitation. Tensiometers on the slopes are much less correlated to runoff than those in the converging zones of the valley bottom. Daily cycles of water uptake by tree roots are reflected by 10 out of 12 tensiometers.

The rapidity and simultaneity of the response of the hydraulic potential in the saturated and nearly-saturated zone was evident during the period of wetting up a portion of the catchment for the steady-state LiBr tracer experiment. On three occasions, the irrigation was interrupted, and that interruption was observed in all of the tensiometers within 20 minutes, including in those more than 15 m outside of the irrigation area (Figure 5.5).

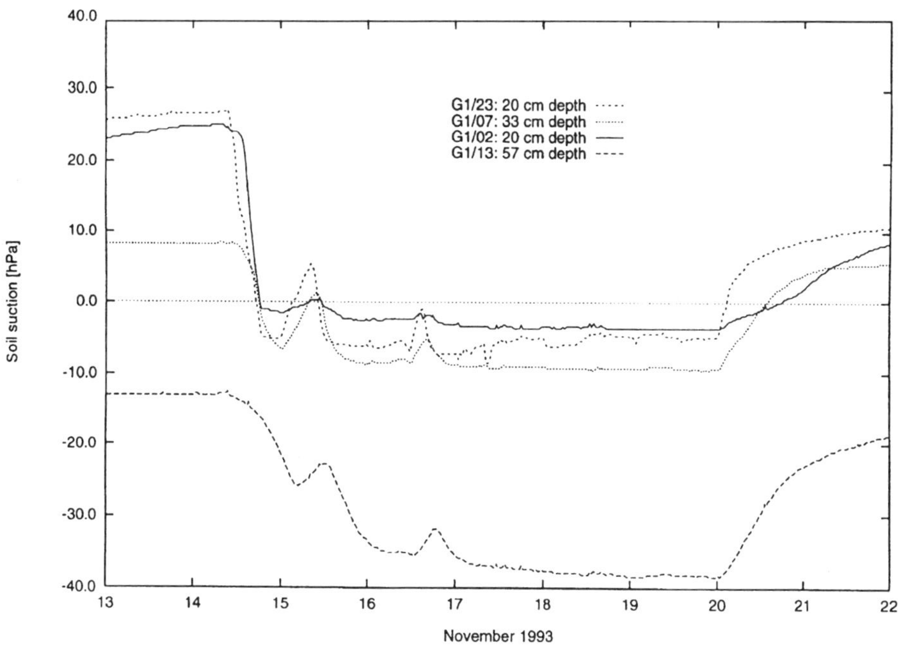


**Figure 5.5.** The time series of soil suction during the first steady-state LiBr tracer experiment. Depths of the tensiometers are given in the legend. Only one of the tensiometers (G1/23) is in the area sprinkled during the experiment. The other tensiometers are 10 to 20 m downslope from the sprinkled area. Increases in soil suction on 15 and 16 November result from interruptions in the sprinkling

## Catchment water storage

The theoretical maximum water storage in the G1 ROOF catchment with all soil fully saturated was 290 mm (1830 $m^3$). The water storage in the catchment estimated from water table data when water was flowing at the outlet ranged between 174 and 243 mm, and was well correlated to flow (Figure 5.6). Since the water storage model did not account for evapotranspiration losses during dry periods, the storage was inferred indirectly at several such occasions. The minimum inferred storage between April 1991 and March 1994 was 125 mm (Bishop *et al.*, 1995a). The mean storage during this 3 year period was 191 mm, with 30 mm held in the mor layer, 20 mm in the Ae soil horizon, 90 mm in the B horizon, and 51 mm in the C horizon. Given the mean output of 1.78 mm $d^{-1}$ (runoff and evapotranspiration combined), the turnover time $T$ (Equation 5.2) was 107 days.

The relationship between flow and water table can be described by a combination of linear equations (Table 5.3). The amount of water in the saturated zone

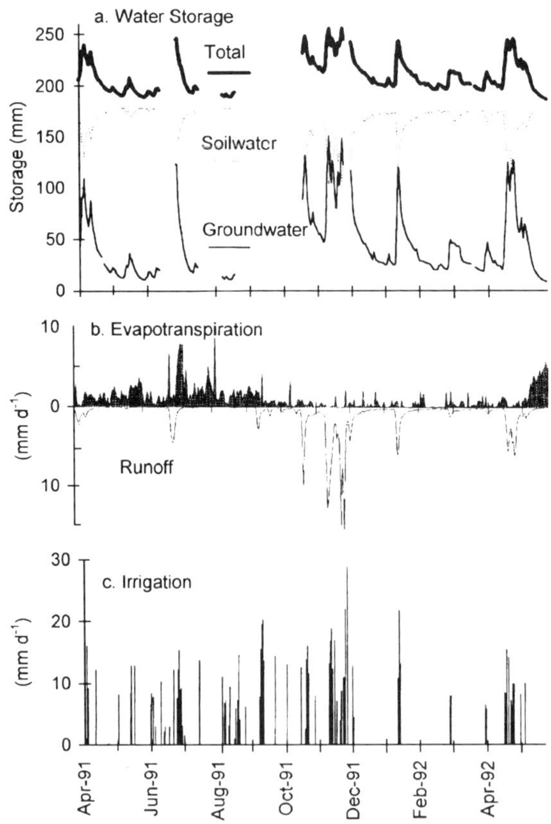


**Figure 5.6.** (a) Estimated daily water storage (total, unsaturated, groundwater), (b) runoff and modelled evapotranspiration and (c) irrigation, for a 14-month period starting in April 1991 (from Bishop *et al.*, 1995a)

**Table 5.3.** Catchment water storage as a function of runoff rate (from Bishop *et al.*, 1995a)

| Runoff ($Q$ (mm d$^{-1}$) | Groundwater storage (mm) | Total water storage (mm) |
|---|---|---|
| $0 \leqslant Q < 0.75$ | $GW = 61.4Q + 14.3$ | $WS = 40.5Q + 190.5$ |
| $0.75 \leqslant Q < 4.5$ | $GW = 15.3Q + 48.9$ | $WS = 6.2Q + 216.2$ |
| $4.5 \leqslant Q \leqslant \sim 15$ | $GW = 3.4Q + 102.4$ | $WS = 1.3Q + 238.2$ |

(groundwater) varied from 10 mm during dry periods to a maximum of 150 mm. The average groundwater storage was 51 mm (or 25% of the average total water storage). The amount of unsaturated water actually decreased with an increasing total amount of catchment water storage, but that is a result of transient saturation of soil layers which "convert" soil water from the unsaturated zone into groundwater.

On 23 April 1993, the water storage was measured with TDR at 60 regularly spaced points, and the total catchment storage was estimated to be 189 mm (Figure 5.7). This value was within 10% of the water storage predicted by the model based on water tables. In this TDR mapping, the range of volumetric water content was 50%, with a standard deviation of 9%. Despite the large variability, the spatial distribution of water content had a fair correlation to the topographic index (Nyberg, 1996), which is reflected in the higher water contents in the central valley where that index is higher. This finding was corroborated by the TDR operating during steady-state flow conditions in June 1995. Along the slope, the water content in the upper 15 cm was 36–40% during steady-state high flow (infiltration rate 3.1 mm h$^{-1}$), whereas TDR results in the valley bottom exceeded 60%.

## Tracer studies

In the $\delta^{18}O$ step-shift experiment, the transition from the pre-roof water to the post-roof irrigation water in the catchment runoff continued from the completion of the

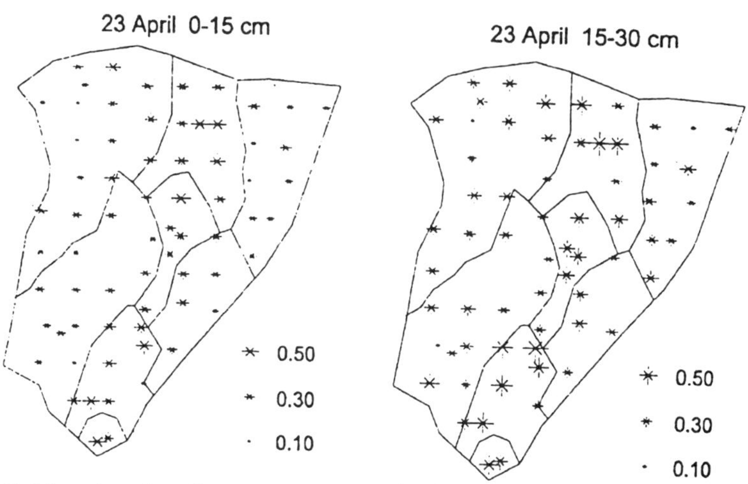


**Figure 5.7.** Time domain reflectometry survey of soil water content at 0–15 cm and 15–30 cm depths on 23 April 1993 (from Nyberg, 1996)

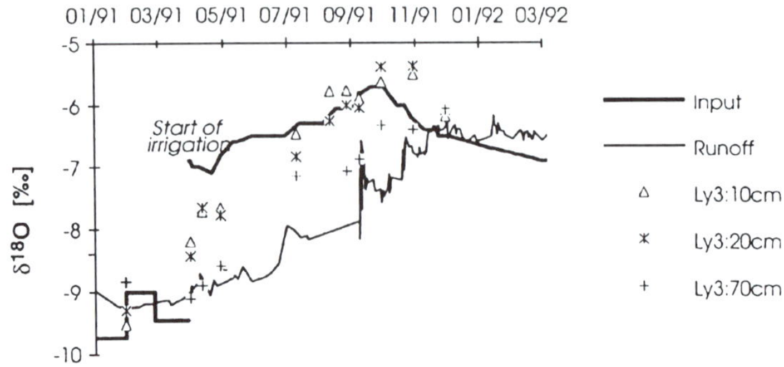


**Figure 5.8.** The transition in the $\delta^{18}O$ of runoff (and the LY3 set of lysimeters in the central valley) from the period prior to the roof becoming operational to values reflecting the more enriched $\delta^{18}O$ of the sprinkler irrigation water. The shift goes most rapidly in conjunction with the runoff episodes, and in the two more superficial lysimeters than in the deepest lysimeter (from Nyberg, 1995b)

roof in April 1991 until November (Figure 5.8). Some 544 mm of sprinkling water was added to the catchment during this period, and 163 mm had flowed out before almost all of the runoff water was post-roof water (Rodhe *et al.*, 1997). Much of the shift occurred in conjunction with periods of high flow.

During this transition period, 60 mm of the runoff was "old" water from before the roof was built. A further 126 mm of "old" water was estimated to have left the catchment in evapotranspiration. The total isotope estimate of the "old" water in the catchment (186 mm) at the time the roof was built in April 1991, is similar to the 205 mm estimate arrived at independently by the water storage model described in the previous section.

The breakthrough curve of the new water (after correction for variation in flow rate) was used to estimate a time-invariant transit time distribution of water leaving the catchment (Rodhe *et al.*, 1997). The mean, flow-corrected transit time corresponded to a runoff of 54 mm. This amount of runoff occurred in periods ranging from 5 days to 192 days during the 3 years after the roof was built.

Lysimeter and groundwater sampling of local $\delta^{18}O$ concentrations within the catchment provided information about the spatial distribution of the transition from pre-roof to post-roof water within the catchment. Some lysimeter water and groundwater made the transition in phase with the runoff. Most of those sampling locations were located in the central valley that were not at the very top or bottom of the soil profile. The transition occurred more quickly at other locations, mostly on the flanks of the catchment and at the top of the valley, while some of the deepest lysimeters in the valley bottom responded more slowly than the runoff. Among these "slow" sampling points was a 100 cm deep lysimeter at the very outlet of the catchment.

The injections of $^{3}H$ and $^{36}Cl$ provided more information about specific flowpaths. In the first such injection, which was made at the groundwater surface *c.* 10 cm below the ground surface 25 m from the outlet in the central valley just prior to a runoff episode, the first sign of both the $^3H$ and $^{36}Cl$ arrived at the outlet less than 4 hours after the injections (Figure 5.9; Nyberg *et al.*, 1993). This corresponds to a travel velocity of 6 m h$^{-1}$. The maximum concentration arrived 10 hours after the injections. Despite this fast response during the runoff episode, the concentrations of both tracers remained well above background for 3 months and 50 mm of runoff after the injection. The relative decrease in the concentration of $^{36}Cl$ was greater than that of $^3H$, which suggests some immobilization of the $^{36}Cl$.

In the second experiment with these tracers, the injection was made *c.* 40 m from the outlet, with the $^{36}Cl$ injected on the water table surface at a depth of 30 cm, and the $^3H$ was injected below the water table at a depth of 60 cm. The $^{36}Cl$ arrived at the outlet first after 30 h, and $^3H$ after 34 h (Nyberg, 1995b). This corresponds to travel velocities of *c.* 1 m h$^{-1}$. These initial arrivals were seen in the soil profile in the upper 30 cm just above the outlet. The maximum concentration of $^{36}Cl$ occurred after 75 h, and after 96 h for $^3H$. After 28 days and 100 mm of runoff, the $^{36}Cl$ concentrations had fallen below the detection limit and almost all of the $^3H$ had left the catchment (i.e. *c.* 96% recovery).

The two tracer experiments with LiBr conducted under steady-state high flow produced breakthrough curves which were significantly different from one another. The input concentrations, however, was about 10 times higher in the second experiment (Figure 5.10). In both experiments, only a small fraction (13.7 and 22.9%) of the applied Br was recovered. Solutions sampled from groundwater wells and soil column studies indicated that Li was strongly retained in the O and A horizons. The loss or retardation of Br is probably due to interference with this Li turnover as these ions comprised most of the ionic charge balance with the tracer pulse (Callenberg *et al.*, 1995).

Groundwater samples showed further that most of the Br transport is confined to the upper 20 cm of the soil. Lateral transport velocities for both experiments were in the range of 1–2 m h$^{-1}$, which are similar to those found by Nyberg (1995b). Tracer

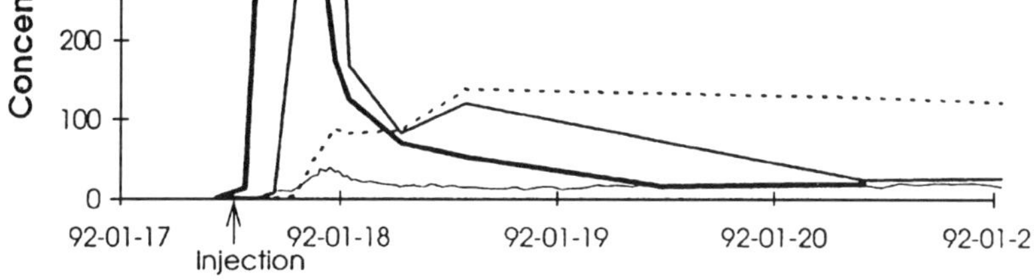


**Figure 5.9.** Concentrations of $^3H$ in water from three groundwater wells and in runoff water after the first injection of $^3H$ in January 1992 (from Nyberg, 1995b)

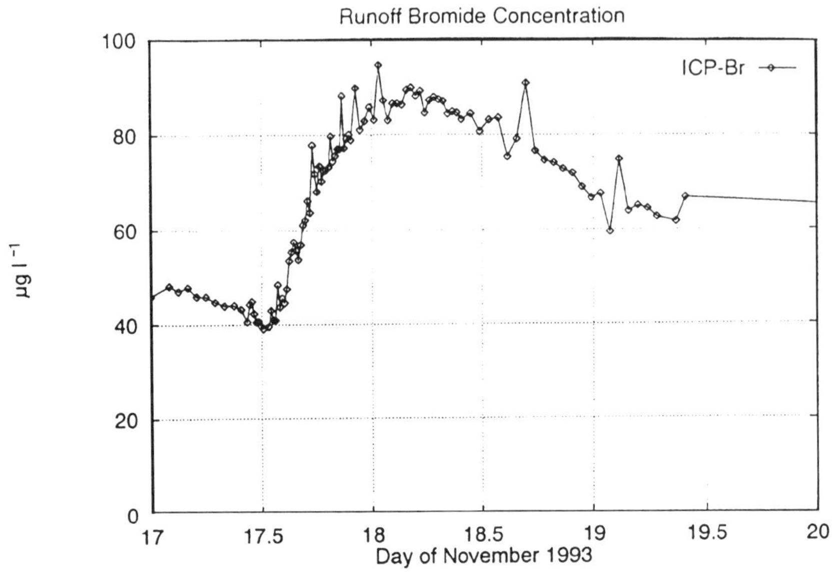


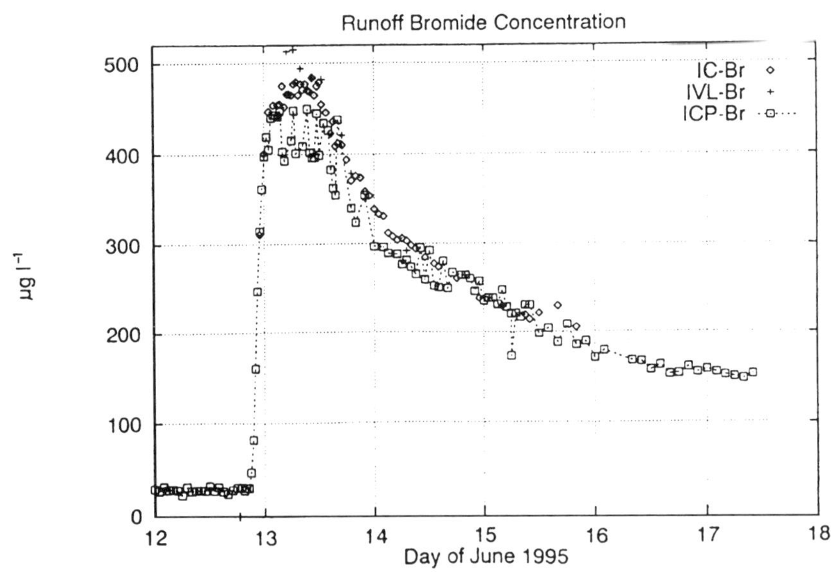


**Figure 5.10.** Breakthrough curves from the two LiBr tracer experiments conducted under steady-state flow conditions (from Callenberg *et al.*, 1995)

concentrations in runoff approached background levels after about 70 mm of runoff in the 1993, whereas in the 1995 experiment concentrations were still at 25% of peak height after 150 mm of runoff. It is unclear to what extent the relatively large difference is due to the higher flow rate in 1993, or due to the higher input concentration used in 1995. In terms of cumulative runoff, both steady-state experiments show a faster breakthrough than was observed under the transient flow conditions in the $\delta^{18}O$ step-shift experiment (Rodhe *et al.*, 1997).

In addition to the loss of Br, a significant portion of the water added to the catchment during the steady-state experiment was not recovered at the weir. During 1993, when the irrigation rate was $3.8\,m^3\,h^{-1}$, the unaccounted water amounted to $0.68\,m^3\,h^{-1}$. The loss was almost identical, $0.69\,m^3\,h^{-1}$, in the 1995 experiment when the irrigation rate was only $3.1\,m^3\,h^{-1}$. During these steady-state experiments, leakage was observed from fractures in a bedrock outcrop below the outlet dam. While the amount of this flow could not be quantified, the Br concentration of that water was similar to background levels in the runoff.

## DISCUSSION

### Runoff response and mechanisms

Despite the switch from natural precipitation to sprinkler irrigation with a uniform intensity of *c.* $2.6\,mm\,h^{-1}$, the pattern of runoff and even the soil moisture conditions in the G1 ROOF catchment appear essentially similar to the natural situation at Gårdsjön. The sprinkling of irrigation water when it is not raining (and hence the trees may be transpiring, which is not usually the case when it rains) has not created a change in the proportion of throughfall/irrigation water that leaves the catchment as evapotranspiration and runoff.

However, the water yield, 48% of sprinkler irrigation, is much less than on the neighbouring catchments. Some of this difference may result from "leakage" through bedrock fracture systems as observed during the steady-state flow experiment. In both steady-state irrigation experiments, sprinkling of the subcatchment at G1 also raised G2 runoff. This may only partly be traced back to nozzles which spilled water directly into G2. The G1 and G2 catchments lie on a zone of older, strongly foliated and frequently veined granodiorite near one of the pronounced fracture systems in the NE–SW direction (Olsson *et al.*, 1985). Thus it cannot be excluded that part of the runoff occurs in fractures parallel to runoff in the valleys.

However, Moldan *et al.* (Chapter 4, this volume), in a fuller discussion of the treatment's effect on precipitation rates and runoff, observed that the catchment input–output budget for Cl was within 5% of being balanced after 5 years. As there are no internal sources or long-term sinks for Cl, the catchment input–output budget for Cl will be balanced only if the water balance is correct. Since soils are shallow on the sideslopes, though, the ion concentration of ungauged water moving through the fracture system might have a Cl content more similar to that of sprinkler irrigation than runoff. The lower mean Cl concentration in sprinkler irrigation (about half of that in runoff) means that Cl loss from the catchment due to fracture flow could be less than the corresponding water loss. None the less, Moldan *et al.* (Chapter 4, this volume) conclude that the lower runoff yield from G1 ROOF is largely a result of greater evapotranspiration resulting from the greater above-ground biomass on G1

ROOF (Andersson *et al.*, Chapter 2, this volume) relative to the adjacent catchments G2 NITREX and F1 CONTROL.

The runoff response to precipitation is rapid, with increases in runoff that are strongly correlated to increasing water tables. The dominant mechanism of runoff generation can be described as "transmissivity feedback" in which the increase in the layer of transiently saturated soil increases the transmissivity of the soil, thus allowing for the associated increases in observed runoff without large changes in the lateral hydraulic potential gradients.

Two important alternative mechanisms of runoff generation, both of which involve significant bypass of the soil matrix by runoff, do not appear to be of major significance in the G1 ROOF catchment. The infiltration capacity of the mineral soil is too high to support infiltration excess overland flow (Horton, 1933), and saturation to the soil surface, which will result in saturation excess overland flow (Dunne and Black, 1970), is restricted to larger flow events when portions of the central valley constitute the partial contributing area (*c.* 5% of the total catchment area). Therefore, while overland flow makes some contribution, it is more a complement to the larger runoff events than a necessary component of runoff events in general.

It is more difficult to test for one other possible form of bypass flow, namely macropore flow, because that involves identification of two different hydraulic potential fields within the soil (Beven and Germann, 1982). The rapid and relatively uniform response of tensiometers to precipitation inputs, however, does not suggest a significant role for macropore flow. A large subsurface stream identified during the steady-state flow experiments also exhibited a breakthrough signature similar to the surface water in the valley bottom and runoff (Hauhs, pers. comm.). On the other hand, some of the injected $^{3}H$ and $^{36}Cl$ tracers arrived at the outlet very rapidly, which would be consistent with macropore flow (Nyberg, 1995b).

## Flowpaths

Laboratory values for the saturated hydraulic conductivity show a one order of magnitude increase from the bottom to the top of the mineral soil profile. The water table–runoff relationship, however, suggested a two order of magnitude increase existing between the base of the soil profile and the upper levels of the water table in that profile, which is a range closer to that reported in other Scandinavian till catchments (Lundin, 1990).

The effect of the vertical gradient in $K_s$ on the cumulative distribution of flow (Figure 5.11) was calculated from the variation in daily mean flow and water table over the course of a year across groundwater well GV182 assuming that lateral subsurface flow was confined to the saturated zone (Figure 5.1; Bishop *et al.*, 1995a). Much of the lateral flow during periods of high flow is transmitted by a 2–3 dm band of transiently saturated superficial soils immediately beneath the water table, which have been termed "spate-specific" flowpaths (Bishop *et al.*, 1990).

While the spatial concentration of flow in the vertical dimension is striking, it fails to capture the temporal concentration of the flow to episodes. Given the assumption that lateral flow occurs in the saturated zone, the much greater lateral flow in the

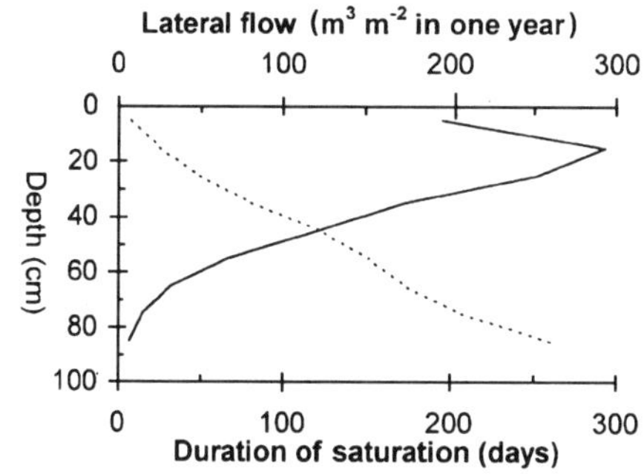


**Figure 5.11.** The amount of lateral flow across different 10 cm layers in the soil near groundwater well GV128 during 1991 (solid line), and the number of days in that year during which each layer was saturated (dashed line) (from Bishop *et al.*, 1995b)

superficial layers is restricted to a period of under 2 weeks, while the lower amount of lateral flow deeper in the soil profile is spread out over much more of the year (Figure 5.11).

The relationship between water table and lateral flow in the vicinity of this soil profile is similar to that which can be inferred from the observations in most of the other groundwater wells, though the absolute rates of flow vary. Thus the pattern of strong vertical and temporal differentiation in lateral flow is expected to be repeated across much of the catchment. The importance of the vertical concentration of flows in the profile, however, is reduced by the shallow soils in the catchment (Bishop *et al.*, 1995b). A 3 dm layer of soil where flow is concentrated is not deep, but still more than half of the mean soil depth of 43 cm.

Nyberg (1995b) has calculated that the lateral flux across the 10 m wide topographic bottleneck in subarea 6 is 500 times the amount of water infiltrating vertically into the soil. There is evidence that this converging flow in the central valley has exerted a strong influence on the long-term development of soil properties within the valley. The soil profiles at the head of the valley have porosities between 75 and 85% in the B horizon, and extremely high bulk organic density. The ranges of both parameters, reported independently by different research groups working on different soil profiles in that area (Borg, 1993; Nyberg, 1995a) are the largest reported to date for mineral soils in Scandinavia. High rates of primary weathering of the minerals, and an associated transport of dissolved organic matter from the side slopes that was deposited in the central valley, may have generated these exceptional soils.

Giesler (pers. comm.) has also reported that a soil profile examined in the central valley had the appearance and chemistry of leached Ae horizon which extended the entire length of the soil profile, which was close to 1 m deep. Swedish forest soils normally exhibit a strong vertical zonation of physical and chemical characteristics, as is indeed seen at G1 ROOF in the podzol soils outside the bottom of the central valley. This zonation results from podzolization, a complex interaction of biological and chemical processes in which the rapid downward infiltration of water when it is raining, punctuated by long periods of slow movements (either downwards or upwards) may play an important role in this development of the characteristic podzol soil profile. The large lateral flux of water in the valley bottom, however, may have obscured the vertical differentiation in the soil profile which the vertical infiltration of water normally generates. Such a relatively undifferentiated soil profile in the valley bottom has been suggested as an explanation for the weak relationship between runoff chemistry and runoff rate at G1 ROOF, both prior to the construction of the roof and during the subsequent recovery from acidification (Bishop *et al.*, 1995b).

### Water storage and residence time

For the purposes of hydrochemical modelling, the water residence time of most interest is the time water has to react with each of the different chemical environments encountered during that water's passage through the catchment (i.e. a local residence time). Assessment of the amount of water in the different chemical environments is a useful place to begin the calculation of local residence times for the G1 ROOF catchment since this, combined with the information on lateral flow within a parcel of soil, can be used to calculate such residence times. A simple illustration of this is an estimate of the local residence time for water moving laterally in a 1 m wide swath of soil at groundwater well GV182 (Figure 5.1). From the porosity of the soil and the calculated lateral fluxes (Figure 5.10), the amount of time that water spends in contact with a 1 m wide swath of a particular 10 cm thick soil layer, when that soil layer is saturated, can be calculated (Figure 5.12).

This calculation ignores the complexities inherent in the potential for mobile–immobile water interactions (cf. Hauhs *et al.*, Chapter 21, this volume), or the spatial and temporal variation in flow through a catchment. The uncertainties associated with such calculations are compounded as one extends calculations from a metre wide band of soil to encompass a catchment, thus requiring quantitative continuity modelling to address the problem of reconciling information on catchment flowpaths and residence times.

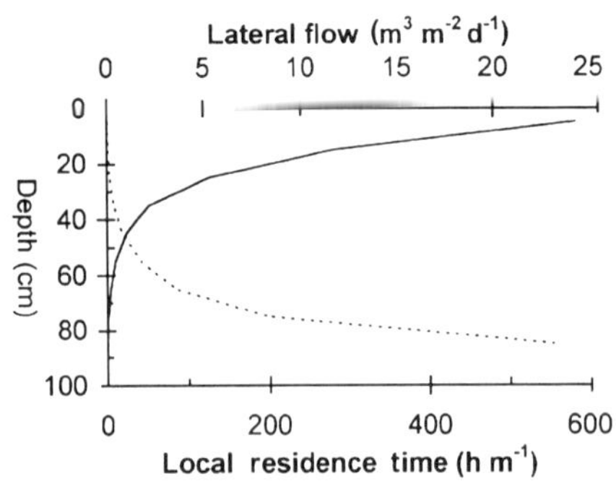


**Figure 5.12.** The lateral flow per day when each layer was saturated in the soil near groundwater well GV128 during 1991 (solid line), and the length of time a 1 m wide swath of each soil layer was in contact with water moving laterally under saturated conditions (dashed line) (from Bishop *et al.*, 1995b)

The modelled storages within the catchment as a whole were well correlated with flow (which was expected since the water table is correlated to flow). Once the amount of water in the catchment exceeds *c.* 200 mm (of the 290 mm maximum storage possible in soil of the catchment), runoff increases rapidly with the amount of water storage (Figure 5.13). During periods with large amounts of water storage, each additional millimetre of water in the catchment (less than half a per cent increase in storage) is accompanied by a *c.* 10% increase in runoff rate (0.32 mm $d^{-1}$, cf. Table 5.3). These large changes in runoff rate result from the transmissivity feedback mechanism in which small amounts of water are sufficient to saturate new soil layers, as can be seen from the increase in the proportion of groundwater as water storage increases (Figure 5.6). Runoff recedes quickly once the precipitation inputs cease. Runoff falls to below 0.01 mm $d^{-1}$ or ceases when storage is below 174 mm.

The large amount of water held in the catchment even at low flow can be overlooked when basing a model of catchment hydrology on the responses of runoff to rainfall alone, since the large store of water at low flow is not evident from that data (Stone and Seip, 1989). This large store of water results in runoff episodes consisting of a large amount of pre-event water displaced from the catchment, a feature which has also been observed at Gårdsjön (Rodhe *et al.*, 1997).

The large displacement of pre-event waters is one of the principal pieces of evidence in support of runoff mechanisms that involved flow through the soil matrix (e.g. the transmissivity feedback mechanism) as opposed to bypass mechanisms which would shunt large amounts of new precipitation through the catchment. Distinguishing the component of event water in runoff, however, is only a very small step towards the characterization of residence time distributions for purposes of hydrochemical modelling.

## Hydrological tracers and residence times

Due to the many assumptions involved in calculating residence times from flowpaths and storages alone, hydrological tracers were employed to provide independent

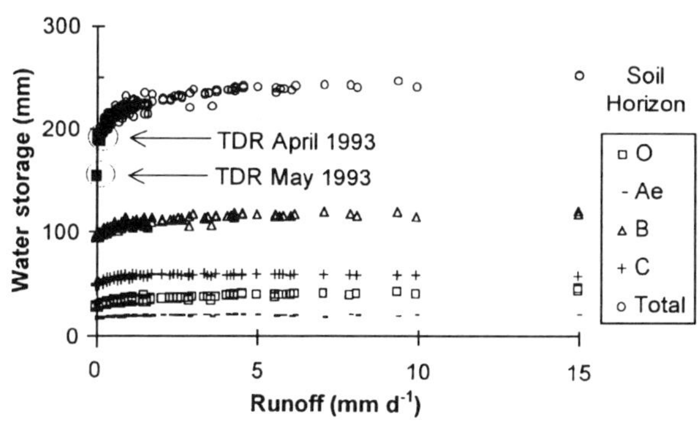


**Figure 5.13.** Water storage in different horizons as a function of runoff during the modelled period between April 1991 and May 1992. The TDR estimates of total catchment water storage made in the spring of 1993 are also plotted (from Bishop *et al.*, 1995a)

information on residence times. The experiments with the hydrological tracers LiBr, $^{36}Cl$ and $^{3}H$ made it apparent that the importance of matrix flow involved in the transmissivity feedback mechanism does not mean that all particles move at uniform rates. In each of these experiments, the first signs of tracer appearance occurred at the outlet in a matter of hours. This rate of movement could be achieved in subsurface flow only through a high degree of dispersion in which some water particles move quickly while others move slowly. This dispersion was complemented by the very rapid rate of superficial flow and, in the cases of the $^{36}Cl$ and $^{3}H$ injected below 50 cm depth, a tendency for these tracer plumes to migrate quickly to superficial layers (Nyberg, 1995b).

The steady-state runoff conditions under which the LiBr experiment was conducted allowed for more definite calculation of the amount of water actively involved in lateral flow, as well as the amount of water in which the tracer was mixed. Assuming that the $Br^-$ tracer which was not recovered was retained irreversibly on the soil, the response of the LiBr indicated that only a fraction of the water stored in the soil was mixed with the tracer. This, and the confinement of much of the Br to the upper 20 cm of the soil, are consistent with the strong concentration of flow to spate-specific flowpaths indicated by the transmissivity feedback runoff mechanism.

The steady-state experiment did, however, indicate shorter residence times than the $\delta^{18}O$ step-shift experiment. The different tracer experiments also revealed different patterns of residence times inside the catchment. A comparison between the $\delta^{18}O$ and the Br experiment has the obvious caveat that in the latter case only a subcatchment was used. It is unlikely, though, that the active flow region within the steady-state experiment (about a third of G1 ROOF) is an unrepresentative sample of the whole catchment at high runoff rates, as indicated by, for example, soil depth distributions. Rather it is the conditions under which the tracer experiments were performed which are more difficult to compare. Whereas the $\delta^{18}O$ signal represents a step-shift experiment under transient flow conditions, the Br experiment is an example of an instantaneous (i.e. Dirac) pulse under steady-state conditions. In the steady-state case, it is mainly water near the surface which transports the tracer signal and takes part in the mixing process. Therefore, it is to be expected that the resulting residence times are smaller in the steady-state case than under the transient conditions of the $\delta^{18}O$ step-shift experiment where the superficial lateral flowpaths may be inactive (i.e. unsaturated) for long periods.

Even though the travel time can vary to different groundwater wells a few metres away from tracers injected on the same occasion, the overall distribution of residence times can be described quite simply as the result of ideal mixing in a single reservoir. This was evident from the $\delta^{18}O$ step-shift experiment (Rodhe *et al.*, 1997) in which the transition from pre-roof water to post-roof water was modelled well as a single, well-mixed reservoir (Figure 5.14).

That does not mean, however, that the internal structure of the distribution of ages within the catchment is consistent with the model of a single, well-mixed reservoir. The rapid shift of the upslope lysimeters to the $\delta^{18}O$ of the irrigation water, the slower, simultaneous transition in runoff in many lysimeters located in the central valley, together with the delayed transition in the deepest lysimeters,

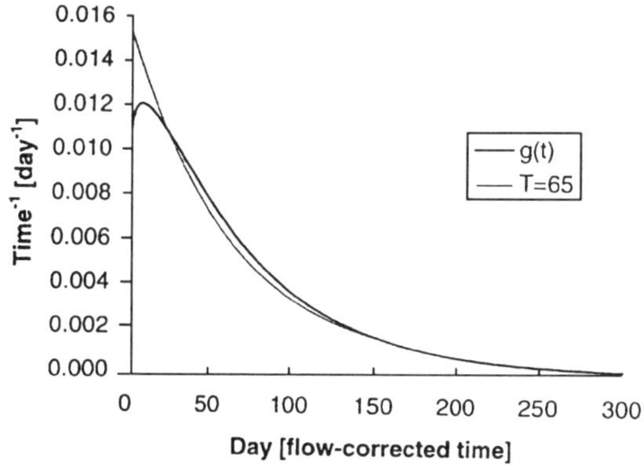


**Figure 5.14.** Transit time distribution obtained from the breakthrough curve for new water in the $\delta^{18}O$ step–shift experiment. An exponential transit time distribution with the mean transit time $T = 65$ days in flow corrected time is also shown (from Rodhe *et al.*, 1997)

including those in the central valley, all suggest a differentiated spatial pattern of age distributions at different locations within the catchment. The spatial pattern of ages may be consistent with the spatial distribution of flowpaths and water storage surmised from the hydrometric observations. Quantitative concordance of flowpaths and residence time information must be achieved through models. The status of the attempts to realize that modelling goal is described by Hauhs *et al.* (Chapter 21, this volume).

## CONCLUSION

In many ways, the hydrology of the G1 ROOF catchment appears typical of the hydrology in till soils which are widespread in Scandinavia. The catchment responds quickly to precipitation inputs, and the mechanism of runoff response can, to a large extent, be described as transmissivity feedback in which precipitation inputs saturate progressively more superficial layers of the soil profile which exhibit a strong increase in $K_s$ towards the soil surface. Overland flow from a limited portion of the catchment contributes to peak flows, but is not a necessary condition for runoff episodes.

The large changes in transmissivity and runoff rate result from the relatively small amounts of infiltrating precipitation needed to create the transient layers of saturated soil which serve as the spate-specific flowpaths that conduct much of the catchment runoff. A model of this soil water storage has been presented which can be used to predict the amount of water held in both the saturated and unsaturated zone over time in the catchment.

Once infiltrating water reaches the water table, the flow direction changes from vertically downwards to laterally downslope. The confinement of much lateral flow to a 2 to 3 dm band is common in many headwater till catchments, and of importance for the hydrogeochemistry of runoff as a result of the strong chemical gradients found in the soil profile. The significance of this vertical concentration of flow may be reduced at Gårdsjön, however, by two factors. The first is the shallow soils which truncate the lower part of the profile where less conductive soils with low flow rates (and hence longer residences times) are found. The second is the lateral concentration of flow across the catchment. The topography creates a strong convergence of flow that makes the central valley transmit a much larger amount of water than the sideslopes. This appears to have been manifested in both the high porosity and organic content of the soils at the head of the valley, as well as in a reduction in the vertical variation of the soil profile chemistry in the valley bottom.

When considering the flow pathways of water, one should also not forget the large flux of water out of the catchment in the form of evapotranspiration. This "flowpath" is taken by half of the water reaching the soil as throughfall. That amount of evapotranspiration is higher than seen on neighbouring catchments, but appears to be largely attributable to the larger amount of biomass per unit area on the G1 ROOF catchment.

The responsive character of the roofed catchment is reflected in the tracer experiments. In each of these, the first signs, or even the peaks of tracer concentration appeared within hours at the outlet. The residence time of water within the catchment can be characterized by an exponential distribution. This can be modelled as mixing of all water within a single reservoir, but the spatial distribution of the response indicates a much more spatially differentiated structure which should be related to the spatial structure of flowpaths. That may, in turn, be related to catchment topography. Athough most of the water leaving the catchment as runoff can be characterized as pre-event water displaced from the catchment, the high conductivity of superficial, spate-specific flowpaths grading into overland flow, as well as dispersion, result in some of the water moving very quickly during periods of high flow.

It remains to be seen whether the detailed information on flowpaths and residence times can be combined into a comprehensive model of catchment hydrology. An even greater challenge, however, is to use this information to improve the hydrological foundation for modelling the hydrochemical response of catchments to changes in acid deposition.

## ACKNOWLEDGEMENTS

This project was supported financially by the European Network of Catchments Organized for Research on Ecosystems (Contract STEP-CT90-0113 Subproject #21), the Swedish Natural Science Research Council, the Swedish Environmental Protection Agency, Great Britain's National Power and PowerGen, as well as the Swedish national power industry (Vattenfall and Elforsk). The prescient comments of an anonymous reviewer were greatly appreciated.

## REFERENCES

Beven, K. and Germann, P. (1982) Macropores and water flow in soils. *Water Resour. Res.*, **18**(5), 1311–1325.

Bishop, K. H. (1991) Episodic increase in stream acidity, catchment flowpaths and hydrograph separation. *Doctoral Dissertation, University of Cambridge*, 246 pp.
Bishop, K. H. and Hultberg, H. (1995) Reversing acidification in a forest ecosystem: The Gårdsjön Covered Catchment Experiment. *Ambio*, **24**(2), 85–91.
Bishop, K. H., Grip, H. and Piggott, E. (1990) Spate-specific flowpaths in an episodically acid stream. In: B. J. Mason (ed.), *The Surface Water Acidification Programme*. Cambridge University Press. Cambridge, pp. 107–120.
Bishop K. H., Nyberg, L. and Seibert, J. (1995a) Spatial and temporal dynamics of water storage and flow in the till soils of the Gårdsjön Covered Catchment in Southern Sweden. In: L. Nyberg (ed.), Soil- and groundwater distribution, flowpaths and transit times in a small till catchment. *Doctoral Dissertation, Uppsala University*.
Bishop, K. H., Nyberg, L., Moldan, F. and Hultberg, H. (1995b) The hydrochemical response of runoff episodes to an experimental reduction in acid deposition. In: A. Jenkins, R. C. Ferrier and C. Kirkby (eds), *Ecosystem Manipulation Experiments: Scientific Approaches, Experimental Method Design and Relevant Results*. Commission of the European Communities, Ecosystems Research Report No. 20, pp. 119–128.
Borg, G. (1993) *Soil Physics and Hydrology at Lake Gårdsjön*. Swedish Environmental Research Institute (IVL) B-report, Gothenburg, Sweden.
Callenberg, R., Hauhs, M., Hoch, R., Köstner, B., Lange, H., Lischeid, G., Müller, M., Prechtel, A., Wörlen, C. and Zimmermann, F. (1995) The second tracer experiment at Gårdsjön G1 during steady state stormflow conditions. (Unpublished.)
Dunne, T. and Black, R. D. (1970) Partial area contributions to storm runoff in a small New England watershed. *Water Resour. Res.*, **6**(5), 1296–1311.
Horton, R. E. (1933) The role of infiltration in the hydrologic cycle. *Trans. Am. Geophys. Union*, **14**, 446–460.
Lange, H., Lischeid, G., Hoch, R. and Hauhs, M. (1996) Water flowpaths and residence times in a small headwater catchment at Gårdsjön (Sweden) during steady state stormflow conditions. *Water Resour. Res.*, **32**, 1689–1698.
Lundin, L. (1990) Saturated hydraulic conductivity of Nordic till soils. In: S. Haldorsen (ed.), *Hydrological Properties of Nordic Tills*. Nordic Hydrological Report No. 25, pp. 51–55.
McDonnell, J. J., Stewart, M. K. and Owens, I. F. (1990) Effect of catchment-scale subsurface mixing on stream isotopic response. *Water Resour. Res.*, **27**, 3065–3073.
NISK (Norwegian Forest Research Institute) (1990) Soil depth measurements and profile descriptions in the G1 catchment at Gårdsjön. (Unpublished manuscript.)
Nyberg, L. (1995a) Water flowpath interactions with soil hydraulic properties in till soil at Gårdsjön, Sweden. *J. Hydrol.*, **170**, 255–275.
Nyberg, L. (1995b) Soil- and groundwater distribution, flowpaths, and transit times in a small till catchment. *Doctoral Dissertation, University of Uppsala*, 158 pp.
Nyberg L. (1996) Spatial variability of soil water content in the covered catchment at Gårdsjön, Sweden. *Hydrol. Proc.*, **10**, 9–103.
Nyberg, L., Bishop, K. and Rodhe, A. (1993) Importance of hydrology in the reversal of acidification in the soils, Gårdsjön, Sweden. *Appl. Geochem.*, **2**, 61–66.
Olsson, B., Hallbäcken, L., Johansson, S., Melkerud, P.-A., Nilsson, S. I. and Nilsson, T. (1985) The Lake Gårdsjön area — physiographical and biological features. *Ecol. Bull.*, **37**, 10–28.
Rodhe, A. (1987) The origin of streamwater traced by oxygen-18. *Doctoral Dissertation, University of Uppsala*, UNGI Report Series A No. 41, 260 pp.
Rodhe, A., Nyberg, L. and Bishop, K. (1997) Water transit times and flowpaths in a microcatchment from a step shift in the $^{18}O$ content of water input. *Water Resour. Res.*, **32**, 3497–3511.
Seibert, J. (1993) *Water Storage and Flux in a Micro-catchment at Gårdsjön, Sweden*. Uppsala University, Institution of Geosciences, Hydrology Section Report, 46 pp.
Stone, A. and Seip, H. M. (1989) Mathematical models and their role in understanding water acidification: An evaluation using the Birkenes Model as an example. *Ambio*, **18**(3), 192–199.

# 6 Changes in the Soil Solution Chemistry After Exclusion of Acid Deposition

REINER GIESLER, ARNE O. STUANES, ULLA S. LUNDSTRÖM,
O. JANNE KJØNAAS & FILIP MOLDAN

The exclusion of acid precipitation in whole-catchment experiments can provide valuable information to further our understanding of recovery processes of acidified soils. In this study we focused on the reversibility of acidification-induced changes in the soil solution in different soil horizons as measured by suction lysimeters and centrifugation. Sulfate concentrations in the uppermost horizons, i.e. the O and E horizons, responded rapidly to the decreased input, and lower concentrations were found within some months after the treatment was initiated. In the Bs horizon, the $SO_4$ concentrations decreased more gradually, being between 20 and 50% of pre-treatment levels in 1993, after 2.5 years of treatment. In 1993, lower concentrations of Al were found in the soil solution. The most pronounced decrease was found in lysimeter solutions close to the outlet whereas no decrease compared to the pre-treatment period was detected in lysimeters 35 m upstream. Higher pH values in the soil solution in 1993 coincided with a decrease in the total ion concentration during the same period. A general decrease in Ca and Mg concentrations was also observed in the soil solution during the treatment period. These changes are ascribed to both a reduced input of sea salts and to reduced sulfate concentrations in the soil solution. So far, the acid exclusion experiment at Gårdsjön shows that some changes in the soil solution chemistry can be achieved after a relatively short time period.

## INTRODUCTION

Long-term exposure to acidifying air pollution has induced several changes in the soil chemistry in many of those areas affected by acid deposition. Losses of base cations, leaching of aluminium and decreased pH are some of those features commonly found in acidified soils. In the long term, the depletion of exchangeable base cations can result in decreased base saturation and the input of anthropogenic $SO_4$ to an accumulation of $SO_4$ in the Bs horizon. The degree of impact of acid deposition on soil chemical processes will be crucial for the time-scale of recovery.

At Gårdsjön, acid deposition inputs were cut down to an assumed pre-industrial level in April 1990 by means of the roof experiment. Changes in the soil solution chemistry might thus give an early indication of ongoing recovery processes. An earlier manipulative roof experiment at Risdalsheia in Norway (Wright *et al.*, 1988a)

*Experimental Reversal of Acid Rain Effects: The Gårdsjön Roof Project.*
Edited by Hans Hultberg and Richard Skeffington. 

resulted in a rapid decrease in $SO_4$ concentrations in the runoff. In a small roof experiment in the Netherlands, decreased inputs of nitrate resulted in a quick response in the soil solution (Boxman *et al.*, 1994). Neither of these results are, however, possible to generalise to Gårdsjön. In the Risdalsheia case the soil was a shallow, organic-rich podzol where the Bs horizon accounted for a very small part of the soil profile. The Bs horizon has a large ability to adsorb $SO_4$ (Harrison *et al.*, 1989) and is thus important since desorption of $SO_4$ is likely to delay the recovery and counteract decreased inputs of $SO_4$. At the Gårdsjön site the Bs makes up 55% of the average soil depth (Chapter 2, this volume) and might therefore influence the time-scale of the recovery. In the roof study in the Netherlands, $NO_3$ was the major anion leaching through the soil whereas in the runoff from the roofed catchment at Gårdsjön, $SO_4$ and marine Cl are the major constituents and $NO_3$ just a minor part of the anions (Chapter 7, this volume).

To facilitate measurements of soil solution chemistry, both suction lysimeters and a centrifuge drainage technique have been used to follow the recovery process. Whereas runoff averages the effects from the whole catchment, study of the soil solution can provide information on specific processes within the soil, deepening our understanding of the recovery from soil acidification.

## MATERIAL AND METHODS

### Soil water sampling

To monitor the soil solution composition, both tension lysimeters and a centrifugation technique have been used. In the G1 ROOF catchment five nests of tension lysimeters were installed (Figure 6.1) and in the F1 CONTROL two nests of tension lysimeters. The lysimeters were under constant suction at 20–30 kPa, provided by a vacuum pump. The porous cups were composed of "Teflon" (Prenart), but in lysimeter nest 4 ceramic lysimeters were also installed at 0.50 m depth for comparison with the "Teflon" lysimeters. In each nest, lysimeters were installed at different depths, from 0.05 to 1.40 m depending on the soil depth. The installation depths for each lysimeter nest are given in Figure 6.1. Two to three lysimeters at each depth were connected to the same sampling bottle. Lysimeters at 0.05 m depth were within the O horizon, lysimeters at 0.10 m were placed in the upper mineral soil and lysimeters at 0.20 m and deeper in the Bs or B/C horizon. A more detailed description of the lysimeter installations is given in Stuanes *et al.* (1995). Water samples for chemical analysis were collected and volume recorded, fortnightly (Stuanes *et al.*, 1995).

At a separate plot in the G1 ROOF (Figure 6.1) and F1 CONTROL catchments soil samples were taken in a soil profile and the soil solution extracted with a centrifuge drainage technique using field-moist soil samples (Giesler and Lundström, 1993). Soil samples were taken from the F- and H-layer of the O horizon, the E horizon, the upper 0.05 m of the Bs horizon and then every 0.10 m down to 0.70 m. The soil samples were taken from the sides of a pit that was refilled after each sampling. The pit was extended approximately 0.20 m beyond the previous sampling location on each new sampling occasion. The soil samples were centrifuged within 36 hours.

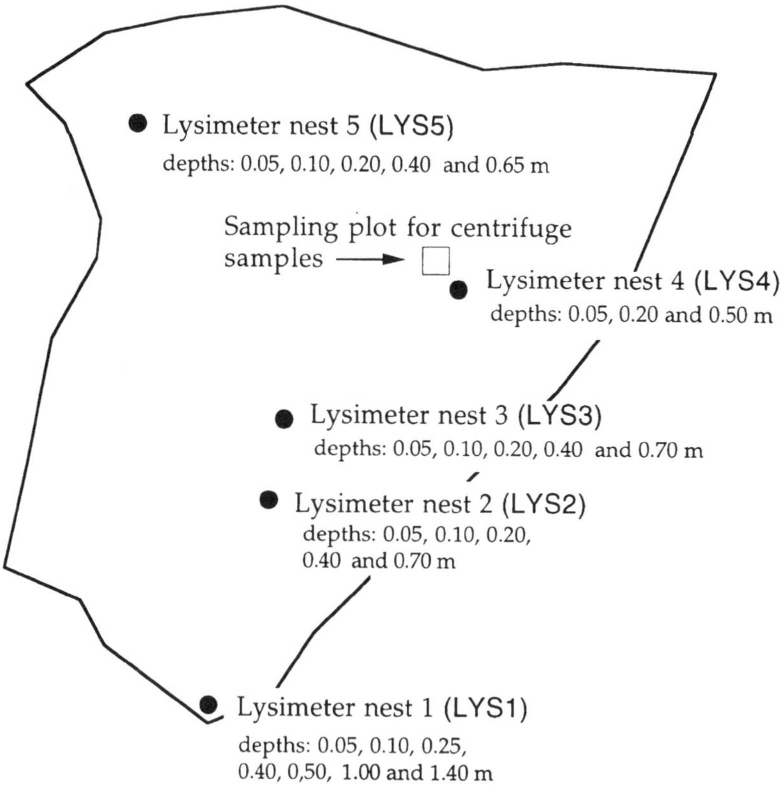


**Figure 6.1.** The placing of lysimeter nests (filled circles) within the G1 ROOF catchment. The outlet is close to lysimeter nest 1. The unfilled square denotes the sampling plot from where soil samples for centrifugation were obtained

### Analyses

Water samples from the lysimeters were analysed at the Norwegian Forest Research Institute (NISK) for Al, Fe, Mn, Mg, Ca, K, Na, $SO_4$, $PO_4$, $NO_3$ and Cl (Ogner *et al.*, 1991). Water samples from the centrifugation were analysed at the Swedish University of Agricultural Sciences (SLU) in Umeå for Al, Fe, Mn, Mg, Ca, K, Na, $SO_4$, $PO_4$, F, $NO_3$ and Cl (Giesler *et al.*, 1996). Speciation of aluminium was performed on selected soil solution samples according to Driscoll (1984) and Barnes (1975).

## RESULTS AND DISCUSSION

### Sulfate decrease

$SO_4$ concentrations have decreased noticeably in the upper horizons (e.g. O, E and upper Bs) since the start of the roof treatment. The response in the centrifuged soil

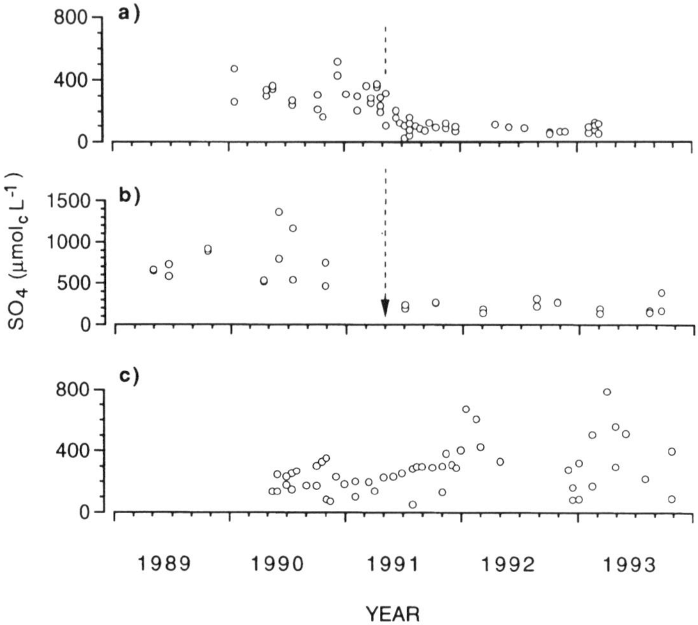


**Figure 6.2.** Sulfate concentrations in the soil solution from the O and E horizons from (a) LYS1 and LYS3 (0.05 and 0.10 m depth), (b) centrifuged soil solutions, and (c) LYS1 (0.05 m depth) in the F1 reference catchment. The arrow and the dotted line denote the time of onset of the ROOF treatment

solution in these horizons could already be detected some months after initiating the experiment (Figure 6.2). The decrease in $SO_4$ concentrations in the centrifuged soil solution from the O horizon was about 75% compared with average pre-treatment concentrations. In the E horizon the corresponding decrease was 63%. Soil solution from the lysimeters installed in the O and E horizons showed a similar quick response in the centrifuged soil solution (Figure 6.2). After the initial decrease, concentrations of $SO_4$ have been relatively constant in the soil solution from the O and E horizons. Another indication of the influence of the changed input is shown in a comparison of the relationship between Na and $SO_4$ in the centrifuged soil solution before and after treatment (Figure 6.3). Na and $SO_4$ were positively correlated in the soil solution of the O and E horizon ($r = 0.76$, $p < 0.001$) during the pre-treatment period. The ratio found in the soil solution corresponded to the ratio found in the throughfall solution. After the exclusion of the non-marine $SO_4$ in the sprinkling solution, a change corresponding to the changed input was found in the ratio between Na versus $SO_4$ in the centrifuged soil solution and the positive correlation remained ($r = 0.71$, $p < 0.001$). In lysimeter solutions from the O horizon a similar rapid shift in the $Na/SO_4$ ratio was observed. Both before and after treatment the $SO_4$ concentrations were still higher in the centrifuged soil solution than in the incoming sprinkling solution. Concentration effects due to evapotranspiration can explain part of the difference. The relative increase in the soil solution concentration

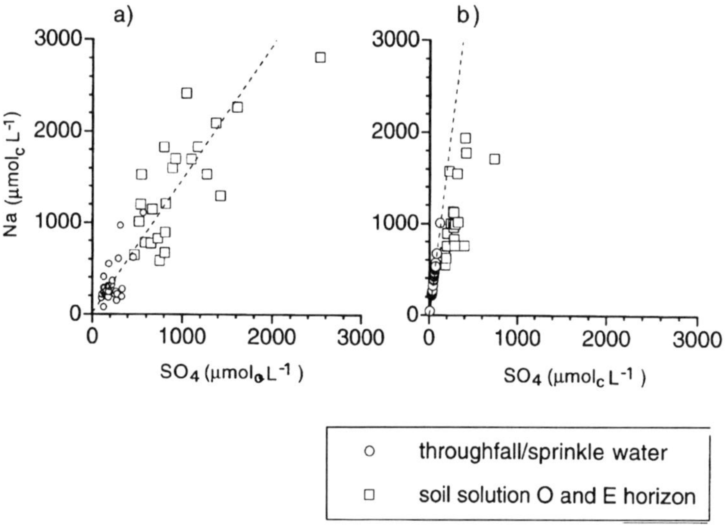


**Figure 6.3.** The relationship between sodium and sulfate in the throughfall/sprinkling solution and in the soil solution from the O and E horizons during (a) the pre-treatment period and (b) the treatment period. The regression lines are for the input solutions (i.e. throughfall/sprinkling solution). Reproduced by permission of the American Society of Agronomy, Crop Science Society of America, and Soil Science Society of America published in *Journal of Environmental Quality* 25(1):110–119

of $SO_4$ in the O and E compared with the sprinkling solution is, however, larger than for Cl. This indicates that additional sources contribute to the elevated $SO_4$ concentrations (assuming that Cl is conservative, i.e. the increase in Cl concentration in the soil solution is only due to evapotranspiration).

Biological processes such as immobilization or mineralization have been reported to be of importance in $SO_4$ dynamics in the O horizon (cf. David and Mitchell, 1987). They reported that additions of $SO_4$ are quickly immobilized but there is also a concomitant mineralization of organic S. Thus, the balance between immobilization and mineralization processes might explain part of the higher $SO_4$ concentrations in the O horizon. Isotopic studies of $^{34}S$ and $^{18}O$ in $SO_4$ from soil solution and runoff samples indicated that approximately 30% of the $SO_4$ might have been involved in mineralization or immobilization reactions (Mörth and Torssander, 1995; Chapter 8, this volume). Whether this reflects a steady state or whether there is a net loss from the O horizon is however still undetermined. A net mineralization of $SO_4$ in the O horizon would thus further delay the decrease of the $SO_4$ output from the catchment.

In the Bs horizon, the decrease has been more gradual (Figure 6.4), following the pattern observed in the runoff (Chapter 7, this volume). The degree to which concentrations were reduced differed depending on the position of the lysimeters within the catchment, but there is a similar decreasing trend for most lysimeters and the centrifuged soil solution. $SO_4$ concentrations in the Bs horizon soil solutions from the centrifugation were approximately 50% of the pre-treatment levels in 1993,

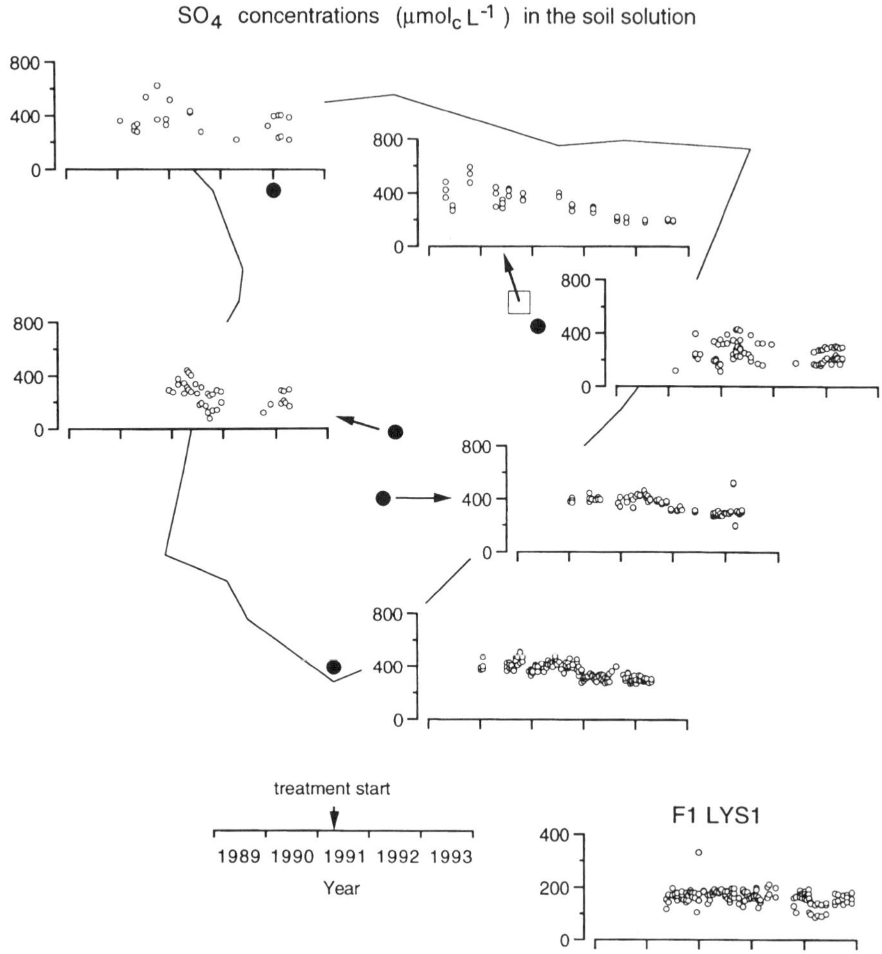


**Figure 6.4.** Sulfate concentrations in lysimeter and centrifuged soil solution from the B horizon (depth > 0.10 m) in the G1 ROOF catchment. Sulfate concentrations from the F1 reference catchment (lysimeter nest 1, depth > 0.10 m) are presented at the bottom of the figure for comparison. On the time-scale at the bottom of the figure, the arrow denotes the onset of the ROOF treatment. The filled circles denote the position of the lysimeters in the G1 catchment and the open square denotes the position of the soil pit from which samples for centrifugation were taken

after 2.5 years of treatment. In the lysimeter soil solution the corresponding decrease is approximately 20% comparing the average concentration before the treatment and the average concentration in spring 1993 for all lysimeters deeper than 0.05 m. The differences between soil horizons in response to the decreased input of $SO_4$ show that there are different processes controlling $SO_4$ concentrations in the soil solution in different horizons. This is also suggested by the absence of a positive correlation between Na and $SO_4$ in the Bs horizon in contrast to the uppermost horizons (see above). The adsorption capacity in the O and E horizon is low (Freiesleben, 1988; David *et al.*, 1991), and thus the ability to retain $SO_4$ through adsorption is very limited. In contrast, several researchers have shown that adsorption and desorption processes play a central part in controlling $SO_4$ concentrations in the Bs horizon (Nodvin *et al.*, 1986; Harrisson *et al.*, 1989; David *et al.*, 1991). In the ROOF catchment, the largest amount of adsorbed $SO_4$ is present in the Bs horizon (Chapter 2, this volume). This is a potential resource supplying the soil solution with $SO_4$ through desorption even after the input of anthropogenic $SO_4$ is cut off; as seen in the maintenance of the high concentrations in the Bs horizon. That this most likely is the case was also suggested by isotopic studies of $^{34}S$ in the input, runoff and different soil pools in the G1 ROOF catchment (Mörth and Torssander, 1995; Chapter 8, this volume).

### Nitrogen

No changes in solute concentrations for $NH_4$ were found in the soil solution after the exclusion of anthropogenic inputs of nitrogen to the G1 ROOF catchment. Ammonium concentrations were generally $< 10$ $\mu$mol $L^{-1}$. $NO_3$ concentrations of up to 100 $\mu$mol $L^{-1}$ could be found in the soil solutions especially in the O horizon and upper mineral soil prior to the ROOF treatment, but no such occasions occurred after the treatment start. Generally, concentrations of $NO_3$ in the lysimeter soil solution were less than 10 $\mu$mol $L^{-1}$ both before and after treatment initiation. In the centrifuged soil solutions $NO_3$ concentrations were somewhat higher, the average concentration being about 30 $\mu$mol $L^{-1}$. Even though average concentrations of $NO_3$ were low, there was a significant decrease in some lysimeters after the initiation of the ROOF treatment (Stuanes and Kjønaas, 1997). The generally low concentrations in the uppermost horizons (i.e. O and E horizons) indicate that inorganic N is quickly withdrawn from the soil solution and thus that there is a tight cycling of inorganic N. This further implies that this ecosystem is still nitrogen-limited.

### Aluminium and pH

Aluminium concentrations of up to 290 $\mu$mol $L^{-1}$ were measured in the lysimeter soil solution, although average values were less, around 90 $\mu$mol $L^{-1}$. Al concentrations in the centrifuged soil solution showed a clear vertical stratification within the soil profile with the highest values in the Bs horizon (Figure 6.5). This was less pronounced in the lysimeter solutions. Al speciation in centrifuged soil solution samples showed that a large part of the aluminium in the Bs horizon was in monomeric inorganic form (Figure 6.5). In the E horizon $> 50\%$ was present as organically complexed Al. This is in accordance with the presence of organic acids leaching from the mor layer. Concentrations of total organic carbon (TOC), as well as the anion deficit, drop rapidly in the upper Bs horizon and thus the organically complexed forms of Al also drop.

No general decrease in Al concentrations could be seen initially in the soil solution after the exclusion of the acid precipitation. In 1993, however, a decrease in concentrations of Al was found in some soil solutions compared with the previous

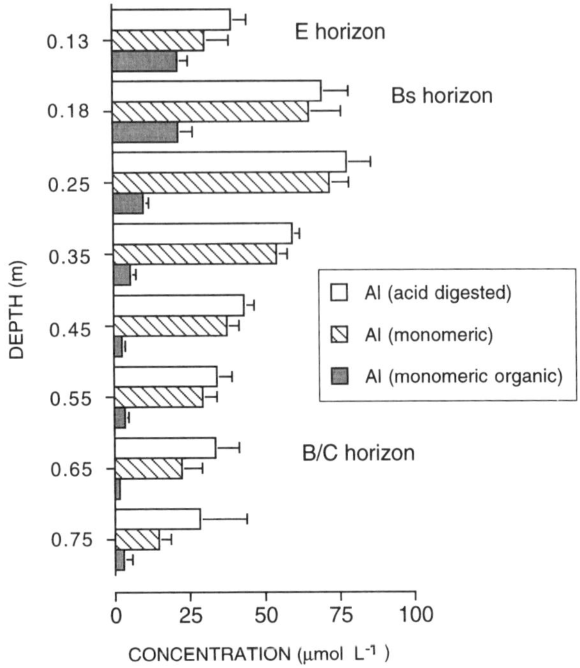


**Figure 6.5.** The distribution of aluminium species in centrifuged soil solution in the soil profile (average values from the pre- and post-treatment period). Highest concentrations of monomeric Al were found in the Bs. The difference between monomeric Al and organically complexed Al (Al monomeric organic) represents the monomeric inorganic Al fraction (Driscoll, 1984). Reproduced by permission of the American Society of Agronomy, Crop Science Society of America, and Soil Science Society of America published in *Journal of Environmental Quality* **25**(1):110–119

year (Figure 6.6). However, only in LYS1 at 0.25 and 0.40 m depths were 1993 concentrations of Al clearly lower than all the previous years. A similar trend was however also found in the centrifuged soil solution (Figure 6.6). In LYS2, 35 m upstream of the outlet, no such decreasing trend was observed. Interestingly, Al concentrations in LYS1 at 0.25 m depth were significantly correlated to the runoff Al concentrations ($r = 0.77$, $p < 0.001$) and the pronounced decrease in 1993 was similarly observed in the runoff. This was not the case for Al concentrations measured at LYS2. Lysimeter data thus indicates that the decrease in runoff Al is related to soil processes in the vicinity of the outlet rather than reflecting the whole catchment. An exception is, however, the results from the centrifuged soil solution where a significant decrease was found 2.5 years after the onset of the ROOF treatment (Figure 6.6, Giesler *et al.*, 1996).

The decrease in Al concentrations in 1993 coincided with a decrease in the concentrations of sea salts (Figure 6.7). A significant positive correlation between Al and Na, Mg and Cl was also found in both suction-lysimeter soil solution and in the soil solution extracted by centrifugation (average $r$ values for Al versus Mg, Na and Cl; 0.67, 0.75 and, 0.68 respectively, using different lysimeter nests and depths). The

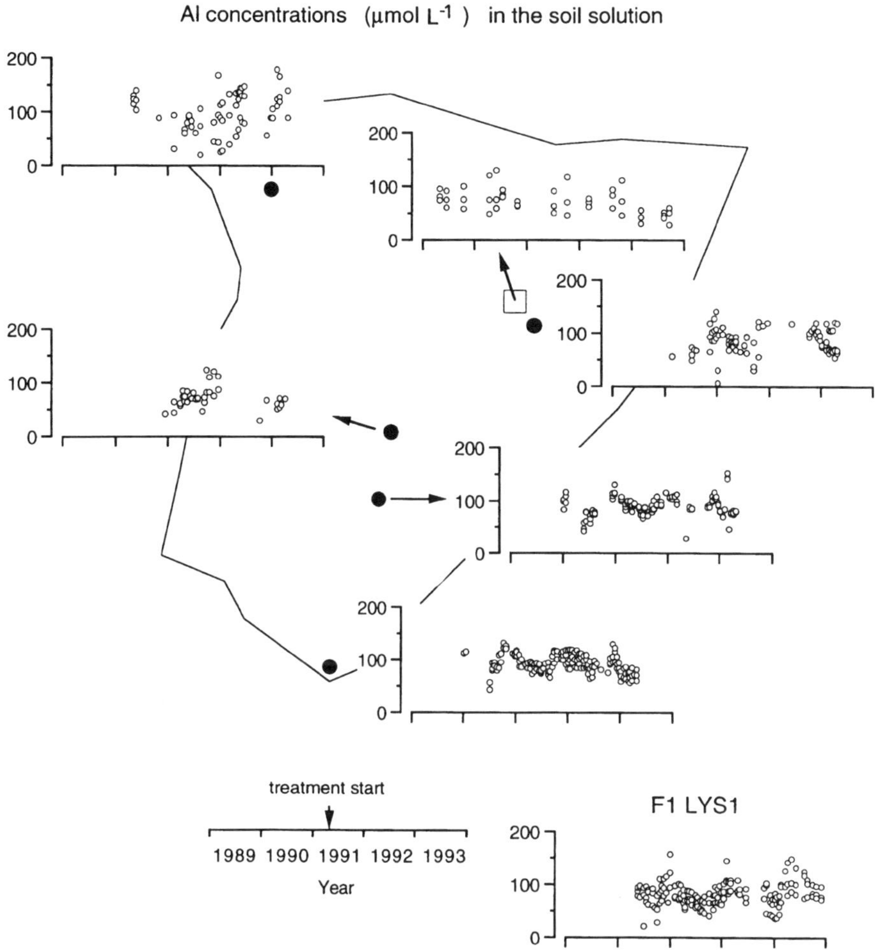


**Figure 6.6.** Aluminium concentrations in lysimeter and centrifuged soil solutions from depths >0.10 m in the G1 ROOF catchment. Aluminium concentrations from the F1 reference catchment (lysimeter nest 1, depth >0.10 m) are presented at the bottom of the figure for comparison. The filled circles denote the position of the lysimeters in the G1 catchment and the open square denotes the position of the soil pit from which samples for centrifugation were taken. On the time-scale at the bottom of the figure, the arrow denotes the onset of the ROOF treatment

strong correlation between Al and these ions show that annual fluctuations in the mainly sea-salt-derived ions are well reflected in soil solution Al concentrations. Changes in Al concentrations in response to fluctuations in the input of sea salts have also been reported earlier from coastal areas (Wright *et al.*, 1988b; Kahl *et al.*, 1992; Hindar *et al.*, 1994; Pedersen and Bille-Hansen, 1995). Pedersen and Bille-Hansen (1995) also showed that a combination of high sea-salt deposition and low amounts of percolating water can exaggerate the effect on Al concentrations in the

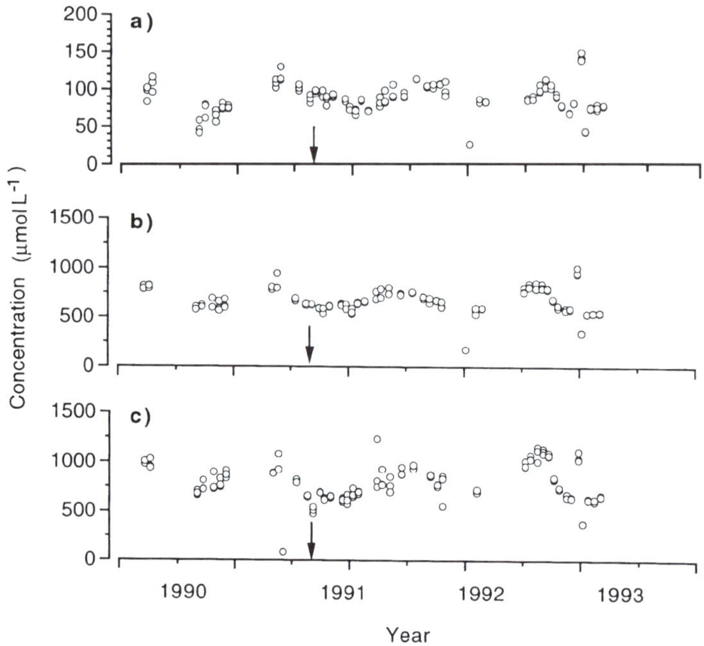


**Figure 6.7.** Time trends for concentrations of (a) Al, (b) Na and (c) Cl in lysimeter soil solution from LYS2 (0.20, 0.40 and 0.70 m depths). The arrow denotes the onset of the ROOF treatment

soil solution. The marked peak in the Cl concentrations (reflecting the influence of sea-salt-derived ions) in the beginning of October 1992 (Figure 6.7) coincided with a period of drought and almost no runoff during June to August 1992. This suggests that the peak was a concentration effect due to low flow and high evapotranspiration. Seasonal fluctuations in Cl concentrations including the peak in 1992 were also markedly well reflected in all lysimeters at depths $>0.20$ m, further supporting the conclusion that the peak was a general effect of the prevailing soil water balance conditions in the whole catchment. The drop in Cl concentration from $>1000\ \mu mol_c\ L^{-1}$ at the peak occasion in 1992 to around 500–600 $\mu mol_c\ L^{-1}$ in spring 1993 is thus most likely a combined effect of dilution due to increased precipitation and decreased evapotranspiration during the winter period, and a decreased input of sea salts starting in June 1992.

Fluctuations as described above will have a large impact on the total ion concentration in the soil solution. This in turn can impact the composition of the soil solution, and particularly the relative amount of Al, especially if ion exchange is a major factor controlling soil solution cation composition (Reuss, 1983; Johnsson, 1992; Richter *et al.*, 1992). An increase in the total ion concentration in the soil solution will according to exchange reactions increase the Al solution fraction of the total equivalent cation concentration in the short term, assuming only minor changes on the exchange complex. Richter *et al.* (1992) showed experimentally that changes in the total ion concentration can increase Al concentrations in the soil solution in soils with high Al saturation (>70%). No such obvious effect was, however, found in the soil solution composition when relating the Al fraction in the soil solution to the total cation concentration. It is noteworthy, however, that the total cation concentration ranged from approximately 1.0 to 1.8 $mmol_c\ L^{-1}$ in the soil solution whereas in the experiment of Richter *et al.* (1992) it spanned a range from 0 to 17 $mmol_c\ L^{-1}$. Changes in the range of 1.0 to 1.8 $mmol_c\ L^{-1}$ were relatively small (approximately 5%) in their experiment and might thus not be detected under field conditions.

The relationship between pH and Al in the soil solution was scattered and no clear relationship indicating equilibrium with a $Al(OH)_3$ (s) solid phase pool could be distinguished from the soil solution data. The chemical equilibrium model ALCHEMI (Schecher and Driscoll, 1988) was applied to Al speciation data from the centrifuge solution to analyse for possible solid phase equilibrium with gibbsite (cf. Giesler *et al.*, 1996). In the Bs and B/C horizons many of the data fell near the solubility line for natural gibbsite (log $Ks = 8.77$; May *et al.*, 1979) or between this and that for microcrystalline gibbsite (log $Ks = 9.35$; Hem *et al.*, 1973). The relationship between pH and Al within each sampling depth was scattered, however, indicating that dissolution of a solid phase Al pool is not a major factor controlling Al solubility.

The largest part of the neutral salt exchangeable Al pool (Chapter 2, this volume) as well as solid phase Al pools is found in the Bs horizon (Giesler, 1996), which is a general feature of podzolic soils. In addition to a relatively large neutral salt exchangeable Al pool in the Bs of approximately 8000 $mmol_c\ m^{-2}$ (Al, 62.7 $mmol_c\ kg^{-1}$; bulk density 610 kg $m^3$ and soil depth 0.22 m), there is also a large pool of pyrophosphate-extractable Al (Gielser, 1996) roughly 20 times larger than the salt exchangeable pool. This shows that there is a large amount of organically bound Al in addition to the neutral salt exchangeable Al pool in the Bs horizon. The presence of different Al pools in the Bs horizon, also including different $Al(OH)_3$ (s) forms, complicates the understanding of Al solubility and might explain the scattered data when studying field soil solution data. The presence of a large organically bound Al pool is most likely of great importance in understanding the Al chemistry in the soil solution. The importance of organically bound Al in the mineral soil has recently also been pointed out by Berggren and Mulder (1995).

No general changes in pH have been seen in the soil solution (Figure 6.8) during the first 2.5 years of treatment. The lysimeter soil solution pH ranges between 3.86 and 4.52 (averaged values for different lysimeters and depths) with the lowest pH values found in the upper part of the mineral soil profile. A similar distribution is found in centrifuged soil solution with a pH decreasing further in the O horizon to about 3.5. An increase in the soil solution pH was found in the LYS1 to LYS3 at depths >0.20 m (Figure 6.8). This was especially pronounced in the beginning of 1993 and coincided with a general decrease in the total ion concentration in the soil solution. In most of these lysimeter soil solutions there was also a significant negative correlation between the total ion concentration and soil solution pH. This suggests that the pH changes reflected changes in the ionic strength of the soil solution (cf. Richter *et al.*, 1988; Ross and Bartlett, 1993). The *dpH* (i.e. the change in pH by

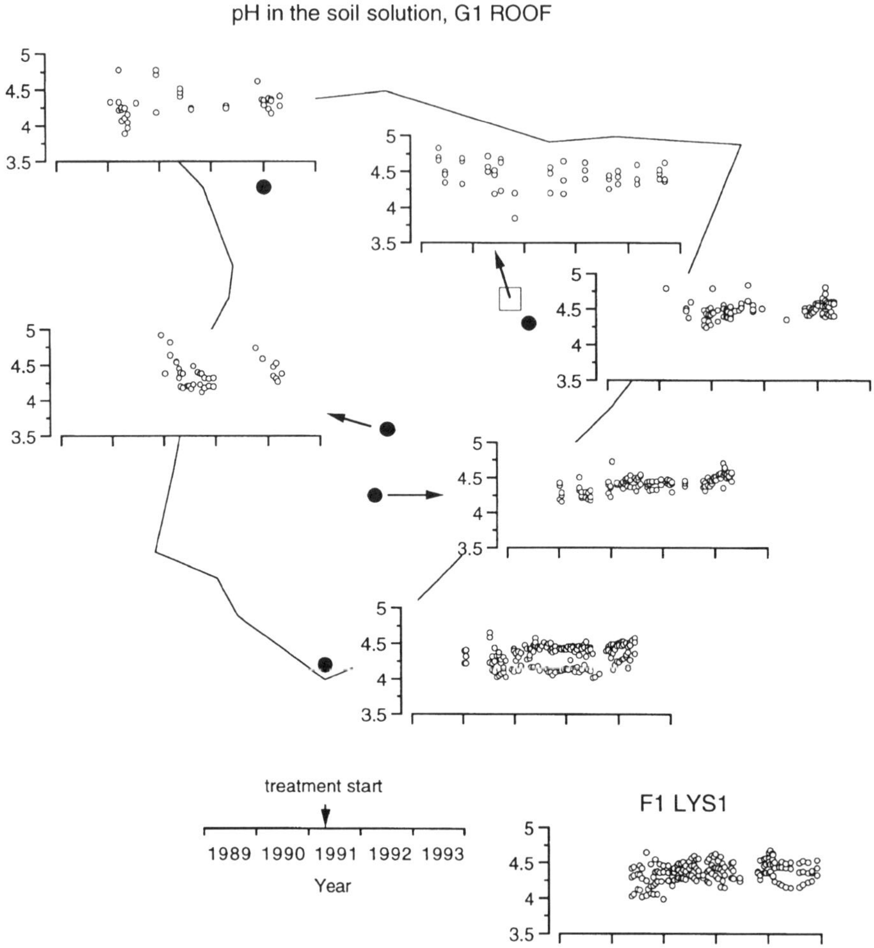


**Figure 6.8.** Soil solution pH in lysimeter and centrifuged soil solution from depths $>0.10$ m in the G1 ROOF catchment. Soil solution pH from the F1 reference catchment (lysimeter nest 1, depth $>0.10$ m) is presented at the bottom of the figure for comparison. On the time-scale at the bottom of the figure, the arrow denotes the onset of the ROOF treatment. The filled circles denote the position of the lysimeters in the G1 catchment and the open square denotes the position of the soil pit from which samples for centrifugation were taken

increasing the ionic strength ($I$) 1 mmol $L^{-1}$ estimated as the slope of the regression with $I$ as independent variable and $pH$ as dependent variable) was approximately 0.18, which is close to those values reported by Richter *et al.* (1988) from different mineral soils.

Even though the input of external acidity is excluded, there is still a large buffering capacity in the Bs horizon that can maintain a low pH in the soil solution in a short-term perspective. In an artificial acidification/recovery experiment in Norway (Stuanes *et al.*, 1992) there were still higher fluxes of H, Al and $SO_4$ found in the soil solution, 10 years after the exclusion of acid treatment. Desorption of $SO_4$ in the Bs might also counteract an increase in pH and thus prolong the recovery since desorption of $SO_4$ can contribute to H ions in the soil solution. In the long term, however, weathering and the decreased input of $SO_4$ might lead to a change on the exchange sites towards an increased base saturation. It should, however, be remembered that the main part of the acidity in the Bs horizon is represented by Al as a result of a long-term accumulation due to the podzolization process and that the base saturation thus can be suspected to be naturally low.

### Base cations

The closeness to the coast and thereby the large input of sea salts such as Na, Mg and Cl gives a clear signature on the soil solution composition. Sodium and magnesium account for approximately 68% of the cations (percentage of equivalent amount). Similarly, Cl is the dominating anion followed by $SO_4$. The relative proportion of Ca in the soil solution (lysimeter) is less than 10% whereas K accounts for less than 3% in the soil solution (lysimeter). A similar relative distribution of the base cations is found in the centrifuged soil solution within the mineral soil. In the O horizon, however, K accounts for approximately 15% of the total cation content.

The average concentrations of Ca and Mg have decreased during the first 2.5 years of treatment in the lysimeter and centrifuged soil solutions (Figure 6.9). This is also the case in the runoff (Chapter 7, this volume). An expected improvement from the exclusion of acid precipitation is a reduced leaching of balancing cations due to the decreased input of $SO_4$. The decline in divalent cations in the soil solution might suggest this, but it does not necessarily follow. The decrease of $SO_4$ in the soil solution/runoff so far is equivalent to about 100 $\mu mol_c\ L^{-1}$. Considering that the total concentration of cations in the soil solution ranges between approximately 600 and 1500 $\mu mol_c\ L^{-1}$, a decrease of 100 $\mu mol_c\ L^{-1}$ is just a small part of the annual fluctuations in cation concentration (Figure 6.10). As has been discussed above, the

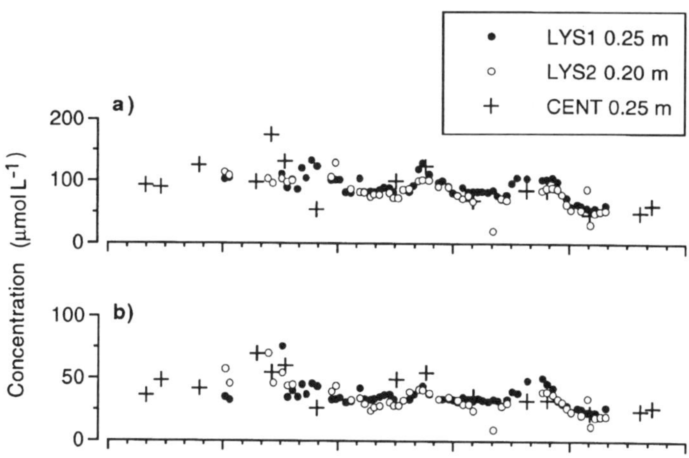


**Figure 6.9.** Time trends for concentrations of (a) Mg and (b) Ca in lysimeter and centrifuge soil solutions

input of sea-salts was less during 1993 than during previous years which probably explains part of the decrease of Mg in the soil solution. Calcium input during the first 2 years of treatment was reduced from 56 $mmol_c\ m^2\ yr^{-1}$ to 18 $mmol_c\ m^{-2}\ yr^{-1}$ (Chapter 7, this volume). This probably also contributed to the reduced concentrations of calcium in the soil solution. Earlier studies in the Gårdsjön area have also shown that annual fluctuations in the input of sea-salts are well reflected in the runoff from the catchments in the area (Hultberg and Grennfelt, 1992). All together, the decrease in base cations might partly be due to the decreased input of $SO_4$, but large variations in the input of base cations, especially Na and Mg, will mask at least part of these effects. A comparison of two occasions with the same Cl concentrations, prior to and after the initiation of the ROOF treatment and assuming Cl to be conservative and thus equal seasalt influence at the two occasions, indicated that a large part of the $SO_4$ reduction was accompanied by reduction of base cations in the soil solution (Figure 6.10).

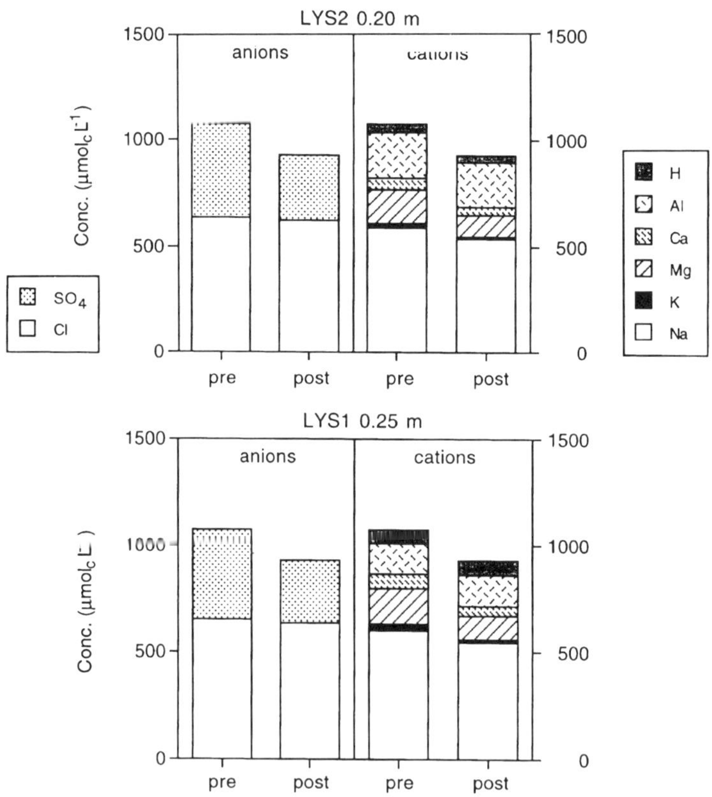


**Figure 6.10.** Comparison of solute composition in lysimeter soil solutions from one sampling occasion in March 1991 (before the start of treatment) and in March, about 2 years after the start of treatment. The sampling occasions were chosen so that Cl concentrations were equal at both occasions

Beside fluctuations in the input there is a large pool of exchangeable base cations that can alter the composition of cations in the soil solution compared with the input. Soil solution concentrations can be considered as intensity measurements whereas the exchangeable pool represents a capacity factor. Exchange reactions can be expressed as in Equation 6.1, here exemplified for Ca and Mg. The term X represents the equivalent fraction of the counterion on the exchanger. $K_{ex}$ is the selectivity coefficient which can be looked at as the equilibrium constant for the interchange of ions. Parentheses signify molar activities for the relevant ions in the soil solution.

$$K_{ex} \times [CaX]/[MgX] = (Ca^{2+})/(Mg^{2+}) \tag{6.1}$$

In the above expression the capacity factors CaX and MgX exceed by far the intensity factors; for example, the amount of Ca and Mg present in the soil solution is much less than the amount of Ca and Mg on the exchange sites. For example, the average exchangeable pool of Ca in the O horizon is about 1400 $mmol_c\ m^{-2}$ (Chapter 2, this volume). The average concentration of Ca in the O horizon (centrifuged soil solution) is about 220 $\mu mol_c\ L^{-1}$. Converting the concentration to the amount of Ca present in the soil solution in the O horizon gives approximately 10 $mmol_c\ m^{-2}$ of Ca in the soil solution, which is less than 1% of the exchangeable pool. Even yearly fluxes of Ca from the catchment are only a minor part of the total exchangeable pool (20 $mmol_c\ m^{-2}\ yr^{-1}$ vs. 2000 $mmol_c m^{-2}$). For Mg in the O horizon the difference is somewhat less, soil solution Mg accounting for about 2 % of the exchangeable Mg pool. Changes in the ratio between Ca and Mg in the soil solution should change the ratio on the exchange complex according to Equation 6.1. In the sprinkling solution the ratio between Ca and Mg changed compared to the pre-treatment period since non-marine Ca was excluded. No change in the ratio between Ca and Mg in the soil solution was, however, detected. Considering the large difference between the capacity and the intensity in the case of Ca and Mg, there is a large ability to "buffer" changes in the soil solution in a short-term perspective. This means that according to the exchange equilibrium, part of Ca on the exchange complex should be replaced by Mg. On the other hand, for Na, which is a monovalent ion with a low affinity to the exchange sites, the yearly fluxes (200 to 300 $mmol_c\ m^{-2}\ yr^{-1}$) are close to the amount on the exchange sites (450 $mmol_c\ m^{-2}$). This means that the relative influence of ion-exchange processes is much less for Na compared, for example, with Ca. This is also indicated by the strong correlation between Na and Cl in the soil solution, the ratio being close to the ratio found in marine seawater and in the throughfall/sprinkling solution.

**Ca/Al ratios**

The lowest molar Ca/Al ratios in the soil solution (0.4 to 0.5) were found in the mineral Bs horizon (Figure 6.11). In the O horizon ratios were much higher ($>1.0$). No general trend over time for the Ca/Al ratio in the soil solution was found. In some cases in the lysimeter soil solutions, the ratio has decreased after the roof application due to decreased Ca concentrations but unchanged Al concentrations (e.g. in LYS2). The Ca/Al ratio (or the BC/Al ratio expressed as (Ca + Mg + K)/Al) of $<1.0$ has been used as a measure of the critical limit at which reduced forest

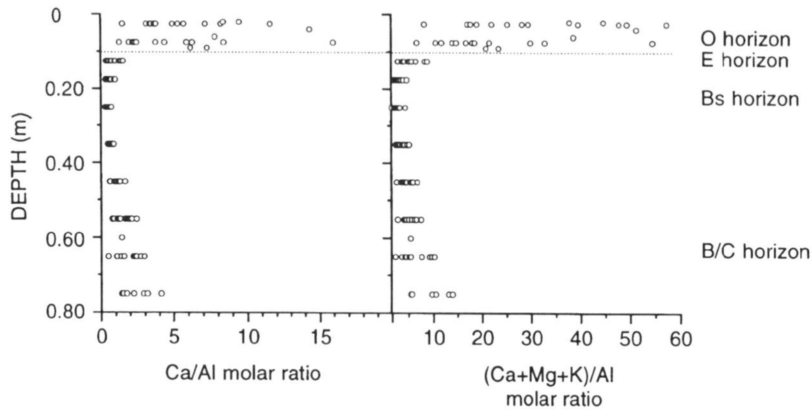


**Figure 6.11.** The distribution of Ca/Al and (Ca+Mg+K)/Al ratios in centrifuged soil solutions in the soil profile. Lowest ratios were found in the Bs horizon. No values less than 1 were found for the (Ca+Mg+K)/Al ratio

growth is assumed to occur (cf. Sverdrup and Warfvinge, 1993). In the O horizon, where much of the nutrient uptake can be assumed to occur since the bulk of fine roots were present in this horizon (Persson, 1993), no values of less than 1.4 were found in the centrifuged soil solution. It should also be mentioned that the ratios here are based on total Al and the ratio would increase if only inorganic Al was used and also if activities rather than concentrations were used (see above and Figure 6.5). The validity of this ratio under field conditions has also been seriously questioned (cf. Högberg and Jensen, 1994; Lundström and Giesler, 1995; Nygaard and Eldhuset, 1996, Lökke *et al.*, 1996). The Ca/Al ratios in the Bs horizon at Gårdsjön were similar to those found in soil solution at 0.15 m depth from a Norwegian survey on forest damage (Nygaard and Eldhuset, 1996). No connection was found between forest vitality and the Ca/Al ratio in these data. Similar observations were reported from a study in southern Sweden by Örhlander *et al.* (1995) where forest growth and BC/Al ratios were compared.

## CONCLUSIONS

The exclusion of anthropogenic $SO_4$ resulted in a rapid decrease in $SO_4$ concentrations in the soil solution of the upper horizons (i.e. the O and E horizons). In the Bs horizon, with a large store of adsorbed $SO_4$, the decrease has been more gradual. The more gradual decrease of $SO_4$ in the Bs also corresponds well to the observed decrease in the runoff. No effect of the exclusion of acid deposition by means of the roof experiment can be traced on $NH_4$ in the soil solution. These concentrations were, however, already low before the roof. For $NO_3$, peak concentrations in the soil solution no longer appeared in the soil solution after the initiation of the ROOF treatment. Decreased Al concentrations in the soil solution in

1993 might partly be ascribed to the combined effect of a lower sea-salt input and dilution during this period. Comparisons of the Al concentrations in the runoff and the lysimeter soil solution 35 m upstream of the catchment exit also indicate that decreased Al concentrations in the runoff are linked to processes at the area close to the outlet. The decreased concentrations of base cations, especially Mg and Ca, are partly related to a reduced input of non-marine constituents and to a decrease in the sea-salt input but are also partly an effect of decreased $SO_4$ concentrations. So far, the acid exclusion experiment at Gårdsjön shows that changes in the soil solution chemistry can be achieved after a relatively short time period. Long-term research is, however, still needed to evaluate the impact of the excluded acid deposition both in the soil solution and on exchangeable pools.

## ACKNOWLEDGEMENTS

This work was financially supported by the Swedish Environmental Protection Agency, the Norwegian Research Council (Programme Committee on Transport and Effects of Airborne Pollutants, TVLF), the Commission of the European Communities (STEP-CV90-0056, EV5V-CT93-0264, EV5V-CT94-0436), and the Norwegian Forest Research Institute. We thank Magne Huse and Ingvar B. Andersson for assistance in the field.

## REFERENCES

Barnes, R. B. (1975) The determination of special forms of aluminium in natural water. *Chemical Geology*, **15**, 177–191.

Berggren, D. and Mulder, J. (1995) The role of organic matter in controlling aluminium solubility in acidic mineral soil horizons. *Geochimica et Cosmochimica Acta*, **59**, 4167–4180.

Boxman, A. W., van Dijk, H. F. G. and Roelofs, J. G. M. (1994) Soil and vegetation response to decreased atmospheric nitrogen and sulphur inputs into a Scots pine stand in the Netherlands. *Forest Ecology and Management*, **68**, 39–45.

David, M. B. and Mitchell, M. J. (1987) Transformation of organic and inorganic sulfur: Importance to sulfate flux in an Adirondack forest soil. *Journal of Air Pollution Control Association*, **37**, 39–44.

David, M. B., Fasth, W. J. and Vance, G. F. (1991) Forest soil response to acid and salt additions of sulfate: I. Sulfur constituents and net retention. *Soil Science*, **151**, 136–145.

Driscoll, C. T. (1984) A procedure for the fractionation of aqueous aluminium in dilute acid waters. *International Journal of Environmental and Analytical Chemistry*, **16**, 267–283

Freiesleben, N. E. V. (1988) Sulfate mobility in an acid Danish forest soil. *Journal of Environmental Quality*, **17**, 278–284.

Giesler, R. (1996) Chemistry of soil solution extracted by centrifugation — Methodology and field applications. PhD thesis, Swedish University of Agricultural Sciences, Umeå, Sweden.

Giesler, R. and Lundström, U. (1993) Soil solution chemistry: Effects of bulking soil samples. *Soil Science Society America Journal*, **57**, 1283–1288.

Giesler, R., Moldan, F., Lundström U., and Hultberg, H. (1996) Reversibility of soil acidification in a forested catchment after exclusion of acid precipitation. *Journal of Environmental Quality*, **25**, 110–119.

Harrison, R. B., Johnson, D. W. and Todd, D. E. (1989) Sulfate adsorption and desorption reversibility in a variety of forest soils. *Journal of Environmental Quality*, **18**, 419–426.

Hem, J. D., Roberson, C. F., Lind, C. J. and Polzer, W. L. (1973) Chemical interactions of aluminum with aqueous silica at 25°C. *US Geological Survey Water Supply Paper* 1827-E.

Hindar, A., Henriksen, A., Tørseth, K. and Semb, A. (1994) Acid water and fish death. *Nature*, **372**, 327–328.

Högberg, P. and Jensen, P. (1994) Aluminium and uptake of base cations by tree roots, a critique of the model proposed by Sverdrup *et al. Water, Air, Soil Pollution*, **75**, 121–125.

Hultberg, H. and Grennfelt, P. (1992) Sulphur and seasalt deposition as reflected by throughfall and runoff chemistry in forested catchments. *Environmental Pollution*, **75**, 215–222.

Johnson, D. W. (1992) Base cation distribution and cycling. In: D. W. Johnson and S. E. Lindberg (eds), *Atmospheric Deposition and Forest Nutrient Cycling*. Ecological studies 91, pp. 467–494.

Kahl, J. S., Norton, S. A., Haines, T. A., Rochette, E. A., Heath, R. H. and Nodvin, S. C. (1992) Mechanisms of episodic acidification in low-order streams in Maine, USA. *Environmental Pollution*, **78**, 37-44.

Lökke, H., Bak, J., Falkengren-Grerup, U., Finlay, R. D., Ilvesniemi, H., Nygaard, P. H. and Starr, M. (1996) Critical loads of acid deposition for forest soils — is the current approach adequate? *Ambio*, **25**, 510–516.

Lundström, U. S. and Giesler, R. (1995) Use of aluminium species composition in soil solution as an indicator of acidification. *Ecological Bulletins (Stockholm)*, **44**, 114–122.

May, H. M., Hlemke P. A. and Jackson, M. L. (1979) Gibbsite solubility and thermodynamic properties of hydroxo-aluminum ions in aqueous solution 25°C. *Geochimica et Cosmochimica Acta*, **43**, 861–868.

Mörth, C.-M. and Torssander, P. (1995) Sulfur and oxygen isotope ratios in sulfate during an acidification reversal study at Lake Gårdsjön, Western Sweden. *Water, Air, and Soil Pollution*, **79**, 261–278.

Nodvin, S. C., Driscoll, C. T. and Likens, G. E. (1986) The effect of pH on sulfate adsorption by a forest soil. *Soil Science*, **142**, 69–75.

Nyberg, L. (1995) Water flow path interactions with soil hydralic properties in till soil at Gårdsjön, Sweden. *Journal of Hydrology*, **170**, 155–175.

Nygaard, P. H. and Eldhuset, T. D. (1996) The ratio between base cations and aluminium in soil solution as a criterion for critical load in forest soil. Research paper of Skogforsk, Norwegian Forest Research Institute, Department of Forestry, Agricultural University of Norway, No. 1. (In Norwegian.)

Ogner, G., Opem, M., Remedios, G., Sjøveit, G. and Sørlie, B. (1991) *The Chemical Analysis Program of Norwegian Forest Research Institute*, 1991. Norwegian Forest Research Institute, Ås, Norway.

Öhrlander, G., Westling, O. and Pettersson, P. (1994) *Markvattnets innehåll av baskatjoner och aluminium och dess påverkan på tillväxt och kådflöde i kraftigt försurad granskog*. Institutet för Vatten-och Luftvårdsforskning (IVL), Sweden, Rapport B 1155. (In Swedish.)

Pedersen, L. P. and Bille-Hansen, J. (1995) Effects of airborne sea salts on soil water acidification and leaching of aluminium in different forest ecosystems in Denmark. *Plant and Soil*, **168-169**, 365–372.

Persson, H. A. (1993) The distribution and productivity of fine roots in boreal forests. *Plant and Soil*, **170**, 87–101.

Reuss, J. O. (1983) Implications of the Ca–Al exchange system for the effect of acid precipitation on soils. *Journal of Environmental Quality*, **12**, 591–595.

Richter, D. D., Comer, P. J., King, K. S., Sawin, H. S. and Wright, D. S. (1988) Effects of low ionic strength solutions on pH of acid forest soils. *Soil Science Society America Journal*, **52**, 261–264.

Richter, D. D., Johnson, D. W. and Dai, K. H. (1992) Cation exchange reactions in acid forest soils: effects of atmospheric pollutant deposition. In: D. W. Johnson and S. E. Lindberg (eds), *Atmospheric Deposition and Forest Nutrient Cycling*. Ecological studies 91, pp. 341–358.

Ross, D. S. and Bartlett, R. J. (1993) Ionic strength effects on acidity and cations leached from forest floor cores. *Soil Science Society America Journal*, **56**, 1796–1799.

Schecher, W. D. and Driscoll, C. T. (1988) An evaluation of the equilibrium calculations within acidification models: The effect of uncertainty in measured chemical components. *Water Resources Research*, **24**, 533–540.

Stuanes A. O. and Kjønaas, O. J. (1997) Soil solution effects after four years' addition of $NH_4NO_3$ to a forested catchment at Gårdsjön, Sweden. *Forest Ecology and Management* (in press).

Stuanes, A. O., Van Miegroet, H., Cole, D. W. and Abrahamsen, G. (1992) Recovery from acidification. In: D. W. Johnsson and S. E. Lindberg (eds), *Atmospheric Deposition and Forest Nutrient Cycling*. Ecological studies 91, pp. 467–494.

Stuanes A. O., Kjønaas, O. J. and Van Miegroet, H. (1995) Soil solution response to experimental addition of nitrogen to a forested catchment at Gårdsjön, Sweden. *Forest Ecology and Management*, **71**, 99–110.

Sverdrup, H. and Warfinge, P. (1993) *The Effect of Soil Acidification on the Growth of Trees, Grass and Herbs as Expressed by the (Ca+Mg+K)/Al Ratio*. Reports in Ecology and Environmental Engineering, 2:193, Lund University, Department of Chemical Engineering, Sweden.

Wright, R. F., Lotse, E. and Semb, A. (1988a) Reversibility of acidification shown by whole-catchment experiments. *Nature*, **334**, 670–675.

Wright, R. F., Norton, S. A., Brakke, D. F. and Frogner, T. (1988b) Experimental verification of episodic acidification of freshwater by seasalts. *Nature*, **334**, 422–424.

# 7 Recovery from Acidification in the Forested Covered Catchment Experiment at Gårdsjön: Effects on Biogeochemical Output Fluxes and Concentrations

HANS HULTBERG, FILIP MOLDAN, B. INGVAR ANDERSSON & RICHARD SKEFFINGTON

The Covered Catchment experiment at Gårdsjön was performed to study recovery of acid forested ecosystems due to ongoing and future reductions of sulphur and nitrogen emissions in Europe. Anthropogenic sulphur and nitrogen deposition to the forest has been experimentally reduced by more than 95% since April 1991. At the same time, ambient S deposition to the control catchment has reduced by almost 50% due to emission reductions. This chapter documents the effects of these reductions on output fluxes and concentrations from both experimental and control catchments.

Compared with the control catchment, the accumulated output fluxes of non-marine sulphur and nitrate decreased by about 45 and 60% respectively over the 5 years of treatment. The fluxes of inorganic aluminium and manganese decreased by 29 and 34% while calcium, magnesium, sodium and potassium decreased by 20, 28, 17% and 15% respectively. These changes were affected by a strong decrease in the ambient sea-salt input between the pre-treatment and experimental periods. The calculated charge balance showed that the 45% decrease in the concentration of non-marine sulphate ($*SO_4^{2-}$) in runoff from the experimental catchment was balanced by a decrease in the concentrations of non-marine calcium (19% of the decrease in $*SO_4^{2-}$) and magnesium (30% of the decrease in $*SO_4^{2-}$): an F-factor of about 0.5. Inorganic aluminium concentration showed the greatest decrease (67% of the decrease in $*SO_4^{2-}$), mostly in the latter part of the experimental period. The decrease in concentration of $H^+$ was small, and did not differ between control and experimental catchments. It will probably not decrease until the flux of mobile inorganic aluminium decreases to near zero. Output water quality improved steadily as judged by a fish toxicity function, though it is still far from being suitable for fish survival and growth.

## INTRODUCTION

The Covered Catchment experiment at Gårdsjön was started to study biological and chemical recovery of acid, forested ecosystems due to ongoing and future reductions of

*Experimental Reversal of Acid Rain Effects: The Gårdsjön Roof Project.*
Edited by Hans Hultberg and Richard Skeffington. 

sulphur and nitrogen emissions in Europe. Since April 1991, non-marine sulphur and inorganic nitrogen deposition to the Covered Catchment has been experimentally reduced by more than 95%. This chapter reports on the effects this reduction has had on the fluxes and concentrations of substances leaving the catchment. To quantify these effects was one of the major aims of the Project (see Chapter 1, this volume).

Other work on recovery from acidification is reviewed in Chapter 1 of this volume. It is worth mentioning here two other attempts to manipulate S inputs at Gårdsjön, and their effects on output fluxes. Clear-cutting of a coniferous forest in a catchment at Lake Gårdsjön in Sweden resulted in a decrease in sulphate flux and also a decrease in $H^+$ (Hultberg, 1992). This permanent (> 10 years) decrease was due to lower dry deposition inputs of sulphur in combination with low desorption of adsorbed sulphate from the soil (Hultberg and Grennfelt, 1992). This contrasted with treatments in the Hubbard Brook Catchments in north-east USA, in which hardwood forest removal by either clear-cutting or whole-tree harvesting reduced S flux but increased $H^+$ flux (Nodvin *et al.*, 1988). It was suggested that the S decrease was directly attributable to increased sulphate adsorption. After tree removal there was an acceleration in nitrification and concomitant generation of $H^+$, the latter depressing the soil pH and increasing the sulphate adsorption potential (Nodvin *et al.*, 1988; Mitchell *et al.*, 1989). For both the Hubbard Brook and Lake Gårdsjön experiments, different interpretation of the results was supported by ancillary experiments (Nodvin *et al.*, 1988; Hultberg *et al.*, 1990).

Other catchments at Gårdsjön were subjected to increases in S input (Hultberg *et al.*, 1990). After treatment with elemental sulphur, a sharp increase in the sulphate flux occurred in August, 9 months after the first treatment. The $SO_4$ flux in runoff increased three times and continued with high fluxes during the following years. An oxidation of the applied sulphur had occurred during the summer months which had produced sulphuric acid. The $H^+$ flux increased in runoff and pH decreased from 4.0–4.1 to 3.8–3.9. The same effect occurred with the inorganic Al flux which increased by about 100%. The high acidity also resulted in high discharge of base cations. The Ca flux was two times higher and Mg two to three times higher than the reference catchment. Also K and Mn increased by one and a half to two times their pre-treatment fluxes. Iron increased by 10 to 30% compared with pre-treatment. The flux of organic Al, sodium and chloride were unaffected by the elemental sulphur treatment. If acidification is reversible, it might be expected that the results of the covered catchment experiment would be the inverse of this.

## SITE DESCRIPTION AND METHODS

### The Covered Catchment in relation to other catchments at Gårdsjön

Detailed descriptions of site characteristics and treatments in the Covered Catchment are given in other chapters of this book by Andersson *et al.* (Chapter 2), Bishop *et al.* (Chapter 5) and Moldan *et al.* (Chapter 4) and only some additional information is presented here. At the start of the Gårdsjön Project three catchments were studied: F1 (3.7 ha) is the long-term study catchment and was used as a reference for the experiments; catchment F2 (2.8 ha) was limed in 1984; and catchment F3 (3.3 ha) was clear-cut in 1984. Catchment: L1 (2.5 ha) was treated with sodium sulphate and catchment F5 (3.1 ha) was treated with elemental sulphur in 1985 and 1986 respectively. Each of these catchments was equipped with a water level recorder and V-notch weir for continuous monitoring of the water flow. The experiments with sulphur were performed to study effects of increased loading of air pollutants like neutral sulphur and acid sulphur on the biogeochemical turnover and transport of elements (Hultberg *et al.*, 1990). The Covered Catchment G1 ROOF (0.63 ha) experiment was started together with the NITREX catchment G2 (0.52 ha) during the winter of 1988/89 (Hultberg *et al.*, 1993a).

These catchments are similar in being vegetated by mixed coniferous stands dominated by Norway spruce (*Picea abies* (L.) Karst.) and usually have a discharge area at the low end with *Sphagnum* spp. Similarities in landscape between catchments, such as percentage forest cover and presence of forest edges, result in similar deposition of air pollutants (Hultberg, 1985b; Olsson *et al.*, 1985; Hasselrot and Grennfelt, 1987; Hultberg and Grennfelt, 1992). Characteristic features of the Covered Catchment (G1) are gneiss granodiorite bedrock and thin podzolic soil (average thickness 0.4 m) (Melkerud, 1977). Small pockets of thicker organic soil or peat occur in the *Sphagnum* dominated valley bottom. Catchment hydrology is characterized by a very high hydraulic conductivity within the upper B horizon determining the water flow pathways (Chapter 5, this volume).

**Treatments, measurements and calculations**

A roof of 7000 $m^2$ located underneath the tree canopy was built to intercept the acidic throughfall which is substituted by "clean throughfall" distributed by a sprinkling system (Hultberg *et al.*, 1993b). The "clean throughfall", with added natural sea salt at current atmospheric deposition amounts and potassium chloride at amounts circulated by the trees, was distributed at the end of precipitation events. Litterfall, also intercepted by the roof, was redistributed in the catchment twice a year. A more detailed description of treatments and logistics is given by Bishop and Hultberg (1995), and in Chapters 1 and 4, this volume, and the latter paper also contains a critique of the extent to which the treatment goals have been fulfilled.

In order to follow the expected changes, water samples were collected from the sprinkling system, from lysimeters at depths between 5 and 70 cm, from groundwater tubes and discharge via the gauged dam. Sprinkling water, lysimeters and groundwater were sampled monthly. Small catchments, such as the roof catchment G1, are very flashy and the majority of transport occurs after precipitation events. To enable an adequate assessment of the runoff fluxes at high discharges, volume-weighted sampling of runoff was performed using a flow-proportional automated ISCO sampler. Sub-samples from each millimetre of runoff were mixed and analysed every 2 weeks to give a 2-week, volume-weighted output concentration.

The volume-weighted runoff concentrations measured at the outlets of catchments F1 and G1 were combined with runoff volumes to calculate monthly runoff fluxes and concentrations. These are described below for each catchment separately, aggregated on a calendar year basis. To attempt to assess whether the responses observed in the G1 ROOF catchment were due to the acid reduction treatment or to concomitant environmental fluctuations, the concentrations and fluxes were also compared with those from the F1 CONTROL catchment. A relationship between monthly fluxes in F1 and G1 during the pre-treatment period was established using regression analysis. These regressions were used to calculate predicted monthly fluxes of each element from G1 using the measured post-treatment fluxes from the F1 CONTROL catchment. These predicted fluxes were compared with the actual measurements in G1, the null hypothesis being that there was no difference between F1 and G1 in which case changes in G1 are not due to the acid reduction treatment. The quantity (observed flux in G1−predicted flux in G1) is known as the "extra" flux (e.g. Tables 7.2 and 7.3) and is thus negative where the flux was less than predicted. The limits of this procedure for assessing the G1 treatment are discussed below (see Discussion). Moldan *et al.* (1995) used randomized intervention analysis (Carpenter *et al.*, 1989) to test whether there was a statistically significant difference between the pre- and post-treatment monthly weighted mean concentrations in the F1 and G1 catchments up to April 1995.

## RESULTS

### Output concentrations

Annual volume-weighted mean concentrations of major ions in runoff from the G1 ROOF and F1 Reference (CONTROL) catchments, averaged over calendar years, are shown in Figure 7.1.

#### *Anions*

Chloride is the major anion in runoff from both catchments (Figure 7.1a). The concentration in runoff from both F1 and G1 has fluctuated markedly during the measurement period, and in particular there has been a steep decline in G1 since the roof was constructed. This decline has to be taken into account when interpreting the results. Total S (Figure 7.1b) and non-marine S (Figure 7.1c) show similar patterns. A decline in S concentration followed construction of the roof over G1, from an initially higher concentration than F1 to a similar value, but since 1992 concentrations in F1 have also been declining. Nitrate concentrations were much lower (Figure 7.1d) and were apparently reduced to even lower values in G1 after roof construction, though the difference is probably within experimental error as nitrate is so close to the limit of detection.

#### *Base cations*

Base cation concentrations are shown in Figures 7.1e–h. Sodium is the predominant base cation (Figure 7.1e), and follows the chloride pattern closely, though not exactly. For the remaining base cations, the pattern in the F1 catchment is for fluctuating concentrations with a reduction towards the end of the period. In runoff from the G1 ROOF catchment, concentrations of Na, Mg and Ca show a sharp and continuing decline after the roof was constructed, whereas the decline in K, the concentration of which was already low, appears to halt in 1993.

*Acidic cations*

Runoff from both catchments is very acid, with a high aluminium concentration. For F1, the highest concentrations of $H^+$ (Figure 7.1i) and $Al_i$ (Figure 7.1j) are found in 1984 and between 1989 and 1993, with an evident decline in 1994 and 1995. Prior to the roof treatment G1 had lower $H^+$ but higher $Al_i$. After the roof was constructed, $H^+$ fell slightly but rose again in 1994 before falling again (Figure 7.1i). Aluminium in contrast remained unaffected initially but then fell sharply (Figure 7.1j). Manganese and iron show no obvious trends related to roof construction (Figures 7.1k and 7.1l respectively). Manganese concentrations fall sharply before roof construction in G1 but fluctuate in F1. Iron concentrations are much lower in F1 than G1, and rise in the final year.

Ammonium concentrations are very low, but increase in both catchments after roof construction, more sharply in G1 (Figure 7.1m).

*Acid neutralizing capacity (ANC)*

The calculated ANC, defined as $(Ca + Mg + Na + K + NH_4{}^+)-(SO_4{}^{2-} + Cl^- + NO_3{}^-)$, is shown in Figure 7.1n. ANC rises over time in both catchments, but apparently more quickly in G1 ROOF.

*Relationships with sulphate*

Charge balance must be maintained in the output solutions, so when $SO_4{}^{2-}$ concentration in G1 declined due to treatment, net cation concentrations must decline, or another anion increase to compensate. Figure 7.2 shows the change in mean annual volume-weighted concentration of ions relative to those in the second pre-treatment year (1990). The effects of changes in sea-salt inputs have to be evaluated concomitantly. In the initial two treatment years, the decline in $*SO_4{}^{2-}$ concentration is largely compensated for by the excess of $Na^+$ over $Cl^-$, a decline in $*Mg^{2+}$ being the only other significant compensating factor. From 1993 on, the decline in inorganic Al becomes increasingly important, the other ions declining to a much smaller extent.

**Output fluxes**

Table 7.1 shows the yearly fluxes of water and major elements in runoff from the reference catchment F1 CONTROL and the Covered Catchment G1 ROOF during the 2 years before treatment (1989 and 1990) and 5 years after start of the treatments in April 1991. For each element, the percentage change in flux comparing the mean yearly flux before treatment and the fluxes during 1995 is also shown. These percentage changes are shown for the purpose of comparison between elements rather than between years: because 1995 was a dry year the percentage change from the pre-treatment years is much larger than it would have been in the wet year 1994. The yearly fluxes of Na, Mn, $NH_4$-N, non-marine S, Cl and $NO_3$-N were very similar between the catchments both before and after the start of treatment. The fluxes of $H^+$, Ca, Mg, K and Fe were higher in F1 CONTROL, but inorganic Al was more than two times higher in G1 ROOF before treatment. During the 5 years from 1991

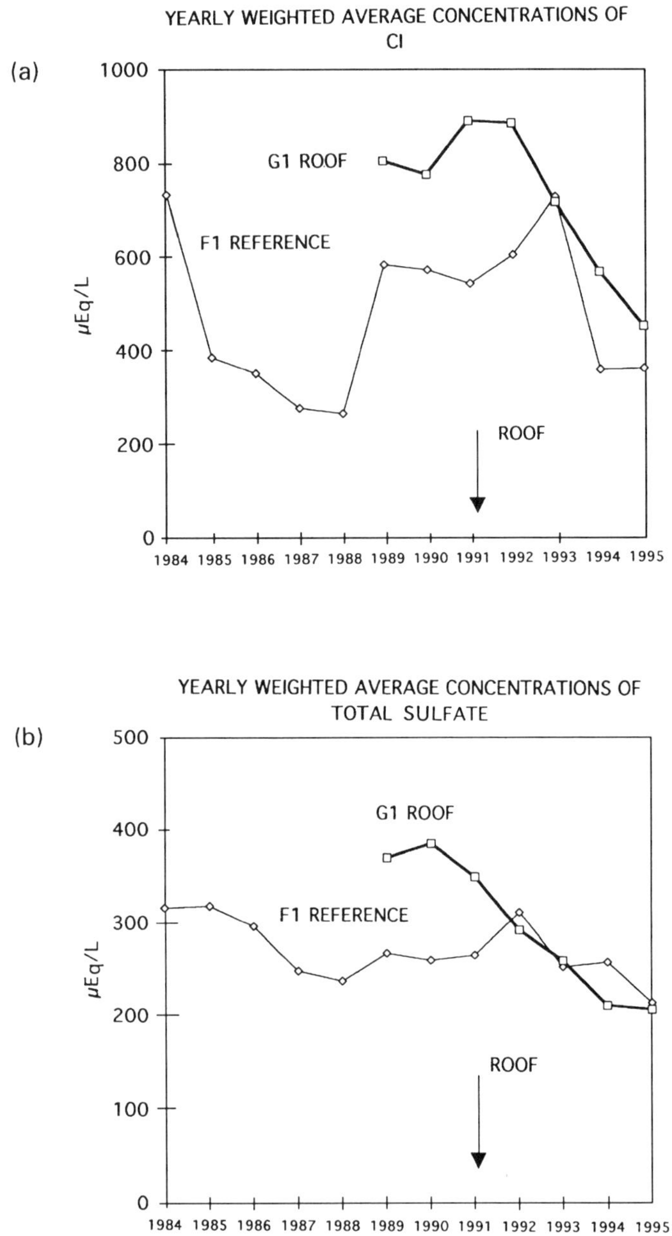


**Figure 7.1.** Annual volume-weighted mean concentrations of major ions in runoff from the G1 ROOF and F1 Reference (CONTROL) catchments, averaged over calendar years. The first two points of the G1 ROOF curve are for pre-treatment years, and the third point incorporates both pre-treatment (to 31 March 1991) and experimental (from 1 April 1991) periods. The remaining points represent periods with the roof in place

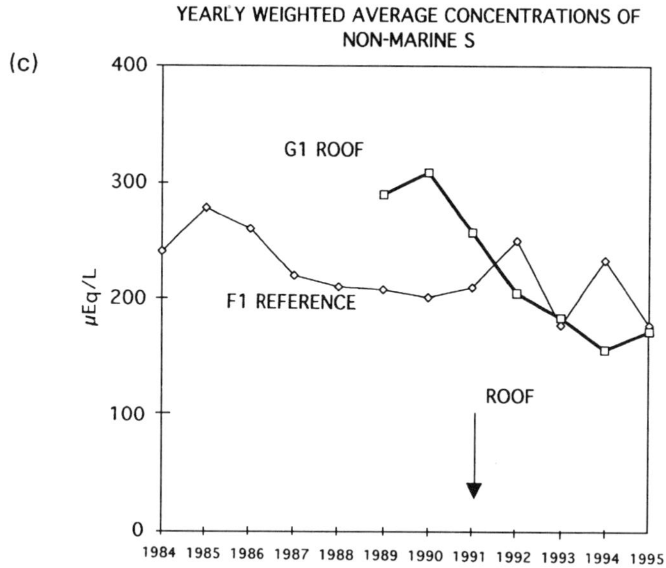


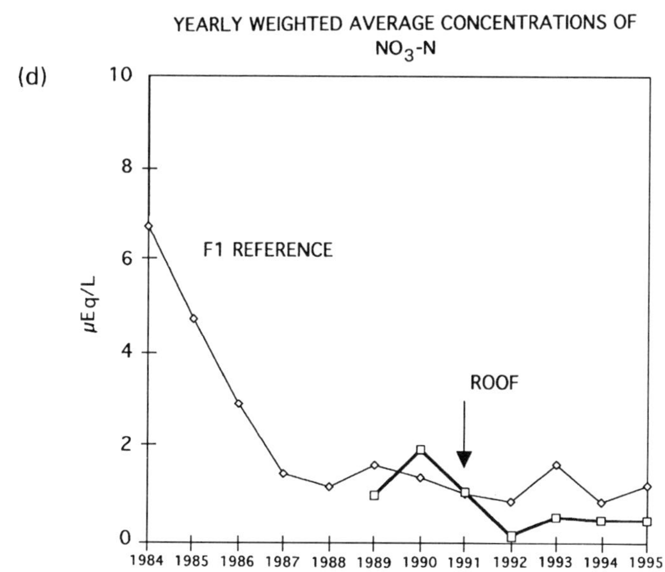


**Figure 7.1.** *continued*

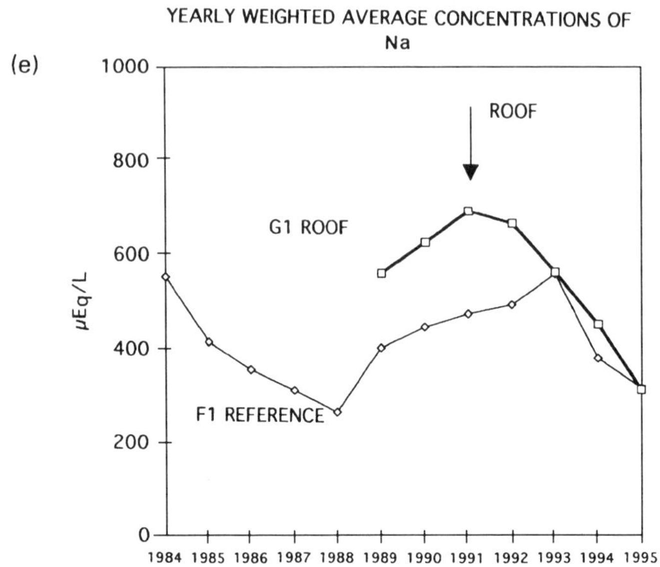


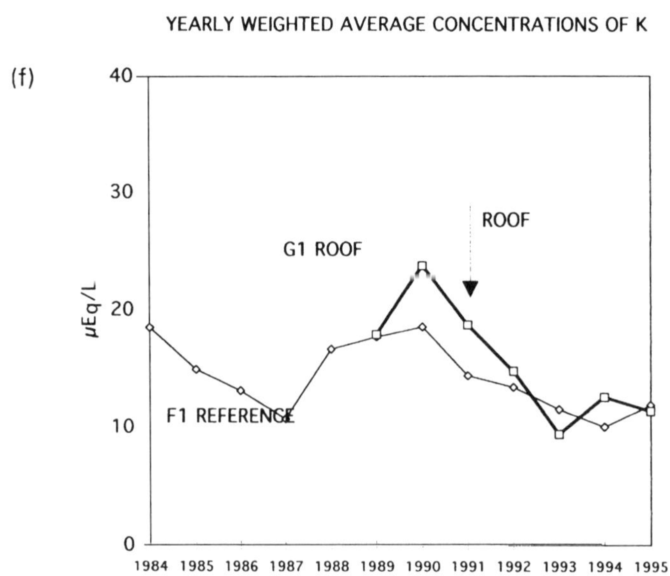


**Figure 7.1.** *continued*

(g)

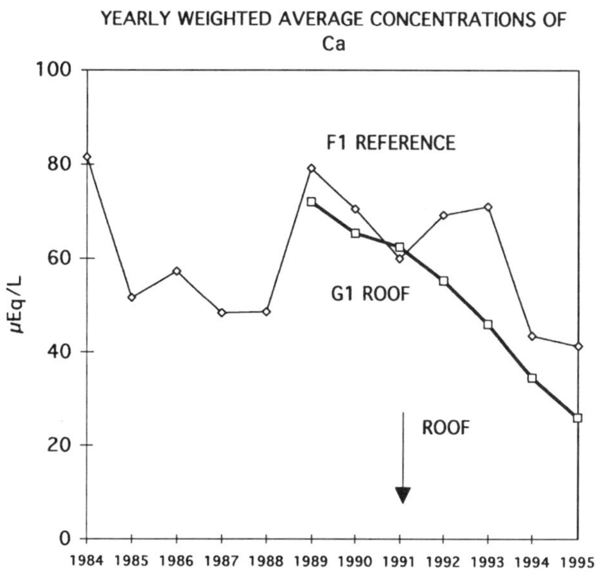


(h)

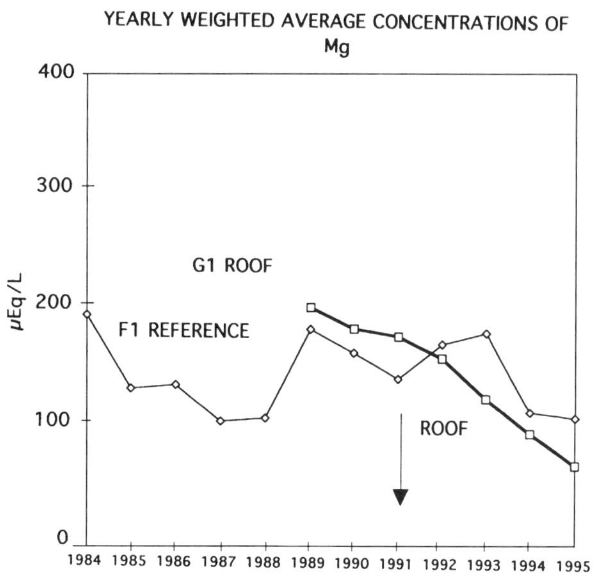


**Figure 7.1.** *continued*

(i)

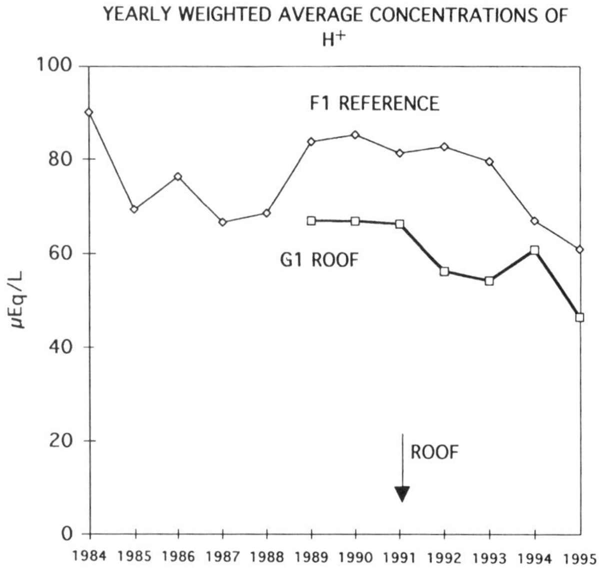


(j)

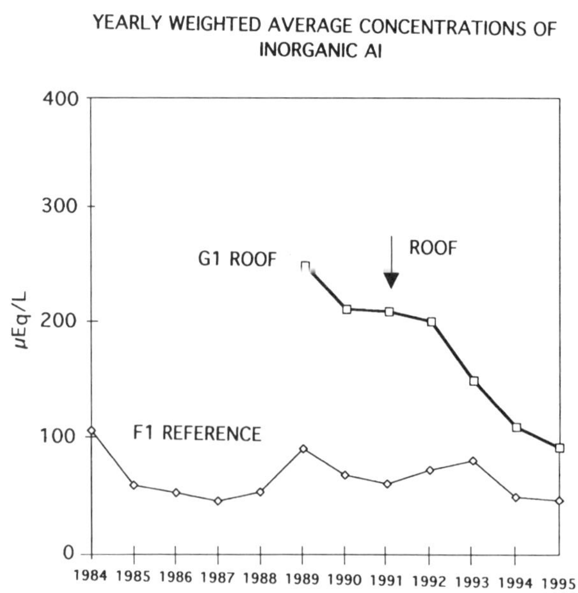


**Figure 7.1.** *continued*

(k)

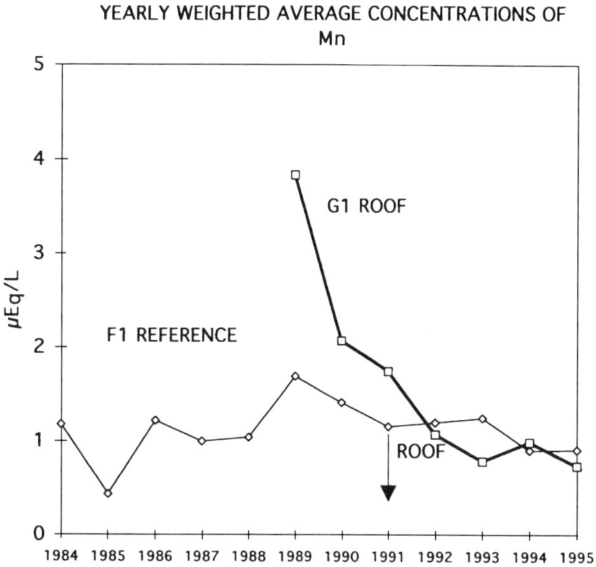


(l)

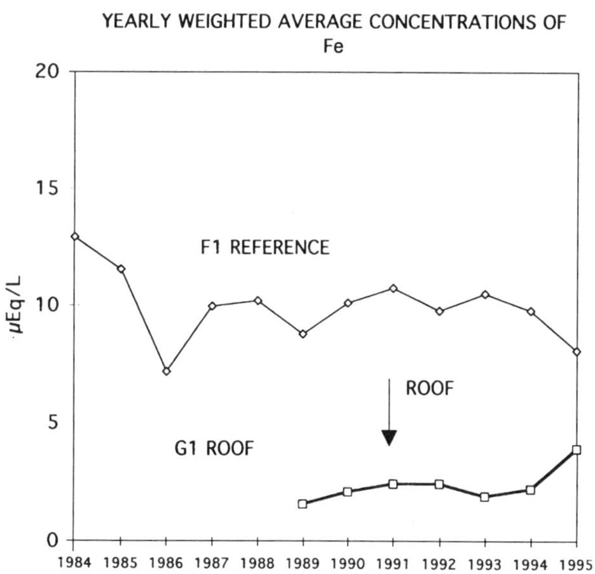


**Figure 7.1.** *continued*

(m)

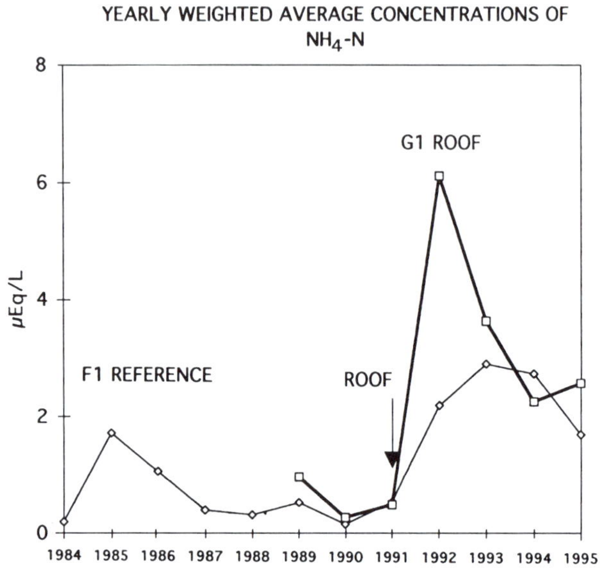


(n)

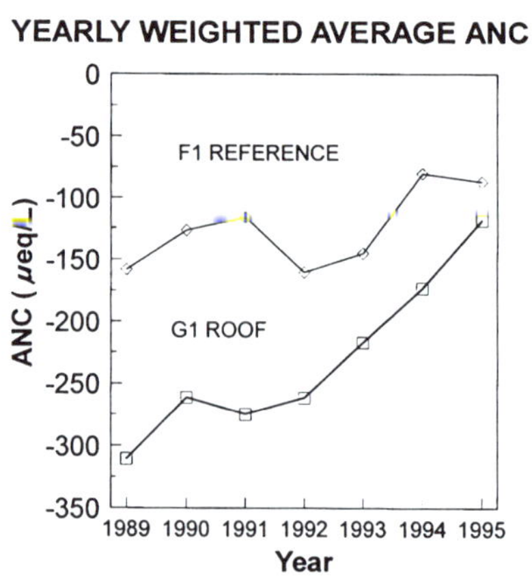


**Figure 7.1.** *continued*

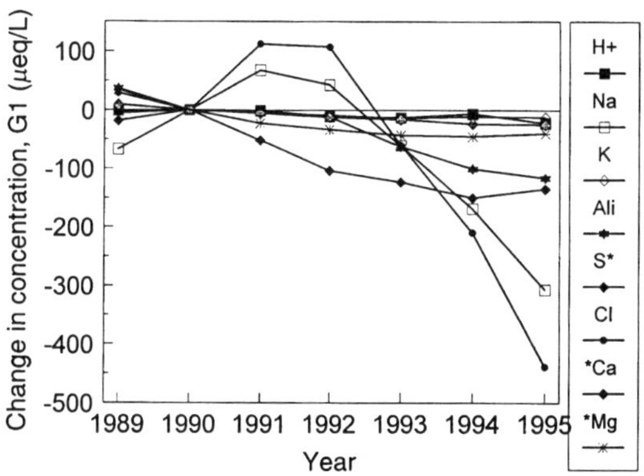


**Figure 7.2.** Change in annual volume-weighted mean cation concentrations of the most significant ions in the output from the G1 ROOF catchment, relative to the concentrations in the last pre-treatment year (1990). * indicates non-marine concentration calculated with respect to Na

to 1995 most fluxes decreased in both the F1 CONTROL and G1 ROOF catchments. The output of water varied from year to year, but showed no trend over this period.

To determine whether there was a treatment effect in the G1 ROOF catchment, the pre-treatment regressions of monthly fluxes of water and major elements between the F1 and the G1 catchments were used to calculate "predicted" monthly fluxes in the G1 catchment after start of the treatments, using the post-treatment monthly fluxes in F1. These "predicted" fluxes can be regarded as an estimate of what the fluxes would have been if the roof was not in place, assuming a linear relationship between F1 and G1. The difference between "predicted" monthly flux and measured monthly flux was termed the *extra* flux resulting from the treatment. Table 7.2 shows accumulated fluxes and extra fluxes for the G1 catchment in both kg/ha and keq/ha, and Table 7.3 shows accumulated extra fluxes expressed as a percentage of the predicted fluxes.

### *Water and anions*

Figures 7.3 and 7.4 show the estimated and measured accumulated monthly fluxes of water and chloride in runoff, the difference being expressed as a percentage in Table 7.3. These show that the experimental treatments in the G1 catchment have been performed with only small differences in water or sea-salt additions compared with the natural changes in the reference catchment F1. This is an important background for evaluating the treatment effects in the G1 catchment. In absolute terms, $Cl^-$ flux has decreased by about 30% in F1 and 40% in G1 compared with the pre-treatment period (Table 7.1).

Figure 7.5 shows decreased yearly fluxes of non-marine sulphur in the G1 ROOF catchment compared with F1, after roof construction. Figure 7.6 and Table 7.3 show that annual fluxes of non-marine S have declined by about 45% compared with what would have been expected from F1. On a more detailed temporal basis, the

**Table 7.1.** Yearly output fluxes in runoff from the reference catchment F1 and the Covered Catchment G1 in kg/ha. The table also includes the calculated change as a percentage between the fluxes in 1995 and the average fluxes in 1989 and 1990 before the start of treatments in G1. All yearly fluxes are based on full years starting on 1 January each year

| Year | $H_2O$ | $H^+$ | Ca | Mg | Na | K | Al Inorg | Mn | Fe | $NH_4$-N | $S^*$ | Cl | $NO_3$-N |
|---|---|---|---|---|---|---|---|---|---|---|---|---|---|
| **Reference Catchment F1** | | | | | | | | | | | | | |
| 1989 | 423 | 0.35 | 6.7 | 9.2 | 39.7 | 2.9 | 3.5 | 0.19 | 1.04 | 0.03 | 14.2 | 87.1 | 0.10 |
| 1990 | 578 | 0.49 | 8.2 | 11.2 | 60.3 | 4.2 | 3.7 | 0.22 | 1.63 | 0.02 | 18.8 | 116.6 | 0.11 |
| 1991 | 520 | 0.42 | 6.3 | 8.7 | 57.7 | 2.9 | 3.0 | 0.16 | 1.56 | 0.04 | 17.6 | 99.5 | 0.07 |
| 1992 | 525 | 0.44 | 7.3 | 10.7 | 60.7 | 2.7 | 3.6 | 0.17 | 1.43 | 0.16 | 21.2 | 111.8 | 0.06 |
| 1993 | 437 | 0.35 | 6.2 | 9.4 | 57.1 | 2.0 | 3.3 | 0.15 | 1.28 | 0.18 | 12.6 | 112.2 | 0.10 |
| 1994 | 653 | 0.44 | 5.7 | 8.8 | 58.8 | 2.6 | 3.2 | 0.16 | 1.79 | 0.25 | 24.6 | 82.1 | 0.08 |
| 1995 | 561 | 0.34 | 4.7 | 7.2 | 42.6 | 2.6 | 2.6 | 0.14 | 1.27 | 0.14 | 16.2 | 70.8 | 0.09 |
| % change 1995/89–90 | | −19 | −37 | −29 | −15 | −27 | −27 | −32 | −5 | 460 | −1 | −30 | −14 |
| **Covered Catchment G1** | | | | | | | | | | | | | |
| 1989 | 300 | 0.20 | 4.3 | 7.2 | 39.0 | 2.1 | 6.8 | 0.31 | 0.13 | 0.04 | 14.0 | 85.3 | 0.04 |
| 1990 | 413 | 0.28 | 5.4 | 9.0 | 60.1 | 3.8 | 8.0 | 0.23 | 0.24 | 0.02 | 20.5 | 113.2 | 0.11 |
| 1991 | 375 | 0.25 | 4.7 | 7.9 | 60.4 | 2.7 | 7.2 | 0.18 | 0.26 | 0.03 | 15.5 | 117.7 | 0.05 |
| 1992 | 300 | 0.17 | 3.3 | 5.7 | 46.6 | 1.7 | 5.5 | 0.09 | 0.20 | 0.26 | 9.9 | 93.6 | 0.01 |
| 1993 | 291 | 0.16 | 2.7 | 4.3 | 38.3 | 1.1 | 4.0 | 0.06 | 0.16 | 0.15 | 8.7 | 73.6 | 0.02 |
| 1994 | 504 | 0.31 | 3.5 | 5.7 | 53.8 | 2.5 | 5.2 | 0.13 | 0.31 | 0.16 | 12.9 | 100.6 | 0.03 |
| 1995 | 371 | 0.17 | 2.0 | 2.9 | 27.8 | 1.6 | 3.3 | 0.07 | 0.41 | 0.14 | 10.4 | 43.9 | 0.02 |
| % change 1995/89–90 | | −29 | −60 | −64 | −44 | −45 | −56 | −74 | 122 | 367 | −39 | −41 | −73 |

S*=non-marine sulphur.

**Table 7.2.** Yearly accumulated total and "extra" non-marine fluxes of elements in kg/ha and keq/ha from the G1 ROOF catchment. The "extra" fluxes represent the difference between the measured flux from G1 and the expected flux calculated from the pre-treatment regression with F1, calculated as described in the Methods section. "Extra" chloride (in parentheses) is also calculated in this way. The yearly fluxes for 1991 are calculated from the start of the experiment on 1 April 1991, and all other years are full years starting on 1 January each year

| Year | $H^+$ | Ca | Mg | Na | K | $NH_4$-N | Al Inorg. | Mn | Fe | $SO_4$-S | CL (extra Cl) | $NO_3$-N | ΣCation | ΣAnion | Anion− Cation | % Anion | % Org. |
|---|---|---|---|---|---|---|---|---|---|---|---|---|---|---|---|---|---|
| **Accumulated fluxes in kg/ha** | | | | | | | | | | | | | | | | | |
| 1991 | 0.18 | 3.2 | 5.5 | 41.9 | 1.7 | 0.03 | 5.0 | 0.1 | 0.2 | 13.5 | 86.6 | 0.05 | | | | | |
| 1992 | 0.34 | 6.5 | 11.2 | 88.5 | 3.4 | 0.3 | 10.5 | 0.2 | 0.4 | 27.6 | 180.3 | 0.06 | | | | | |
| 1993 | 0.50 | 9.2 | 15.5 | 126.7 | 4.4 | 0.4 | 14.6 | 0.3 | 0.5 | 39.8 | 253.9 | 0.08 | | | | | |
| 1994 | 0.81 | 12.7 | 21.2 | 180.5 | 6.9 | 0.6 | 19.8 | 0.4 | 0.9 | 57.0 | 354.5 | 0.11 | | | | | |
| 1995 | 0.99 | 14.7 | 24.1 | 208.3 | 8.6 | 0.7 | 23.0 | 0.5 | 1.3 | 69.5 | 398.4 | 0.13 | | | | | |
| **Extra accumulated non-marine fluxes in kg/ha** | | | | | | | | | | | | | | | | | |
| 1991 | 0.04 | 0.1 | −0.8 | −7.0 | −0.2 | −0.01 | 0.7 | 0.0 | 0.02 | −2.6 | 0 (31.6) | 0.04 | | | | | |
| 1992 | −0.03 | −1.1 | −2.8 | −11.2 | −0.6 | 0.2 | −1.5 | −0.1 | 0.02 | −15.5 | 0 (13.7) | 0.00 | | | | | |
| 1993 | −0.06 | −2.1 | −4.2 | −10.4 | −0.5 | 0.3 | −4.8 | −0.2 | −0.01 | −21.2 | 0 (−18.7) | −0.08 | | | | | |
| 1994 | 0.01 | −2.8 | −6.7 | −19.1 | −0.7 | 0.3 | −6.7 | −0.2 | 0.03 | −34.6 | 0 (−2.2) | −0.11 | | | | | |
| 1995 | 0.00 | −3.5 | −7.7 | −16.8 | −0.8 | 0.4 | −9.4 | −0.2 | −0.25 | −42.1 | 0 (−7.7) | −0.18 | | | | | |
| **Accumulated fluxes in keq/ha** | | | | | | | | | | | | | | | | | |
| 1991 | 0.18 | 0.2 | 0.5 | 1.8 | 0.0 | 0.00 | 0.6 | 0.00 | 0.01 | 0.8 | 2.4 | 0.00 | 3.2 | 3.3 | 0.1 | 102 | −2 |
| 1992 | 0.34 | 0.3 | 0.9 | 3.8 | 0.1 | 0.02 | 1.2 | 0.01 | 0.01 | 1.7 | 5.1 | 0.00 | 6.7 | 6.8 | 0.1 | 101 | −1 |
| 1993 | 0.50 | 0.5 | 1.3 | 5.5 | 0.1 | 0.03 | 1.6 | 0.01 | 0.02 | 2.5 | 7.2 | 0.01 | 9.5 | 9.7 | 0.1 | 101 | −1 |
| 1994 | 0.81 | 0.6 | 1.7 | 7.8 | 0.2 | 0.04 | 2.2 | 0.01 | 0.03 | 3.6 | 10.0 | 0.01 | 13.5 | 13.6 | 0.1 | 100 | 0 |
| 1995 | 0.99 | 0.7 | 2.0 | 9.1 | 0.2 | 0.05 | 2.6 | 0.02 | 0.05 | 4.3 | 11.2 | 0.01 | 15.7 | 16.0 | 0.4 | 102 | −2 |
| **Extra accumulated non-marine fluxes in keq/ha** | | | | | | | | | | | | | | | | | |
| 1991 | 0.04 | 0.0 | −0.1 | −0.3 | −0.01 | 0.00 | 0.1 | 0.00 | 0.00 | −0.2 | 0 | 0.00 | −0.2 | −0.2 | 0.1 | | |
| 1992 | −0.03 | −0.1 | −0.2 | −0.5 | −0.02 | 0.01 | −0.2 | 0.00 | 0.00 | −1.0 | 0 | 0.00 | −1.0 | −1.0 | 0.0 | 99 | 1 |
| 1993 | −0.06 | −0.1 | −0.3 | −0.5 | −0.01 | 0.02 | −0.5 | −0.01 | 0.00 | −1.3 | 0 | −0.01 | −1.5 | −1.3 | 0.2 | 89 | 11 |
| 1994 | 0.01 | −0.1 | −0.6 | −0.8 | −0.02 | 0.02 | −0.7 | −0.01 | 0.00 | −2.2 | 0 | −0.01 | −2.3 | −2.2 | 0.1 | 96 | 4 |
| 1995 | 0.00 | −0.2 | −0.6 | −0.7 | −0.02 | 0.03 | −1.0 | −0.01 | 0.01 | −2.6 | 0 | −0.01 | −2.6 | −2.6 | −0.1 | 102 | −2 |

**Table 7.3.** Yearly accumulated measured fluxes of elements and water as a percentage of fluxes from the G1 ROOF catchment predicted from the pre-treatment regression with F1, calculated as described in the Methods section

| Year | $H_2O$ | $H^+$ | Ca | Mg | Na | K | Al Inorg | Mn | Fe | $S^*$ | Cl | $NO_3$-N |
|---|---|---|---|---|---|---|---|---|---|---|---|---|
| 1991 | 12 | 28 | 37 | 32 | 28 | 30 | 18 | 15 | 13 | −22 | 57 | 488 |
| 1992 | −8 | −8 | −9 | −13 | −5 | −9 | −13 | −27 | 4 | −45 | 10 | 4 |
| 1993 | −8 | −10 | −19 | −24 | −16 | −18 | −25 | −38 | −3 | −43 | −7 | −50 |
| 1994 | −2 | 1 | −16 | −23 | −13 | −10 | −25 | −32 | 4 | −46 | 1 | −50 |
| 1995 | −4 | 0 | −20 | −28 | −17 | −15 | −29 | −34 | 25 | −45 | −2 | −57 |

S*=non-marine sulphur.

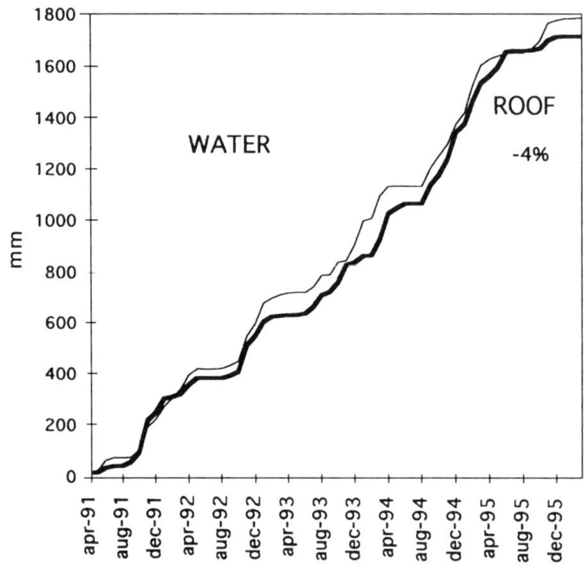


**Figure 7.3.** Accumulated measured monthly flux of water in runoff from the Covered Catchment (thick line) compared with the flux predicted from the pre-treatment relationship of runoff from the F1 CONTROL catchment (thin line)

calculated extra monthly fluxes for both water and non- marine sulphur are shown in Figure 7.7. From the figure it is obvious that the sprinkling of water under the roof generally caused only small differences in water output compared with the runoff predicted from F1. Some differences occurred during the winter periods when freezing of the sprinkler system delayed water application. For non-marine sulphur, there were generally lower monthly fluxes than predicted after the start of treatment. Output from G1 exceeded that predicted from F1 during a few summer months only.

The calculated charge balance of the measured fluxes (Table 7.2) shows that the sum of cations was nearly in exact balance with the inorganic anions, showing that organic anions are unimportant for the transport of cations in the G1 catchment.

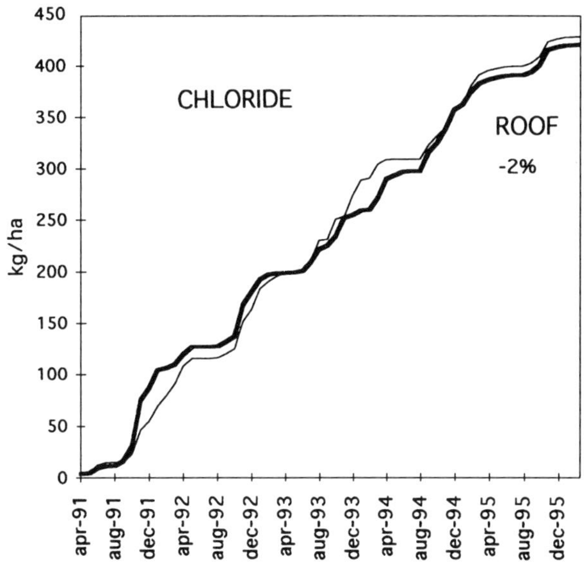


**Figure 7.4.** Accumulated measured monthly flux of chloride in runoff from the Covered Catchment (thick line) compared with the flux predicted from the pre-treatment relationship of runoff from the F1 CONTROL catchment (thin line)

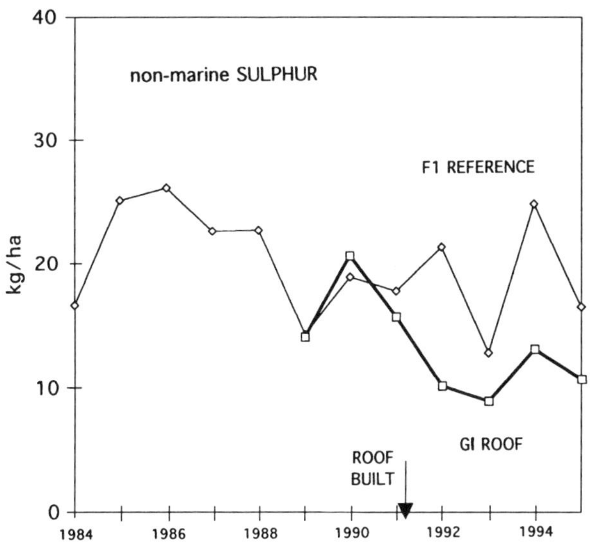


**Figure 7.5.** Yearly fluxes of sulphur in runoff from the F1 CONTROL catchment and the Covered Catchment G1 ROOF between 1984 and 1995

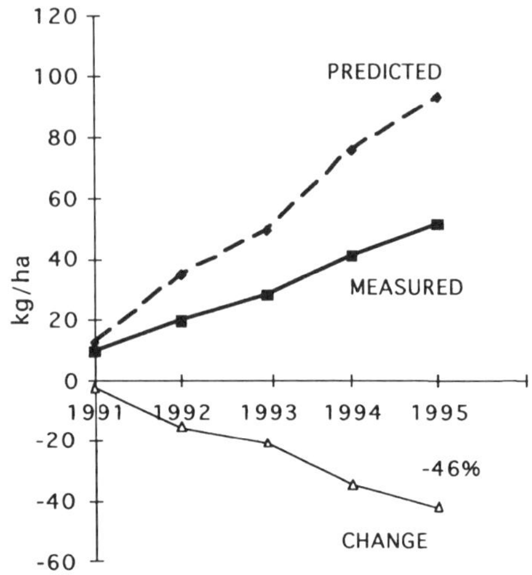


**Figure 7.6.** Accumulated annual non-marine sulphur fluxes in runoff from the G1 ROOF catchment at Gårdsjön. Thick line — measured; dashed line — as predicted from the pre-treatment relationship with the F1 CONTROL catchment. The thin line illustrates the difference between predicted and measured

Nitrate fluxes were very small before roof treatment, and decreased to even smaller amounts after roof construction (Table 7.1).

*Base cations*

Fluxes of divalent base cations decreased by 60–65%, and those of monovalent base cations by about 45%, in the G1 catchment compared with the mean of the two pre-treatment years (Table 7.1). The accumulated percentage change relative to F1 was smaller, varying from −15% for $K^+$ to −28% for $Mg^{2+}$ (Table 7.3).

*Acidic cations*

The flux of $H^+$ decreased by 19 and 29% in F1 and G1 respectively (Table 7.1) compared with the pre-treatment period. However, there was no extra decrease in flux of $H^+$ due to G1 treatment, unlike most ions (Table 7.3). Inorganic Al, in contrast, showed a much larger decline in G1 than F1 compared with pre-treatment values (56 compared with 27%; Table 7.1) and a large extra decline in G1 compared with F1 (Table 7.3). Figure 7.8 shows this decline on a temporal basis, showing that the flux from G1 declined sharply, although still higher than F1. Manganese showed even larger percentage declines (Tables 7.1 and 7.3) although the absolute values of the manganese flux were much smaller.

*Relationships with sulphate*

The "extra" accumulated decrease in non-marine sulphur flux (keq/ha) is plotted against the extra accumulated fluxes of major cations in Figure 7.9. The figure shows

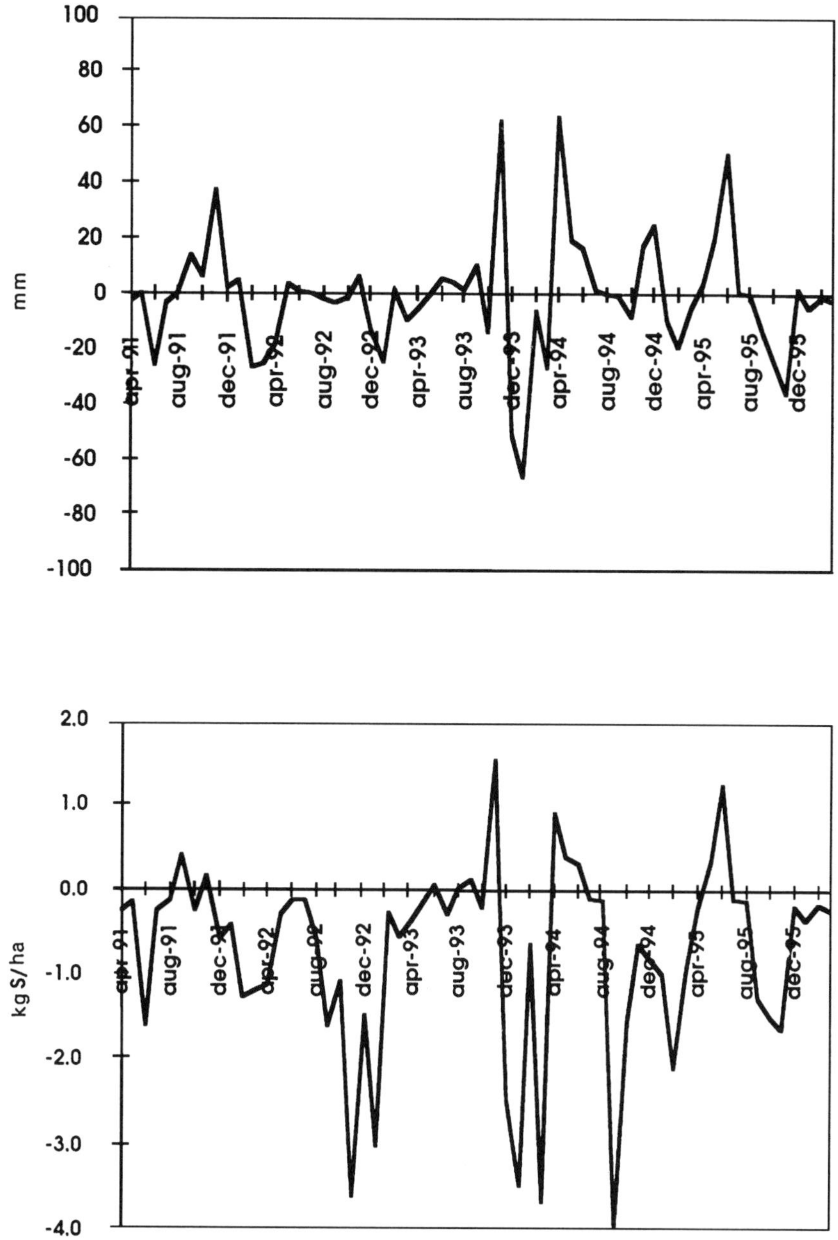


**Figure 7.7.** Monthly output of water and non-marine sulphur from the G1 ROOF catchment calculated as the difference between the flux as predicted from the pre-treatment relationship with the F1 CONTROL catchment and the measured flux

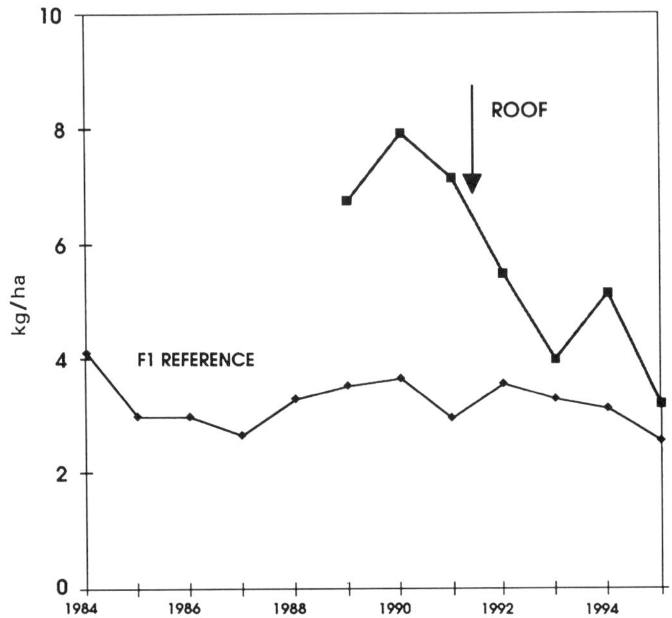


**Figure 7.8.** Yearly fluxes of inorganic Al in runoff from the reference catchment F1 CONTROL and the Covered Catchment G1 ROOF between 1984 and 1995

clearly the increasing importance of Al in balancing the reduction in S flux from the G1 catchment compared with F1.

## DISCUSSION

### Initial state of the Gårdsjön catchments

Long-term monitoring (Figure 7.10) and experiments in catchments at Gårdsjön show that the 10 year means of sulphur fluxes in throughfall and runoff were almost identical (Hultberg and Grennfelt, 1992). Wet deposition over the period 1980 to 1990 was estimated to be 11.6 kg S/ha/yr, and dry deposition to be 10.5 ± 1.7 kg/ha/yr based on throughfall, and 11.6 ± 5.5 kg/ha/yr based on runoff. The sulphur flux in stemflow was measured to about 1 kg/ha/yr which is 5% of the total sulphur input. This amount should be added to the dry deposition estimates based on throughfall. When estimating the dry deposition of sulphur, the delayed hydrological response in runoff compared to throughfall results in a year to year difference, with estimates based on runoff giving much larger variations between years (Hultberg and Likens, 1991; Hultberg and Grennfelt, 1992; Hultberg *et al.*, 1994). The between-year variations in runoff fluxes of sulphur are large as a result of years with droughts like 1982/83, 1989, 1992, 1994 and 1995. Fluxes of the sea-salt ions sodium and chloride during the period 1980 to 1990 were very similar in throughfall and runoff indicating

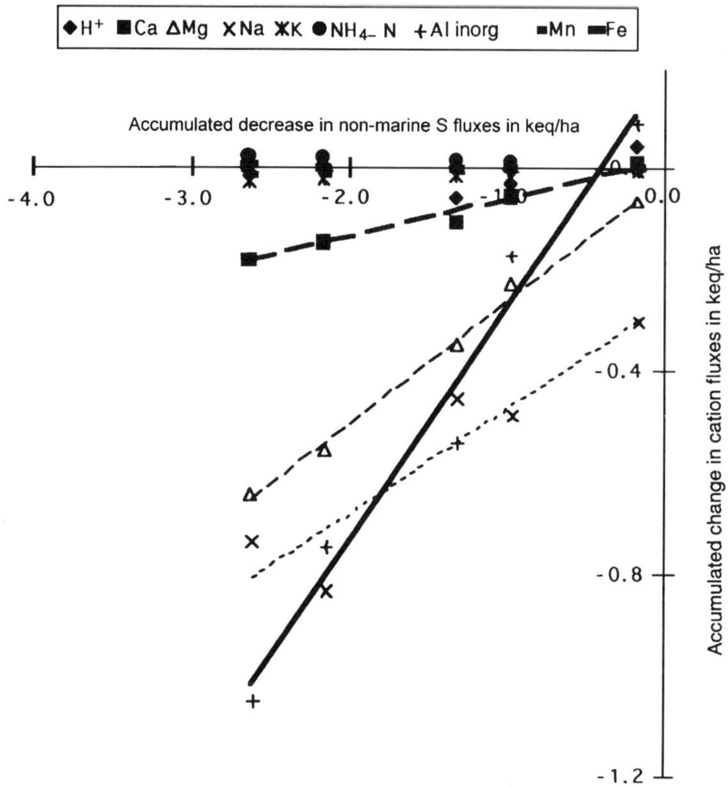


**Figure 7.9.** Accumulated decrease in annual non-marine sulphur fluxes versus accumulated changes in cation fluxes in keq/ha. Time flows from right to left and top to bottom on this diagram

that the ecosystem was in equilibrium with these elements also (Hultberg 1985b; Hultberg and Grennfelt 1992).

Between 1980 and 1990, there was no significant trend in S fluxes in either throughfall or runoff at Gårdsjön, in spite of a 29% reduction in European $SO_2$ emissions (Hultberg and Grennfelt, 1992). Since the end of the old political system in East Germany and other former Eastern European countries in 1989/90, however, S deposition has decreased markedly (Hultberg and Ferm, 1995; Chapter 3, this volume). By 1995 wet and dry sulphur deposition had decreased by about 50% compared to 1989/90. This has resulted in decreased input in throughfall and resulted in decreased output of sulphur from the reference catchment F1 (Figure 7.10). The long-term measurements on F1 have been shown to be more and more valuable with time as a reference to experimental catchment research, but the large changes in deposition have also turned it into an experiment in itself.

With the large changes in atmospheric sulphur deposition occurring during the Covered Catchment experiment it was very important that the project started with two or more years of pre-treatment calibration studies of the G1 ROOF and F1

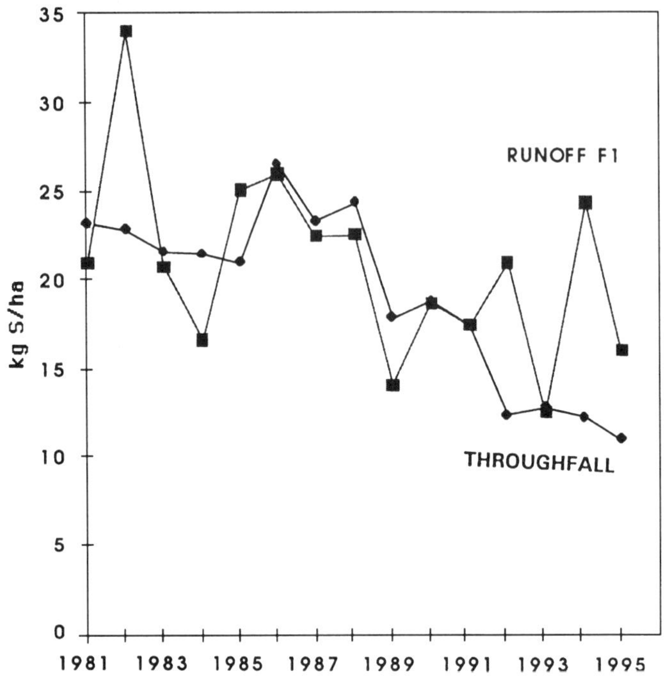


**Figure 7.10.** Yearly non-marine sulphur fluxes in throughfall and runoff at the reference catchment F1 CONTROL at Gårdsjön

CONTROL catchments. These pre-treatment studies now allow for detailed quantitative comparisons between the catchments.

## Changes in fluxes and concentrations

Several considerations need to be borne in mind in interpreting the changes of fluxes and concentrations in the G1 ROOF catchment. The experimental treatment is intended to mimic the reduction of S and N deposition which would be observed if emissions were reduced. Concentrations and fluxes before and after roof construction should therefore be compared. However, these concentrations and fluxes also respond to changes in environmental variables not under the control of the experimenters, such as temperature and rainfall amount and distribution. The F1 catchment will also be exposed to these changes in environmental variables and its response will thus act as a control for the experimental treatments in G1. The F1 catchment is not a perfect control for G1, however, for two main reasons. Firstly, no two natural catchments can be exactly alike and even though the F1 and G1 catchments are within 1 km distance there are differences in microclimate, soils, vegetation and size which are reflected in differences in pre-treatment water chemistry (Chapter 2, this volume). Secondly, ambient S deposition to the F1 catchment has also been decreasing (Chapter 3, this volume) in line with reductions in $SO_2$ emissions, so the responses of F1 also include a response to decreased S deposition, though the decrease is at a different rate to that of G1. Comparisons of F1 and G1 are therefore testing the effect of the *additional* reduction in S deposition under the G1 ROOF, rather than comparing the response of G1 with no S reduction. Calculation of "extra" fluxes (see Methods) implicitly assumes that the relationships between processes in F1 and G1 are linear, which may not be the case. Moldan *et al.* (1995) used randomized intervention analysis (Carpenter *et al.*, 1989) to test whether there was a statistically significant difference between the pre- and post-treatment monthly weighted mean concentrations in the F1 and G1 catchments up to April 1995. This tests the null hypothesis that the same changes have occurred in the treatment and reference catchments. As with all correlative methods, a significant difference between F1 and G1 does not conclusively prove that the difference is caused by the roof treatment, but it adds considerable strength to such a conclusion.

Another consideration is whether the unintended effects of roof construction have been compensated for sufficiently (Chapters 4 and 22, this volume). There are small differences in the timing of water inputs which lead to output differences (Figures 7.3 and 7.7), but the overall differences are small and use of annual means to assess responses should minimize the problems. More importantly, there were some differences between F1 and G1 in sea-salt inputs in the first 2 years (Chapter 4, this volume) and the same authors suggested that more $Ca^{2+}$ should have been added to G1 to compensate for interruption of the return of $Ca^{2+}$ in throughfall. Though these considerations complicate interpretation, in most cases the changes resulting from reductions in S and N deposition due to the roof are sufficiently large to be identifiable with a fair degree of confidence.

*Anions*

Table 7.1, Figure 7.5 and Figure 7.1c show that there has been a marked reduction in both the flux and concentration of non-marine $SO_4^{2-}$ due to the roof treatment. The difference in response between F1 and G1 can be seen in Table 7.2, and was statistically significant ($p < 0.01$; Moldan *et al.*, 1995). Output of non-marine $SO_4^{2-}$ from G1 ROOF was about 45% less than would have been predicted knowing the output of F1 CONTROL (Table 7.3). Figure 7.7 shows clearly that, except for a period in late summer each year, output of non-marine S is consistently lower in G1 than F1, whereas water fluxes show no such effect. Figure 7.1c appears to show the $SO_4^{2-}$ concentration levelling off in 1995, the final year considered here. The catchment is still losing $SO_4^{2-}$ (see Moldan *et al.*, 1995; Chapter 24, this volume), so this does not represent the attainment of a new equilibrium with the reduced S input. It may however represent a temporary steady state which, in view of the large store of S in the soil, may last for some years. The RAIN Project in contrast reached final equilibrium after 6–7 years (Wright *et al.*, 1993; Chapter 1, this volume). According to the "mobile anion hypothesis" (Johnson and Cole, 1980) such reductions in anion output will force corresponding reductions in cation output. The other major anions have first to be considered, however.

Chloride is the dominant anion, and the long-term record from the F1 CONTROL catchment (Figure 7.1a) shows output concentrations fluctuating markedly, due to input fluctuations (Hultberg and Grennfelt, 1992; Chapter 3, this volume). In the first year of treatment, sea salt and thus $Cl^-$ was overapplied to G1 ROOF relative to the amount deposited on F1 (Chapter 4, this volume), being compensated for by underapplication the following year, to reach a balance in the third year. The years 1994 and 1995 had low $Cl^-$ deposition compared with previous years (Chapter 4, this volume) and these two factors together have led to a strong declining trend in $Cl^-$ concentration over the first 5 years of the roof experiment in both G1 and F1 (Figure 7.1a). The reduction in ionic strength over this period may itself have led to a reduction in acidity by the "salt effect" mechanism (e.g. Reuss and Johnson, 1986), and the reduction in mobile anion flux will lead to a reduction in total cation flux, but not of non-marine cation flux provided they do not lag anion fluxes significantly on the yearly time-scale being considered.

Nitrate outputs were low before the roof treatment, and have been reduced to even lower concentrations afterwards (Figure 7.1d). The untreated Gårdsjön catchments appear to be N-limited, and thus $NO_3^-$ fluxes make an insignificant contribution to the total anion flux. The ionic balance of measured cations and anions is very close (Table 7.2) indicating that organic anions also make a very small contribution to anion flux. This contrasts strongly with the Risdalsheia roof (Wright *et al.*, 1993) as described in Chapter 1 of this volume. The difference in response between F1 and G1 was significant only for $SO_4^{2-}$ among the anions (Moldan *et al.*, 1995). Only changes in $SO_4^{2-}$ and $Cl^-$ thus need to be considered as drivers of cation fluxes.

*Base cations*

The concentrations of Na, Mg and Ca show a sharp and continued decline after roof construction (Figures 7.1e, g and h) whereas K shows the same pattern but halting in 1993 (Figure 7.1f). This latter is probably due to an increase in K applied to compensate for interruption of throughfall supplies in the year 1994/95 (Chapter 4, this volume). Of these four base cations, the difference in response between F1 and G1 was significant only for Mg ($p<0.01$: Moldan *et al.*, 1995). However, Table 7.2 shows that the output of all base cations from G1 is progressively less than would have been predicted from the relationship between F1 and G1 before roof construction. The initial overapplication of sea salt led to an increased flux in the first year (Table 7.3) but the pattern after that is consistent. The relative decreases in total cation concentrations are very large—around 60% for the divalent cations and 45% for the monovalent (Figure 7.1; Table 7.1). The F-factor (Henriksen, 1984)—i.e. the change in non-marine base cation concentrations relative to the change in non-marine $SO_4^{2-}$—varies between 0.45 (in 1994) and 0.54 (in 1991). This can be compared with an F-factor of 0.4 after additions of elemental sulphur (Hultberg *et al.*, 1990), and an F-factor of 0.5 for the decreased sulphur loading following clear-felling (Hultberg, 1992) in other Gårdsjön catchments. Though the decline in Ca concentration is particularly sharp, some of it may be due to insufficient compensation for throughfall losses rather than the reduction in S output (Chapter 4, this volume).

*Acidic cations*

As with the base cations, $H^+$, inorganic Al and Mn decrease after the roof treatment. For Al and Mn, but not $H^+$, the decrease in G1 relative to F1 is statistically significant (Moldan *et al.*, 1995). The reduction in Al flux is particularly striking (Figure 7.8), amounting to 38% of the equivalent reduction in non-marine $SO_4^{2-}$ flux (Table 7.2). This contrasts with Risdalsheia, where Al concentrations were much lower and reacted more slowly (Wright *et al.*, 1993; see Chapter 1, this volume). An increase in Al flux in G1 relative to F1 in the first year of treatment (Tables 7.2 and 7.3) was probably due to sea-salt effects. Although $H^+$ concentrations have declined in absolute terms, the change in G1 is no more than would have been expected from the change in F1 (Table 7.2). The slow response of $H^+$ is consistent with other roof projects (Wright *et al.*, 1993; Boxman *et al.*, 1995; Bredemeier *et al.*, 1995). The other acidic cations are quantitatively unimportant, though there is a striking increase in $NH_4^+$ concentrations in the output from both catchments (Figure 7.1m).

### *ANC*

ANC rises due to treatment, randomized intervention analysis showing that the difference between F1 and G1 up to 1994 is significant at $p<0.05$ (Moldan *et al.*, 1995). The almost linear increase in ANC in G1 is encouraging, though it may not continue if the decline in $SO_4^{2-}$ output slows down, or if sea-salt deposition increases again.

### *Overall*

Removal of 95% of the $SO_4^{2-}$ input to the G1 catchment has apparently caused the $SO_4^{2-}$ concentration of the output water to decline steadily over 4 years by a total of about 43%. There are signs in the final year that this decline has levelled out. Sulphate concentrations in the output water are still high, however, at around 215 $\mu$eq/L total or 175 $\mu$eq/L non-marine. These $SO_4^{2-}$ concentrations are being maintained by desorption and mineralization of S stored in the catchment soils (Chapters 8 and 9, this volume). The length of time this process can continue is a key question for the Covered Catchment Experiment (Chapter 24, this volume).

From the point of view of reducing toxicity to aquatic biota, it would be advantageous if the decline in $SO_4^{2-}$ was accompanied by a decline in toxic aluminium and $H^+$ ions rather than $Ca^{2+}$ or $Mg^{2+}$ ions. The calculated F-factor (Henriksen, 1984) of around 0.5 implies that both will be reduced about equally. The increasing importance of Al over time in balancing the reduction in $SO_4^{2-}$ is encouraging. However, there has been a considerable reduction in ionic strength in both G1 and F1 catchments due to declining sea-salt inputs. This "inverse salt effect" would be expected to reduce concentrations of $H^+$ and Al both absolutely and relative to $Ca^{2+}$ and $Mg^{2+}$, irrespective of the fate of $SO_4^{2-}$ (Reuss and Johnson, 1986). It will be important to measure whether the decline of Al will be maintained if sea-salt inputs stabilize or rise again. Comparison of the reactions of F1 and G1 by computing "extra" fluxes supports the assertion that Al concentrations have responded more strongly to the reduction in $SO_4^{2-}$ than have base cations (Table 7.2). The same is not true of $H^+$, however, which has declined to a similar small extent in both F1 and G1, suggesting it may be responding only to the reduction in sea-salt inputs. The high concentration of inorganic Al implies the presence of an aluminium buffer system of some sort: in that case the concentration of Al will probably have to decline to quite low values before there is much reduction in $H^+$ flux or concentration.

### Fish toxicity

In Chapter 1 of this volume an approximate fish toxicity function due to Sadler (1994) was applied to the results from the covered catchment at Risdalsheia in the RAIN Project to evaluate whether reduction in acid deposition was improving the prospects for fish — i.e. whether the reduction in Al and $H^+$ was sufficient to overcome the reduction in base cation concentrations. The same function has been applied to the output from G1 and F1 (Figure 7.11). The result is very similar to that at Risdalsheia in that treatment is slowly but steadily improving the prospects for fish, but that the water quality after 5 years is still well short of that necessary for fish survival. As at Risdalsheia, however, water suitable for fish could be produced if this runoff water were mixed with sufficient more alkaline water, as might happen if the water flowed into a lake or stream with some groundwater input (such as Lake Gårdsjön). The improvement observed thus remains relevant.

## CONCLUSIONS

Reducing acid inputs to the covered catchment has led to a steady improvement in the quality of the output water. The output flux of toxic aluminium ions has been considerably reduced, though it remains to be seen how much of this reduction is due to a reduction in sea-salt inputs. The output of base cations has also been reduced, though to a smaller extent, and the overall effect judged by the fish toxicity function remains positive. The flux of non-marine $SO_4^{2-}$ and inorganic Al remains high, however, in spite of 5 years of elimination of acid deposition. The rate of recovery of the catchment will now depend on the rate of release of stored S from the catchment soils.

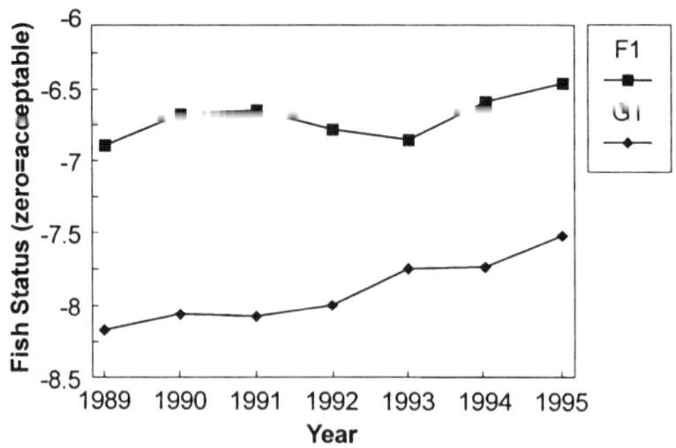


**Figure 7.11.** Calculated fish status in the G1 ROOF and F1 CONTROL catchments, using discriminant functions described by Sadler (1994) based on pH, Al and Ca. All waters are unsuitable for fish, but the less negative the value of the fish status indicator, the better the water quality. All values are based on annual means

## ACKNOWLEDGEMENTS

We greatly appreciate grants for performing this study from Vattenfall and later Elforsk AB, the National Power and Power Gen (UK), the CEC-STEP programme (ENCORE-project), the Swedish Environmental Research Institute (IVL) research programme, the Council of Nordic Ministers (NMR). We would also like to thank the many field assistants, laboratory technicians and scientists from many countries who all have made important contributions to the scientific results within the project.

## REFERENCES

Bishop, K. and Hultberg, H. (1995) Reversing acidification in a forest ecosystem: The Gårdsjön Covered Catchment. *Ambio*, **24**, 85–91.

Boxman, A. W., Vandam, D., Vandijk, H. F. G., Hogervorst, R. F. and Koopmans, C. J. (1995) Ecosystem responses to reduced nitrogen and sulphur inputs into two coniferous forest stands in The Netherlands. *Forest Ecology and Management*, **71**, 7–29.

Bredemeier, M., Blanck, K., Lamersdorf, N. and Wiedey, G. A. (1995) Response of soil water chemistry to experimental "clean rain" in the NITREX roof experiment at Solling, Germany. *Forest Ecology and Management*, **71**, 31–44.

Carpenter, S. R., Frost, T. M., Heisey, D. and Kratz, T. K. (1989) Randomized intervention analysis and the interpretation of whole-ecosystem experiments. *Ecology*, **70**, 1142–1152.

Hasselrot, B. and Grennfelt, P. (1987) Deposition of air pollutants in a wind-exposed forest edge. *Water, Air, and Soil Pollution*, **34**, 135–143.

Henriksen, A. (1984) Changes in base cation concentrations due to freshwater acidification. *Verh. Internat. Verein. Limnol.*, **22**, 692–698.

Hultberg, H. (1985a) Changes in fish populations and water chemistry in Lake Gårdsjön and neighbouring lakes during the last century, *Ecological Bulletins (Stockholm)*, **37**, 64–72.

Hultberg, H. (1985b) Budgets of base cations, chloride, nitrogen and sulphur in the acid Lake Gårdsjön catchment, SW Sweden. *Ecological Bulletins (Stockholm)*, **37**, 133–157.

Hultberg, H. (1992) Biochemical cycling of sulphur and its effects on surface water chemistry in an acid catchment in SW Sweden. Proceedings of International Symposium on *Terrestrial Ecosystem Effects*, Commission of the European Communities, May 1991, Florence.

Hultberg, H. and Ferm, M. (1995) Measurements of atmospheric deposition and internal circulation of base cations to a forested catchment area. *Water, Air, and Soil Pollution*, **85**, 2235–2240.

Hultberg, H. and Grennfelt, P. (1992) Sulphur and sea salt deposition as reflected by throughfall and runoff chemistry in forested catchments. *Environmental Pollution*, **75**, 215–222.

Hultberg, H. and Likens, G. E. (1991) Sulphur deposition to forested catchments in Northern Europe and North America — Large scale variations and long term dynamics. In: S. E. Schwartz and W. G. N. Slinn (eds), Proceedings of 5th International Conference on *Precipitation Scavenging and Atmosphere–Surface Exchange*, Vol. 3, Hemisphere Publ. Corp., USA, pp. 1343–1365.

Hultberg, H., Lee, Y. H., Nyström, U. and Nilsson, S. I. (1990) Chemical effects on surface-, ground- and soil-water of adding acid and neutral sulphate to catchments in south-west Sweden. In: B. J. Mason (ed.), *The Surface Water Acidification Programme SWAP*. Cambridge University Press, Cambridge, pp. 167–182.

Hultberg, H., Dise, N. B., Wright, R. F., Andersson, B. I. and U. Nyström, U. (1993a) Nitrogen saturation induced during winter by experimental $NH_4NO_3$ addition to a forested catchment. *Environmental Pollution*, **84**, 145–147.

Hultberg, H., Andersson, B. I. and Moldan, F. (1993b) The covered catchment—An experimental approach to reversal of acidification in a forest ecosystem. In: L. Rassmusen, T. Brydges and P. Mathy (eds), *Experimental Manipulations of Biota and Biogeochemical Cycling in Ecosystems*. Ecosystems Research Report, 4, CEC, Brussels, pp. 46–54.

Hultberg, H. *et al.* (1994) Sulphur. In: B. Moldan and J. Cherny, (eds), *Biogeochemistry of Small Catchments: A Tool for Environmental Research. SCOPE 51*. John Wiley & Sons Ltd, Chichester, pp. 229–254.

Johnson, D. W. and Cole, D. W. (1980) Anion mobility in soils: relevance to nutrient transport from terrestrial ecosystems. *Environment International*, **3**, 79–90.

Melkerud, P.-A. (1977) Samband mellan morän och berggrund. En travers över Östergötlands kambrosilurområde. PhD thesis, Department of Quaternary Research, University of Stockholm, 163 pp. (In Swedish with English summary.)

Mitchell, M. J., Driscoll, C. T., Fuller, R. D., David M. B. and Likens, G. E (1989) Effect of whole-tree harvesting on the sulfur dynamics of a forest soil. *Soil Science Society of America Journal*, **53**, 933–940.

Moldan, F., Hultberg, H. and Andersson, I. (1995) Covered Catchment Experiment at Gårdsjön: Changes in runoff chemistry after four years of experimentally reduced acid deposition. *Water, Air, and Soil Pollution*, **85**, 1599–1604.

Nodvin, S. C., Driscoll, C. T. and Likens, G. E. (1988) Soil processes and sulphate loss at the Hubbard Brook Experimental Forest. *Biogeochemistry*, **5**, 185–199.

Olsson, B., Hallbäcken, L., Johansson, S., Melkerud, P. A., Nilsson, S. I. and Nilsson, T. (1985) The Lake Gårdsjön area—physiographical and biological features. *Ecological Bulletins (Stockholm)*, **37**, 10–28.

Reuss, J. O. and Johnson, D. W. (1986) *Acid Deposition and the Acidification of Soils and Waters*. Springer-Verlag, New York.

Sadler, K. (1994) *Damage Functions for Predicting Fishery Status from Water Chemistry*. Report PT/94/330068/M. PowerGen, Ratcliffe-on-Soar, UK.

Wright, R. F., Lotse, E. and Semb, A. (1993) RAIN Project: results after 8 years of experimentally reduced acid deposition to a whole catchment. *Canadian Journal of Fisheries and Aquatic Sciences*, **50**, 258–268.

# 8 Sulfur Dynamics in the Roof Experiment at Lake Gårdsjön Deduced from Sulfur and Oxygen Isotope Ratios in Sulfate

PETER TORSSANDER & CARL-MAGNUS MÖRTH

Sulfur concentration and isotope composition were determined in deposition, runoff, groundwater and soils of the G1 Roof Catchment at Lake Gårdsjön during one pre-treatment and two post-treatment years. The $\delta^{34}S$ values of the deposition prior to sprinkling were almost identical for bulk deposition (+7.4‰) and throughfall (+7.5‰). Two sources of sulfur in the atmospheric deposition were identified: anthropogenic sulfur and seawater spray. Sulfur isotope compositions of sulfate in runoff, groundwater and soil extractions collected prior to the start of the experiment were similar, which suggests that the inorganic sulfate pool within the catchment has a homogeneous $\delta^{34}S$ value of about +5.5‰. This implies that sulfate exchanges readily and that sulfate adsorption is reversible. Sprinkler water spiked with small amounts of seawater sulfate has a sulfur $\delta^{34}S$ value of +19.5‰. Since the sprinkling started in April 1991, the sulfate concentration in runoff has decreased by 30%, but the $\delta^{34}S$ value has increased only by 0.5‰. This suggests mixing of sprinkler water with the large inorganic sulfur reservoir in the catchment. This pool is sufficiently large to control the sulfur isotope composition of the runoff. Oxygen isotopes in sulfate suggest that around one-third of sulfate transfer occurs through the A-horizon organic sulfur pool. Heavy sulfur isotope values for total sulfur in the B-horizon suggest that organic sulfur is accumulating and that sulfur cycling is not important, because the only heavy sulfur source is seawater spray in the deposition. Most of the accumulation must have occurred in pre-acidification times. The conclusion is that sorption processes in the B-horizon are the main sulfur flux mechanism in the catchment. Acidification processes will continue for 25–45 years after a reduction of sulfate input, because adsorption of sulfate is a reversible process. Mineralization of organic sulfur could, however, prolong acidification processes for centuries.

## INTRODUCTION

Sulfur emissions have decreased over the last decade by 20–30% in Europe and a similar decrease has been observed in the sulfur deposition (Hedin *et al.*, 1994).

*Experimental Reversal of Acid Rain Effects: The Gårdsjön Roof Project.*
Edited by Hans Hultberg and Richard Skeffington. 

Reduction of atmospheric acid deposition towards natural levels has raised the questions of whether the decrease in acid deposition will result in recovery of the acidified soils and waters, and what would be the time-scale for a full recovery. Sulfate is the predominant anion associated with acid rain. The mobility of sulfate is of great importance in acidified waters, because sulfate facilitates base cation flow (Likens *et al.*, 1977; Christophersen and Wright, 1981). Reduction in sulfate flux reduces the flux of base cations (Wright *et al.*, 1988). Mitchell *et al.* (1991) have shown that retention of sulfate can reduce the acidification of streams. Conversely, during reversal of acidification, sulfate desorption may prolong the time for recovery. If sulfate adsorption is a reversible process, sulfate and base cations will desorb from the soil until a new steady state is reached (Reuss and Johnson, 1986). As sulfur deposition declines, sulfate outputs may exceed inputs, so sulfate and base cations will be leached from the soil to a greater extent than the reduction in acid deposition predicts.

Whole-catchment manipulation experiments began at Lake Gårdsjön in 1984 (Andersson and Olsson, 1985; Hultberg and Grennfelt, 1992). One catchment, F1, has been used as a reference catchment since 1979 for more than 10 sub-catchments, which are and were exposed to various investigations and experiments (Hultberg and Grennfelt, 1992). Several of these experiments have dealt with sulfur flux dynamics. The results from these experiments have indicated that the retention time for sulfur in Lake Gårdsjön soils is short. Sodium sulfate was spread in one area in 1985 and after 3 years almost all sulfur added (95%) could be accounted for in runoff by mass balance calculations. Further evidence for short retention times derives from clear-felling of a catchment thereby reducing dry sulfur deposition. The sulfate flux in runoff quickly reduced to a value corresponding to open field collector deposition (Hultberg and Grennfelt, 1992).

Construction of a roof over the forested catchment G1 (Figure 8.1) allowed us to test the hypothesis of short retention times and to evaluate the reversibility of sulfate retention and acidification processes in general. The roof enables researchers to control depositional variables and evaluate the potential of using isotope tracers. The deposition within the roofed catchment (G1) is controlled by a sprinkler system depositing sulfur from seawater sulfate. We can therefore use stable isotopes in sulfate as environmental tracers, because the $\delta^{34}$S value of marine sulfate is +20.7‰ and the $\delta^{18}$O value is +9.5‰.

Sulfur isotope ratios have been a useful tool for investigating sulfur cycling in catchments and soils (Krouse and Case, 1981, Fuller *et al.*, 1986a; Krouse and Tabatabai, 1986; Van Stempvoort *et al.*, 1990, 1992; Stam *et al.*, 1992; Mayer *et al.*, 1995). However, until now, very few catchment manipulations using stable sulfur and oxygen isotopes in sulfate as tracers have been performed to study sulfur dynamics in soils. Mayer (1993) and Prietzel *et al.* (1995) investigated sulfur dynamics in soil during an irrigation experiment where sulfur was added with a high $\delta^{34}$S value compared with the naturally occurring sulfur in the soil. Sulfur isotope ratios from the roofed catchment have been reported in Mörth and Torssander (1993, 1995a,b). The purpose of this paper is to present isotope monitoring data from one year before the experiment and from the first two years of experimental treatment, with the aim determining the source of sulfur in the deposition and of investigating the sulfur fluxes and retention time within the catchment.

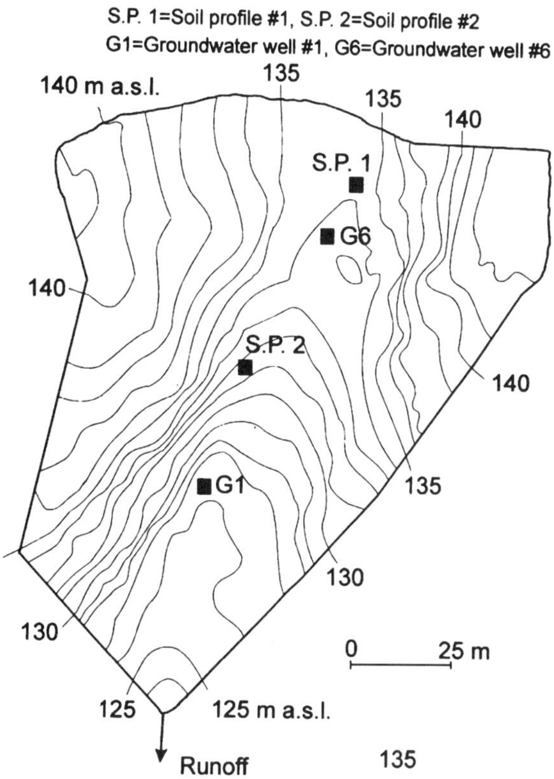


**Figure 8.1.** Map of sampling stations in the roof catchment (G1)

## Site description

Detailed descriptions on geology, topography, climate, soils, vegetation, etc. are given by Andersson *et al.* (Chapter 2), hydrology by Bishop *et al.* (Chapter 5), experimental design by Moldan *et al.* (Chapter 4), deposition by Ferm *et al.* (Chapter 3) and element fluxes by Hultberg *et al.* (Chapter 7, this volume). Only information specific to this chapter is given here.

Under the roof a sprinkler system is installed which replaces the original throughfall deposition. The sprinkler water contains no nitrogen compounds and has much lower concentrations of sulfur, 18–29 $\mu$M, compared with 125 $\mu$M (IVL, 1991) in natural throughfall in the area. The artificial deposition water is taken from a nearby lake. This water is deionized and seawater sulfate, which is isotopically distinct from the natural input, is added to the catchment in water sprinkled under the roof. Seawater sulfate has very well-defined $\delta^{34}S$ and $\delta^{18}O_{SO4}$ values of +20.7‰ (measured Kattegat water) and +9.5‰ (Claypool *et al.*, 1980) respectively. The $\delta^{34}S$ value of the sprinkler water is +19.5 and the $\delta^{18}O_{SO4}$ value is +9.3‰.

# METHODS

## Sampling

Sprinkling started under the roof in G1 in April 1991. However, sampling started in February 1990, 14 months prior to sprinkling. Samples were collected from bulk deposition (open field collectors), throughfall, runoff, groundwater and soils. Samples were also collected from the control catchment F1. Runoff was initially collected at predetermined time intervals and later by automatic sampling proportional to the runoff. All the waters were sampled once a month, although dry conditions did not always allow sampling. The samples were stored in 2 L polyethylene bottles and filtered with 0.45 $\mu$m membrane filters before processing. The soil samples were sampled prior to the experiment and 18 months after sprinkling started. The soil centrifugates were collected twice: after 1 and 2 years of experimental control.

## Analytical techniques

### *Chemical analyses*

Collected water samples were analysed for sodium, chloride and sulfate concentrations and the soil extracts for sulfate concentrations. The analyses were made by anion chromatography.

### *Extraction of sulfate in water samples*

Dissolved sulfate was converted to $BaSO_4$ for the isotopic analyses (Mörth and Torssander, 1993). An ion exchange system modified from Nehreing *et al.* (1977) and Hesslein *et al.* (1988) was used to concentrate sulfate and to ensure pure precipitates of $BaSO_4$.

### *Extraction of sulfur in soil samples*

Sulfur in the soil samples was extracted with various techniques for chemical and isotopical analyses. Dried soil samples were reacted with concentrated $HNO_3$ and Br to extract total sulfur as sulfate (Krouse and Tabatabai, 1986). Soil solutions were obtained from thawed wet soil samples by a centrifuge technique modified by Giesler and Lundström (1993). Extractions of adsorbed $SO_4$ were made on wet soil samples with 0.01 M $CaCl_2$ (Tabatabai, 1982) and 0.04 M $NaHCO_3$ (Van Stempvoort *et al.*, 1990). The dissolved sulfate obtained from all these extractions was quantitatively converted to $BaSO_4$ for isotope analyses.

### *Isotope analyses*

Sulfur dioxide used for sulfur isotope analyses was prepared by burning a mixture of 10 mg $BaSO_4$ with 200 mg $V_2O_5$ and $SiO_2$ at 950°C in a vacuum line (Yanagisawa and Sakai, 1983). Carbon dioxide prepared for oxygen isotope analyses in sulfate was made in a similar manner by mixing of 10 mg of $BaSO_4$ with 10 mg graphite and burning the mixture at 1200°C under vacuum (Rafter, 1967).

The $SO_2$(g) and $CO_2$(g) were analysed with a modernized and rebuilt Micromass 602 mass spectrometer at the Laboratory of Isotope Geology of the Naturhistoriska riksmuseet in Stockholm. The sulfur isotope composition is defined as a deviation in ‰ of the ratio $^{34}S/^{32}S$ between a sample and a standard, expressed in the conventional $\delta^{34}S$ notation relative to Cañon Diablo Troilite (CDT). The accuracy of the measurements based on standard measurements was better than ±0.2‰ for all samples but generally around ±0.1‰. The oxygen isotope composition is defined as a deviation in ‰ of the ratio $^{18}O/^{16}O$ expressed in the conventional $\delta^{18}O$ notation relative to SMOW (Standard Mean Ocean Water). The accuracy of the measurements was better than ±0.4‰.

## RESULTS

### Deposition

The sodium and chloride concentration in bulk deposition and throughfall prior to the experiment varied from 80 to 560 $\mu$M and from 80 to 600 $\mu$M respectively (Tables 8.1 and 8.2). The sulfur isotope composition of bulk deposition and throughfall varied between +4.6‰ and +12.6‰, and the weighted mean $\delta^{34}S$ values for the pre-treatment period were very similar between bulk deposition and throughfall, +7.3‰ and +7.5‰ respectively. Anthropogenic $\delta^{34}S$ values, obtained by calculating excess

**Table 8.1.** The pre-roof bulk deposition concentration of sodium, chloride and sulfate and the $\delta^{34}S$ and $\delta^{34}S$ excess value at Lake Gårdsjön. Concentrations are given in $\mu$M and isotope values are given in ‰. — = Data not available. Excess $\delta^{34}S$ values are calculated for the anthropogenic component as described in the text

| Date | $[Na^+]$ | $[Cl^-]$ | $[SO_4^{2-}]$ | $\delta^{34}S$ | $\delta^{34}S$, excess |
|---|---|---|---|---|---|
| 11 Apr 90 | — | — | — | 6.58 | — |
| 29 May 90 | 64 | 46 | 38 | — | — |
| 28 Jun 90 | 32 | 22 | 32 | 4.82 | 4.23 |
| 30 Jul 90 | 70 | 55 | 31 | 7.09 | 5.71 |
| 30 Aug 90 | 73 | 54 | 28 | 6.14 | 4.54 |
| 27 Sep 90 | 76 | 60 | 22 | 6.01 | 3.58 |
| 30 Oct 90 | 117 | 104 | 23 | 8.53 | 4.80 |
| 29 Nov 90 | 54 | 41 | 28 | 6.37 | 5.20 |
| 02 Jan 91 | 131 | 123 | 24 | 10.12 | 6.38 |
| 04 Feb 91 | 214 | 221 | 25 | 12.63 | 5.90 |
| 04 Mar 91 | 51 | 52 | 44 | 6.35 | 5.42 |
| 02 Apr 91 | 52 | 46 | 43 | 4.87 | 3.93 |
| Mean | | | | | |
| Mass flux weighted | 90 | 80 | 30 | 7.25 | 5.02 |
| Arithmetic | 85 | 75 | 31 | 7.23 | 4.97 |

**Table 8.2.** The pre-roof throughfall concentration of sodium, chloride and sulfate and the $\delta^{34}S$ and $\delta^{34}S$ excess value in catchments G1 and F1. Concentrations are given in $\mu M$ and isotope values are given in ‰. — = Data not available

| Date | $[Na^+]$ | $[Cl^-]$ | $[SO_4^{2-}]$ | $\delta^{34}S$ | $\delta^{34}S$, excess |
|---|---|---|---|---|---|
| **G1 catchment** | | | | | |
| 11 Apr 90 | — | — | — | 10.47 | — |
| 30 May 90 | 182 | 221 | 85 | 5.82 | 3.50 |
| 28 Jun 90 | 83 | 81 | 62 | 5.50 | 4.39 |
| 30 Jul 90 | 203 | 260 | 66 | 8.84 | 5.81 |
| 30 Aug 90 | 232 | 268 | 56 | 8.17 | 4.00 |
| 27 Sep 90 | 193 | 235 | 46 | 7.73 | 3.01 |
| 30 Oct 90 | 288 | 350 | 55 | 9.00 | 3.22 |
| 29 Nov 90 | 278 | 291 | 160 | 5.34 | 3.74 |
| 02 Jan 91 | 274 | 324 | 76 | 7.48 | 3.73 |
| 04 Feb 91 | — | — | — | — | — |
| 04 Mar 91 | — | — | — | — | — |
| 02 Apr 91 | — | — | — | — | — |
| Mean | | | | | |
| Weighted | 215 | 253 | 65 | 7.36 | 3.96 |
| Arithmetic | 217 | 254 | 76 | 7.59 | 3.93 |
| **F1 catchment** | | | | | |
| 11 Apr 90 | — | — | — | 10.56 | — |
| 30 May 90 | 207 | 210 | 91 | 6.38 | 4.42 |
| 28 Jun 90 | 105 | 87 | 58 | 5.88 | 4.64 |
| 30 Jul 90 | 231 | 247 | 67 | 8.38 | 5.46 |
| 30 Aug 90 | 317 | 307 | 66 | 8.43 | 4.57 |
| 27 Sep 90 | 277 | 266 | 58 | 7.87 | 3.86 |
| 30 Oct 90 | 366 | 366 | 58 | 8.63 | 2.75 |
| 29 Nov 90 | 422 | 368 | 158 | 5.55 | 3.48 |
| 02 Jan 91 | 386 | 471 | 112 | 6.25 | 2.23 |
| 04 Feb 91 | 559 | 555 | 75 | 9.91 | 3.26 |
| 04 Mar 91 | 267 | 229 | 141 | 5.63 | 4.25 |
| 02 Apr 91 | 189 | 170 | 129 | 4.66 | 3.49 |
| Mean | | | | | |
| Weighted | 312 | 316 | 86 | 6.88 | 2.95 |
| Arithmetic | 302 | 298 | 92 | 7.34 | 3.17 |

sulfur using Cl as a conservative element and then assigning the marine sulfur a $\delta^{34}S$ of +20.7‰, were a little different for bulk deposition and throughfall: the weighted mean average $\delta^{34}S$ values were +5.0‰ and +3.5‰ respectively.

Table 8.2 shows a close resemblance between throughfall $\delta^{34}S$ values from catchments G1 and F1 prior to the onset of the experiment. Seasonal variation in the $\delta^{34}S$ values in bulk deposition and throughfall has been reported from other investigations (Nriagu and Coker, 1978; Caron *et al.*, 1986; Andersson *et al.*, 1992) but cannot be identified in these data, whether or not the marine input is subtracted from $\delta^{34}S$ by excess sulfate calculation (Figure 8.2).

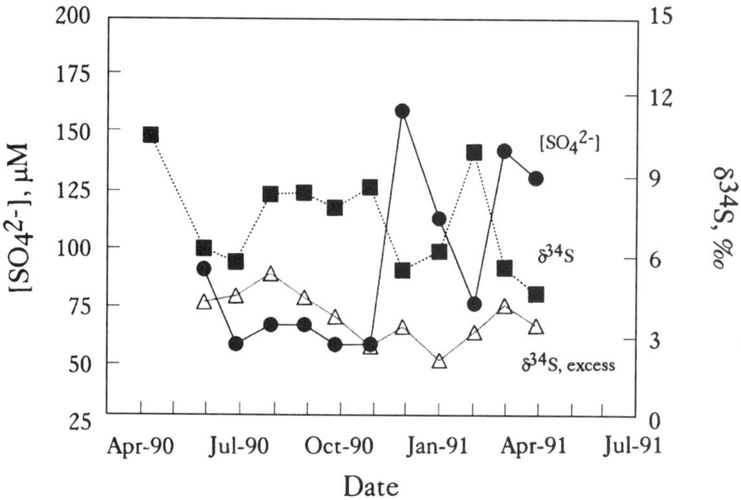


**Figure 8.2.** The throughfall concentration of sulfate and $\delta^{34}S$ value in sulfate for the pre-roof sampling period. Data from the F1 catchment (roof control catchment)

### Runoff and groundwater

The sulfur concentration and $\delta^{34}S$ in runoff in G1 are listed in Table 8.3 and shown in Figure 8.3. The sulfur concentration had decreased from the pre-treatment concentrations of around 200 $\mu M$ down to around 130 $\mu M$ after 2 years of sprinkling. At the same time, the $\delta^{34}S$ had increased by only 0.5‰ from +5.49‰ ($\pm$0.14 at 95% confidence interval for unweighted samples) to around +6.0‰. The $\delta^{34}S$ of the groundwaters does not show any statistically significant trend with time (Tables 8.4 and 8.5) although the decrease in sulfate concentration was around 15%. Groundwater well #6 had slightly higher $\delta^{34}S$ values than simultaneous samples taken from well #1 and runoff. However, in September 1992, an episodic decrease in $\delta^{34}S$ was connected to an increase in sulfate concentration (Figure 8.4).

### Soil samples

The results from the sulfate extractions of the soil profiles are listed in Tables 8.6 and 8.7. The samples from the two soil profiles sampled prior to treatment had large variation in sulfur concentration. The $\delta^{34}S$ values were almost identical between the non-specifically and specifically adsorbed sulfate (extraction with $CaCl_2$ and $NaHCO_3$ respectively). The $\delta^{34}S$ values in each soil profile were similar irrespective of depth but differ some 0.5 between the two profiles. The $\delta^{34}S$ values of the soil profiles were similar to the groundwater well in the vicinity of the soil profile. After 18 months of treatment another soil profile was sampled at location #1 (SP #1B). The $\delta^{34}S$ values increased by 0.5‰ to 2‰ (Table 8.7). Soil waters were sampled after 1 and 2 years of sprinkling from location #1 by centrifugation of soil samples (Table 8.8). The $\delta^{34}S$ values of the soil water were higher than pre-roof sampled soil profiles

**Table 8.3.** Pre-roof and roof experiment concentration of chloride and sulfate and the $\delta^{34}S$ value in runoff sulfate from catchment G1. Concentrations are given in $\mu$M and isotope values are given in ‰. Partially reproduced from Mörth and Torssander (1995a). Reprinted by permission of Kluwer Academic Publishers

| Date | $[Cl^-]$ | $[SO_4^{2-}]$ | $\delta^{34}S$ | $\delta^{34}S$, moving five mass flux weighted mean |
|---|---|---|---|---|
| 27 Feb 90 | 551 | 185 | 5.77 | — |
| 11 Apr 90 | 685 | 203 | 5.72 | — |
| 27 Jun 90 | 807 | 199 | 5.36 | 5.57 |
| 26 Sep 90 | 996 | 193 | 5.41 | 5.57 |
| 31 Oct 90 | 943 | 190 | 5.59 | 5.07 |
| 28 Nov 90 | 923 | 186 | 5.50 | 5.08 |
| 04 Jan 91 | 744 | 192 | 4.59 | 5.06 |
| 30 Jan 91 | 692 | 195 | 5.28 | 5.00 |
| 27 Feb 91 | 671 | 202 | 5.26 | 5.05 |
| 27 Mar 91 | 655 | 195 | 5.55 | 5.40 |
| ROOF ON | — | — | — | — |
| 06 May 91 | 661 | 196 | 5.78 | 5.56 |
| 27 Jun 91 | 770 | 199 | 5.33 | 5.66 |
| 10 Jul 91 | 799 | 189 | 5.72 | 5.74 |
| 24 Jul 91 | 774 | 188 | 5.99 | 5.66 |
| 13 Sep 91 | 1175 | 178 | 6.16 | 5.72 |
| 25 Sep 91 | 1006 | 186 | 5.51 | 5.69 |
| 02 Oct 91 | 1032 | 187 | 5.50 | 5.82 |
| 31 Oct 91 | 1084 | 166 | 5.68 | 5.82 |
| 20 Nov 91 | 1049 | 148 | 5.96 | 5.77 |
| 22 Nov 91 | 995 | 138 | 6.61 | 5.78 |
| 23 Dec 91 | 914 | 153 | 5.69 | 5.79 |
| 08 Jan 92 | 954 | 157 | 5.93 | 5.73 |
| 29 Jan 92 | 861 | 156 | 5.73 | 5.70 |
| 26 Feb 92 | 854 | 159 | 5.62 | 5.70 |
| 15 Apr 92 | 825 | 163 | 5.81 | 5.76 |
| 29 Apr 92 | 761 | 142 | 6.04 | 5.70 |
| 13 May 92 | 735 | 142 | 5.64 | 5.83 |
| 09 Sep 92 | 1084 | 153 | 5.27 | 5.84 |
| 18 Nov 92 | 906 | 128 | 5.93 | 5.81 |
| 02 Dec 92 | 702 | 119 | 5.86 | 5.84 |
| 16 Dec 92 | 660 | 132 | 5.83 | 5.95 |
| 30 Dec 92 | 670 | 133 | 5.98 | 5.96 |
| 27 Jan 93 | 596 | 122 | 6.12 | 6.01 |
| 10 Feb 93 | 570 | 125 | 5.91 | 6.06 |
| 01 Mar 93 | 593 | 132 | 5.99 | 6.04 |
| 10 Mar 93 | 605 | 136 | 5.90 | 5.89 |
| 21 Apr 93 | 598 | 147 | 5.57 | — |
| 05 May 93 | 666 | 149 | 5.62 | — |

and similar to post-roof sampled soil samples. The distribution of $\delta^{34}S$ values in centrifugates and SP #1B were similar, with isotopically heavier sulfate in the A-horizon than in B-horizon where the shift compared with earlier sampled soils was smaller. The soil samples were also measured for their $\delta^{34}S$ of total sulfur (Table 8.7). The $\delta^{34}S$ for all soil profiles were around +6‰ in the A-horizon and between +12

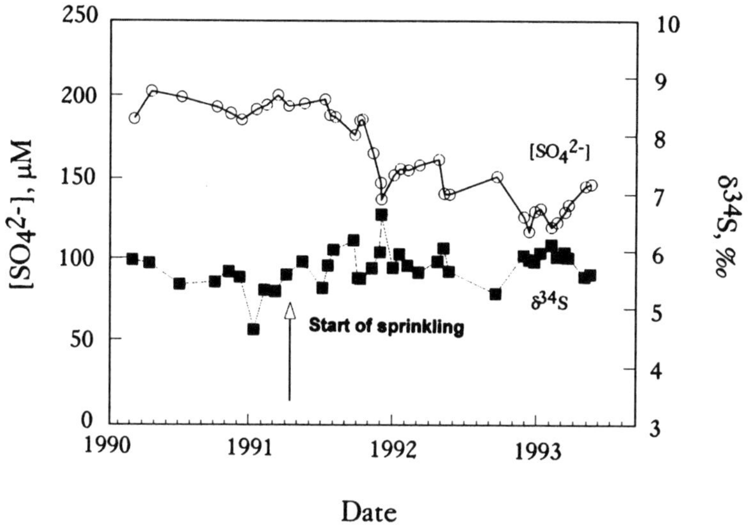


**Figure 8.3.** The sulfate concentration and the $\delta^{34}S$ value in runoff from the roof experiment catchment. Redrawn from Mörth and Torssander (1995a). Reprinted by permission of Kluwer Academic Publisher.

and +13‰ in the B-horizon. The total sulfur concentration in A- and B-horizon samples measured was an order of magnitude larger than that measured for extracted sulfate.

### Oxygen isotopes

Oxygen isotope measurements have been performed on sulfate in bulk deposition, throughfall, runoff and groundwater (Figure 8.5). The throughfall $\delta^{18}O$ values vary between +9.5 and 12‰ compared with the mean $\delta^{18}O$ in runoff of +5.4‰. No change in the runoff is expected from sprinkling as the $\delta^{18}O_{SO4}$ values are almost identical in depositional and sprinkler sulfate.

## DISCUSSION

### Deposition

Chloride and sodium in the deposition at Lake Gårdsjön originate from sea spray (Hultberg and Grennfelt, 1992; Mörth and Torssander, 1993), and are the dominant ions in the runoff from the catchments in the area. Calculation on the basis of all Na and Cl being derived from sea salt will give very similar excess sulfur values.

Sulfur deposition to the Lake Gårdsjön area is best described as a mixture between a seawater source and anthropogenic sources. This is clearly seen in Figure 8.6, where $Cl/SO_4$ ratios are plotted against $\delta^{34}S$ values in deposition. The regression line

**Table 8.4.** Pre-roof and roof experiment concentration of chloride and sulfate and the $\delta^{34}S$ value in groundwater well #1 from catchment G1. Concentrations are given in $\mu$M and isotope values are given in ‰. Partially reproduced from Mörth and Torssander (1995a). Reprinted by permission of Kluwer Academic Publishers

| Date | $[Cl^-]$ | $[SO_4^{2-}]$ | $\delta^{34}S$ | $\delta^{34}S$, moving five mean |
|---|---|---|---|---|
| 27 Feb 90 | 506 | 182 | 5.85 | — |
| 11 Apr 90 | 529 | 176 | 5.62 | — |
| 02 Jul 90 | 599 | 171 | 5.29 | 5.56 |
| 29 Sep 90 | 689 | 168 | 5.71 | 5.49 |
| 31 Oct 90 | 779 | 163 | 5.31 | 5.43 |
| 28 Nov 90 | 757 | 161 | 5.54 | 5.39 |
| 04 Jan 91 | 699 | 187 | 5.31 | 5.36 |
| 01 Feb 91 | 642 | 182 | 5.07 | 5.33 |
| 27 Feb 91 | 632 | 188 | 5.56 | 5.30 |
| 27 Mar 91 | 615 | 186 | 5.15 | 5.36 |
| ROOF ON | — | — | — | — |
| 06 May 91 | 632 | 186 | 5.42 | 5.47 |
| 12 Jun 91 | 653 | 190 | 5.62 | 5.54 |
| 17 Jul 91 | 662 | 178 | 5.60 | 5.69 |
| 28 Aug 91 | 710 | 191 | 5.92 | 5.73 |
| 18 Sep 91 | 1135 | 181 | 5.88 | 5.67 |
| 02 Oct 91 | 1037 | 170 | 5.62 | 5.80 |
| 30 Oct 91 | 1061 | 157 | 5.32 | 5.75 |
| 27 Nov 91 | 1055 | 137 | 6.24 | 5.67 |
| 31 Dec 91 | 955 | 152 | 5.67 | 5.64 |
| 29 Jan 92 | 844 | 154 | 5.48 | 5.67 |
| 12 Feb 92 | 838 | 158 | 5.49 | 5.48 |
| 27 Feb 92 | 846 | 162 | 5.48 | 5.46 |
| 25 Mar 92 | 826 | 161 | 5.30 | 5.47 |
| 29 Apr 92 | 783 | 149 | 5.56 | 5.33 |
| 27 May 92 | 764 | 158 | 5.53 | 5.02 |
| 02 Sep 92 | 1181 | 177 | 4.78 | 5.05 |
| 30 Sep 92 | 838 | 170 | 3.94 | 5.01 |
| 04 Nov 92 | 999 | 140 | 5.42 | 4.99 |
| 25 Nov 92 | 794 | 149 | 5.36 | 5.11 |
| 29 Dec 92 | 688 | 144 | 5.44 | 5.42 |
| 27 Jan 93 | 626 | 143 | 5.39 | 5.48 |
| 26 Feb 93 | 608 | 142 | 5.49 | 5.50 |
| 10 Mar 93 | 622 | 141 | 5.70 | 5.48 |
| 30 Mar 93 | 631 | 142 | 5.49 | — |
| 27 Apr 93 | 675 | 156 | 5.32 | — |

indicates a $\delta^{34}S$ value of +18.85‰ for the $Cl/SO_4$ ratio in sea water, compared with the measured value of 19.32. The anthropogenic end member is at the intercept with a $\delta^{34}S$ value around +4.3‰, in relatively good agreement with the calculated mean excess $\delta^{34}S$ values (Tables 8.1 and 8.2). Interestingly, excess sulfate calculations show that around 75% of the deposition is of anthropogenic origin for the analysed time period.

**Table 8.5.** Pre-roof and roof concentration of chloride and sulfate and the $\delta^{34}S$ value in groundwater well #6. Concentrations are given in $\mu M$ and isotope values are given in ‰. Partially reproduced from Mörth and Torssander (1995a). Reprinted by permission of Kluwer Academic Publishers

| Date | $[Cl^-]$ | $[SO_4^{2-}]$ | $\delta^{34}S$ | $\delta^{34}S$, moving five mean |
|---|---|---|---|---|
| 27 Feb 90 | 647 | 224 | 6.08 | — |
| 11 Apr 90 | 632 | 212 | 5.84 | 5.86 |
| 04 Jan 91 | 877 | 219 | 5.65 | 5.74 |
| ROOF-ON | — | — | — | — |
| 13 Sep 91 | 1191 | 173 | 5.72 | 5.72 |
| 30 Oct 91 | 1320 | 184 | 5.78 | 5.90 |
| 27 Nov 91 | 1081 | 169 | 6.19 | 5.97 |
| 29 Apr 92 | 730 | 156 | 5.95 | 5.99 |
| 02 Sep 92 | 1162 | 140 | 5.82 | 5.93 |
| 04 Nov 92 | 1122 | 137 | 6.01 | 5.94 |
| 25 Nov 92 | 885 | 142 | 5.98 | 6.02 |
| 27 Jan 93 | 580 | 131 | 6.06 | — |

The small difference between the $\delta^{34}S$ values of bulk deposition and throughfall indicates that the internal circulation of S in these trees is small. If internal circulation was important, excess sulfur in throughfall would show $\delta^{34}S$ values closer to those of the soil reservoirs. This is consistent with earlier work, for example Lindberg and Garten (1988) showed that less than 10% of the sulfur in throughfall was internally circulated by labelling soil sulfur with $^{35}S$. The finding is consistent with other studies in Sweden at Lake Mjösjön (Andersson *et al.*, 1992).

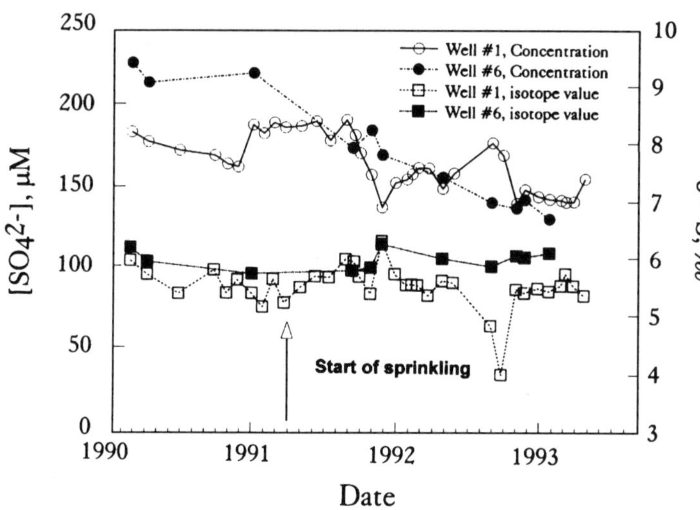


**Figure 8.4.** The sulfate concentration and the $\delta^{34}S$ value in the two groundwater wells G1 and G6 (from the roof experiment catchment). Redrawn from Mörth and Torssander (1995a). Reprinted by permission of Kluwer Academic Publishers

**Table 8.6.** Soil concentration of sulfur from two different soil profiles, one profile was taken pre-roof and post-roof and one only pre-roof. Concentrations are given in $\mu$mol S 100 $g^{-1}$ dry soil. Soil depth is given from the forest floor. — = Data not available

| Sample | Depth (cm) | Total sulfur | Inorganic sulfur, $NaHCO_3$-extractable | Inorganic sulfur, $CaCl_2$-extractable |
|---|---|---|---|---|
| Profile #1, pre-roof (1 June, 1990) | | | | |
| 1 | 5 | 819 | 37 | 21 |
| 2 | 22.5 | 2106 | 306 | 49 |
| 3 | 43 | 1676 | 95 | 25 |
| 4 | 72 | 1101 | 112 | 20 |
| Profile #1, post-roof (20 Oct 1992) | | | | |
| 1 | 3 | 4313 | 114 | — |
| 2 | 7.5 | 447 | 27 | — |
| 3 | 14 | 153 | — | — |
| 4 | 22.5 | 911 | 191 | — |
| 5 | 35 | 2402 | 303 | — |
| Profile #2, pre-roof (1 June 1990) | | | | |
| 1 | 3 | 4651 | 155 | 89 |
| 2 | 18 | 3748 | 519 | 75 |
| 3 | 35 | 4285 | 495 | 76 |
| 4 | 68.5 | 2653 | 193 | 48 |

**Table 8.7.** The $\delta^{34}S$ value from two different soil profiles, one profile pre-roof and post-roof and one profile only pre-roof. The $\delta^{34}S$ value is reported in ‰. Soil depth is given from the forest floor. — = Data not available

| Sample | Depth (cm) | Total sulfur | Inorganic sulfur, $NaHCO_3$-extractable | Inorganic sulfur, $CaCl_2$-extractable |
|---|---|---|---|---|
| Profile #1, pre-roof (1 June 1990) | | | | |
| 1 | 5 | 7.06 | — | — |
| 2 | 22.5 | 12.41 | 6.35 | 6.39 |
| 3 | 43 | 11.35 | 6.18 | 6.33 |
| 4 | 72 | 12.61 | 5.69 | 6.49 |
| Profile #1, post-roof (28 Oct 1992) | | | | |
| 1 | 3 | 5.39 | 8.15 | — |
| 2 | 7.5 | 7.97 | 6.61 | — |
| 3 | 14 | — | — | — |
| 4 | 22.5 | 11.30 | 8.39 | — |
| 5 | 35 | 12.84 | 6.63 | — |
| Profile #2, pre-roof (1 June 1990) | | | | |
| 1 | 3 | 5.97 | — | — |
| 2 | 18 | 11.94 | 5.48 | 5.59 |
| 3 | 35 | 13.56 | 5.09 | 5.46 |
| 4 | 68.5 | 10.48 | 5.52 | 5.40 |

**Table 8.8.** The $\delta^{34}S$ value in soluble sulfate as shown by soil centrifugation. Concentrations are given in $\mu M$ and isotope values are given in ‰. Soil depth is given from the forest floor

| Sample | Depth (cm) | $\delta^{34}S$ |
|---|---|---|
| Soluble sulfate, centrifugated from soil (2 March 1992) | | |
| 1 | 5 | 6.94 |
| 2 | 10.5 | 8.37 |
| 3 | 22.5 | 6.42 |
| Soluble sulfate, centrifugated from soil (20 May 1993) | | |
| 1 | 0 | 7.52 |
| 2 | 5 | 7.36 |
| 3 | 10.5 | 8.36 |
| 4 | 15 | 8.98 |
| 5 | 22.5 | 6.99 |
| 6 | 35 | 6.90 |
| 7 | 43 | 6.71 |

It has been suggested that seasonal variation in $\delta^{34}S$ values is a result of different reaction pathways in the atmosphere during winter and summer (Saltzman *et al.*, 1983). A homogeneous gas phase reaction will dominate during summer months giving lower $\delta^{34}S$ values in the formed sulfate due to a kinetic isotope effect. A heterogeneous reaction will dominate during winter and give a temperature-dependent equilibrium fractionation (Eriksen, 1972), with a fractionation factor $\alpha = 1.0109$ at 25°C, when dissolving $SO_2$ in water and a small kinetic isotope effect

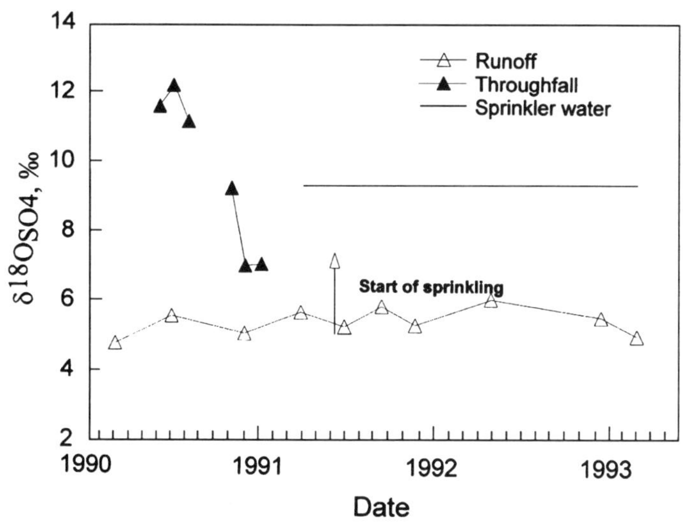


**Figure 8.5.** The $\delta^{18}O_{SO4}$ value in throughfall, sprinkler water and runoff (from the roof experiment catchment)

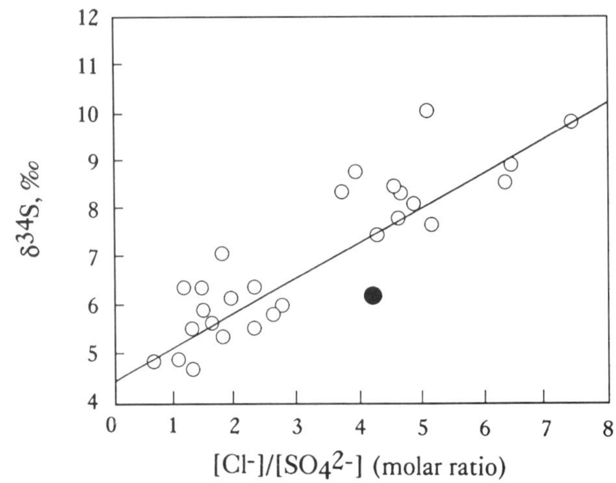


**Figure 8.6.** The $\delta^{34}S$ value plotted against the chloride/sulfate ratio in throughfall and bulk deposition. Data from the pre-roof period

(Saltzman *et al.*, 1983), $\alpha = 0.996$, when oxidizing sulfite to sulfate in water and higher $\delta^{34}S$ values. An explanation of the lack of a seasonal variation at Lake Gårdsjön could be that a large proportion of the sulfur is oxidized to sulfate before it is deposited. Nearly all deposited sulfate at Lake Gårdsjön is long-range transported. In that case the seasonal variation in $\delta^{34}S$ will be small, whereas near the emitting source seasonal variation is expected among the sulfur compounds as a small fraction of sulfur dioxide or sulfite will be oxidized.

There are many studies that support the use of throughfall data as an estimate of total deposition for sulfur in forested catchments (Lindberg and Garten, 1988; Hultberg and Grennfelt, 1992; Mitchell *et al.*, 1992). Hultberg and Grennfelt (1992) showed that throughfall data from a 10 year time span are needed to balance the amount of sulfate deposited and the amount of sulfate in runoff at Lake Gårdsjön. The annual variation was large, showing both a negative and positive sulfur mass balance in individual years. The 14 month pre-treatment sampling period showed that sulfur deposited to the catchment (bulk deposition and throughfall) had a significantly higher $\delta^{34}S$ value than the runoff, about 7.5‰ compared with 5.5‰. Sulfate deposition has decreased by 25% since 1970, due to reduced emissions of $SO_2$ (Hedin *et al.*, 1994). This decrease in depositional sulfates indicates that the $\delta^{34}S$ values of depositional sulfate could be presently higher than before, as only the proportion of anthropogenic sulfate in total sulfate deposition has decreased in concentration and its $\delta^{34}S$ values are lower than those of depositional sulfate $\delta^{34}S$ values. Mörth and Torssander (1993) calculated excess sulfate in throughfall over the time period 1980–90 from Hultberg and Grennfelt's (1992) Lake Gårdsjön measurements. They found that 10 year average excess sulfate in throughfall amounts to 88% of total sulfate, which corresponds to a $\delta^{34}S$ value of around +6‰. Thus, the higher sulfur deposition in earlier decades corresponds to an $\delta^{34}S$ value

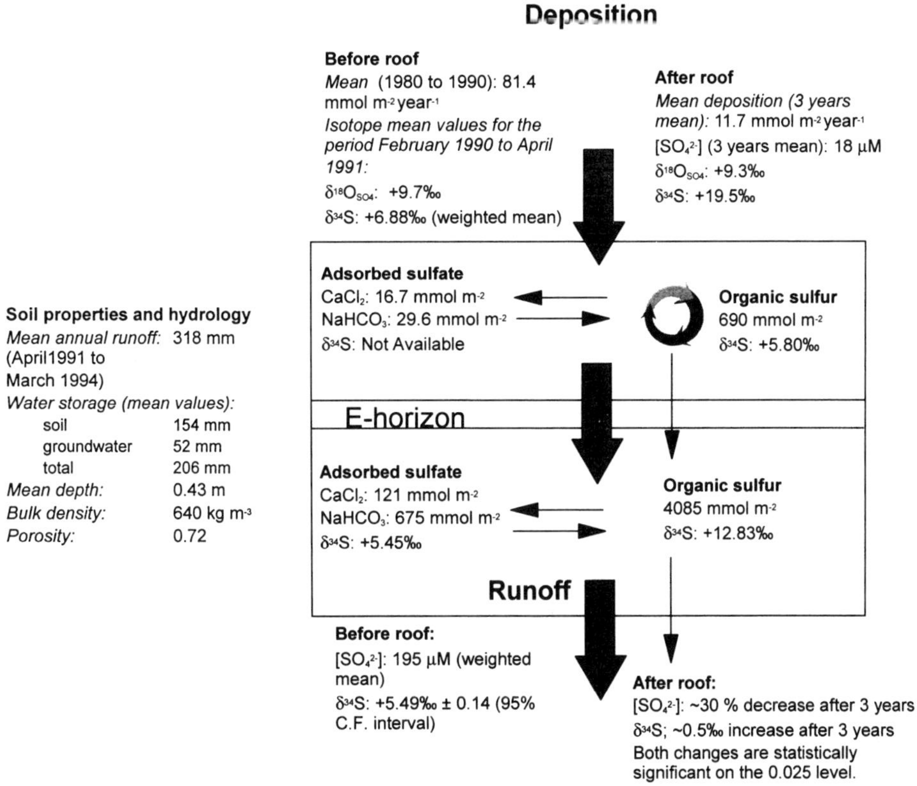


**Figure 8.7.** Summary of sulfur pool sizes, sulfur fluxes and their $\delta^{34}S$ values. Note that only about 4% of the adsorbed sulfate is available in the upper soil (A-horizon). To recover the adsorbed sulfate the soil was extracted with two different solutions, $CaCl_2$ and $NaHCO_3$. Data from Hultberg and Grennfelt (1992), Mörth (1995), Mörth and Torssander (1995a), Mörth and Torssander (1995b) and Nyberg (1995)

between +5‰ and +6‰ confirming that deposition over the last few decades is the source of sulfur in the catchment runoff.

## Sulfur pools in the soil

Runoff water and groundwater showed little variation over time in $\delta^{34}S$ values before the experiment started. These $\delta^{34}S$ values were similar to the soil profile $\delta^{34}S$ values for inorganic sulfur which suggests that the inorganic sulfur pool has a homogeneous $\delta^{34}S$ value of about +5.5‰. Sulfur in the soil could be present as inorganic sulfide, inorganic sulfate or organic sulfur. Sulfides are not present in the G1 soil and measurements show the organic sulfur pool in G1 to be about 10 times the size of the inorganic sulfur pool (Table 8.6). The organic pool is much larger than the inorganic pool in most soils (David *et al.*, 1983; Mitchell *et al.*, 1989). In the G1 soil the organic sulfur reservoir in just the A-horizon is about the same size as the inorganic pool in the whole soil profile. The $\delta^{34}S$ value of the soil sulfur as a whole will therefore reflect that of the organic pool.

### Dissimilatory sulfate reduction

Dissimilatory sulfate reduction by bacteria is the only process in soils that can fractionate sulfur isotopes at Earth surface conditions to any large degree (Van Stempvoort *et al.*, 1990; Krouse and Grinenko, 1991). Sulfur isotope studies have identified this process in terrestrial environments; in peat areas (Morgan, 1992), in watersheds (Hesslein *et al.*, 1988; Mörth, 1995) as well as during low discharge in a small creek with no peat, mire or bog surroundings (Andersson *et al.*, 1992). An indication of dissimilatory sulfate reduction, especially during low discharge periods, is a decrease in sulfate concentration in combination with an increase in the $\delta^{34}S$ values, because dissimilatory sulfate reduction enriches $^{32}S$ in the formed $H_2S$ that commonly precipitates as sulfides. No such changes have been observed in the G1 catchment except during high discharge periods when the sprinklers were run to simulate rainstorms. Up to 25% of the runoff has consisted of sprinkler water during single sprinkler events (Nyberg *et al.*, 1993). However, oxidation of reduced sulfur produced by bacteria will lead to lighter sulfur and a high sulfur content in the water. This oxidation has been identified in other Swedish catchments after dry summer periods (Mörth, 1995) and was shown to occur in the Storgama catchment in Norway (Christophersen *et al.*, 1983/84).

### Sorption processes

The similar $\delta^{34}S$ values of the extracted sulfate from the soil profiles (irrespective of method) and the groundwater suggest that there is exchange between the soluble sulfate and the adsorbed sulfate. These results also confirm that sulfur isotope fractionation is negligible during sorption processes in soils, as also shown by Fuller *et al.* (1986a) and Van Stempvoort *et al.* (1990). The latter authors also showed that oxygen isotopes in sulfate do not change isotopic composition during adsorption/desorption processes in soils. Sulfate in soils is adsorbed on Al- and Fe-oxides or hydroxides (Hingston *et al.*, 1968; Rajan, 1978). Kinetic studies of the adsorption and desorption rate have shown the sorption processes to be rapid (Rajan, 1979; Singh, 1984; Houle and Carignan, 1995). High recovery of adsorbed sulfate during laboratory desorption studies has been found (Alewell and Matzner, 1993; Karltun 1994). As both the non-specifically and the specifically adsorbed sulfate in the G1 soil samples must have had the same isotopic composition, it is suggested that all the adsorbed sulfate is available for desorption. Thus, the decrease in sulfur concentration in runoff and the small increase in $\delta^{34}S$ suggest that the sprinkled seawater sulfate is mixed with a large sulfate reservoir in the soil (Mörth and Torssander, 1995a). The small shift in $\delta^{34}S$ values in runoff is predicted by the fact that the pool of adsorbed sulfate is very large compared with the relatively small input of sulfate from the sprinkler system.

The sulfur isotope composition of groundwater and runoff in the roofed catchment reflects the isotopic composition of deposited sulfur. Short-term sulfur isotope variations in throughfall and sprinkler water are not detectable in runoff because of mixing with a large reservoir of adsorbed sulfur. Variation and response time in runoff to changes in the deposited $\delta^{34}S$ values must reflect transfers and mixing among different sulfur pools in the catchment. However, a more rapid response time and less smoothing of the throughfall signature is reflected by the slightly higher $\delta^{34}S$ in groundwater well #1 and soil profile 1A because the site is mostly a recharge area. This has been even more accentuated by the up to 2‰ increase in $\delta^{34}S$ values in soil profile 1B samples and the centrifugates after almost 2 years of sprinkling.

## Organic sulfur cycling

By organic sulfur cycling we mean immobilization by assimilatory sulfate reduction followed by mineralization, which is oxidation to sulfate. Little or no difference in $\delta^{34}S$ values between biomass, organic and inorganic sulfur in soils were found by Kusakabe *et al.* (1976) Consequently, incorporation of sulfur into biomass as organic sulfur causes small or no sulfur isotope fractionation. This result is supported by the studies of Krouse and Tabatabai (1986) and Van Steempvoort *et al.* (1992). Likewise, mineralization causes little or no fractionation among sulfur isotopes (e.g. Krouse and Grinenko, 1991).

Organic sulfur cycling has been shown to be important in the upper soil horizons, but less important in the B-horizon (Johnson and Lindberg, 1992). The large organic sulfur pool in the B-horizon at Lake Gårdsjön implies that organic sulfur is accumulating in the B-horizon. The high $\delta^{34}S$ values of around +13‰ imply that this accumulation occurred a long time ago since no heavy sulfur source is present in the area apart from sea spray. The organic sulfur pool in the A-horizon has a $\delta^{34}S$ value (+6‰) identical to the inorganic sulfate, which reflects atmospheric deposition during later part of the 20th century. This implies that most of the organic sulfur in the B-horizon has an origin from sea spray and accumulated earlier, before the $\delta^{34}S$ values of deposition and A-horizon organic sulfur changed.

Substantial amounts of sulfur can be cycled through the upper soil horizons (Fuller *et al.*, 1986b). Sulfate has been observed to be quickly transformed to organic sulfur in a study by Schindler *et al.* (1986). Only 4% of the inorganic sulfur pool is absorbed in the A-horizon. The extracted sulfate and soil water sulfate in samples collected about 2 years after the onset of the experiment had isotopic compositions that alternatively were heavy (from sprinklers) and light (from organic S mineralization). It is implied that these low $\delta^{34}S$ values could arise from mineralization of organic sulfur and that transformation of deposited sulfate to organic sulfur is not complete (high $\delta^{34}S$ values).

Oxygen isotopes in sulfate have been used not only as evidence of organic sulfur cycling but to determine the extent of organic sulfur cycling. This is because oxygen isotope exchange in sulfate is a slow process at low temperatures, even if the pH is low (Lloyd, 1967; Chiba and Sakai, 1985) and the fractionation is negligible during sorption processes (Van Stempvoort *et al.*, 1990). Thus, the change in $\delta^{18}O_{SO4}$ values between bulk deposition, throughfall or sprinkler water and runoff to around 5‰ lighter values is caused by sulfate formed within the catchment. Oxygen isotopes of sulfate in soils can be modelled as a mix between deposited sulfate and a sulfate reservoir that has undergone transformations in the soil, i.e. secondary sulfate (Van Stempvoort *et al.*, 1992). The oxygen isotope shift has been used to calculate the amount of organic sulfur oxidized (Caron *et al.*, 1986; Gélineau *et al.*, 1989) from a formula originally used to calculate oxidation of mine waste given by Taylor *et al.* (1984) and Van Everdingen and Krouse (1985). The shift in the $\delta^{18}O$ values is dependent on the oxygen source, whether it is from water or air. Depending on reaction pathway the shift may vary (Toran and Harris, 1989). Assuming that assimilatory reduction does not cause any oxygen isotope fractionation and that the origin of the oxygen is the water, a calculation shows that less than one-third of the sulfate in runoff is secondary formed within the soil profile (Mörth and Torssander, 1995a). Thus, sulfate is partly transferred through the organic sulfur pool in the A-horizon, but this does not mean that a net mineralization of sulfur is occurring in the catchment. However, recently Mayer *et al.* (1995) have pointed out that the calculated oxygen percentage by this method is unreliable, because the oxygen's origin will be dependent on the reaction pathway.

In a soil study by Fitzgerald and Autry (1992) it was shown that increased concentration of sulfate in solution increased the immobilization. In a clear-cutting experiment at Hubbard Brook, the organic sulfur pool remained constant (Mitchell *et al.*, 1989), although Hultberg and Grennfelt (1992) have shown that clear-cutting reduces sulfur deposition drastically.

At Risdalsheia in Norway a roof catchment experiment has been undertaken where net mineralization was suggested to have occurred because the sulfate pool could not account for the total sulfur mass flux after 8 years of experimental treatment (Wright *et al.*, 1993). Thus, it is unclear if reductions in sulfate deposition will result in net mineralization of organic sulfur. To evaluate the importance of mineralization at G1, a longer time series or a sulfur tracer experiment are needed, because at present we do not have any indication of net mineralization.

**Acidification reversal**

Mass balance calculations after one year of sprinkling suggest that 25% of the sulfate in runoff had its origin in the sprinkled sulfate, but the $\delta^{34}S$ does not support this hypothesis (Mörth and Torssander, 1993). If the sulfate in runoff is a mixture of sprinkler water sulfate and that already fixed within the catchment, almost all sulfate in runoff is derived from the fixed sulfate. It is implied that desorption of sulfate (dependent on adsorption isotherms) may prolong acidifying processes by enhanced base cation leaching from soils for a considerable time. If net mineralization of organic sulfur occurs then acidification processes are even more prolonged, accentuated by the size of the A-horizon active organic sulfur pool.

How should we define an acidification reversal? In the soil water and the runoff we need to reach pre-acidification element concentrations before the reversal is complete. The decrease in sulfate concentrations down to pre-acidification values is one of the keys to understanding the reversal in time and space. If the catchment was in a steady state (i.e. inputs of sulfate in deposition balance outputs in runoff) under a high sulfur load, then to reach a steady state under low sulfur load, stored sulfur will have to be leached from the catchment. The sulfur concentration in the output water is controlled by adsorption isotherms, and the sulfur isotope composition by sorption processes and to some extent by organic sulfur cycling. Modelling of the reversal of acidification in G1 runoff, assuming that sulfate output is controlled purely by desorption of inorganic sulfate, predicted that a new steady state was reached in 25–45 years depending on whether Freundlich or Langmuir isotherms were used (Mörth, 1995). If the organic sulfur cycling is not negligible, however, a new steady-state will take much longer to attain. One of the strengths of the sulfur isotope approach is that mass fluxes and transformations of the soil sulfur pool can be determined and compared with the changes predicted by modelling.

## CONCLUSIONS

In Figure 8.7, we have summarized some of the data we have used in this paper. The conclusions are as follows:

1. Two sources of sulfur are identified in the deposition at Lake Gårdsjön: sea spray and a dominant anthropogenic source. Most of the sulfur deposition is long-range transported, which explains why no seasonal variation in $\delta^{34}S$ has been found. The origin of sulfur in the area is the deposition.
2. Sulfur isotope fractionation between sulfate in runoff and the adsorbed pool is non-existent. The similar $\delta^{34}S$ values in dissolved and adsorbed sulfate imply that sulfate exchanges readily and that sulfate adsorption is reversible.
3. The $\delta^{34}S$ values of sulfate in runoff and groundwater reflect the atmospheric sulfur deposition and mixing with the inorganic sulfate pool in the area. This pool is sufficiently large to control the sulfur isotope composition of the runoff. The $\delta^{34}S$ values change very slowly through time. Thus, sorption processes in the B-horizon are the main flux mechanism.
4. In the A-horizon less than one-third of sulfate transfer occurs through the organic sulfur pool, while in the B-horizon organic sulfur is accumulating. Thus, mineralization of organic sulfur in the B-horizon is not currently important as a source of sulfate in runoff.
5. Acidification processes continue for 25–45 years if the organic sulfur cycling is assumed to be negligible, because adsorption of sulfate is a reversible process. Mineralization of organic sulfur could prolong acidification processes for centuries.
6. Sulfur isotope ratios are an excellent tool for determination of mass fluxes and transformation in and out of sulfur pools. Used in connection with soil and water concentration, predictions of changes between reservoirs can be made through time.

## ACKNOWLEDGEMENTS

We are grateful to Dr R. Giesler for supplying the centrifugates and to K. Hajnal for preparation of the isotope samples. Financial support was obtained from Swedish Natural Science Research Council.

## REFERENCES

Alewell, C. and Matzner, E. (1993) Reversibility of soil solution acidity and of sulfate retention in acid forest soils. *Water, Air, and Soil Pollution*, **71**, 155–165.

Andersson, F. and Olsson, B. (eds) (1985) *Lake Gårdsjön, An Acid Forest Lake and its Catchment. Ecological Bulletins (Stockholm)*, **37**.

Andersson, P., Torssander, P. and Ingri, J. (1992) Sulphur isotope ratios and oxygen isotopes in water from a small watershed in Central Sweden. *Hydrobiology*, **235/236**, 205–217.

Caron, F., Tessier, A., Kramer, J. R., Schwarcz, H. P. and Rees, C. E. (1986) Sulfur and oxygen isotopes of sulfate in precipitation and lake water, Quebec, Canada. *Applied Geochemistry*, **1**, 601–606.

Chiba, H. and Sakai, H. (1985) Oxygen isotope exchange rate between dissolved sulfate and water at hydrothermal temperatures. *Geochimica et Cosmochimica Acta*, **29**, 993–1000.

Christophersen, N. and Wright, R. F. (1981) Sulfate budget and a model for sulfate concentrations in stream water at Birkenes, a small forested catchment in southernmost Norway. *Water Resources Research*, **17**, 377–389.

Christophersen, N., Dymbe, L. H., Johannssen, M. and Seip, H. M. (1983/84) A model for sulphate in streamwater at Storgama, Southern Norway. *Ecological Modelling*, **21**, 35–61.

Claypool, G. E., Holser, W. T., Kaplan, I. R., Sakai, H. and Zak, I. (1980) The age curves of sulfur and oxygen isotopes in marine sulfate and their mutual interpretation. *Chemical Geology*, **28**, 199–260.

David, M. B., Schindler, S. C., Mitchell, M. J. and Strick, J. E. (1983) Importance of organic and inorganic sulfur to mineralization processes in a forest soil. *Soil Biology and Biochemistry*, **15**, 671–677.

Eriksen, T. E. (1972) Sulfur isotope effects. I. The isotopic exchange coefficient for the sulfur isotopes $^{34}S$–$^{32}S$ in the system $SO_2(g)$–$HSO_3^-(aq)$ at 25, 35 and 45°C. *Acta Chemica Scandinavia*, **26**, 573–580.

Fitzgerald, J. W. and Autry, A. R. (1992) Organic sulfur dynamics including mineralization and immobilization of various organic fractions. In: D. W. Johnson and S. E. Lindberg (ed.), *Atmospheric Deposition and Forest Nutrient Cycling*. Ecological Studies 91, Springer-Verlag, Heidelberg, pp. 118–129.

Fuller, R. D., Mitchell, M. J., Krouse, H. R., Wyskowski, B. J. and Driscoll, C. T. (1986a) Stable sulfur isotope ratios as a tool for interpreting ecosystem sulfur dynamics. *Water, Air, and Soil Pollution*, **28**, 163–171.

Fuller, R. D., Driscoll, C. T., Schindler, S. C. and Mitchell, M. J. (1986b) A simulation model of sulfur transformations in forested spodosols. *Biogeochemistry*, **2**, 313–328.

Gélineau, M., Carignan, R. and Tessier, A. (1989) Study of the transit of sulfate in a Canadian shield lake watershed with stable oxygen isotope ratios. *Applied Geochemistry*, **4**, 195–201.

Giesler, R. and Lundström, U. (1993) Soil solution chemistry: effects of bulking soil samples. *Soil Science Society of America Journal*, **57**, 1283–1288.

Hedin, L. O., Granat, L., Likens, G. E., Buishand, T. A., Galloway, J. N., Butler, T. J. and Rodhe, H. (1994) Steep declines in atmospheric base cations in regions of Europe and North America. *Nature*, **367**, 351–354.

Hesslein, R. H., Capel, M. J. and Fox, D. E. (1988) Sulfur isotopes in sulfate in the inputs and outputs of a Canadian shield watershed. *Biogeochemistry*, **5**, 263–273.

Hingston, F. J., Atkinson, R. J., Posner, A. M. and Quirk, J. P. (1968) Specific adsorption of anions on goethite. *Transactions International Congress of Soil Science*, **1**, 669–678.

Houle, D. and Carignan, R. (1995) Role of $SO_4$ adsorption and desorption in the long-term S budget of a coniferous catchment on the Canadian shield. *Biogeochemistry*, **28**, 161–182.

Hultberg, H. and Grennfelt, P. (1992) Sulphur and sea salt deposition as reflected by throughfall and runoff chemistry in forested catchments. *Environmental Pollution*, **75**, 215–222.

IVL (1991) *Miljöatlas*. Resultat från IVLs undersökningar i miljön 1991, Institutet för Vatten och Luftvårdsforskning, Göteborg, Sweden. (In Swedish.)

Johnson, D. W. and Lindberg, S. E. (1992) *Atmospheric Deposition and Forest Nutrient Cycling*. Ecological studies 91. Springer-Verlag, Heidelberg.

Karltun, E. (1994) The determination of adsorbed sulfate with isotopic dilution of $^{35}S$ sulfur compared with calcium dihydrogen phosphate extraction. *Communications in Soil Science and Plant Analysis*, **25**, 207–214.

Krouse, H. R. and Case, J. W. (1981) Sulphur isotope ratios in water, air, soil and vegetation near Tepee Creek gas plant, Alberta. *Water, Air, and Soil Pollution*, **15**, 11–28.

Krouse, H. R. and Grinenko, V. A. (1991) *Stable Isotopes Natural and Anthropogenic Sulphur in the Environment. Scope 43.* John Wiley and Sons Ltd, Chichester.

Krouse, H. R. and Tabatabai, M. A. (1986) Stable sulfur isotopes. In: M. A. Tabatabai (ed.), *Sulfur in Agriculture.* Amer. Soc. Agron.—Crop Sci. Soc. Amer.—Soil Sci. Soc. Amer., Madison, Wisconsin.

Kusakabe, M., Rafter, T. A., Stout, J. D. and Collie, T. W. (1976) Sulfur isotope variations in nature. Part 12. Isotopes of sulfur extracted from some plants soils, and related materials. *New Zealand Journal of Science*, **19**, 433–440.

Likens, G. E., Bormann, F. H., Pierce, R. S., Eaton, J. S. and Johnson, N. M. (1977) *Biogeochemistry of a Forested Ecosystem.* Springer-Verlag, New York.

Lindberg, S. E. and Garten, Jr. C. T. (1988) Sources of sulfur in forest canopy throughfall. *Nature*, **336**, 148–151.

Lloyd, R. M. (1967) Oxygen-18 composition of oceanic sulfate. *Science*, **154**, 1228–1231.

Mayer, B. (1993) *Untersuchungen zur Isotopengeochemie des Schwefels in Waldbödeln und neu gebildetem Grundwasser unter Wald.* GSF-Bericht, 2/93. Neuherberg, Germany.

Mayer, B., Fritz, P., Prietzel, J. and Krouse, H. R. (1995) The use of stable sulfur and oxygen isotope ratios for interpreting the mobility of sulfate in aerobic forests soils. *Applied Geochemistry*, **10**, 161–173.

Mitchell, M. J., Driscoll, C. T., Fuller, R. D., David, M. B. and Likens, G. E. (1989) Effect of whole-tree harvesting on the sulfur dynamics of a forest soil. *Soil Science Society of America Journal*, **53**, 933–940.

Mitchell, M. J., David, M. B. and Harrison, R.B. (1991) Sulfur cycling in forest ecosystems. In: R. W. Howarth and J. W. B. Stewart (eds), *Sulfur Cycling in Terrestrial Systems and Wetlands.* John Wiley and Sons Ltd, New York.

Mitchell, M. J., Harrison, R. B., Fitzgerald, J. W., Johnson, D. W., Lindberg, S. E., Zhang, Y. and Autry A. (1992) In: D. W. Johnson and S. E. Lindberg (eds), *Atmospheric Deposition and Forest Nutrient Cycling*, Ecological Studies 91. Springer-Verlag, Berlin.

Morgan, M. D. (1992) Sulfur pool sizes and stable isotope ratios in HUMEX peat before and immediately after the onset of acidification. *Environmental International*, **18**, 545–553.

Mörth, C.-M. (1995) Sulfur isotopes used as a tracer of acidification reversal, dispersion of acid mine drainage and sulfur dynamics in small and large catchments. *Meddelanden från Stockholms Universitets Institution för Geologi och Geokemi*, no. 290.

Mörth, C.-M. and Torssander, P. (1993) Sulfur isotopes in a roof covered forested catchment at Lake Gårdsjön, Western Sweden. *IAH Selected Papers*, **4**, 63–75.

Mörth, C.-M. and Torssander, P. (1995a) Sulfur and oxygen isotope ratios in sulfate during an acidification reversal study at Lake Gårdsjön, Western Sweden. *Water, Air, and Soil Pollution*, **79**, 261–278.

Mörth, C.-M. and Torssander, P. (1995b) Sulfur and oxygen isotope ratios in sulfate used as tracers of sulfur dynamics in an acidification reversal study. In: A. Jenkins, R. C. Ferrier and C. Kirby (eds), *Ecosystem Manipulation Experiments: Scientific Approaches, Experimental Design and Relevant Results.* Ecosystems Research Report 20, CEC, Windermere, UK, 16–21 October 1994, pp. 264–270.

Nehring, N. L., Bowen, P. A. and Thrusdell, A. T. (1977) Techniques for the conversion to carbon dioxide of oxygen from dissolved sulfate in thermal waters. *Geothermics*, **5**, 63–66.

Nriagu, J. O. and Coker, R. D. (1978) Isotopic composition of sulfur in precipitation within the Great Lakes Basin. *Tellus*, **30**, 365–375.

Nyberg, L. (1995) Soil and groundwater distribution, flowpaths, and transit times in a small till catchment, PhD thesis, Uppsala University, Uppsala, Sweden.

Nyberg, L., Bishop, K. and Rodhe, A. (1993) Importance of hydrology in the reversal of acidification in till soils, Gardsjoen, Norway. *Applied Geochemistry, Suppl. Issue*, **8**(2), 61–66.

Olsson, B., Hallbäcken, L., Johansson, S., Melkerud, P.-A., Nilsson, S. I. and Nilsson, T. (1985) The Lake Gårdsjön area—physiographical and biological features. *Ecological Bulletins (Stockholm)*, **37**, 10–28.

Prietzel, P., Mayer, B., Krouse, H. R., Rehfuess, K. E. and Fritz, P. (1995) Transformations of simulated wet sulfate deposition in forest soils assessed by a core experiment using stable sulfur isotopes. *Water, Air, and Soil Pollution*, **79**, 243–260.

Rafter, T. A. (1967) Oxygen isotopic composition of sulfates. Part 1. A method for the extraction of oxygen and its quantitative conversion to carbon dioxide for isotope radiation measurements. *New Zealand Journal of Science*, **10**, 493–510.

Rajan, S. S. S. (1978) Sulphate adsorption on hydrous alumina, ligands displaced and changes in surface charge. *Soil Science Society of America Journal*, **42**, 39–44.

Rajan, S. S. S. (1979) Adsorption and desorption of sulphate and charge relationships on allophanic clays. *Soil Science Society of America Journal*, **43**, 65–69.

Reuss, J. O. and Johnson, D. W. (1986) *Acid Deposition and the Acidification of Soils and Waters*. Ecological Studies 59. Springer-Verlag, Heidelberg.

Saltzman, E. S., Brass, G. W. and Price, D. A. (1983) The mechanism of sulfate aerosol formation; Chemical and sulfur isotopic evidence. *Geophysical Research Letters*, **10**, 513–516.

Schindler, S. C., Mitchell, M. J., Scott, T. J., Fuller, R. D. and Driscoll, C. T. (1986) Incorporation of $^{35}S$ sulfate into inorganic and organic constituents of two forested soils. *Soil Science Society of America Journal*, **50**, 457–462.

Singh, B. R. (1984) Sulphate sorption by acid forest soils. 3. Desorption of sulphate from adsorbed surfaces as a function of time, desorbing ion, pH and amount of adsorption. *Soil Science*, **138**, 346–353.

Stam, A. C., Mitchell, M. J., Krouse, H. R. and Kahl, J. S. (1992) Stable sulfur isotopes of sulfate in precipitation and stream solutions in a northern hardwood watershed. *Water Resources Research*, **28**, 231–236.

Tabatabai, M. A. (1982) Sulfur. In: A. L. Page, R. H. Miller and D. R. Keeney (eds), *Methods of Soil Analysis, Part 2, Chemical and Microbial Properties. Agronomy*, number 9 (part 2), American Society of Agronomy, Madison, Wisconsin, pp. 501–538.

Taylor, B. E., Wheeler, M. C. and Nordstrom, D. K. (1984) Isotope composition of sulfate in acid mine drainage as measure of bacterial oxidation. *Nature*, **308**, 538–541.

Toran, L. and Harris, R. F. (1989) Interpretation of sulfur and oxygen isotopes in biological and abiological sulfide oxidation. *Geochimica et Cosmochimica Acta*, **53**, 2341–2348.

Van Everdingen, R. O. and Krouse, H. R. (1985) Isotope composition of sulfates generated by bacterial and abiological oxidation. *Nature*, **315**, 395–396.

Van Stempvoort, D. R., Reardon, E. J. and Fritz, P. (1990) Fractionation of sulfur and oxygen isotopes in sulfate by soil sorption. *Geochimica et Cosmochimica Acta*, **54**, 2817–2826.

Van Stempvoort, D. R., Fritz, P. and Reardon, E. J. (1992) Sulfate dynamics in upland forest soils, central and southern Ontario, Canada; stable isotope evidence. *Applied Geochemistry*, **7**, 159–175.

Wright, R. F., Lotse, E. and Semb, A. (1988) Reversibility of acidification shown by whole catchment experiments. *Nature*, **334**, 670–675.

Wright, R. F., Lotse, E. and Semb, A. (1993) Rain project—results after 8 years of experimentally reduced acid deposition to a whole catchment. *Canadian Journal of Fisheries and Aquatic Science*, **50**, 258–268.

Yanagisawa, F. and Sakai, H. (1983) Thermal decomposition of barium sulfate—vanadium pentoxide—silica glass mixtures for preparation of sulfur dioxide in sulfur isotope ratio measurements. *Analytical Chemistry*, **55**, 985–987.

# 9 Relationships Between Sulfate Retention and Release, Solution pH and DOC in the Gårdsjön Soils

G. R. GOBRAN, F. COURCHESNE & A. DUFRESNE

Sulfate ($SO_4$) retention and release are important components of the buffering capacity of soils against acidic deposition. Sequential leaching experiments were undertaken to evaluate the effects of dissolved organic carbon (DOC) and pH on $SO_4$ retention and release from a Swedish podzol. Results show that DOC reduces $SO_4$ retention and that DOC is released when $SO_4$ is retained. The DOC also increases the rate of $SO_4$ release although about 40% of total sorbed $SO_4$ is retained irreversibly. The capacity of DOC to compete with $SO_4$ is pH-dependent, increasing with pH. The ratios of $OH^-$ released to $SO_4$ retained were low ($<0.3$) and suggest that $SO_4$ retention mostly involved the formation of outer-sphere complexes.

The interrelationship $SO_4$–DOC–pH and the irreversible $SO_4$ sorption–desorption behaviour are vital in the podzolic B horizon of the Gårdsjön catchment. As this horizon actually stores a large quantity of both C and S, it may, therefore, play a decisive role in the recovery process in the G1 Roof catchment.

These findings also suggest that models with fully reversible Langmuir or Freundlich approaches need to account for additional irreversible reactions and the competitive behaviour between $SO_4$ and DOC. In addition, other types of model should be considered. These should include kinetic models of the multireaction type and multicomponent approaches that consider reversible retention and irreversible processes.

## INTRODUCTION

### Acidic deposition and $SO_4$ retention

Sulfate retention by soils has received considerable attention because it can delay the detrimental effects of sulfuric acid ($H_2SO_4$) deposition on terrestrial and aquatic ecosystems. The input of $H_2SO_4$ to terrestrial ecosystems has been hypothesized to cause an increase in cation outflux from soils to ground and surface waters through an increase in the ionic strength of the soil solution (Reuss, 1983; Reuss and Johnson, 1986). Moreover, cation transport has been shown to increase more rapidly in soils that do not retain $SO_4$ compared with soils exhibiting a large anion retention

*Experimental Reversal of Acid Rain Effects: The Gårdsjön Roof Project.*
Edited by Hans Hultberg and Richard Skeffington. 

capacity (Huete and McColl, 1984; Gobran and Bosatta, 1988). The mobilization of cations associated with $H_2SO_4$ deposition can have considerable impacts on the chemical composition of soils and surface waters. On the one hand, the acidification of soils can occur if base cations are progressively replaced by H (from $H_2SO_4$) and Al (from Al oxides and hydroxides) on the exchange complex. However, the net rate of base cation depletion ultimately depends on the relative rate of: (1) base cation production (weathering) and cycling (organic matter decomposition) in the soil–plant system and/or (2) acid cation (H and Al) input via atmospheric deposition and mineral dissolution. On the other hand, surface water acidification can also occur following the mobilization of exchangeable Al by the acidity generated by soil-forming processes like podzolization. The overall rate of soil and water acidification can thus be accelerated by $H_2SO_4$ deposition in areas where the acid neutralizing capacity is low (Van Bremen *et al.*, 1983).

**Mechanisms of $SO_4$ retention by forest soils**

The process of $SO_4$ retention is still debated and varies with soil properties which, in turn, are controlled by climate, vegetation, parent material and time since the initiation of pedogenesis (Rochelle *et al.*, 1987). Retention of $SO_4$ has frequently been attributed to adsorption on mineral soil surfaces resulting in the formation of outer- or inner-sphere surface complexes (Parfitt and Smart, 1978; Curtin and Syers, 1990). Outer-sphere complexes refer to $SO_4$ being retained in the diffuse layer with at least one water molecule remaining between the surface and the sorbed $SO_4$ ion (Sposito, 1984). The formation of inner-sphere complexes, or chemisorption, implies a ligand exchange reaction between $SO_4$ and OH or $OH_2$ groups, and thus the creation of ionic bonds with metal atoms at the surface of solids (Charlet *et al.*, 1990). The precipitation of alumino-hydroxy sulfate minerals (alunite, jurbanite) has also been considered to control $SO_4$ and Al concentration in soil solutions (Adams and Rawajfih, 1977; Khanna *et al.*, 1987). However, the solid phases have never been conclusively identified in forest soils affected by acidic deposition (Gobran and Nilsson, 1988). Several studies have indicated that $SO_4$ can be incorporated into soil organic matter (Strickland *et al.*, 1986; Schindler *et al.*, 1986). This phenomenon occurs mostly in surface organic horizons and appears to be much less important in the mineral soil, at least in the short term. Finally, $SO_4$ can be absorbed by plants through root uptake.

The magnitude of $SO_4$ retention has been related to a number of soil properties with the most important being: $SO_4$ concentration in solution (e.g. Chao *et al.*, 1962a; Johnson and Henderson, 1979), aluminium (Al) and iron (Fe) sesquioxides content (Johnson and Todd, 1983; Fuller *et al.*, 1985), solution pH (Nodvin *et al.*, 1986; Courchesne and Hendershot, 1989) and electrolyte concentration and composition (Bolan *et al.*, 1986; Courchesne, 1991; Gobran and Clegg, 1992). While the first two properties enhance $SO_4$ retention, increasing pH and ionic strength (I) is known to decrease $SO_4$ retention.

**Organic matter and $SO_4$ retention**

Organic matter, both in the solid and aqueous phases, also interacts with $SO_4$ retention. Organic substances are ubiquitous in forest soils as a result of the decomposition of organic debris, root exudation and microbial activity (Fox and Comerford, 1990).

Organic matter abundance typically decreases from the surface horizons towards the lower part of soil profiles. The spodosol profile constitutes a notable exception to this general trend since organic matter generally accumulates at two different levels (Sandborn and Lavkulich, 1989). A first peak resulting from *in situ* accumulation is found in surface organic horizons. A second peak is observed in the upper part of the spodic B horizon where eluvial humified organic substances dominate. The amount of soil organic matter is negatively correlated to the quantity of sorbed $SO_4$ in a variety of mineral soil horizons (Johnson and Todd, 1983). It thus seems that the accumulation of organic matter on mineral soil surfaces can efficiently block anion sorption sites. Contrarily, the presence of organic matter could potentially enhance the retention of anions. Huang and Violante (1986) showed that organic matter interferes with the crystallization of Al and Fe oxides, hence favouring the formation of amorphous Fe and Al materials having a higher surface area, and thus a higher anion retention capacity, than well-crystallized oxides.

Organic substances also interact with $SO_4$ through the production of organic anions that can compete, like inorganic anions, with $SO_4$ for sorbing sites (Krug and Frink, 1983). Evans (1986) suggested that the increased mobility of $SO_4$ in the presence of dissolved organic carbon (DOC) was the result of anionic competition for similar surface sites. Gobran and Nilsson (1986, 1988) showed that DOC from forest floor leachates could inhibit $SO_4$ retention by a podzol. Inskeep (1989) demonstrated that the DOC-induced reduction in $SO_4$ sorption was controlled by the amount of oxygen-containing functional groups which are active at the surface of organic ligands. The efficiency of organic acids in reducing $SO_4$ retention was also found to be strongly pH-dependent (Courchesne and Landry, 1994).

In conclusion, the weight of evidence suggests that the presence of DOC plays a decisive role in sulfate retention and mobility in acid forest soils (Gobran and Nilsson, 1988; Kaiser and Zech, 1996).

### Reversibility of $SO_4$ retention

Over the last few years, anthropogenic S emissions to the atmosphere and $SO_4$ deposition to terrestrial and aquatic ecosystems have decreased substantially in North America, Central Europe and Scandinavia (Harriman *et al.*, 1990). Several studies, like the Gårdsjön Roof experiment and the RAIN Project, are addressing the question of how reduced $SO_4$ inputs to forest ecosystems will affect $SO_4$ concentrations in soil solution and in lake and streamwaters (Bishop and Hultberg, 1995; Wright *et al.*, 1993). In this context, the evaluation of the extent and rate of $SO_4$ retention and reversibility in soils is crucial to assess the recovery of soils and surface waters following a decrease in S deposition.

### Objectives

Several aspects of the $SO_4$ retention and release process deserve attention and will be the focus of this chapter: (1) what is the role of organic anions in influencing the rate and extent of $SO_4$ release to solution? (2) what is the impact of $SO_4$ release in soil solution pH? (3) how much of the retained $SO_4$ can be released? and (4) what are the practical implications of the findings of this work and how could they be incorporated in models?

In order to answer the above questions, we attempted to elucidate the relationships between $SO_4$ retention and release, DOC and pH in podzolic soil from the Gårdsjön catchment. Therefore, the specific objectives were: (1) to demonstrate the competitive behaviour between $SO_4$ and organic anions for sorption sites, (2) to examine the effects of solution pH on anionic competition, (3) to assess the extent of $SO_4$ retention and reversibility in the presence of organic anions and (4) to use these results to help projecting $SO_4$ recovery after excluding S in the G1 Roof catchment.

### Approach

Two types of laboratory experiments were conducted. The first was an equilibrium experiment using a sequential leaching technique (SLT). The second (SET) was a kinetic experiment using continuous flow technique. Both experiments examined the relationships between $SO_4$ retention and release, DOC and pH in soils, especially in the Bs1 horizon that is believed to be the most important horizon controlling $SO_4$ retention and release.

## MATERIALS AND METHODS

Soil samples were collected from the Lake Gårdsjön catchment located on the Swedish west coast. The representative soil type in the Gårdsjön area is a haplorthod according to the US Soil Taxonomy classification (Soil Survey Staff, 1995) or ferro-humic and humo-ferric podzol according to the Canadian System (ACECSS, 1987).

### The equilibrium experiment

Soil samples from the G1 Roof catchment at the time of constructing the roof were taken. Soil profiles were dug and the samples were taken by horizon, i.e. forest floor (FF), Ah and upper Bs1 (5 cm) and lower Bs2 (5 cm) from the Bs horizon. These soil profiles are described in Andersson *et al.* (Chapter 2, this volume). Field moist samples of humus (FF) were passed through a 5 mm sieve and mineral soils through a 2 mm sieve. The moist fresh samples were stored in black plastic bags at 3°C. All the sequential equilibration experiments were conducted at a constant temperature ($15 \pm 1$°C).

Due to the importance of including DOC in the soil extractions to assess $SO_4$ sorption (Gobran and Nilsson, 1988) the following leaching solutions were prepared for use in our experiments. First, forest floor leachate (FFL) from the FF sample was obtained by shaking the FF in deionized water (DW) 1:10 (wt./wt.) for 2 h. The suspension was filtered and the filtrate was then diluted with DW to bring the DOC concentration to that observed beneath the FF in the field (Table 9.1). Second, sulfate solutions (FFL-$SO_4$) used for $SO_4$ sorption were prepared by adding different aliquots from a stock solution of $Na_2SO_4$ (prepared in DW) to FFL. The FFL-$SO_4$ contained S up to 775 mg S $L^{-1}$.

**Table 9.1.** Chemical characteristics of input–output soil solution of different horizons from the G1 Roof catchment. Each value represents one observation. From Gobran *et al.* (1997). With kind permission from Kluwer Academic Publishers

| Soil variable | FFL | Ah | Bs1 | Bs2 |
|---|---|---|---|---|
| pH | 4.74 | 4.05 | 4.23 | 4.77 |
| | | 4.04 | 4.35 | 4.77 |
| $SO_4$-S (mg $L^{-1}$) | 0.38 | 2.48 | 4.54 | 6.16 |
| | | 2.16 | 5.05 | 6.26 |
| DOC-C (mg $L^{-1}$) | 10.40 | 74.60 | 58.50 | 38.60 |
| | | 69.10 | 56.40 | 39.70 |

Since we were primarily concerned with desorption processes that may occur in the environment where Cl and $NO_3$ are the major competing anions, release or desorption processes were examined using 0.01 M $NH_4Cl$ solution, which was recommended for determining sulfur in organic horizons of forest soils (Maynard *et al.*, 1987). Therefore, ammonium chloride (FFL-$NH_4$) solution used for $SO_4$ desorption was prepared as with the FFL except that 0.01 M $NH_4Cl$ was used instead of DW. The FF-$NH_4Cl$ was used to estimate the labile or readily desorbed $SO_4$ from the respective soil in the retention sequence (Figure 9.1).

### The sequential equilibration technique

We used a sequential leaching technique (SLT) that mimics leaching under forest field conditions. Figure 9.1 illustrates sulfate retention (I) and release (II) sequences.

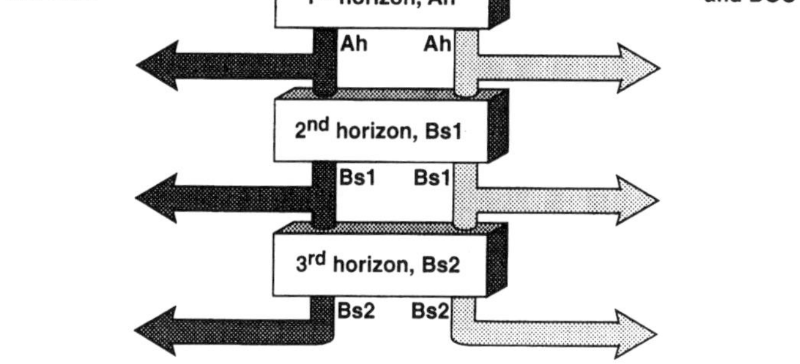


**Figure 9.1.** The sequential leaching technique. From Gobran *et al.* (1997). With kind permission from Kluwer Academic Publishers

For any given $SO_4$ concentration, the equilibrium solution (output) from the first horizon (Ah) was used as a leaching (input) solution for the lower horizon (Bs1) and so forth. As a result, in this sequential leaching technique, the soil system reacts with only one single input variable used, e.g. $SO_4$ in $Na_2SO_4$. Moreover, since the composition and concentration of the electrolyte influences $SO_4$ sorption (Courchesne, 1991), this technique keeps the leaching solutions close to that found under field conditions.

In both retention and release sequences (I and II in Figure 9.1), soil suspensions were shaken end-over-end on a reciprocating shaker for 1 h and left standing overnight, then shaken again for another hour (total period exactly 24 h). The ratio of equivalent dry weight soil to solution was kept constant at 1:5 (soil moisture content was always considered). All equilibrium solutions were obtained by centrifugation for 20 minutes at 3000 rpm. Next, the equilibrium solutions were filtered through a 0.45 $\mu$m filter and analysed for pH, $SO_4$ and DOC. The $SO_4$ was measured by ion chromatography (Dionex 2000i HPLC). The DOC content was determined with a Shimadzu total organic carbon analyser (TOC-5000). All experimental work and analysis were done in duplicate.

**The kinetic experiment (SET)**

A bulk sample of the upper B horizons (Bhf1, Bhf2) was collected from G1 Roof catchment, air-dried, lightly crushed, passed through a 2 mm sieve and stored at room temperature. Soil pH was measured in $H_2O$ and in 0.01 M $CaCl_2$ using a 1:10 soil to solution ratio. Organic C content was analysed by combustion with a LECO C analyser. Iron and Al were extracted with Na-pyrophosphate and acid ammonium oxalate (in the dark); dissolved Fe and Al were measured by atomic absorption spectrophotometry (Varian). The $PO_4$-exchangeable native $SO_4$ content of the soil was estimated by three sequential extractions of triplicate 3 g samples with 30 ml 0.01 M $NaH_2PO_4$ using an equilibration time of 1 h. Solutions were separated by centrifugation at $10\,000 \times g$, filtered through 0.4 $\mu$m polycarbonate membranes and analysed for $SO_4$ by ion chromatography (Waters). The DOC was measured in each extract with a Shimadzu TOC analyser. The amount of native $SO_4$ and DOC were corrected for entrapped solution. The particle size distribution was determined by the hydrometer method and by wet sieving after removal of coatings with $H_2O_2$ and dithionite-citrate-bicarbonate (DCB) following the procedures of Sheldrick (1984). Table 9.2 lists some soil properties of the B horizon from the G1 Roof catchment.

**Table 9.2.** Some soil properties of the B horizon from the G1 Roof catchment. From Courchesne *et al.* (1995). By kind permission of Kluwer Academic Publishers

| pH | | | Pyrophosphate-extractable | | Oxalate-extractable | | | | $PO_4$-exchangeable | |
|---|---|---|---|---|---|---|---|---|---|---|
| $H_2O$ 1:10 w:v | $CaCl_2$ 1:10 w:v | Organic C % | Fe % | Al % | Fe % | Al % | Sand | Clay | $SO_4$ (mmol $kg^{-1}$) | DOC (g $kg^{-1}$) |
| 4.25 | 3.66 | 7.1 | 2.27 | 0.72 | 2.88 | 0.84 | 59 | 12 | 2.2 | 1.1 |

**Table 9.3.** Chemical composition of input solutions used in the kinetic experiment. From Courchesne *et al.* (1995). By kind permission from Kluwer Academic Publishers

| | Retention phase | | Release phase | |
|---|---|---|---|---|
| Solution no. | $SO_4$ ($\mu$mol $L^{-1}$) | DOC (mg $L^{-1}$) | $SO_4$ ($\mu$mol $L^{-1}$) | DOC (mg $L^{-1}$) |
| blank | 0 | 0 | 0 | 0 |
| 1 | 250 | 0 | 0 | 0 |
| 2 | 250 | 0 | 0 | 25 |
| 3 | 250 | 0 | 0 | 50 |
| 4 | 250 | 25 | — | — |

The experiment was conducted at 24 ± 2°C using a continuous flow technique. One gram of < 2 mm soil was loaded into a 25 mm diameter filtration assembly containing a 0.4 $\mu$m membrane. This reaction chamber was attached to a peristaltic pump and a fraction collector. The soil samples were then leached at a flow rate of 0.2 mL $min^{-1}$ with five different input solutions containing $SO_4$ ($Na_2SO_4$) and/or DOC (Aldrich humic acid, Na-salt) at two pH values: 3.2 and 4.2 (Table 9.3). The Aldrich humic acid (HA) was selected as a surrogate for soil DOC, even knowing the limitations of using commercial humic acids (Malcolm and MacCarthy, 1986). We assumed that HA is more representative of the heterogeneous DOC compounds found in forest soils than purified low-molecular-weight organic acids. The pH and the ionic strength ($I$ = 0.01 M) of the input solutions were adjusted to the desired level using 0.01 M HCl and NaCl respectively. Two drops of toluene were added to each litre of solution to reduce microbial activity. The leaching experiment included three successive steps: soil equilibration, $SO_4$ retention and $SO_4$ release. The objective of this procedure was to maintain a constant pH and ionic strength (I) throughout

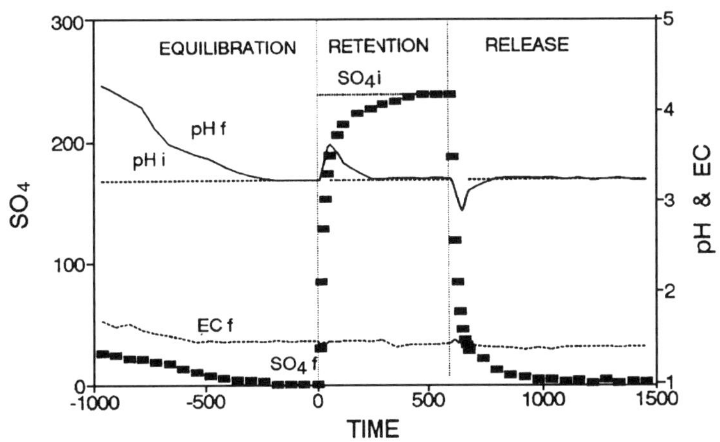


**Figure 9.2.** The design of the kinetic experiment. The pH and ionic strength (I) of the leaching solutions were maintaned throughout the equilibration, retention and release phases. The units for $SO_4$, electrical conductivity (EC) and time are $\mu$mol $SO_4$ $L^{-1}$, mS $cm^{-1}$ and minutes respectively. The symbols i and f refer to initial and final values

the entire experiment (Figure 9.2). During the equilibration phase, the soils were leached with a $SO_4$ and DOC-free 0.01 M NaCl solution adjusted to the appropriate pH until steady-state conditions were reached ($pH_{initial} = pH_{final}$). At the beginning of the retention phase (0 min), the equilibration solution was replaced by a solution of the same pH and I values but containing $SO_4$ with and without added DOC (Table 9.3). Leaching continued up to the point where the input and output $SO_4$ concentrations were the same ($[SO_4]i = [SO_4]f$). At the start of the $SO_4$ release phase (600 min), the input solution was changed and leaching proceeded until $[SO_4]i = [SO_4]f$ (Table 9.3).

Each treatment was repeated eight times (replicates) and the output solutions were bulked prior to chemical analyses. During the retention and release phases, solutions were collected every 10 min for the first hour and then every hour until the end of each phase. All solutions were analysed for $SO_4$ and DOC as indicated before and for pH (Radiometer PHM82) and conductivity (Radiometer CDM83). For each time interval, $SO_4$ and DOC retention and release were calculated as the difference between fluxes into and out of the reaction chamber.

The kinetics of $SO_4$ retention and release were described by the Elovich equation expressed as (Hodges and Johnson, 1987):

$$S_x = (1/\beta)\ln(\alpha\beta) + (1/\beta)\ln t \qquad (9.1)$$

where $S_x$ is the amount of $SO_4$ sorbed or desorbed (mmol $SO_4$ kg $soil^{-1}$) at time $t$ (min), and $\beta$ and $\alpha$ are constants. The Elovich equation was selected because it gave the best representation of $SO_4$ sorption kinetics for a range of podzols from Quebec (Courchesne, 1992). The parameters of the equation and the standard error of the slope ($SE_s$) were determined by regressing ln $t$ against $S_x$. The slope parameter ($1/\beta$) was used as the $SO_4$ retention or release rate. The homogeneity of slopes between different treatments was tested using the covariance module in SYSTAT (Wilkinson, 1989).

## RESULTS AND DISCUSSION

### The equilibrium experiment

Table 9.1 lists the chemical characteristics of the equilibrium soil solution of the field moist soil samples FF, Ah, Bs1 and Bs2 horizons using the sequential equilibration technique. The DW (S free) was used to extract the FF horizon and the equilibrium solution FFL was used to equilibrate the lower Ah horizon and so forth. The values listed in Table 9.1 represent the first observations in the sulfate retention experiments, i.e. native soil solution pH, $SO_4$ and DOC.

Sulfate retention by all soil horizons (Ah, Bs1 and Bs2) increased linearly with increasing $SO_4$ concentration in the soil solution (Figure 9.3). Such a linear relationship suggests that the retained $SO_4$ depends on the $SO_4$ concentration in the solution (e.g. Chao *et al.*, 1962a). Also, this would indicate the possible involvement of several mechanisms of $SO_4$ retention (e.g. Chao *et al.*, 1962b; Hue *et al.*, 1985). Moreover, these results show that the retention of $SO_4$ by the Bs horizons was distinctly greater than for the Ah horizons. This is a characteristic of spodosol

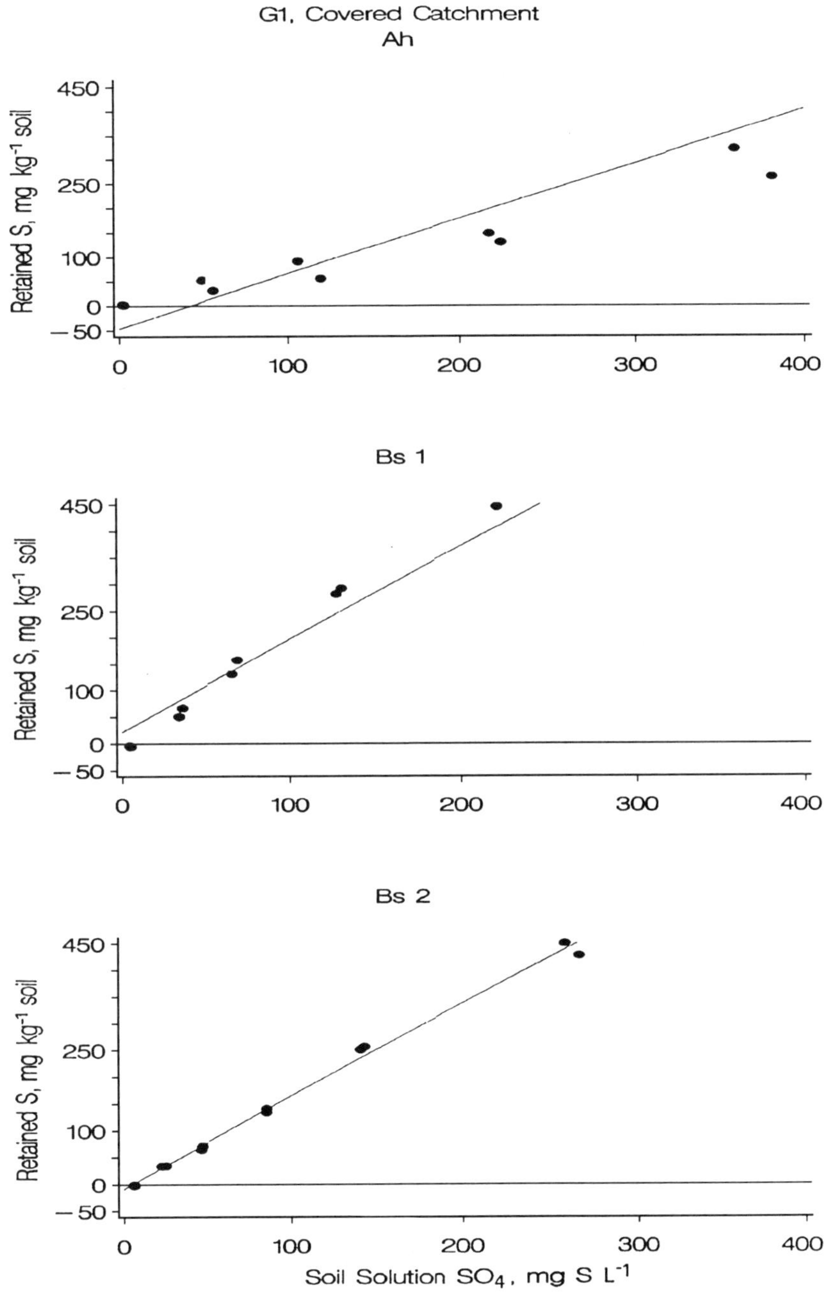


**Figure 9.3.** Relationship between $SO_4$ in solution and $SO_4$ retention of the three different horizons (Ah, Bs1 and Bs2) from the G1 Roof catchment. The straight lines represent the regression lines of the entire range of $SO_4$ used. From Gobran *et al.* (1997). With kind permission from Kluwer Academic Publishers

horizons rich in Al and Fe hydrous oxides, which also retain Si (Saito and Shoji, 1984). These minerals are known specifically to adsorb $SO_4$ through ligand exchange processes (e.g. Hingston *et al.*, 1968, 1972; Parfitt and Smart, 1978; Rajan, 1978). In Scandinavia, minerals such as imogolite and proto-imogolite allphane may be the most important for sulfate adsorption (Gustafsson *et al.*, 1995).

Furthermore, the Bs1 and Bs2 horizons showed a strong capacity for retaining DOC. For example, when the FFL was used to equilibrate Ah, DOC was released (Table 9.1). As the Ah solution was brought into equilibrium with Bs1, the DOC concentration in the solution decreased. Further disappearance (retention) of the DOC from soil solution occurred when the Bs1 solution was used to equilibrate Bs2 (Table 9.1). These observations demonstrate that the Bs soil horizons have a strong capacity to retain both DOC and $SO_4$. The Gårdsjön B horizon is acidic and coarse textured, two features commonly associated with podzols (Table 9.2). The soil material from the podzolic B horizon is rich in organic C and most of the extractable Al and Fe was found to be complexed with organic matter as indicated by the difference between oxalate and pyrophosphate extractions (Table 9.2).

### The sorption–desorption of $SO_4$

The labile or readily desorbed $SO_4$ from the Bs1 horizon is shown in Figure 9.4. The straight line represents the regression line of the sorption data from Figure 9.3, and also would represent the desorption line if desorption was fully reversible. Figure 9.4 shows clearly that at a fixed level of S in equilibrium solution, the desorption experiments resulted in more S remaining on sorption sites than during the sorption phase. On average (within the range of S concentrations used), only 30% of the adsorbed $SO_4$ was desorbed by the FF-$NH_4Cl$ extract from Bs1 of the G1 Roof catchment. In the F1 Control catchment in Gårdsjön, the desorbed $SO_4$ was only about 40% of the adsorbed $SO_4$ (Gobran *et al.*, 1997).

As the experimental duration of adsorption and desorption cycles was equal, the difference between the desorption data points and the adsorption regression line

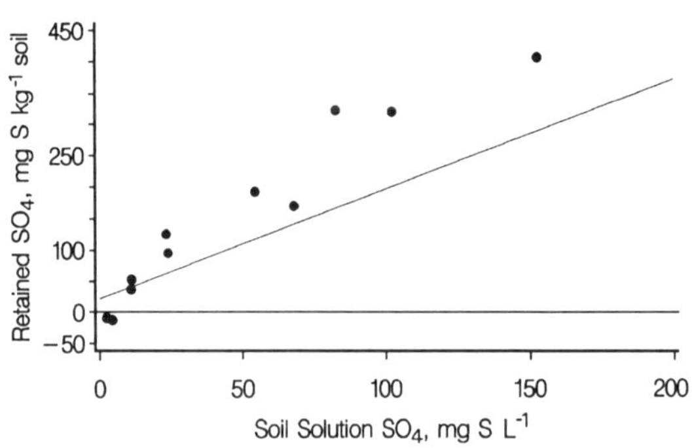


**Figure 9.4.** Desorption of $SO_4$ from $SO_4$-equilibrated Bs1 from the G1 Roof catchment. The straight line represents the regression line of the sorption data from Figure 9.3. From Gobran *et al.* (1997). With kind permission from Kluwer Academic Publishers

(Figure 9.4) reflects the extent of irreversibility for a given $SO_4$ concentration in soil solution. This discrepancy between sorption and desorption pathways is often referred to as hysteresis. Similar results were obtained in soils by Singh (1984) and Harrison *et al.* (1989), and in oxide minerals (Turner and Kramer, 1992). The $SO_4$ sorption–desorption irreversibility behaviour reported here has been observed for anions and organic ions in different soils (for a review see Selim, 1992).

### The effects of $SO_4$ input on DOC

The effect of $SO_4$ retention on the retention of DOC in Bs1 is illustrated in Figure 9.5. It demonstrates clearly a negative relationship between the retention of DOC and $SO_4$. This opposite trend confirms the competitive behaviour of both $SO_4$ and DOC sorption sites on matrix surfaces. These results provide a further support to our hypothesis that dissolved substances from the forest floor play a decisive role for the $SO_4$ retention in forest soils. These findings also suggest that organic ligands possibly were competing with $SO_4$ for ligand sites (Gobran and Nilsson, 1986, 1988). This hypothesis has been supported by other investigators (e.g. Inskeep, 1989; Evans and Anderson, 1990; Vance and David, 1992).

### The $SO_4$–DOC–pH interrelationship

Our datasets from Bs1 of the G1 catchment were further utilized to examine the relative retention of $SO_4$ and DOC versus pH. Figure 9.6 illustrates the relationship between the ratio of retained $SO_4$ and total retained anions ($SO_4$+DOC) and pH. This relationship points to the predominance of $SO_4$ retention at pH $<4$ and DOC retention at pH $>4$. Courchesne and Hendershot (1989) also found that $SO_4$ sorption reached its maximum at a similar pH range (pH 3.8 to 4.0) in the B horizons of two spodosols from Quebec. Jardine *et al.* (1989), working with an ultisol, also observed a maximum DOC sorption at pH 4.5. The increased DOC at pH $>4$ may be due to increased dissociation of humic substances with increasing pH (Stevenson, 1982;

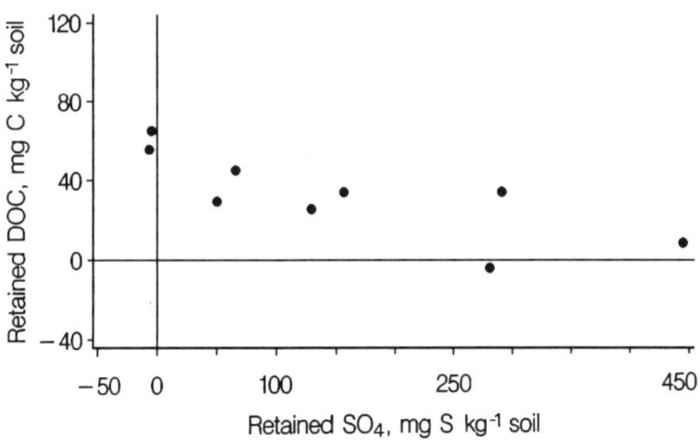


**Figure 9.5.** Relationship between $SO_4$ retention and DOC retention in Bs1 from the G1 Roof catchment. From Gobran *et al.* (1997). With kind permission from Kluwer Academic Publishers

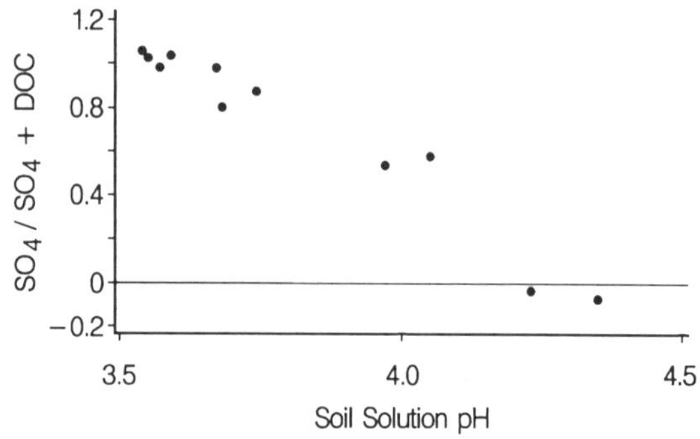


**Figure 9.6.** Relative retention of $SO_4$ to total anion retention ($SO_4$+DOC) as dependent on soil solution pH of Bs1 from the G1 Roof catchment. The $SO_4$-S and DOC-C are expressed in mg $kg^{-1}$ dry soil. Notice, the ratio values $<0$ and $>1$ indicate the released $SO_4$ and DOC respectively. From Gobran *et al.* (1997). With kind permission from Kluwer Academic Publishers

Gobran and Tipping, 1993) and thus a higher competition between the organic anions and the inorganic anion $SO_4$. Lee and Hultberg (1990) estimated that the average charge of carboxyl acid groups (COO) of DOC in streamwaters of the Lake Gårdsjön catchment varies from 1 to 4 ($mmol_c\ g^{-1}$) within a pH range of 3.8–4.3.

It is worth mentioning here that all the results of the G1 Roof catchment mentioned above were rather similar to those of the F1 Control catchment when the equilibrium experiment technique was used (Gobran *et al.*, 1997). This finding was the case just before the construction of the roof. However, the differences between the F1 Control and G1 Roof catchments ought to be large by now and, therefore, await investigation.

### The kinetic experiment

The amount of $PO_4$-exchangeable native $SO_4$ decreased from 2.2 (Table 9.2) to 1.0 mmol $kg^{-1}$ after the equilibration phase. This reduction in $SO_4$ probably results from the partial dissolution of Al and Fe oxides during soil equilibration and from native $SO_4$ being exchanged for $Cl^{-1}$ ions present in the electrolyte.

### DOC effects on $SO_4$ retention

Constant chemical conditions were attained at the end of the equilibration phase and were maintained throughout the $SO_4$ retention and release phases (Figure 9.2 and 9.7). The electrical conductivity (EC) of output solutions ranged from 1.38 to 1.48 mS $cm^{-1}$ for all treatments and was stable over the entire 1200 min period for a given treatment. Figure 9.8 shows that the output solution pH equalled the initial pH ($pH_i$) for all sampling intervals except when $SO_4$ retention and release were at their maximum. The impact of $SO_4$ retention and release on solution pH will be addressed below.

The presence of humic acid (HA) at an initial level of 25 mg DOC $L^{-1}$ significantly decreased the rate of $SO_4$ retention at both pH values (Table 9.4). Sulfate retention rates were lowered by 26 and 42% at pHi of 3.2 and 4.2 respectively. As in the equilibrium experiment (Figure 9.6), the increased competitiveness of DOC towards

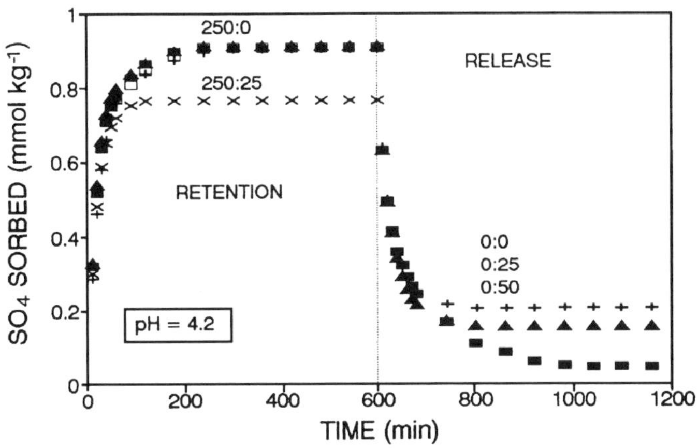


**Figure 9.7.** Sulfate retention (250 $\mu$mol $SO_4$ $L^{-1}$ : 0 mg DOC $L^{-1}$ and 250 $\mu$mol $SO_4$ $L^{-1}$ : 25 mg DOC $L^{-1}$) and release (0 $\mu$mol $SO_4$ $L^{-1}$ : 0, 25, and 50 mg DOC $L^{-1}$) at pHi of 4.2, from the Bs horizon of the G1 catchment

$SO_4$ is considered to result from the higher degree of dissociation of HA at pHi of 4.2 than at pHi of 3.2. The soil released DOC to solution (0–4 mg DOC $L^{-1}$) as $SO_4$ was retained (Figure 9.9). However, DOC was not released during blank runs (0.01 M NaCl with no $SO_4$ or DOC added) conducted after the equilibration phase. This clearly indicates: (1) that $Cl^-$ ions could no longer displace DOC and (2) that the release of DOC resulted from anion exchange reactions triggered by the addition of $SO_4$ to the input solution. The synchronicity between $SO_4$ retention and DOC release also lends support to the anion competition hypthesis. Finally, the extent of $SO_4$ retention, despite the presence of 0.01 M Cl, suggests that the Gårdsjön B horizon had a stronger affinity for $SO_4$ and DOC than for $Cl^-$ ions. This is similar to the order of anion affinity reported by Nodvin *et al.* (1986).

### DOC effects on $SO_4$ release

The ability of HA to displace $SO_4$ from anion sorption sites was pH-dependent, and was similar to what we observed during the $SO_4$ retention phase. At pHi of 3.2, the rate of $SO_4$ release was not affected by the level of DOC addition in the input solution (Table 9.5). However, at pHi of 4.2, $SO_4$ release to solution proceeded at a significantly faster rate as DOC concentration increased from 0 to 50 mg $L^{-1}$ (Table 9.5). The efficiency of DOC in displacing $SO_4$ from surface sites appears to be controlled by the extent of proton dissociation from the functional groups present at the surface of HA. Although pK values are unknown for HA, we can assume that deprotonation increased from pH 3.2 to 4.2, thus increasing the net negative charge of HA molecules together with their capacity to compete with $SO_4$. These results also are in agreement with those obtained from the equilibrium experiment (Figure 9.6).

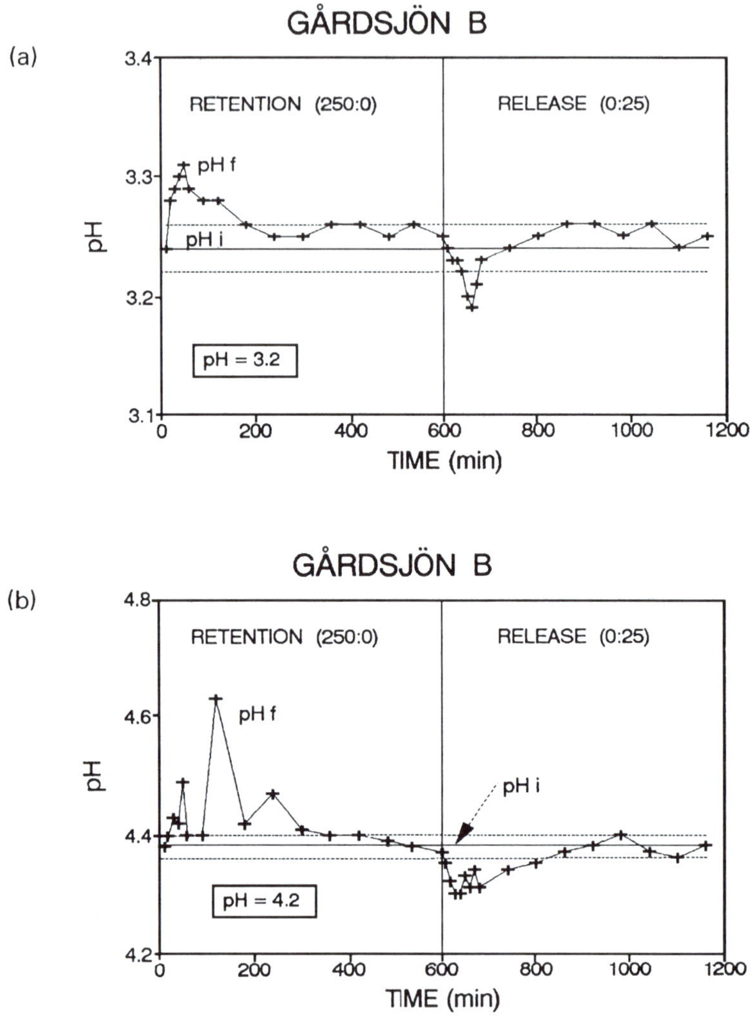


**Figure 9.8.** Output solution pH values during the $SO_4$ retention (250 $\mu$mol $SO_4$ $L^{-1}$ : 0 mg DOC $L^{-1}$) and release (0 $\mu$mol $SO_4$ $L^{-1}$ : 25 mg DOC $L^{-1}$) phases. Input pH (pHi) level is (a) 3.2 and (b) 4.2; dashed lines mark the upper and lower limits of a 95% confidence interval. From Courchesne *et al.* (1995). With kind permission from Kluwer Academic Publishers

The period of time needed for $SO_4$ desorption to cease was longer at higher DOC levels (Figure 9.7). The reverse trend was observed during the retention phase with the duration of the sorption process decreasing as DOC increased (Table 9.4). In both phases, the time differences between DOC levels were larger at pHi of 4.2 than at 3.2. This further indicates that DOC enhances the release of $SO_4$.

The total amount of $SO_4$ released during the desorption phase was equivalent to about 90% of the amount of $SO_4$ retained during the retention phase (200 minutes). However, at least 40% of total retained $SO_4$ ($SO_4$ retained during the

**Table 9.4.** Rate of $SO_4$ retention (±standard error) and duration of the retention phase for two treatments. From Courchesne *et al.* (1995). With kind permission from Kluwer Academic Publishers

| Treatment $SO_4$:DOC ($\mu$mol $L^{-1}$:mg $L^{-1}$) | Retention rate (mmol $kg^{-1}$ ln $min^{-1}$) | Duration (min) |
|---|---|---|
| **pH = 3.2** | | |
| 250: 0 | 0.35a* ±0.02 | 270 |
| 250:25 | 0.26b ±0.01 | 240 |
| **pH = 4.2** | | |
| 250: 0 | 0.24a ±0.01 | 250 |
| 250:25 | 0.14b ±0.02 | 120 |

*Rate values followed by the same letter are not significantly different at the $\alpha = 0.05$ level.

experiment+native $SO_4$ left after the equilibration period) was not released by the Gårdsjön B horizon during the desorption phase (Figure 9.10). It should be remembered that about 50% of the native $SO_4$ was released by 0.01 M NaCl during the equilibrium phase (from 2.2 to 1.1 mol $SO_4$ $kg^{-1}$ (Table 9.2). Obviously, NaCl desorbed some of the native $SO_4$ but only the $SO_4$ ions that were more loosely bound to the surface, i.e. the non-specifically retained $SO_4$ that was mostly involved in the formation of outer-sphere complexes (Hingston *et al.*, 1968; Parfitt and Smart, 1978). Therefore, it is understandable that the newly retained $SO_4$ during the retention phase was non-specifically adsorbed by displacing Cl from the soil surface. During the desorption phase, Cl could only displace the non-specifically sorbed $SO_4$. However, Cl could not displace $SO_4$ that was strongly bound to the surface or

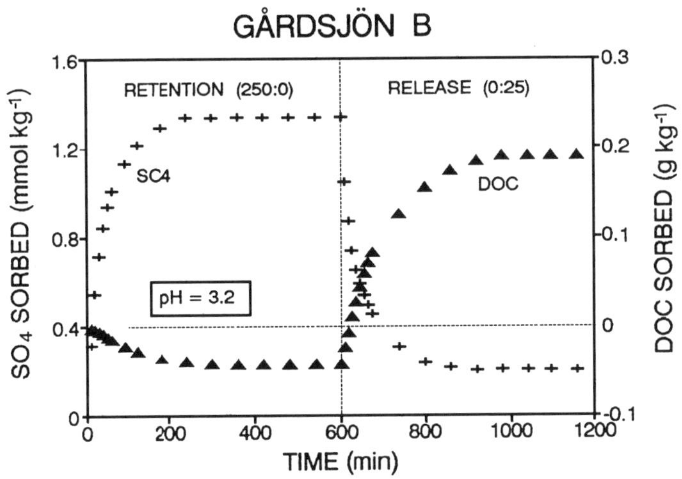


**Figure 9.9.** Sulfate and DOC retention (250 $\mu$mol $SO_4$ $L^{-1}$ : 0 mg DOC $L^{-1}$) and release (0 $\mu$mol $SO_4$ $L^{-1}$ : 25 mg DOC $L^{-1}$) at pHi of 3.2, in the Bs horizon of the Gl catchment. From Courchesne *et al.* (1995). With kind permission from Kluwer Academic Publishers

**Table 9.5.** Rate of $SO_4$ release (± standard error), amount released and duration of the release phase for three treatments. From Courchesne *et al.* (1995). With kind permission from Kluwer Academic Publishers

| Treatment $SO_4$:DOC ($\mu$mol $L^{-1}$:mg $L^{-1}$) | Release rate (mmol $kg^{-1}$ ln $min^{-1}$) | Amount released (%)* | Duration (min) |
|---|---|---|---|
| **pH = 3.2** | | | |
| 0: 0 | 0.29a[†] ±0.01 | 54 | 380 |
| 0:25 | 0.25b ±0.01 | 48 | 380 |
| 0:50 | 0.28ab±0.01 | 51 | 440 |
| **pH = 4.2** | | | |
| 0: 0 | 0.15a ±0.01 | 37 | 200 |
| 0:25 | 0.16ab±0.01 | 40 | 200 |
| 0:50 | 0.19b ±0.01 | 45 | 440 |

*% of total retained $SO_4$ (retained during the experiment + $PO_4$ exchangeable after the equilibration period).
†Rate values followed by the same letter are not significantly different at the $\alpha = 0.05$ level.

specifically adsorbed by the ligand exchange process, i.e. the formation of inner-sphere surface complex.

Harrison *et al.* (1989) reported a similar figure of the irreversibly retained $SO_4$ (36%) for a variety of forest soils that were leached with deionized water. This significant $SO_4$ fraction can be considered to be irreversibly retained, or very slowly available (Table 9.5), at least under conditions where organic anions, Cl and $NO_3$ are the major competing anions. This is the case in most forest soils. Figure 9.10 also shows that at a given time, the $SO_4$ sorbed in the retention phase was higher that found in the desorption phase. The results of Figure 9.10 are in full agreement with those found under equilibrium conditions, which were performed during 24 hours (Figure 9.4). Because of the magnitude of the irreversibly retained $SO_4$ pool,

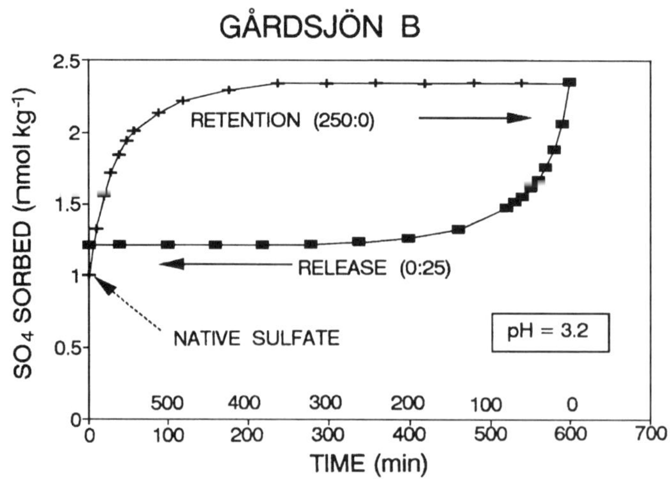


**Figure 9.10.** Sulfate retention (250 $\mu$mol $SO_4$ $L^{-1}$ : 0 mg DOC $L^{-1}$) and release (0 $\mu$mol $SO_4$ $L^{-1}$ : 25 mg DOC $L^{-1}$) at pHi of 3.2, in the Bs horizon of the G1 catchment

knowledge on the extent of temporal dynamics of this pool is critical for modelling $SO_4$ release from soils, should the input of $SO_4$ decrease.

**The effects of $SO_4$ retention and release on solution pH**

The pH of the output solution increased temporarily during the $SO_4$ retention phase while it decreased as $SO_4$ was released (Figure 9.8). This is in agreement with the view that $SO_4$ retention causes a release of OH ions whereas $SO_4$ release consumes OH. For the different treatments, the ratios of OH released to $SO_4$ retained ranged from 0.04 to 0.29, on an equivalent basis. These values compare well with data from other soils (Zhang *et al.*, 1989) but are much lower than the ratios obtained for Al and Fe oxides and allophane (Rajan, 1978; Turner and Kramer, 1992). Moreover, in this study, the lowest $OH/SO_4$ ratios (lowest pH changes) were recorded at pHi of 4.2 and in the presence of DOC in the input solution (Courchesne *et al.*, 1985).

The extent of OH release indicates that most of the $SO_4$ ions retained during the experiment by the Gårdsjön B horizon were involved in exchange reactions with organic anions, OH or Cl. Obviously, such anion exchange reactions do not imply that $SO_4$ is chemisorbed. On the other hand, the incomplete removal of native sorbed $SO_4$ by 0.01 M NaCl and DOC suggests that, to a certain extent, $SO_4$ ions can form stable complexes with metal atoms at the surface of soil particles. Whether these bonds developed upon ageing or following dry periods in the field remains to be demonstrated. None the less, although some of the $SO_4$ ions retained during the experiment may have formed inner-sphere complexes and become chemically co-ordinated to the soil surface, the low values for $OH/SO_4$ tend to support the idea that $SO_4$ ions mostly formed outer-sphere complexes. However, the partial buffering of $SO_4$-induced changes in solution pH by DOC cannot be ruled out and is surely likely. This could potentially affect the interpretation of $OH/SO_4$ ratios.

## PRACTICAL IMPLICATIONS AND MODELLING

The relationship between sulphate and DOC cannot be understood without taking charge effects (including pH and complexation effects) into account. This, however, requires an accurate description of cation exchange equilibria, Al solubility and other processes. Ongoing modelling efforts should be focused on better describing these processes as well as better quantifying the more immediate effects of pH and DOC on $SO_4$ sorption. These processes are vital in the podzolic B horizon of the Gårdsjön catchment, since it has strong capacity to retain both DOC and $SO_4$. This horizon actually stores a large quantity of both C and S (Table 9.2) due to its relatively high anion retention capacity. Indeed, the irreversibly adsorbed $SO_4$ represents a considerable pool of $SO_4$ relatively permanently removed from cycling (Harrison *et al.*, 1989). The podzolic B horizon should, therefore, play a decisive role in the recovery process in the Roof catchment.

A model simulation using the ordinary Freundlich model was made as an attempt to describe our results of the $SO_4$ sorption/desorption from the SET experiment (Gobran, G., Gaston, L. and Selim, H. M., unpublished). Simple Freundlich models could generally approximate $SO_4$ sorption. For certain horizons in the G1 Roof catchment, however, inclusion of pH-dependency improved description of the data. As expected from the results of Figure 9.4, the Freundlich model underestimated the quantity of $SO_4$ remaining sorbed during desorption. When incorporating the pH-dependent Freundlich model into a broader model (Gaston *et al.*, 1992) including cation exchange, Al solubility, $H_2CO_3$ equilibria, $SO_4$ complex formation with Ca, mg and Al, a mixing-cell model was used to approximate convective–dispersive transport of ions, and applied to the sequential batch data, accounting for the pH effect on $SO_4$ sorption led to no better predictions than the use of the ordinary Freundlich sorption model because soil pH was not predicted with sufficient accuracy.

Although the simple Freundlich model was adequate to describe $SO_4$ sorption in the sequential batch experiments SET, this was not the case for input of broad $SO_4$ pulse (2.6 to 50 mg S $L^{-1}$ and displacement through Bs horizon soil (Gobran, G., Gaston, L. and Selim, H. M., unpublished). In this experiment, homogeneous columns of Bs material were water-saturated, then subjected to broad pulses of $SO_4$ solutions with and without accompanying 0.01 M $NH_4Cl$ or $CaCl_2$ (adsorption phase). Sulfate was then recovered by $SO_4$-free solutions ($<2.6$ mg S $L^{-1}$). The simple Freundlich model failed to (qualitatively) describe the shape of the $SO_4$ BTC (break-through curve). Also, sorption tended to be underpredicted at low input concentrations and overpredicted at high input concentrations. In contrast, the pH-dependent Freundlich model offered better descriptions of the BTCs and predictions of net sorption. Discrepancies between experimental data and model simulations may reflect shortcomings in estimates of underlying assumptions on exchange equilibria, Al solubility, etc., all of which strongly influence predicted pH and, therefore, $SO_4$ sorption/desorption. Ongoing modelling efforts are, therefore, focused on better describing the background processes as well as better quantifying the more immediate effects of pH and DOC on $SO_4$ sorption.

It is worth mentioning here that the long-term monitoring study (Hultberg and Grennfelt, 1992) indicated that 10 year average fluxes of S in throughfall and runoff were identical. However, this trend has changed after the construction of the roof, and the $SO_4$-S concentration dropped from about 7 to 4 mg S $L^{-1}$ during the first 2 years followed by a very slow decrease to 3–4 mg S $L^{-1}$ now. This suggests a release of $SO_4$ from the native $SO_4$ pool of the podzolic B horizon after excluding S input in the Roof catchment (G1). According to our preliminary data from SLT, we believe that this level of $SO_4$ in the runoff will remain virtually unchanged for a long time. If the leaching rate of $SO_4$ remains around 3 mg S $L^{-1}$ per year, the storage of the podzolic B horizon might be depleted in about two or three decades. Considering S mineralization and transport through the soil, the depletion of the podzolic B horizon might be delayed. The opposite might happen if the production and decomposition of organic matter and the transport of DOC increases, leading to more desorption of $SO_4$ from the podzolic B horizon. We certainly agree that the recovery of soil chemistry of this soil will take several decades (Sverdrup *et al.*, 1996).

The lack of reversibility also has important implications for models which simulate soil and watershed response to changes in sulfate deposition inputs. According to Figure 9.4, models assuming full reversibility (i.e. same pathway for sorption and desorption) such as those using either the Freundlich model (Gaston *et al.*, 1992) or the Langmuir model (Cosby *et al.*, 1985, 1986), may underestimate the $SO_4$ retention capacity of the soil when the $SO_4$ concentration of the input is reduced. This means that the recovery of $SO_4$ ought to be less than predicted by reversible models.

It should be noted here, however, that deviations between sorption and desorption could also be related to kinetic retention behaviour. Irreversible sorption/desorption and the extent of kinetic reactions has also been quantified using the pressure-jump relaxation method on geothite by Zhang and Sparks (1990). Consistent with earlier findings of other investigators, Zhang and Sparks (1990) found that sulfate adsorption occurs rapidly at initial reaction stages whereas initial desorption is slow and may be considered as a limiting step. Such findings have been observed for other inorganic and heavy metal species present in the soil solution (Selim, 1992).

It is obvious that there is a need for developing process-orientated models to predict future change. This includes investigating the applicability of incorporating reversible and irreversible approaches for the description of the equilibrium and/or kinetic sulfate reactions with organic anions.

## CONCLUSIONS

1. The presence of DOC added as humic acid reduced the rate of $SO_4$ retention whereas it increased the $SO_4$ release rate, suggesting that both anions compete for the same sorption sites. This is supported by the inverse synchronicity of $SO_4$:DOC retention and release.
2. The ability of DOC to compete with $SO_4$ was pH-dependent, increasing with pH. This is due to the progressive deprotonation of functional groups present at the surface of organic acid molecules as solution pH increases.
3. Observed changes in solution pH indicate that $SO_4$ retention mostly releases OH and DOC and that outer-sphere, rather than inner-sphere, complexes are preferentially formed in the Gårdsjön soil.
4. A significant fraction of sorbed $SO_4$ is retained irreversibly in systems where DOC and $Cl^-$ (or $NO_3^-$) are the major competing anions. Knowledge on the extent and temporal dynamics of the strongly retained $SO_4$ pool is critical for assessing $SO_4$ release and soil recovery after a reduction in S inputs to forest ecosystems like the Gårdsjön catchment.
5. Considering the intimate links observed between $SO_4$ and DOC reaction at the solid–liquid interface, we suggest that the processes of DOC retention and release be included in models to improve the prediction of $SO_4$ movement and of cation transport in forest soils.
6. The competition between dissolved cations (Ca, Mg, K, Na, Al, $NH_4$ and H) for negatively charged surface sites has long been recognized and included in predictive models. Such is not the case for anions. However, we feel that the competition between inorganic ($SO_4$, Cl, $NO_3$) and organic anions (humate, fulvate, oxalate) in forest soils is, like cationic competition, an important process affecting soil acidification and recovery. As such, anion competition needs to be considered when assessing, and modelling, changes in forest soils due to atmospheric deposition, perturbations (clear-cutting, fire) and climatic fluctuations.

## ACKNOWLEDGEMENTS

The authors gratefully acknowledge I. Andersson, P. Beauchemin, H. Hultberg, J. Langlois and L. Mathier for their contribution. This research was supported by the Natural Sciences and Engineering Research Council of Canada, Le Fonds FCAR, the Swedish Environmental Protection Agency (SNV) and the Commission of the European Union as part of the REPITISC (Role of Exchange Processes in Controlling Ion Transport in Forest Soils and Catchments) project.

## REFERENCES

ACECSS (Agriculture Canada Expert Committee on Soil Survey (1987) *The Canadian System of Soil Classification*, 2nd edition. Agriculture Canada, Publication no. 1646, 164 pp.

Adams, F. and Rawajfih, A. (1977) Basaluminite and alunite: A possible cause of sulfate retention in acid soils. *Soil Science Society of America Journal*, **41**, 686–692.

Bishop, K. H. and Hultberg, H. (1995) Reversing acidification in a forest ecosystem: the Gårdsjön covered catchment experiment. *Ambio*, **24**, 85–91.

Bolan, N. S., Syers, J. K. and Tillman, R. W. (1986) Ionic strength effects on surface charge and adsorption of phosphate and sulphate in soils. *Journal of Soil Science*, **37**, 379–388.

Chao, T. T., Harward, M. E. and Fang, S. C. (1962a) Movement of $^{35}S$ tagged sulfate through soil columns. *Soil Science Society of America Proceedings*, **26**, 27–32.

Chao, T. T., Harward, M. E. and Fang, S. C. (1962b) Adsorption and desorption phenomena of sulfate ions in soils. *Soil Science Society of America Proceedings*, **36**, 234–237.

Charlet, L., Dise, N. and Stumm, W. (1990) Sulfate adsorption on a variable charge soil and on reference minerals. *Agriculture, Ecosystems and Environment*, **47**, 87–102.

Cosby, B. J., Wright, R. F., Hornberger, G. M. and Galloway, J. N. (1985) Modeling the effects of acid deposition: Estimation of long-term water quality responses in a small forested catchment. *Water Resources Research*, **21**, 1591–1601.

Cosby, B. J., Hornberger, G. M., Wright, R. F. and Galloway, J. N. (1986) Modeling the effects of acid deposition: Control of long-term sulfate dynamics by soil sulfate adsorption. *Water Resources Research*, **22**, 1283–1291.

Courchesne, F. (1991) Electrolyte concentration and composition effects on sulfate sorption by two spodosols. *Soil Science Society of America Journal*, **55**, 1576–1581.

Courchesne, F. (1992) Relationships between soil chemical properties and sulfate sorption kinetics in podzolic soils from Quebec. *Canadian Journal of Soil Science*, **72**, 467–480.

Courchesne, F. and Hendershot, W. H. (1989) Sulfate retention in some podzolic soils of the Southern Laurontians, Quebec. *Canadian Journal of Soil Science*, **69**, 337–350.

Courchesne, F. and Landry, R. (1994) Sulfate retention by spodosols in the presence of organic ligands. *Soil Science*, **158**, 329–336.

Courchesne C., Gobran, G. R. and Dufresne, A. (1995) The role of humic acid on sulfate retention and release in a podzol. *Water, Air, and Soil Pollution*, **85**, 1813–1818.

Curtin, D. and Syers, J. K. (1990) Mechanism of sulphate adsorption by two tropical soils. *Journal of Soil Science*, **41**, 295–304.

Evans, A. Jr (1986) Effects of dissolved organic carbon and sulfate on aluminum mobilization in forest soil columns. *Soil Science Society of America Journal*, **50**, 1576–1578.

Evans, A. Jr and Anderson, T. J. (1990) Aliphatic acids: influence on sulfate mobility in a forest Cecil soil. *Soil Science Society of America Journal*, **54**, 1136–1139.

Fox, T. R. and Comerford, N. B. (1990) Low-molecular-weight organic acids in selected forest soils of the Southeastern USA. *Soil Science Society of America Journal*, **54**, 1139–1144.

Fuller, R. D., David, M. B. and Driscoll, C. T. (1985) Sulfate adsorption relationships in forested spodosols of the Northeastern USA. *Soil Science Society of America Journal*, **49**, 1034–1040.

Gaston, L. A., Mansell, R. S. and Selim, H. M. (1992) Predicting removal of major soil cations and anions during acid infiltration: model evaluation. *Soil Science Society of America Journal*, **56**, 944–950.

Gobran, G.R. and Bosatta, E. (1988) Base cation depletion rate as a measure of soil sensitivity to acidic deposition: theory. *Ecological Modelling*, **40**, 25–36.

Gobran, G.R. and Clegg, S. (1992) Relationship between TOC and $SO_4^{2-}$ in soil and waters. *The Science of the Total Environment*, **117/118**, 449–461.

Gobran, G. R. and Nilsson, S. I. (1986) Effects of forest floor on sulfate mobility in a sandy iron podzol. *Proceedings International Congress of Soil Science*, 13–20 August, 1986, Hamburg, Germany pp. 304–305.

Gobran, G. R. and Nilsson, S. I. (1988) Effects of forest floor leachate on sulfate retention in a Spodosol soil. *Journal of Environmental Quality*, **17**, 235–239.

Gobran, G. R. and Tipping, E. (1993) Modelling the chemistry of humic-rich soil leachates. *Applied Geochemistry, Suppl. Issue*, **2**, 121–124.

Gobran, G. R., Selim, H. M., Hultberg, H. and Andersson, I. (1997) Description of sulfate adsorption–desorption and movement in a Swedish forest soil. *Water, Air, and Soil Pollution* (accepted for publication).

Gustafsson, J. P., Bhattacharya, P., Bain, D. C., Fraser, A. R. and McHardy, W. J. (1995) Podzolisation mechanisms and the synthesis of imogolite in northern Scandinavia. *Geoderma*, **66**, 167–184.

Harriman, R., Ferrier, R. C., Jenkins, A. and Miller, J. D. (1990) Long- and short-term hydrochemical budgets in Scottish catchments. In: B. J. Mason (ed.), *The Surface Water Acidification Programme SWAP*. Cambridge University Press, Cambridge, pp. 31–45.

Harrison, R. B., Johnson, D. W. and Todd, D. E. (1989) Sulfate adsorption and desorption reversibility in a variety of forest soils. *Journal Environmental Quality*, **18**, 419–426.

Hingston, F. J., Atkinson, R. J., Posner, A. M. and Quirk, J. P. (1968) Specific adsorption of anions on geothite. *Transactions of the 9th International Congress on Soil Science*, **1**, 669–678.

Hingston, F. J., Posner, A. M. and Quirk, J. P. (1972) Anion adsorption by goethite and gibbsite. 1. The role of the proton in determining adsorption envelopes. *Journal of Soil Science*, **23**, 177–192.

Hodges, S. C. and Johnson, G. C. (1987) Kinetics of sulfate adsorption and desorbtion by Cecil soil using miscible displacement. *Soil Science Society of America Journal*, **51**, 323–331.

Huang, P. M. and Violante A. (1986) Influence of organic acids on crystallization and surface properties of precipitation products of aluminum. In: P. M. Huang and M. Schnitzer (eds), *Interactions of Soil Minerals with Natural Organics and Microbes*. Soil Science Society of America, Madison, WI.

Hue, N. V., Adams, F. and Evans, C. E. (1985) Sulfate retention by an acid BE horizon of an Ultisol. *Soil Science Society of America Journal*, **49**, 1196–1200.

Huete, A. R. and McColl, J. G. (1984) Soil cation leaching by acid rain with varying nitrate–sulfate ratios. *Journal of Environmental Quality*, **13**, 366–371.

Hultberg, H. and Grennfelt, P. (1992) Sulphur and seasalt deposition as reflected by throughfall and runoff chemistry in forested catchments. *Environmental Pollution*, **75**, 215–222.

Inskeep, W. P. (1989) Adsorption of sulfate by kaolinite and amorphous iron oxide in the presence of organic ligands. *Journal of Environmental Quality*, **18**, 379–385.

Jardine, P. M., Weber, N. L. and McCarthy, J. F. (1989) Mechanisms of dissolved organic carbon adsorption on soil. *Soil Science Society of America Journal*, **53**, 1378–1385.

Johnson, D. W. and Henderson, G. S. (1979) Sulfate adsorption and sulfur fractions in a highly weathered soil under a mixed deciduous forest. *Soil Science*, **128**, 34–40.

Johnson, D. W. and Todd, D. E. (1983) Relationships among iron, aluminum, carbon, and sulfate in a variety of forest soils. *Soil Science Society of America Journal*, **47**, 792–800.

Khanna, P. K., Prenzel, J., Meiwes, K. J., Ulrich, B. and Matzner E. (1987) Dynamics of sulfate retention by acid forest soils in an acidic deposition environment. *Soil Science Society of America Journal*, **51**, 446–452.

Kaiser, K. and Zech, W. (1996) Nitrate, sulfate and biphosphate retention in acid forest soils affected by natural dissolved organic carbon. *Journal of Environmental Quality*, **25**, 1325–1331.

Krug, E. C. and Frink, C. R. (1983) Acid rain on acid soil: A new perspective. *Science*, **221**, 520–525.

Lee, Y. H. and Hultberg, H. (1990) Studies of strong and weak acid speciation in fresh waters in experimental forested catchments. In: B. J. Mason (ed.), *The Surface Waters Acidification Programme*. Cambridge University Press, Cambridge, pp. 183–192.

Malcolm, R. L. and MacCarthy, P. (1986) Limitations in the use of commercial humic acids in water and soil research. *Environmental Science and Technology*, **20**, 904–911.

Maynard, D. G., Kalra, Y. P. and Radford, F. G. (1987) Extraction and determination of sulfur in organic horizons of forest soils. *Soil Science Society of America Journal*, **51**, 801–805.

Nodvin, S. C., Driscoll, C. T. and Likens, G. E. (1986) Simple partitioning of anions and dissolved organic carbon in a forest soil. *Soil Science*, **142**, 27–35.

Parfitt, R. L. and Smart, R. S. C. (1978) The mechanism of sulfate adsorption on iron oxides. *Soil Science Society of America Journal*, **42**, 48–50.

Rajan, S. S. S. (1978) Sulfate adsorbed on hydrous alumina, ligands displaced, and changes in surface charge. *Soil Science Society of America Journal*, **42**, 39–44.

Reuss, J. O. (1983) Implications of the Ca–Al exchange system for the effect of acid precipitation on soils. *Journal of Environmental Quality*, **12**, 591–595.

Reuss, J. O. and Johnson, D. W. (1986) *Acid Deposition and the Acidification of Soils and Waters*. Ecological Studies, **59**, Springer-Verlag, Berlin.

Rochelle, B. P., Church, M. R. and David, M. B. (1987) Sulfur retention at intensively studied sites in the U.S. and Canada. *Water, Air, and Soil Pollution*, **33**, 73–83.

Saito, K. and Shoji, S. (1984) Silica adsorption and dissolution properties of Andosol from Northeast Japan as related to their noncrystalline clay mineralogical composition. *Soil Science*, **138**, 341–345.

Sandbron, P. and Lavkulich, L. M. (1989) Ferro-humic podzols of Coastal British Columbia: I. Morphology, selected properties, and classification. *Soil Science Society of America Journal*, **53**, 511–517.

Schindler, S. C., Mitchell, M. J., Scott, T. J., Fuller, R. D. and Driscoll, C. T. (1986) Incorporation of $^{35}$S-sulfate into inorganic and organic sulfur constituents of two forest soils. *Soil Science Society of America Journal*, **50**, 457–462.

Selim, H. M. (1992) Modeling the transport and retention of inorganics in soils. *Advances in Agronomy*, **47**, 331–384.

Sheldrick, B. H. (1984) *Analytical Methods Manual*. Land Resource Research Institute, 84–30, Agriculture Canada, Ottawa.

Singh, B. R. (1984) Sulfate sorption by acid forest soils: 2. Sulfate adsorption isotherms with and without organic matter and oxides of aluminum and iron. *Soil Science*, **138**, 294–297.

Soil Survey Staff (1975) *Soil Taxonomy: A Basic System of Soil Classification for Making and Interpreting Soil Surveys*. USDA–SCS Agricultural Handbook, 436, US Government Printing Office, Washington DC, USA.

Sposito, G. (1984) *The Surface Chemistry of Soils*. Oxford University Press. New York.

Stevenson, F. J. (1982) *Humus Chemistry*. John Wiley & Sons, Chichester.

Strickland, T. C., Fritzgerald, J. W. and Swank, W. T. (1986) *In situ* measurements of sulfur incorporation into forest floor and soil organic matter. *Canadian Journal of Forest Research*, **16**, 549–553.

Sverdrup, H., Warfvinge, P., Hultberg, H., and Moldan, P. (1996) Modelling acidification and recovery in the roof catchment at Lake Gårdsjön *Water, Air, and Soil Pollution*, **85**, 1753–1758.

Turner, L. J. and Kramer, J. R. (1992) Irreversibility of sulfate sorption on goethite and hematite. *Water, Air, and Soil Pollution*, **63**, 23–32.

Van Breemen, N., Mulder, J. and Driscoll, C. T. (1983) Acidification and alkalinization of soils. *Plant and Soil*, **75**, 283–308.

Vance, G. F. and David, M. B. (1992) Dissolved organic carbon and sulfate sorption by spodosol mineral horizons. *Soil Science*, **154**, 136–144.

Wilkinson, L. (1989) *SYSTAT: The System for Statistics for the PC*. SYSTAT Inc., Evanston, IL, USA.

Wright, R. F., Lotse, E. and Semb, A. (1993) RAIN project: Results after 8 years of experimentally reduced acid deposition to a whole catchment. *Canadian Journal of Aquatic Sciences*, **50**, 258–268.

Zhang, G. Y., Zhang, X. N. and Yu, T. R. (1989) Adsorption of sulphate and fluoride by variable charge soils. *Journal of Soil Science*, **38**, 29–38.

Zhang, P. C. and Sparks, D. L. (1990) Kinetics and mechanisms of sulfate adsorption/desorption on goethite using pressure-jump relaxation. *Soil Science Society of America Journal*, **54**, 1266–1273.

# 10 Estimating the Weathering Rate at Gårdsjön Using Different Methods

**HARALD SVERDRUP, PER WARFVINGE & TONIE WICKMAN**

The weathering rate of the soil in the catchment at Lake Gårdsjön is well known from a number of determinations. The weathering rate is one of the long-term determining factors for a number of processes in the soil: for the chemical state of soils, long-term nutrient supply to forests, the chemical composition of streams and lakes, the chemical composition of groundwater and the sensitivity of soils and surface waters to acidification. Great emphasis has been put on getting a good estimate of the weathering rate at the Lake Gårdsjön catchments. The rate was determined by different methods: (1) Total base cation net soil export by budget calculation, (2) estimation by the Sr-isotope ratio method, (3) correlation to total analysis, (4) geochemical modelling using the PROFILE model, (5) historic rate as estimated with the quartz enrichment/mineral depletion method and (6) by running the SAFE model for 12 500 years.

In the roofed catchment G1, different estimates yield a present weathering rate in the range 0.3–0.4 keq $ha^{-1}$ $yr^{-1}$. The weathering rate determined with PROFILE was 0.3 keq $ha^{-1}$ $yr^{-1}$ in the pre-industrial era, and at present it is calculated as 0.37 keq $ha^{-1}$ $yr^{-1}$ , assuming that the average soil depth is 0.43 m. Budget studies give a weathering rate of 0.36 keq $ha^{-1}$ $yr^{-1}$.

For F1, the present weathering rate obtained with the PROFILE model is 0.56 keq $ha^{-1}$ $yr^{-1}$. Budget studies yield a weathering rate of 0.65 keq $ha^{-1}$ $yr^{-1}$ in 1981, and 0.55 keq $ha^{-1}$ $yr^{-1}$ for 1991–93. The historical average weathering rate was estimated using SAFE to be 0.74 keq $ha^{-1}$ $yr^{-1}$, and the immediate pre-industrial rate to be approximately 0.45 keq $ha^{-1}$ $yr^{-1}$.

## INTRODUCTION

This text deals with the chemical weathering of minerals in the Gårdsjön catchment. By chemical weathering we understand the release of base cations due to chemical dissolution from minerals in the soil matrix, and the neutralization or production of alkalinity connected to this process, as distinguished from the chemical dissolution of rock or mineral mass, which is also called chemical weathering. We strictly separate this from the denudation of base cations from ion exchange complexes in the soil which is termed leaching or cation exchange, as well as the degradation of different

*Experimental Reversal of Acid Rain Effects: The Gårdsjön Roof Project.*
Edited by Hans Hultberg and Richard Skeffington. 

types of organic matter. We also disregard any type of mechanical degradation of soil minerals.

In soils, cations are produced from minerals by chemical weathering. The rate of production is dependent on the chemical and physical properties of the soil, as well as the chemical conditions. The weathering of the mineral matrix will be the major long-term source of base cations to replace removed base cations, and an important factor for determining the chemical status of the soil. Present-day soils are subject to leaching of base cations by acid deposition percolating through the soil profile, as well as base cation removal by uptake caused by forest growth. The base cations removed by these processes are fully or partially replaced by cations produced by weathering reactions in the soil mineral matrix, by degradation of organic material and atmospheric deposition of base cations.

Several weathering rates reported earlier have been revised, since new and better data have become available for the Gårdsjön catchment. Soil thickness has been re-estimated from 0.8–0.7 m in F1 to 0.63 m. Better uptake and cation export rates have become available since earlier estimates in Sverdrup and Warfvinge (1988), which relied mostly on information collected before 1982. Better hydrology and soil moisture values are available today.

## GEOCHEMICAL PROPERTIES OF THE SOILS

The Gårdsjön area is a small watershed with several lakes, located on the Fenno-Scandic granitic shield on the Swedish west coast. The soils in the area are shallow orthic podzols. Lake and soil acidification has been intensively studied in the area in a large multidisciplinary project, and it has been determined that the lake acidified in the early 1960s (Renberg *et al.*, 1985). Since then the lake and its watershed has stayed in a stable acid condition. The acidity of the lake is but a mirror of the chemical conditions in the soil of the watershed, as the runoff water composition is largely formed by chemical reactions occurring when the water passes through the soil matrix. As a part of the lake and soil acidification study, the mineralogy of the soil as well as the bedrock was determined along with parameters concerning the soil texture and chemical status at a number of sites in the watershed (Melkerud 1983; Olsson *et al.*, 1985).

The soil solution has pH values of 3.9–4.3 in the upper 0.30 m of the soil profile, pH 4.3–4.5 in the 0.3–0.6 m layer and pH values of 4.5–5.0 in the lowest layers (Chapter 23, this volume). In 1981 the soil pH was slightly higher (Melkerud, 1983). The till mineral composition (Table 10.1) has been estimated in F1, F2 and F3 using data from total analysis and XRD analysis (Melkerud, 1983). Total analysis for subcatchments F1 and G1 at Lake Gårdsjön, data obtained from Melkerud (1983), are shown in Table 10.2. Site 23:1 is located in the upper part of F1, close to G1.

The soil moisture was set at 0.15 $m^3 m^{-3}$ in the top of the soil profile increasing to 0.35 at 0.6 m depth in G1, according to recent data (Chapter 5, this volume). The average depth of the soil layer is 0.43 m in G1, 0.63 m in F1, and it has been estimated that it is 0.6–0.8 m in F2 and F3. No soil mineralogy is available for the

**Table 10.1.** Till and bedrock composition in three catchments on the north slope of Gårdsjön (different estimates taken from Melkerud, 1983) estimated using a combination of total analysis, XRF and XRD analysis

| Mineral | Bedrock composition (%) | At soil depth 2 m (%) | At soil depth 0.7 m (%) | Average soil composition F1 (%) | F2 (%) | F3 (%) |
|---|---|---|---|---|---|---|
| K-feldspar | 20.0 | 14.0 | 19.0 | 15.0 | 15.0 | 12.0 |
| Plagioclase | 32.3 | 27.0 | 16.0 | 14.0 | 12.0 | 12.0 |
| Hornblende | 2.6 | 2.0 | 1.0 | 0.5 | 0.5 | — |
| Epidote | 4.5 | 2.0 | 2.0 | 0.1 | 0.1 | 1.1 |
| Biotite | 13.0 | 1.0 | 1.0 | 1.2 | 1.0 | — |
| Garnet | 0.1 | 0.1 | — | 0.1 | 0.1 | — |
| Chlorite | 0.4 | 0.4 | — | 0.4 | 0.4 | — |
| Vermiculite | — | 1.1 | 5.0 | 5.0 | 5.0 | 8.0 |
| Apatite | 0.1 | 0.1 | 0.1 | 0.1 | 0.1 | 0.1 |
| Undefined | — | 12.3 | 23.9 | 30.5 | 32.8 | 33.8 |
| Quartz | 27.0 | 40.0 | 28.0 | 28.0 | 28.0 | 28.0 |

**Table 10.2.** Total analysis (%) for soil from subcatchments F1 and G1 at Lake Gårdsjön. Data obtained from Melkerud (1983). Site number after Melkerud (1983)

| Site | CaO | MgO | $Na_2O$ | $K_2O$ | $P_2O_5$ | $SiO_2$ | $Al_2O_3$ |
|---|---|---|---|---|---|---|---|
| F1; 21:1 | 2.11 | 0.88 | 2.8 | 2.76 | 0.06 | 73.3 | 13.9 |
| G1; 23:1 | 2.46 | 1.57 | 2.9 | 2.61 | 0.10 | 69.4 | 15.6 |

roofed catchment, G1, but it was assumed that the soil mineralogy was the same as in catchment F1.

The net amount of biomass cation accumulation was estimated as the difference over time between the measured storage in tree trunks and substantial roots.

## METHODS OF ESTIMATING WEATHERING

### Budget study

There are a number of studies where the base cation release rate due to chemical weathering has been estimated from catchment studies (Balek *et al.*, 1978; Paçes, 1983; Cronan, 1985; Olsson *et al.*, 1985; Velbel, 1985; Olsson and Melkerud, 1990). However, until recently, only a few studies have had access to datasets that permitted the base cation transport rates to be separated into rates of chemical weathering and depletion of the exchange pool in the soil (Åberg *et al.*, 1989; Wickman and Jacks, 1992; Miller *et al.*, 1993; Bain and Bacon, 1994, Bain *et al.*, 1994). Others have made field–laboratory comparisons (Paçes, 1983; Velbel, 1985), but several were unsuccessful because of laboratory kinetic data highly inadequate for the purpose. More recently Sverdrup (1990), Warfvinge and Sverdrup (1992), Sverdrup and Warfvinge (1993) have more systematically succeeded in comparing laboratory rates and field weathering observations, using integrated soil chemistry models.

A mass balance for a base cation keeps account of all sources and sinks of the substance considered in the system, and may be expressed as:

$$\text{influx} + \text{production} = \text{accumulation} + \text{sinks} + \text{outflux} \tag{10.1}$$

where production may be the amount produced by weathering reactions and decomposition of organic matter, and accumulation is reversible storage or depletion in the soil solution or ion exchange reservoir. Sinks may represent irreversible removal of cations from the system, such as base cation uptake through forest growth or permanent accumulation in peat. All sink terms and source terms for a base cation are balanced in a mass balance for each soil horizon. It is assumed that each soil horizon is well mixed. The mass balance equation can be rearranged to an expression for the weathering rate for each base cation:

$$\text{weathering} = \text{leaching} + \text{uptake} - \text{deposition} - \text{base saturation decrease} \tag{10.2}$$

It should be noted that this equation only comprises weathering in the sense that it represents irreversible chemical dissolution of soil primary minerals, expressed as base cation flux due to the dissolution process. In most budget calculations, the base cation flux from change in base saturation cannot be separated from the flux of base cations from weathering. If it is assumed that the catchment is already at steady state, then that reduces the expression for the base cation release rate due to chemical weathering to:

$$\text{weathering} = \text{leaching} + \text{uptake} - \text{deposition} \tag{10.3}$$

The basic assumptions of steady state must be carefully tested in each case in order to use the methodology for determining the weathering rate.

Cation exchange capacity is often assumed not to have changed with time. There is evidence available that shows that the CEC has changed most significantly in many parts of Europe during historic times when climatic changes occurred (Burl, 1976; Champion *et al.*, 1984) but also in the last 50–120 years (Johnston *et al.*, 1986; Tamm and Hallbäcken, 1986; Berden *et al.*, 1987; Falkengren-Grerup, 1987), but no decisive conclusion can be made concerning change of CEC over time. A very small denudation of the cation exchange pool will amount to a significant amount as compared to the weathering state.

**Budget studies using the strontium isotope ratio**

Åberg *et al.* (1989), Jacks *et al.* (1989), Miller *et al.* (1993) and Bain and Bacon (1994) used a modified budget approach where the difference in the ratio between the two naturally occurring strontium isotopes ($^{87}Sr/^{86}Sr$) in the deposition and the bedrock, were used to determine the fraction of the cation export coming from the weathering of soil minerals. The approach assumes the isotopic ratio in the part originating from chemical weathering to be equal to the ratio measured in a citric or salicylic acid leachate, after several extractions to remove exchangeable Sr (Wickman and Jacks, 1992). Base cation release caused by weathering can be calculated from the relation:

$$W = \frac{rBC}{rCa} D \left( \frac{Sr_L - Sr_D}{Sr_W - Sr_R} \right) \tag{10.4}$$

where: W = base cation weathering rate (keq ha$^{-1}$ yr$^{-1}$); D = calcium cation deposition rate (keq ha$^{-1}$ yr$^{-1}$); $r_{BC}/r_{Ca}$ = the base cation to calcium weathering rate ratio; $Sr_D$ = the strontium isotope ratio in the deposition; $Sr_R$ = the strontium isotope ratio in the runoff; $Sr_W$ = the strontium isotope ratio in the mineral matrix.

The method is based on the assumptions that the weathering of strontium from the mineral matrix is linearly correlated with the weathering rate of all base cations. The method is unique in that when combined with a budget study it allows the weathering rate to be separated from leaching from the ion exchange complex.

**The historic weathering rate**

The till in Scandinavia is generally of a geologically recent origin, most of the mineral matrix being formed during the last period of glaciation. One approach may be to assume that, at the end of the glaciation, the mineral matrix of the soil consisted of freshly ground material, which only has been exposed to weathering since then. If the composition of the parent minerals is known, this may be used together with the present composition of the soil mineral matrix to estimate the approximate weathering rate during the last 12 500 years. The mineral composition of the soil and the different parent bedrock materials at Lake Gårdsjön is shown in Table 10.1. Many samples have been taken in the lake catchment area, and several sampling sites are located close to the roofed experiment, but no sample is available from under the roof. The dissolved mass of each mineral is calculated from the difference between the parent material composition, either at the bottom of the profile or bedrock and the present-day till mineral composition. The formula used was:

$$R_W = \sum_{j}^{\text{minerals}} \frac{(C_{REF,j} - C_{soil,j})\, \rho_{SOIL}\, Z\, 10{,}000}{t} ME \tag{10.5}$$

where: $R_W$ = weathering rate; $C_{REF,j}$ = fraction of mineral in reference materials (keq ha$^{-1}$ yr$^{-1}$); $C_{soil,j}$ = fraction of mineral in soil; $\rho_{SOIL}$ = soil density (kg m$^{-3}$); ME = equivalent amount of base cation per mineral mass (keq kg$^{-1}$); t = time since the material was formed, (yr); $Z_{soil}$ = soil depth (m).

Zirconium, quartz or rutile may also be used as the internal inert standard (Olsson and Melkerud, 1990). This approach has several inherent weaknesses. The mineral matrix produced by the glacier some 12 000 years ago probably contained a certain fraction of fine particles which would dissolve very fast in a geological perspective. Studies of moraines in the Arctic and Antarctic indicate that this initial rapid phase lasts from a few hundred years to maybe a thousand years. The effect is somewhat counteracted by the substantially lower temperature in the immediate post-glacial period. Data indicate that a maximum of 25% of the mass dissolved in the total time period will be dissolved in the initial weathering stage. This would imply the rates could be 25% larger. Quartz dissolves very slowly, but is not totally inert. The rate is approximately one order of magnitude slower than the rate of orthoclase dissolution in the natural soil, enough to justify the assumption that quartz is inert.

### Total cation content correlation

The correlation is based on the fact that the base cations magnesium and calcium are mainly associated with easily weatherable minerals, whereas sodium and potassium are associated with very slowly weathering minerals such as feldspars. It also uses the fact that the weathering rate is dependent on the temperature. The empirical correlation (derived using data from Olsson and Melkerud, 1990), applicable to soils of granitic origin, is expressed in keq $ha^{-1}$ $yr^{-1}$ (Sverdrup, 1990; Sverdrup *et al.*, 1990);

$$W_{Ca} = z_{soil}(1.0 \times 10^{-4}\%CaO\,(768 + 104\,T) - 0.08) \tag{10.6}$$

$$W_{Mg} = z_{soil}(2.12 \times 10^{-4}\%MgO\,(768 + 104\,T) - 0.09) \tag{10.7}$$

$$W_{Na} = z_{soil}(0.37 \times 10^{-4}\%Na_2O\,(768 + 104\,T)) \tag{10.8}$$

$$W_{K} = z_{soil}(0.21 \times 10^{-4}\%K_2O\,(768 + 104\,T)) \tag{10.9}$$

where the symbols are: T = average annual temperature (°C); %CaO, . . . = cation content expressed as oxide in the soil (%); $z_{soil}$ = soil depth (m).

The temperature sum is an alternative way of describing the annual average temperature. The two are linked by:

$$TSUM = \sum_{i=1}^{365} T_i > 5°C = (768 + 104\,T) \tag{10.10}$$

A similar empirical expression is obtained by adding the expression for individual base cations and adding the rate of Ca release from apatite (Sverdrup *et al.*, 1990):

$$\begin{aligned} W = z_{soil}\,0.98 \times 10^{-4}\,(768 + 104\,T)\,(\%CaO + 2.12\,\%MgO \\ + 0.37\,\%Na_2O + 0.21\,\%K_2O + 2.7\,\%P_2O_5 - 0.17) \end{aligned} \tag{10.11}$$

The expressions are calibrated for Swedish podzolic soils and only valid for them. For other types of soil, other correlations will presumably apply. The base cation content is based on an analysis of a totally dissolved soil sample. Similar correlations have also been derived for the individual base cations (Olsson and Melkerud, 1990), in keq $ha^{-1}$ $yr^{-1}$, but these are based on a mixture of values from sites with weathering to different soil depths. Those relations cannot be used until they have been normalized with respect to weathering depth. The correlations above were based on the historical weathering rate at multiple sites in Gävleborgs Län in Sweden and the province of Skåne, as derived from Zr depletion studies.

## PROFILE

PROFILE is a steady-state soil chemistry model that includes several processes like weathering, uptake, nitrification, nutrient cycling and solution equilibria (Sverdrup and Warfvinge, 1993, 1995a,b). It circumpasses the transient phenomena considered in its dynamic counterpart SAFE, and directly calculates the chemical steady state that will result from a certain set of conditions.

An important subroutine in PROFILE is the weathering submodel. Several chemical reactions between the mineral and constituents in the liquid solution contribute to the base cation release rate due to chemical weathering of silicate minerals, and the total rate will be the sum of the rates of the individual reactions. For most minerals several reactions have been experimentally identified:

- The reaction with hydrogen ion, aluminium and base cations.
- The reaction with water, aluminium and base cations.
- The reaction with hydroxyl ion and aluminium.
- The reaction with carbon dioxide.
- The reaction with strongly complexing organic acids.

The total base cation release rate by chemical weathering will be the sum of the rate of all parallel simultaneous processes regardless of the molecular mechanism, minus the rate of precipitation of secondary solid phases (Lasaga, 1981; Aagaard and Helgeson, 1982; Helgeson *et al.*, 1984). Base cations can be removed from the aqueous solution by the precipitation of secondary minerals, and the forward reaction rate can under certain circumstances be affected by dissolved species which interfere with the decay mechanism of the activated complexes.

The rate is proportional to the exposed surface area of the mineral, $A_w$:

$$R_W = r\,A_W \tag{10.12}$$

The principal form of the rate equation derived is consistent with the theories developed by Aagaard and Helgeson (1982) and by Chou and Wollast (1985), and a general rate equation was derived from the transition state theory. The synthesis of data and theory led to the following rate law incorporated in the PROFILE and SAFE models (Warfvinge and Sverdrup, 1992):

$$r = k_{H+}\,[H^+]^n\,\frac{1}{f_H} + k_{H_2O}\,\frac{1}{f_{H_2O}} + k_{CO_2}\,P_{CO_2}^m\,\frac{1}{f_{CO_2}} + k_{org}\,[R^-]^{0.5}\,\frac{1}{f_{org}} \tag{10.13}$$

where the retardation factors $f_H, f_{H_2O}, f_{Al}, f_{CO_2}$ and $f_{org}$ are given by:

$$f_H = \left(1 + \left(\frac{[BC]}{k_{BC}}\right)^{x_{BC}}\right)\left(1 + \left(\frac{[Al^{3+}]}{x_{Al}}\right)^{x_{Al}}\right) \tag{10.14}$$

$$F_{CO_2} = f_{H_2O} = \left(1 + \left(\frac{[BC]}{k_{BC}}\right)^{z_{BC}}\right)\left(1 + \left(\frac{[Al^{3+}]}{k_{Al}}\right)^{z_{Al}}\right) \tag{10.15}$$

$$f_{org} = \left(1 + \left(\frac{[R^-]}{k_R}\right)^{0.5}\right) \tag{10.16}$$

where: $k_{H+}$ = rate coefficient in the $H^+$ reaction; $k_{H_2O}$ = rate coefficient in the $H_2O$ reaction; $k_{CO_2}$ = rate coefficient in the $CO_2$ reaction; $k_{org}$ = lumped rate coefficient in organic acid reactions; $\kappa_{org}$ = organic acid reaction saturation constant; r = reaction rate ($kmol\,m^{-2}\,s^{-1}$); x = base cation reaction order in the $H^+$ reaction; u = Al reaction order in the $H_2O$ reaction; m = reaction order in the $CO_2$ reaction; $P_{CO_2}$ = partial pressure $CO_2$ in the soil solution (atm); [BC] = base cation concentration ($kmol\,m^{-3}$); [org] = organic acid concentration (Oliver eq.) ($kmol\,m^{-3}$).

In soils the availability of moisture may be limited, causing the reactivity of the exposed mineral surface to be less than unity. The reactions only take place on wetted surfaces, and the degree of surface wetting is taken to be proportional to soil moisture. All surfaces which participate in the reactions must be wetted, but there must also be sufficient soil solution present for the weathering process to respond to and exchange mass with other soil processes.

Soil moisture saturation is calculated by combining the densities of the solid ($2700\,kg\,m^{-3}$) wet ($1000\,kg\,m^{-3}$) and gaseous phases ($0.010\,kg\,m^{-3}$) and the bulk soil density $\rho_{soil}$, with the volumetric water content $\Theta$:

$$\theta = \frac{2700\Theta}{2700 + 1000\Theta - \rho_{soil}} \tag{10.17}$$

The total rate for the soil horizon is obtained by repeating the calculation additively for all minerals present:

$$R_W = \sum_{j=1}^{\text{minerals}} r_j\, A_{tot}\, x_j\, \frac{\theta}{\theta_m}\, z \tag{10.18}$$

where: $A_W$ = exposed surface area of the mineral fraction ($m^2\,ha^{-1}$); $x_j$ = the surface area fraction of mineral j in the soil; $\theta$ = the soil moisture content ($kg\,m^{-3}$); $\theta_m$ = soil porosity as kg water per volume unit ($kg\,mg^{-3}$); z = soil layer thickness.

The total exposed area ($A_{tot}$) can be estimated using the formula:

$$A_{tot} = (8.0\,x_{clay} + 2.2\,x_{silt} + 0.3\,x_{sand})\,\frac{\rho_{soil}}{1000} \tag{10.19}$$

where: $x_{clay}$ = less than 2 $\mu$m size fraction; $x_{silt}$ = 2–60 $\mu$m size fraction; $x_{sand}$ = 60–250 $\mu$m size fraction; $\rho_{soil}$ = soil bulk density.

The weathering rate in the natural soil is calculated with rate coefficients taken from laboratory studies carried out at 25°C, and adjusted to soil temperatures using an Arrhenius relationship:

$$\log_{10}\left(\frac{r_1}{r_2}\right) = A\left(\frac{1}{T_2} - \frac{1}{T_1}\right) \tag{10.20}$$

where: A = Arrhenius factor; T = the absolute temperature (Kelvin); r = chemical weathering rate in a specific reaction ($kmol/m^2\,s$).

Sverdrup (1990) made a critical review of the literature concerning mineral dissolution, and a re-evaluation of some of the experimental data presented in the literature. The results were integrated with the experimental data of the authors to determine rate coefficients for the base cation release rate from a large number of rock-forming primary minerals. The rate coefficients are presented in Table 10.3.

PROFILE is a steady-state model that ignores development over time. It is used to calculate soil chemistry and soil weathering at steady state under a specified set of conditions. SAFE is a dynamic model used for acidification assessments, and calculates changes in soil chemistry and weathering over time. A special version of the model has been developed to calculate changes during the Holocene period, and make predictions for mineralogical and texture changes with time (Rietz, 1995).

## RESULTS

### Budget method

Hultberg (1985) determined weathering plus ion exchange plus net mineralization to 1.20 keq $ha^{-1} yr^{-1}$ for 1981, and cited a weathering rate from an aluminium budget in 1985 as approximately 0.6 keq $ha^{-1} yr^{-1}$. Lundström (1987) estimated weathering plus ion exchange plus net mineralization to be 0.85 keq $ha^{-1} yr^{-1}$.

The base cation release rate due to chemical weathering of silicate minerals was calculated for the Gårdsjön watersheds using data available in the literature (Melkerud, 1983; Hultberg, 1985; Olsson *et al.*, 1985) and new data from the monitoring program. Table 10.4 shows the results obtained using flux data for two periods, 1981 and 1991–93. Using input values for 1991–93 yields similar weathering rates for Ca, lower for Mg, higher for Na and a very uncertain value for K as compared with 1981.

The results involve large uncertainties, and very different values are obtained, depending on which year the data were collected. At Gårdsjön, base cation uptake was estimated to be 0.6–0.7 keq $ha^{-1} yr^{-1}$. Leaching in 1991 was estimated to be in the range 0.65–0.75 keq $ha^{-1} yr^{-1}$ for Ca+Mg+K, and for Na it was 2.2–2.4 keq $ha^{-1} yr^{-1}$. The amounts of cations which may potentially be removed from the soil by biological processes will be dependent on whether there is a net accumulation of biomass in the system or a net breakdown of biomass. If the canopy is harvested, the removal becomes irreversible. The high rates for Mg in 1981 may be suspected to be caused by acidity leaching of Mg from the ion exchange complex. The uptake numbers are rather uncertain, and uptake could possibly be only half of the indicated amount. Base cation deposition values are also notoriously difficult to verify, leaving room for much uncertainty as well. That would give a range in the calculated weathering rate of 0.3–0.7 keq $ha^{-1} yr^{-1}$ for F1 and 0.1–0.4 keq $ha^{-1} yr^{-1}$ for G1. The throughfall value for K is higher than the deposition value due to canopy leaching. Deposition of K was set at 0.2 keq $ha^{-1} yr^{-1}$ for F1 and 0.1–0.4 keq $ha^{-1} yr^{-1}$ for G1. In F1, atmospheric inputs are a little lower than at G1. This could imply that the reservoir of K is being depleted, since the roof was installed, but the number is very uncertain. In 1991–93 the acid deposition load was approximately 40% lower

**Table 10.3.** Approximate laboratory rate coefficients for the chemical weathering rate for minerals expressed as the flux of base cations (Ca, Mg, Na, K) related to a total rate expressed as keq $m^2 s^{-1}$ at 8°C, applied in the PROFILE model. Values within parentheses are interpolations and estimates from Madelung site energies and other crystallographic data (modified after Sverdrup, 1990). Muscovite includes muscovite and secondary illite; biotite includes biotite, glauconite, primary illite and phlogopite; hornblende includes hornblende, tremolite and glaucophane; and pyroxene includes minerals of the pyroxene group and talc, and has the properties of augite. Garnet includes all nesosilicates and anorthite; vermiculite includes do-octahedral vermiculite, secondary illite, smectite and montmorillonite

| Mineral | $pk_H$ | $n_H$ | $C_{Al}$ | $y_{Al}$ | $C_{BC}$ | $x_{BC}$ | $pk_{H_2O}$ | $z_{Al}$ | $z_{BC}$ | $pk_{CO_2}$ | $n_{CO_2}$ | $pk_{org}$ | $C_R$ |
|---|---|---|---|---|---|---|---|---|---|---|---|---|---|
| K-feldspar | 14.7 | 0.5 | 0.5 | 0.4 | 300 | 0.15 | 17.2 | 0.14 | 0.15 | 16.8 | 0.6 | 15.0 | (5) |
| Plagioclase | 14.6 | 0.5 | 0.5 | 0.4 | 300 | 0.2 | 16.8 | 0.14 | 0.15 | 15.9 | 0.6 | 14.7 | (5) |
| Albite | 14.5 | 0.5 | 0.5 | 0.4 | 300 | 0.2 | 16.5 | 0.14 | 0.15 | 15.9 | 0.6 | 14.7 | 5 |
| Hornblende | 13.3 | 0.7 | 5 | 0.4 | 200 | 0.3 | 16.3 | (0.3) | 0.3 | (15.9) | (0.6) | 14.4 | (5) |
| Pyroxene | 12.3 | 0.7 | (500) | (0.2) | 200 | (0.3) | 17.5 | 0.1 | 0.3 | 15.8 | 0.6 | 14.4 | (5) |
| Epidote | 14.0 | 0.5 | (50) | (0.3) | 200 | (0.2) | 17.2 | 0.2 | 0.2 | (16.2) | 0.6 | (14.4) | (5) |
| Garnet | 12.4 | 1.0 | 300 | 0.4 | 500 | (0.2) | 16.9 | 0.2 | 0.2 | (15.8) | (0.6) | (14.7) | (50) |
| Biotite | 14.8 | 0.6 | 10 | 0.3 | (500) | 0.2 | 16.7 | 0.2 | 0.2 | 15.8 | (0.5) | 14.8 | (50) |
| Muscovite | 15.2 | 0.5 | 4 | 0.4 | (500) | 0.1 | 17.5 | 0.2 | 0.1 | 16.5 | (0.5) | 15.3 | (5) |
| Fe-chlorite | 14.8 | 0.7 | (50) | (0.2) | (500) | (0.2) | (17.0) | (0.1) | (0.1) | (16.2) | (0.5) | (15.0) | (5) |
| Mg-chlorite | 14.3 | 0.7 | (50) | (0.2) | (200) | (0.2) | (16.7) | (0.1) | (0.1) | 15.8 | (0.5) | 14.5 | (5) |
| Fe-vermiculite | 15.2 | 0.6 | (4) | (0.4) | (500) | 0.2 | 17.6 | 0.1 | (0.1) | (16.5) | (0.5) | (15.6) | (5) |
| Mg-vermiculite | 14.8 | 0.6 | 4 | 0.4 | 200 | 0.2 | 17.2 | 0.1 | (0.1) | (16.2) | (0.5) | (15.2) | (5) |
| Apatite | 12.8 | 0.7 | 100 | — | 300 | 0.4 | 15.8 | — | 0.2 | 15.8 | 0.6 | (19.5) | (5) |
| Kaolinite | 15.1 | 0.7 | 2 | 0.4 | 500 | 0.4 | 17.6 | 0.2 | 0.2 | (16.5) | (0.5) | 19.5 | (5) |
| Calcite | 13.6 | 1.0 | 5000 | 0.4 | 1000 | 0.4 | 15.2 | — | 0.2 | 13.2 | 1.0 | 13.2 | (5) |

**Table 10.4.** Results from two budget studies. The weathering plus ion exchange rate (in keq $ha^{-1} yr^{-1}$) for the three "F" subcatchments at Gårdsjön (Sverdrup and Warfvinge, 1988) using input data from Hultberg (1985) for 1981 is shown in the upper part, and from 1991–93 for F1 and G1 in the lower part

| Catchment | Ca | Mg | K | Na | Sum |
|---|---|---|---|---|---|
| F1 1981 | 0.21 | 0.42 | 0.11 | 0.01 | 0.75 |
| F2 1981 | 0.18 | 0.21 | 0.07 | 0.02 | 0.48 |
| F3 1981 | 0.19 | 0.19 | 0.07 | 0.02 | 0.47 |
| Average | 0.19 | 0.27 | 0.08 | 0.02 | 0.56 |
| F1 1991–93 | 0.25 | 0.16 | 0.10 | 0.13 | 0.54 |
| G1 1991–93 | 0.13 | 0.10 | 0.05 | 0.08 | 0.36 |

than in 1981. This could explain the lower Mg values; the acid rain is no longer depleting the ion exchange reservoir to the same extent.

### Strontium method

Water samples were collected in April 1992 in G1. This was in the beginning of the hydrological year 1992/93, when 0.09 keq Ca $ha^{-1}$ $yr^{-1}$ was sprinkled under the roof. In the hydrological year before, more calcium was applied (0.13 keq Ca $ha^{-1}$ $yr^{-1}$). The average value of 0.11 keq Ca $ha^{-1}$ $yr^{-1}$ was used in the present calculation. It was assumed that the Sr-isotope ratio stayed constant in the sprinkler and runoff water, as well as in the exchangeable pool in the time period considered here.

Extraction with salicylic acid gave a mineral Sr-isotope ratio of 0.7270. A check indicated that some exchangeable Sr remained after extraction, so this Sr-isotope ratio may be too low. The results obtained for G1 were:

| | | |
|---|---|---|
| Deposition | $^{87}Sr/^{86}Sr =$ | 0.7097 |
| Exchangeable | $^{87}Sr/^{86}Sr =$ | 0.7158–0.7230 |
| Soil water at 0.5 m depth | $^{87}Sr/^{86}Sr =$ | 0.7191 |
| Groundwater | $^{87}Sr/^{86}Sr =$ | 0.7196 |
| Mineral | $^{87}Sr/^{86}Sr =$ | 0.7270 |
| Runoff | $^{87}Sr/^{86}Sr =$ | 0.7196 |

Using mineral $^{87}Sr/^{86}Sr = 0.7270$ and the average Ca deposition of 0.11 keq $ha^{-1}$ $yr^{-1}$, we get a weathering rate of 0.147 $keq^{-1}$ Ca $ha^{-1}$ $yr^{-1}$. In the PROFILE calculations, the ratio of Ca weathering to total base cation weathering is 2.7. In the budget calculations the ratio of Ca weathering to total base cation weathering is 3.07 during 1981 and 2.46 for 1991; the average is 2.76. The value 2.7 was used to scale up from Ca weathering to total weathering. The obtained value for the base cation weathering rate is thus 0.39 keq $ha^{-1}$ $yr^{-1}$ for G1, with an uncertainty range from 0.33 keq $ha^{-1}$ $yr^{-1}$ to 0.47 keq $ha^{-1}$ $yr^{-1}$, depending on the Ca deposition rate.

For F1, no measurement of Sr isotopes are available. Thus, only a rough estimate of the weathering rate can be discussed, if we assume that the ratios in the minerals are the same in F1 as in G1. This is not unreasonable since the total analysis and the bedrock mineralogy in both catchments appear to be similar, though G1 has a slightly higher MgO content (Melkerud, 1983). This would tend to make this estimate for F1 slightly too large.

F1 lies to the east of G1, behind a small hill, so the deposition may be slightly lower there. The deposition estimate of 0.09 keq Ca $ha^{-1}$ $yr^{-1}$ is thus used. Using the average value of Sr-isotope ratios measured ($^{87}Sr/^{86}Sr = 0.7135$) in C-layer samples of soil water in G1 at approximately 0.5–0.6 m depth, gives a weathering rate of 0.18–0.23 C keq $ha^{-1}$ $yr^{-1}$, corresponding to base cation weathering rate of 0.5–0.6 keq $ha^{-1}$ $yr^{-1}$. This could serve as an indicative value for F1 since that catchment has an average soil depth of 0.63 m. It is consistent with other estimates for F1.

## Historic rate

The investigations by Olsson *et al.* (1985) in the Gårdsjön acidification research area may be used to estimate the average weathering rate of the past. The rates obtained are shown in Table 10.5.

The present matrix may be a mixture of minerals weathered before the glaciation and after the glaciation. For the first case where we use the bedrock values we get a weathering rate of 1.31–0.81 keq $ha^{-1}$ $yr^{-1}$ for subcatchment F1, assuming average soil depth to be 0.6 m. Using the values from the bottom of the profile we get a past weathering rate of 0.53 keq $ha^{-1}$ $yr^{-1}$. The range in rates arises from uncertainties in

**Table 10.5.** The dissolved amounts of mineral in the upper metre of soil matrix, since the glaciation in the soil from catchment F3, and the corresponding weathering rates. For catchment F1 it would be approximately the same, but with a lower initial biotite content, maybe only 2–3%. Thus for F1 the average historic rate would be 0.87 keq $ha^{-1}$ $yr^{-1}$. ME is the equivalents base cations per kilogram mineral. Soil age *t* was set to 12 500 years

| | | Mineralogy | | | | Weathering rate | | |
|---|---|---|---|---|---|---|---|---|
| Mineral | ME | Gneiss bedrock IDQ 211 (%) | Granite bedrock IDQ 43 (%) | Profile bottom F1 (%) | Average soil mineral (%) | Gneiss ref. (keq $ha^{-1}$ $yr^{-1}$) | Granite ref. | Bottom ref. |
| Microcline | 279 | 19.5 | 35.4 | 14 | 15 | 0.10 | 0.46 | — |
| Plagioclase | 260 | 32.8 | 29.8 | 27 | 14 | 0.42 | 0.35 | 0.28 |
| Hornblende | 70 | 2.6 | — | 2 | 0.5 | 0.18 | — | 0.18 |
| Epidote | 145 | 4.2 | — | 2 | 0.1 | 0.14 | — | 0.07 |
| Biotite | 142 | 12.9 | 1.1 | 1 | 1.2 | 0.47 | — | — |
| Weathering | | | | | | 1.31 | 0.81 | 0.53 |

the estimates of mineralogy and soil density. The average historic rate was estimated using SAFE to be 0.74 keq $ha^{-1}$ $yr^{-1}$, but that may be a low estimate.

Instead of using individual minerals as the conservative substance, quartz, zircon or rutile may be used as the conservative tracer. They are relatively resistant to chemical dissolution (Nickel, 1973). However, Ti may be incorporated in biotite, pyroxenes and amphiboles in a less resistant form, which may offset the methodology. Another confounding factor occurs when the soil is layered with different geological origins of the layers.

### Total analysis correlation

The total analysis correlation was applied to catchments F1, F2, F3 and G1. For F1, 0.53 keq $ha^{-1}$ $yr^{-1}$ was obtained by using the simplified expression, 0.44 keq $ha^{-1}$ $yr^{-1}$ using the full set of equations, and the results for individual cations using the complete expressions are shown in Table 10.6. Data corresponding to Melkerud's sample 21:1 was used for F1, and 23:1 was used for G1. Assuming the same total analysis for G1, but applying a soil depth of 0.43 m, would yield a base cation weathering rate of 0.42 keq $ha^{-1}$ $yr^{-1}$ using the simplified expression and 0.38 using the full set of equations. In G1, the soil density may be somewhat less than in F1 since, in a thinner soil, the organic horizons make up a larger part of the whole soil profile. Assuming the average catchment soil density for G1 to be 870 kg $m^{-3}$ and for F1 to be 1000 kg $m^{-3}$, would yield a base cation weathering rate of 0.31 keq $ha^{-1}$ $yr^{-1}$ for G1. The results are compared with those for F2 and F3, and with results generated by other methods in Table 10.11. For F2 and F3, base cation weathering rates of 0.6 keq $ha^{-1}$ $yr^{-1}$ and 0.61 keq $ha^{-1}$ $yr^{-1}$ were obtained. If soil depth is less than estimated in 1983, then these rates must be changed correspondingly.

## PROFILE

The input data requirements for a model calculation of the weathering rate in the three watersheds at Gårdsjön are different from the requirements of a budget calculation, but the required information for the watersheds is available in the literature (Melkerud, 1983; Hultberg, 1985; Olsson *et al.*, 1985) and is summarized in Table 10.7.

**Table 10.6.** The base cation weathering rates (in keq $ha^{-1}$ $yr^{-1}$) for subcatchments F1 and G1 using the total analysis correlation method. Equations 10.6–10.9 were used for the upper part of the table, and Equation 10.11 for the lower part

| Catchment | Ca | Mg | K | Na | Sum |
|---|---|---|---|---|---|
| F1 | 0.15 | 0.14 | 0.05 | 0.10 | 0.44 |
| G1 | 0.12 | 0.17 | 0.03 | 0.06 | 0.38 |
| F1 | | | | | 0.55–0.52 |
| G1 | | | | | 0.43–0.41 |

**Table 10.7.** SAFE and PROFILE input data for the F1 Catchment at Gårdsjön

| Parameter | Unit | 1 | 2 | 3 | 4 |
|---|---|---|---|---|---|
| Morphological characterization | | O | A/E | B | B/C |
| Soil layer thickness | m | 0.09 | 0.05 | 0.3 | 0.23 |
| Moisture content | $m^3 m^{-3}$ | 0.15 | 0.25 | 0.3 | 0.35 |
| Soil bulk density | $kg m^{-3}$ | 300 | 800 | 700 | 1250 |
| Specific surface area | $m^2 m^{-3} \times 10^{-6}$ | 0.48 | 1.2 | 1.11 | 1.98 |
| Cation exchange capacity (CEC) | $keq\ kg^{-1} \times 10^{-6}$ | 352 | 22 | 61 | 13 |
| $CO_2$ pressure | times ambient | 2 | 5 | 20 | 30 |
| Dissolved organic carbon | $mg\ l^{-1}$ | 35 | 30 | 10 | 5 |
| log Gibbsite eq. constant | $kmol^2\ m^{-3}$ | 6.5 | 7.5 | 8.5 | 9.5 |
| 1978–92 inflow | % of precipitation | 100 | 64 | 54 | 54 |
| 1978–92 percolation | % of precipitation | 64 | 54 | 54 | 54 |
| Mg+Ca+K uptake | % of total max. | 50 | 40 | 10 | 0 |
| N uptake | % of total max. | 50 | 40 | 10 | 0 |
| Soil mineralogy | % of total | | | | |
| K-feldspar | | 15 | 15 | 18 | 19 |
| Oligoclase | | 14 | 14 | 15 | 16 |
| Albite | | 0 | 0 | 0 | 0 |
| Hornblende | | 0.1 | 0.5 | 1.5 | 1.5 |
| Pyroxene | | 0 | 0 | 0 | 0 |
| Epidote | | 0.1 | 0.5 | 0.75 | 1.0 |
| Garnet | | 0 | 0.1 | 0.1 | 0.1 |
| Biotite | | 0 | 0.5 | 0.5 | 0.5 |
| Muscovite | | 0 | 0 | 0 | 0 |
| Chlorite | | 0 | 0.4 | 0.4 | 0.4 |
| Vermiculite | | 0 | 3 | 15 | 5 |
| Apatite | | 0 | 0.1 | 0.2 | 0.3 |
| Kaolinite | | 0 | 0 | 5 | 0 |

The soil matrix particle size distribution was determined at different levels in the soil profile at three locations, in all 26 samples (Melkerud, 1983), and from this material the surface area was calculated as indicated earlier. The present base cation release rate was calculated for the Gårdsjön watersheds using the PROFILE model, based on observed particle size distributions, soil pH values, soil texture, water content and weathering rate coefficients determined in laboratory studies.

PROFILE was used with data for the soils in catchments G1, F1, F2 and F3, obtained from Olsson *et al.* (1985), and the results are shown in Tables 10.8 and 10.9. The calculated rate values correlate well with the observed values, and the observed difference is within the uncertainty of the values.

The weathering rate per mineral is shown in Table 10.10. It can be seen that K-feldspar, plagioclase, hornblende, epidote, vermiculite and apatite account for almost all of the weathering in the profile. Of the individual minerals, plagioclase feldspar appears to be the most important mineral at this site. The feldspars account for 60% and hornblende and epidote account for 20% of the weathering rate. All mafic minerals account for approximately 30% of the weathering rate. Clay mineral and layer silicates account for less than 10% in such soils, and only have a marginal

**Table 10.8.** The weathering rate in keq $ha^{-1}$ $yr^{-1}$ for catchment F1 calculated with PROFILE

| Depth (m) | Ca | Mg | K | Na | BC | P |
|---|---|---|---|---|---|---|
| 0–0.05 | 0.002 | 0.000 | 0.002 | 0.003 | 0.007 | 0.000 |
| 0.05–0.11 | 0.009 | 0.003 | 0.006 | 0.009 | 0.027 | 0.001 |
| 0.11–0.37 | 0.049 | 0.028 | 0.034 | 0.039 | 0.151 | 0.009 |
| 0.37–0.59 | 0.107 | 0.043 | 0.052 | 0.078 | 0.280 | 0.021 |
| 0.59–0.64 | 0.041 | 0.014 | 0.017 | 0.030 | 0.101 | 0.009 |
| Sum | 0.209 | 0.088 | 0.111 | 0.158 | 0.566 | 0.040 |

**Table 10.9.** The weathering rate in keq $ha^{-1}$ $yr^{-1}$ for catchment G1 calculated with PROFILE

| Depth (m) | Ca | Mg | K | Na | Layer sum | P |
|---|---|---|---|---|---|---|
| 0–0.05 | 0.003 | 0.000 | 0.003 | 0.004 | 0.009 | 0.000 |
| 0.05–0.11 | 0.008 | 0.002 | 0.006 | 0.008 | 0.024 | 0.001 |
| 0.11–0.33 | 0.046 | 0.027 | 0.033 | 0.038 | 0.144 | 0.008 |
| 0.33–0.46 | 0.072 | 0.030 | 0.036 | 0.054 | 0.191 | 0.013 |
| Sum | 0.129 | 0.059 | 0.077 | 0.103 | 0.369 | 0.022 |

**Table 10.10.** The weathering rate for each mineral in the soil in keq $ha^{-1}$ $yr^{-1}$ for catchment G1, as calculated with PROFILE

| | Weathering in layer | | | | | |
|---|---|---|---|---|---|---|
| Mineral | 1 | 2 | 3 | 4 | Sum | % of total |
| K-feldspar | 0.003 | 0.006 | 0.025 | 0.029 | 0.062 | 16.8 |
| Plagioclase | 0.006 | 0.014 | 0.060 | 0.082 | 0.161 | 43.6 |
| Hornblende | — | 0.001 | 0.015 | 0.018 | 0.034 | 9.2 |
| Epidote | — | 0.001 | 0.009 | 0.028 | 0.038 | 10.3 |
| Garnet | — | — | 0.001 | — | 0.001 | 0.3 |
| Biotite | — | — | 0.002 | 0.002 | 0.004 | 1.1 |
| Chlorite | — | 0.001 | 0.004 | 0.004 | 0.009 | 2.4 |
| Vermiculite | — | 0.001 | 0.017 | 0.007 | 0.024 | 6.5 |
| Apatite | — | — | 0.013 | 0.023 | 0.035 | 9.5 |
| Sum | 0.009 | 0.024 | 0.144 | 0.191 | 0.369 | 100 |

importance for the weathering rate. The role of biotite for the present weathering rate is insignificant. Apatite accounts for almost 10% of the weathering rate according to the calculations, but it must be kept in mind that the kinetic coefficients for apatite dissolution are rather uncertain at present.

**The HISTORIC-SAFE model**

The SAFE model was applied to the whole Gårdsjön catchment and run for the 12 500 years that has elapsed since the last glaciation. The model was set up to subtract each year's weathering from the previous year's amount of each mineral. At the same time the particle size was recalculated, as a shrinking sphere. This means that the particle size distribution changes over time, depending on how the different minerals were initially distributed over the different particle size fractions. The model calculates the development of soil chemistry, soil mineralogy and soil texture over time, and partly reconstructs soil development. Both the consumption of primary minerals and the production of weathering residues and secondary minerals are considered. The reconstruction of the weathering rate at Lake Gårdsjön starts with a mineral matrix composed of crushed bedrock material, of the same texture as the soil at 2 m depth.

The results are shown in Figure 10.1. The model calculates present weathering, 12 500 years after ice-out, to 0.39 keq $ha^{-1}$ $yr^{-1}$. At present, the model also underpredicts present soil texture, yielding a slightly too low present weathering rate. This is still close to the observed present value of 0.58 keq $ha^{-1}$ $yr^{-1}$ as estimated by other methods. The general declining trend is caused by the disappearance of a large part of the finest particles, and a depletion of the mineralogy. The sharp increases and decreases are caused by historical variations in temperature and soil moisture. The model calculation suggests that climate has a strong effect on the weathering rate, and that significant climatic changes could also change lake chemistry significantly. The "Holocene-SAFE" model is still under development and the estimates with it are preliminary. The calculated present soil content of K-feldspar, plagioclase, hornblende and vermiculite are closely reconstructed by the model (Rietz, 1995). This is an indication that laboratory kinetics do indeed apply to field conditions.

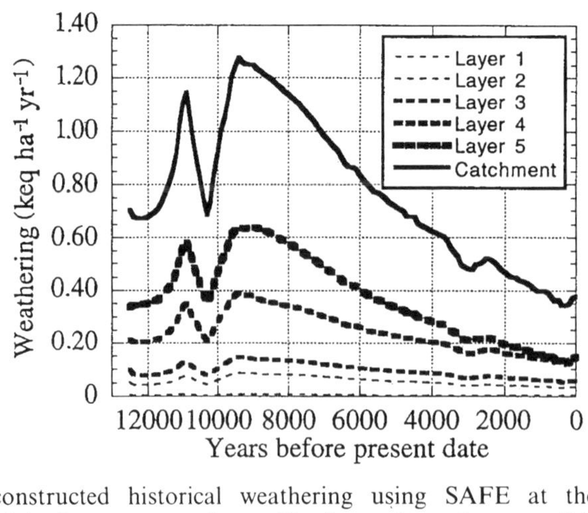


**Figure 10.1.** Reconstructed historical weathering using SAFE at the Lake Gårdsjön catchment. Layer 1 is the uppermost layer. The layers have the same thickness as in Table 10.7 plus an extra layer to make up to 0.65 m soil depth

# DISCUSSION

There are several different weathering rate estimates available for the catchments in the Gårdsjön area. The weathering rates appear to be consistent between measurements even when different methods have been used. The budget method seems likely to give less accuracy, since the role of ion exchange may still be significant in the catchment, and because this rate cannot be readily measured.

For the roofed catchment G1, the weathering rate in keq $ha^{-1}$ $yr^{-1}$ at any point can be calculated using an empirical expression derived in this study:

$$W = a\, z_{O-\text{layer}} + b\, z_{A-\text{layer}} + c\, z_{B-\text{layer}} + d\, z_{C-\text{layer}} \qquad (10.21)$$

where $a=0.1$, $b=0.36$, $c=0.48$ and $d=1.33$. For catchment F1, the weathering rate in keq $ha^{-1}$ $yr^{-1}$ is given by $a=0.1$, $b=0.4$, $c=0.52$ and $d=1.42$. For the whole Lake Gårdsjön catchment, we suggest using the coefficients $a=0.1$, $b=0.38$, $c=0.5$ and $d=1.4$.

A summary of the different weathering rates obtained by different methods is shown in Table 10.11.

# CONCLUSION

The different estimates of the weathering rate appear to be consistent with each other, despite different possibilities for uncertainties and weaknesses in the different methods. It can be concluded that the G1 catchment weathering rate is in the range 0.3–0.4 keq $ha^{-1}$ $yr^{-1}$. For F1 and the whole lake catchment the weathering rate is probably in the range 0.5–0.6 keq $ha^{-1}$ $yr^{-1}$.

G1 has a lower rate than F1 because it has a smaller mean soil depth (0.47 m) than F1 (0.67 m). This is amplified by the fact that the organic horizon, which has a low mineral weathering rate, constitutes a larger fraction of the whole soil in shallower soils. These are typical weathering rates for Sweden. The average weathering rate in

**Table 10.11.** Summary of base cation release rates due to chemical weathering in keq $ha^{-1}$ $yr^{-1}$ at different catchments at Lake Gårdsjön, Sweden, using different methods. It can be seen that the different methods give approximately the same results

| F1 | F2 | F3 | G1 | Method |
|---|---|---|---|---|
| 0.54 | 0.61 | 0.59 | 0.36 | Budget study |
| 0.57 | 0.60 | 0.67 | 0.37 | Laboratory kinetics, PROFILE |
| 0.53 | 0.60 | 0.61 | 0.38 | Total analysis correlation |
| 0.59 | — | — | 0.39 | Sr-isotope ratio method |
| 0.56 | 0.60 | 0.64 | 0.38 | Average present value |
| 0.74–0.81 | — | 0.81–1.31 | 0.35 | Historic rate, bedrock mineralogy reference |
| 0.53 | — | — | 0.30 | Historic rate, C-layer mineralogy reference |
| 0.74 | — | 1.08 | 0.39 | Historic rate, laboratory kinetics, SAFE |
| 0.47 | 0.47 | 0.48 | 0.37 | Historic conditions, present soil, PROFILE |

the upper 0.5 m of soil was 0.42 keq $ha^{-1}$ $yr^{-1}$ according to measurements at 1884 points evenly spread over the whole of Sweden (Sverdrup and Warfvinge, 1995b).

## REFERENCES


Aagaard, P. and Helgeson, H. (1982) Thermodynamic and kinetic constraints on reaction rates among minerals and aqueous solutions: I. Theoretical considerations. *American Journal of Science*, **282**, 237–285.

Åberg, G., Jacks, G. and Hamilton, J. (1989) Weathering rates and $^{87}Sr/^{86}Sr$ ratios: An isotopic approach. *Journal of Hydrology*, **109**, 65–78.

Bain, D. and Bacon, J. (1994) Strontium isotopes as indicators of mineral weathering in catchments. *Catena*, **22**, 201–214.

Bain, D., Mellor, A., Wilson, J. and Duthie, D. (1994) Chemical and mineralogical rates and processes in an upland granitic till catchment in Scotland. *Water, Air, and Soil Pollution*, **73**, 11–27.

Balek, J., Moldan, B., Paçes, T. and Skorepa, J. (1978) Hydrological and geochemical mass balances in small forested and agricultural basins. *Modelling the Water Quality of the Hydrological Cycle*. IAHS–AISH Publication No. 125, pp. 50–59.

Berden, M., Nilsson, S., Rosen, K. and Tyler, G. (1987) *Soil Acidification, Extent, Causes and Consequences—An Evaluation of Literature Information and Current Research*. National Swedish Environment Protection Board. Report 3292.

Burl, A. (1976) *The Stone Circles of the British Isles*. Yale University Press, Yale.

Champion, T., Gamble, C., Shennan, S. and Whittle, A. (1984) *Prehistoric Europe*. Academic Press, London.

Chou, L. and Wollast, R. (1985) Steady state kinetics and dissolution mechanisms of albite. *American Journal of Science*, **285**, 963–993.

Cronan, C. S. (1985) Chemical weathering and solution chemistry in acid forest soils: Differential influence of soil type, biotic processes and $H^+$ deposition. In: J. Drever (ed.), *The Chemistry of Weathering*. D. Reidel Publishing Company, Dordrecht, pp. 175–195.

Falkengren-Grerup, U. (1987) Long-term changes in pH of forest soils in southern Sweden. *Environmental Pollution*, **43**, 79–90.

Helgeson, H., Murphy, W. and Aagaard, P. (1984) Thermodynamics and kinetic constraints on reaction rates among minerals and aqueous solutions: II. Rate constants, effective surface area and the hydrolysis of feldspar. *Geochimica et Cosmochimica Acta*, **48**, 2405–2432.

Hultberg, H. (1985) Budgets of base cations, chloride, nitrogen and sulphur in the acid Lake Gårdsjön. *Ecological Bulletins (Stockholm)*, **37**, 133–157.

Jacks, G., Åberg, G. and Hamilton, P. J. (1989) Calcium budgets for catchments as interpreted by strontium isotopes. *Nordic Hydrology*, **20**, 85–96.

Johnston, A., Goulding, K. and Poulton, P. (1986) Soil acidification during more than 100 years under permanent grassland and woodland at Rothamsted. *Soil Use and Management*, **2**(1), 3–10.

Lasaga, A. (1981) Transition state theory. In: A. Lasaga and R. Kirkpatrick (eds), *Kinetics of Geochemical Processes*. Reviews in Mineralogy, Vol. 8, Mineralogical Society of America, pp. 135–170.

Lundström, U. (1987) Studies of chemical weathering by laboratory batch experiments and lysimeter investigations. In: B. J. Mason (ed.) *Surface Water Acidification Programme*. The Royal Society, The Royal Swedish Academy of Sciences, The Norwegian Academy of Science and Letters. Mid-term review conference, Bergen, Norway, 22–26 June, pp. 137–146.

Melkerud, P. (1983) *Quaternary Deposits and Bedrock Outcrops in an Area around Gårdsjön, South-West Sweden, with Physical, Mineralogical and Geochemical Investigation*. Reports on Forest Ecology and Forest Soils No. 40, Swedish University of Agricultural Sciences, Uppsala, Sweden.

Miller, E., Blum, J. and Friedland, A. (1993) Determination of soil exchangeable cation loss and weathering rates using Sr isotopes. *Nature*, **362**, 438–441.
Nickel, E. (1973) Experimental dissolution of light and heavy minerals in comparison with weathering and intrastitial solution. *Contributions in Sedimentology*, **1**, 1–68.
Olsson, B., Hallbäcken, L., Johansson, S., Melkerud, P.-A. Nilsson, I. and Nilsson, T. (1985) The Lake Gårdsjön area—physiogeographical and biological features. *Ecological Bulletins (Stockholm)*, **37**, 10–28.
Olsson, M. and Melkerud, P. A. (1990) Determination of weathering rates based on geochemical properties of the soil. In: E. Pulkinene (ed.) *Proceedings of the Conference on Environmental Geochemistry in Northern Europe*. Symposium Series no. 34, Geological Survey of Finland, pp. 45–61.
Paçes, T. (1973) Steady-state kinetics and equilibrium between ground water and granitic rock. *Geochimica et Cosmochimica Acta*, **37**, 2641–2663.
Paçes, T. (1978) Reversible control of aqueous aluminium and silica during the irreversible evolution of natural waters. *Geochimica et Cosmochimica Acta*, **42**, 1487–1493.
Paçes, T. (1983) Rate constants of dissolution derived from the measurements of mass balances in catchments. *Geochimica et Cosmochimica Acta*, **47**, 1855–1863.
Renberg, I., Hellberg, T. and Nilsson, M. (1985) Effects of acidification on diatom communities as revealed by analyses of lake sediments. *Ecological Bulletins (Stockholm)*, **37**, 219–223.
Rietz, F. (1995) *Modeling Mineral Weathering and Soil Acidity During Post-glacial Period.* Reports in Ecology and Environmental Engineering 1:1995. Department of Chemical Engineering, Lund University, Sweden.
Sverdrup, H. (1990) *The Kinetics of Base Cation Release Due to Chemical Weathering*. Lund University Press, Lund, or Chartwell Bratt Ltd, London, 246 pp.
Sverdrup, H. and Warfvinge, P. (1988) Weathering of primary silicate minerals in the natural soil environment in relation to a chemical weathering model. *Water, Air, and Soil Pollution*, **38**, 387–408.
Sverdrup, H. and Warfvinge, P. (1993) Calculating field weathering rates using a mechanistic geochemical model—PROFILE. *Journal of Applied Geochemistry*, **8**, 273–283.
Sverdrup, H. and Warfvinge, P. (1995a) Estimating field weathering rates using laboratory kinetics. In: A. F. White and S. Brantley (eds), *Chemical Weathering Rates of Silicate Minerals*. Reviews in Mineralogy, Vol. 31. Mineralogical Society of America, Washington, DC, pp. 485–541.
Sverdrup, H. and Warfvinge, P. (1995b) Critical loads of acidity for Swedish forest ecosystems. *Ecological Bulletins (Stockholm)*, **44**, 75–89.
Sverdrup, H. U., de Vries, W. and Henriksen, A. (1990) *Mapping Critical Loads*. Nordic Council of Ministers, Geneva. Miljõrapport 1990:15, Nord 1990:98.
Tamm, C. and Hallbäcken, L. (1986) Changes in soil pH over a 50-year period under different canopies in SW Sweden. *Water, Air, and Soil Pollution*, **36**, 337–341.
Velbel, M. (1985) Geochemical mass balances and weathering rates in forested watersheds in Southern Blue Ridge. *American Journal of Science*, **285**, 904–930.
Warfvinge, P. and Sverdrup, H. (1992) Calculating critical loads of acid deposition with PROFILE—a steady-state soil chemistry model. *Water, Air, and Soil Pollution*, **63**, 119–143.
Warfvinge, P., Falkengren-Grerup, U. and Sverdrup, H. (1993) Modeling long-term base supply to acidified forest stands. *Environmental Pollution*, **80**, 1–13.
Wickman, T. and Jacks, G. (1992) Strontium isotopes in weathering budgeting. In: Y. Kharaka and A. Mast (eds), *Water–Rock Interaction*. A. A. Balkema Publishing Company, pp. 611–614.

# 11 Trace Metals in the Lake Gårdsjön Catchments

**BENNY LYVÉN, VIVEKA PALM, CONNY HARALDSSON & PER ÖSTLUND**

The trace elements arsenic, cadmium, cobalt, copper, iron, lead, manganese, nickel, vanadium and zinc were studied, from March 1991 to October 1995, in the roofed catchment (G1) and in the reference catchment (F1) at the Gårdsjön area. Over this period of time a statistically significant decrease in the ratio G1/F1 of the metal concentrations was observed for cobalt, manganese, vanadium and zinc at the $2\sigma$ level, and for arsenic, cadmium and copper at the $1\sigma$ level. No effect was found for iron and lead.

When studying the speciation of some trace metals in this area, the dominating species were found to be the easily bioavailable forms.

## INTRODUCTION

There are two principal sources of metals in a soil, either successively mobilised by weathering processes of primary minerals in the soil material itself, or imported to the soil, e.g. due to atmospheric deposition. The retention of metals in soils is severely affected by the ongoing deposition of acidifying substances, especially in areas with limited soil buffering capacity. Both the mobility and the bioavailability of many metals are increased by acidification. Thus it is expected that most metal fluxes in a forest catchment will decrease, if the pH in the catchment is increased. Both mobility and bioavailability are a consequence of interactions between the element in its chemical/physical form, speciation, and the chemical/physical properties of dissolved and solid phases (soil particles) in its vicinity.

The speciation of elements in a soil solution, either as free ions or as different complexes, from relatively weak electrostatic associations to strong covalent bonds, depends on the composition and concentration of the involved ions and ligands. the surfaces of soil particles are chemically active; surfaces are either neutral or positively or negatively charged. Ions with opposite charges in the soil solution are attracted to these surfaces. Also, secondary minerals are formed depending on the concentration and form of the involved elements, as well as the composition of the soil solution. The most important of these secondary minerals are oxides, hydroxides, oxyhydroxides and carbonates. Secondary minerals not only retain the elements involved in the formation of the actual mineral, but also offer charged surfaces which can scavenge

*Experimental Reversal of Acid Rain Effects: The Gårdsjön Roof Project.*
Edited by Hans Hultberg and Richard Skeffington. 

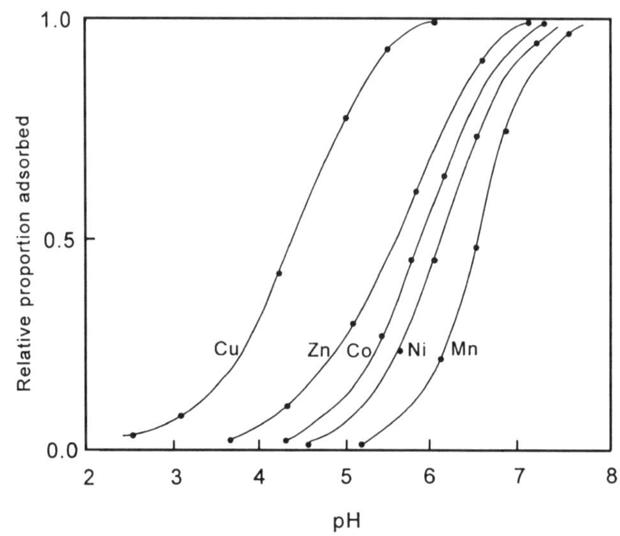


**Figure 11.1.** Relative retention of some metals on goethite. Modified from McKenzie (1980)

other elements. Abundant minerals such as oxides and oxyhydroxides of Al, Mn and Fe are of importance in these processes. This can be exemplified by the relative retention of some metals on goethite, a secondary ironhydroxyoxide which may be formed when reduced iron is oxidised (Taylor and McKenzie, 1966) (Figure 11.1). The surface charge of a mineral is a function of the soil solution pH and may shift from positive to negative over an environmentally relevant pH interval. Also the retention of metals forming oxyanions, such as Mo, V and Cr, may be drastically pH-dependent due to changes in oxidation states.

Organic matter like humic and fulvic acids offers binding sites, thereby affecting the retention of metals. All the processes involved in metal retention are dependent on the soil properties and especially the alkalinity or acidity of the soil. Reduction of inputs of acid deposition, by means of the Gårdsjön Roof, might thus be expected to reduce the mobility, and perhaps change the speciation and bioavailability for a number of metals. A study was thus initiated to:

- explain and quantify the main mechanisms behind fluxes of metals in such a system;
- quantitatively measure metal speciation in soil runoff, under the influence of reduced deposition of acidifying substances.

## METHODS

The strategy for the study was to:

- Repeatedly sample and analyse filtered and unfiltered soil runoff water in the parallel catchments (G1 ROOF, G2 NITREX and F1 CONTROL) in order to detect metal concentration trends versus time.

- Study trends in the chemical/physical properties of metals in the soil runoff water. This strategy is dependent on the expected similarities between the catchments, or on the consistency in time of the differences.

**Sampling**

Sampling and analysis of runoff water in the F1, G1 and G2 parallel catchments were started at the end of March 1991. Sampling bottles made of polyethylene are extensively cleaned by leaching new bottles in 6 M hydrochloric acid at 50°C for 3 weeks. Following this, the bottles are rinsed with Milli-Q water and filled with 1% V/V nitric acid and stored until use; however, not less than 1 month. The samples are collected, transported to the laboratory and immediately filtered through pre-cleaned 0.45 $\mu$m pore size cellulose acetate filters (Millipore HA). Aliquots of the filtered and unfiltered samples (10 ml) are taken for determination of the metal content and the samples are acidified to 0.15 M $HNO_3$ and stored until analysis. The remaining parts of the collected samples are used for speciation studies. These samples are ultrafiltered or pumped over the adsorbents, Chelex and DEAE, within 8 hours after collection.

In order to obtain reference data, samples have also been collected at two places in Gothenburg, 50 km south of Lake Gårdsjön. The two stations are Säveån (SÅ), a small river, and Delsjöbäcken (DB), a small stream. Although the geological context is different in these areas it is interesting to compare the results from Lake Gårdsjön with results from a less acidified area. Chemical data used for characterisation of the different sampling stations are presented in Table 11.1.

**Laboratory procedures**

In all trace metal analysis, contamination by reagents, sample bottles and air-borne particles, etc. is a problem that has to be considered. All experimental work has therefore been performed in a clean room laboratory (a positive pressure laboratory with Class-100 filtered air supply). High-purity acids are prepared in the clean lab by sub-boiling-point quartz distillation of analytical grade HCl and $HNO_3$. Water used for cleaning bottles and other equipment is prepared using a Milli-Q (Millipore) system.

A plasma source mass spectrometer (ICP-MS)(VG Elemental) is used for the metal determinations. The operating conditions are described in detail by Haraldsson *et al.* (1993). Aluminium (Al), arsenic (As), cadmium (Cd), chromium

**Table 11.1.** pH, dissolved organic carbon, sulfate and chlorine in the samples used for trace metal speciation

| Station | Date | pH | DOC (mg/l) | $SO_4$-S(mg/l) | Cl(mg/l) |
|---|---|---|---|---|---|
| SÅ | 92/12/17 | 6.4 | 3.85 | 3.24 | 10.3 |
| DB | 92/12/17 | 6.86 | 4.71 | 3.68 | 14.3 |
| F1 | 92/12/09 | 4.12 | | 4.95 | 17.6 |
| G1 | 92/12/09 | 4.22 | | 4.57 | 26.1 |
| G2 | 92/12/09 | | 15.4 | | |

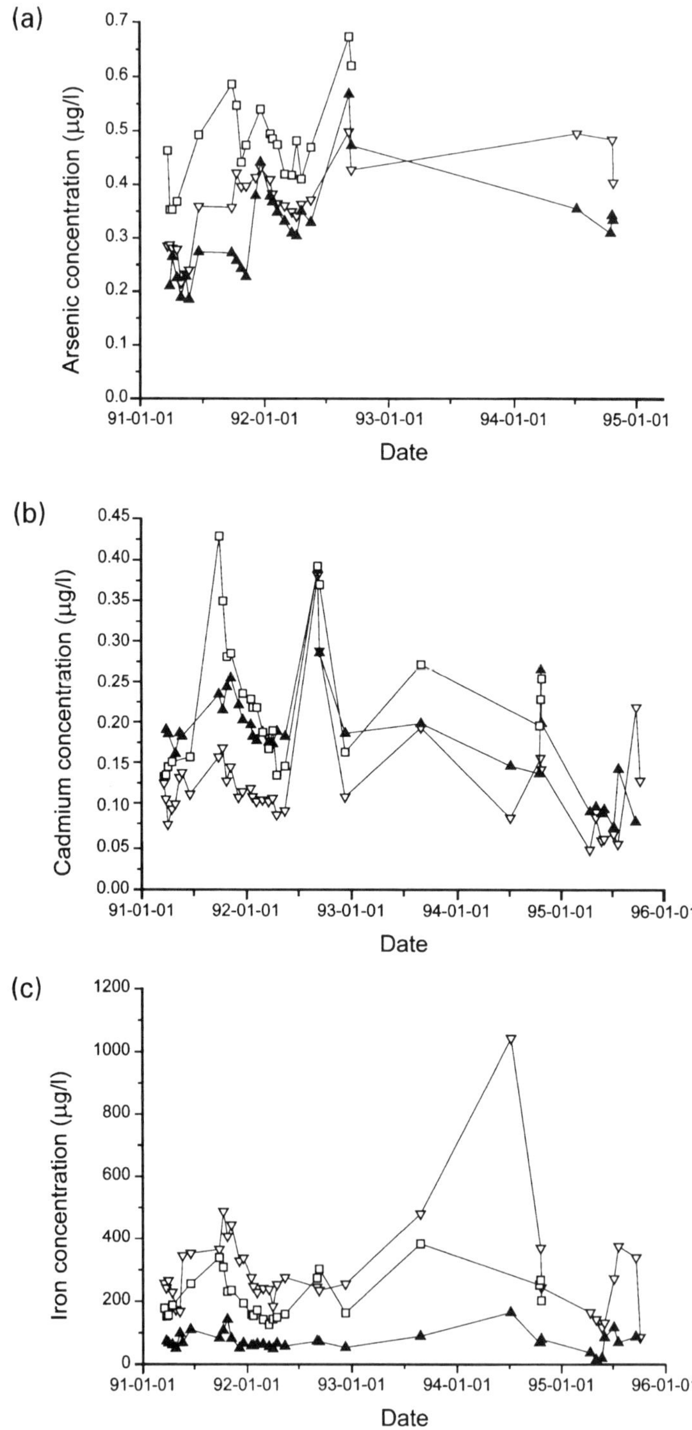
(a)
Arsenic concentration (µg/l)
0.0
0.1
0.2
0.3
0.4
0.5
0.6
0.7
91-01-01
92-01-01
93-01-01
94-01-01
95-01-01
Date
(b)
Cadmium concentration (µg/l)
0.00
0.05
0.10
0.15
0.20
0.25
0.30
0.35
0.40
0.45
91-01-01
92-01-01
93-01-01
94-01-01
95-01-01
96-01-01
Date
(c)
Iron concentration (µg/l)
0
200
400
600
800
1000
1200
91-01-01
92-01-01
93-01-01
94-01-01
95-01-01
96-01-01
Date

(d)

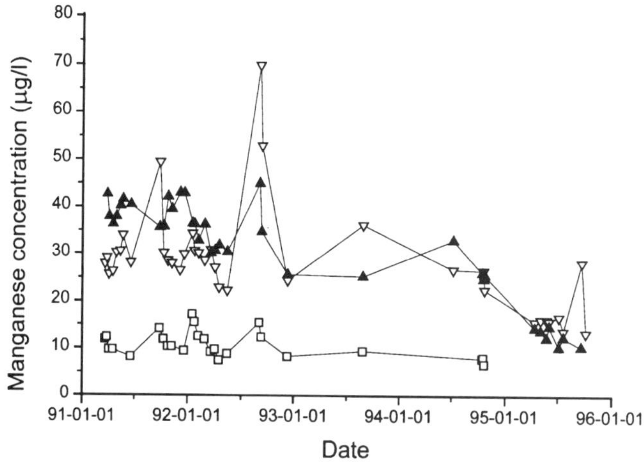


(e)

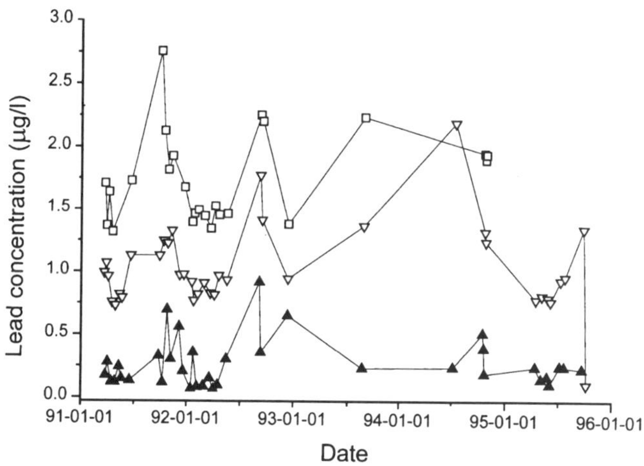


**Figure 11.2.** The metal concentration ($\mu$g/l) versus time of (a) arsenic, (b) cadmium, (c) iron, (d) lead and (e) manganese. ▽ = F1; ▲ = G1; □ = G2

(Cr), cobalt (Co), copper (Cu), iron (Fe), manganese (Mn), nickel (Ni), lead (Pb), vanadium (V) and zinc (Zn) are analysed.

## Speciation

Speciation was studied as described by Haraldsson *et al.* (1993). The dissolved fraction as defined by filtration with 0.45 $\mu$m cellulose acetate filters was divided into groups in order to assess the bioavailable fraction. Chelex-100 (200–400 mesh pore size, BIO-RAD) adsorbs free, positively-charged metal ions, including metal ions bound to weak complexes which dissociate during the contact time between the sample and the adsorbent, giving an estimate of the easily bioavailable fraction of the metals. This was complemented by the use of DEAE (Fractogel TSK DEAE-650 with 200–400 mesh pore size, Merck) to determine the fraction of the metals bound to organic matter, and thus significantly less bioavailable. Ultrafilters with a nominal molecular weight cutoff at 10 000 (Mini Ultrasette, Filtron) were used to study the amount of metals associated with colloids passing through the ultrafilter. Together the three methods give a consistent description of mobility and bioavailability in an acidified forest ecosystem. In this paper, results from the speciation study are exemplified by Al and Pb.

## RESULTS AND DISCUSSION

### Metal concentrations in runoff

The concentrations of As, Cd, Fe, Pb and Mn versus time from 20 March 1991 are presented in Figures 11.2a–e. In the beginning samples were taken weekly, but as the changes in trace metal concentrations in the roofed catchment runoff were similar to those in the parallel catchments, sampling was made less frequently. The trends are similar from year to year, with more or less pronounced concentration peaks each autumn, after the long drought of the summer and a following period of rain. The lowest concentrations were registered between February and May.

Differences in metal concentrations between the three catchments: F1 (the reference catchment), G1 (the roofed catchment) and G2, (the NITREX catchment) can be seen in Figures 11.2a–e. Variables that are commonly thought to govern the metal concentration in runoff, i.e. pH, DOC and major anions, also differ between the catchments (Table 11.2). As expected, G2 is the one most affected by acidification, whereas G1 is less acid than the reference catchment.

The trace metal mobility in the roofed catchment has decreased for a number of metals. As seen in Figure 11.3, the concentration in runoff of cadmium, cobalt, manganese, nickel and vanadium has decreased significantly during the time the roof has existed. In order to assess the influence of the roof, results from the roofed catchment should be compared with the reference catchment (Figure 11.4). This figure shows the metal concentration ratios between the catchments. These results indicate a beneficial effect of the roof so far on all metals except lead and iron for which no significant changes can be observed.

In F1 there are strong correlations between Mn and Ni ($r = 0.90$), Mn and Co ($r = 0.93$) and Mn and Cd ($r = 0.84$). A strong correlation between Fe and Pb ($r = 0.76$) is also found. Trace elements are often adsorbed on secondary minerals of oxidised Fe and Mn. In a situation where the chemical environment varies between oxidising and reducing and consequently affects the formation/dissolution of these

**Table 11.2.** Median values in the runoff from the catchments, 1992 (From Molden *et al.*, pers. comm.)

| Station | pH | DOC (mg/l) | $SO_4$-S(mg/l) | Cl(mg/l) | Runoff(mm/month) |
|---|---|---|---|---|---|
| F1, reference | 4.1 | 9 | 4 | 22 | 46 |
| G1, roof | 4.4 | 7 | 5 | 31 | 49 |
| G2, nitrogen added | 3.9 | 13 | 5 | 22 | 46 |

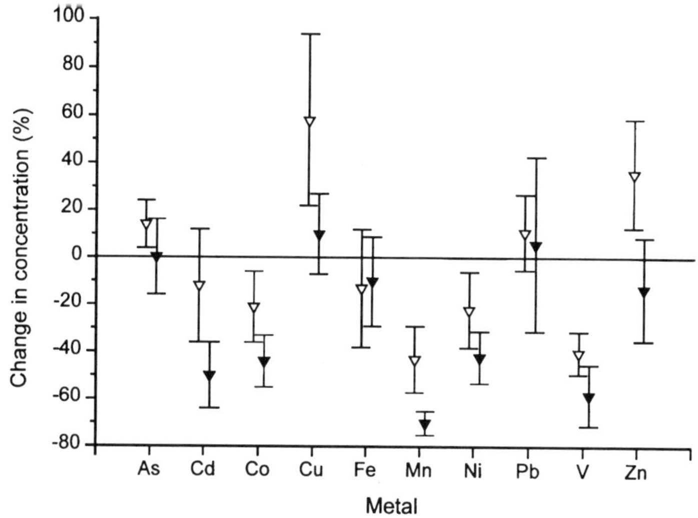


**Figure 11.3.** Relative change in metal concentration calculated by least squares fitting of a straight line to all determined concentrations of each metal. The error bars denote the standard deviation of the slope calculated at the 1$\sigma$ level. ▽ = F1; ▼ = G1

secondary minerals, correlations between Fe/Mn and trace elements should be expected.

In G1, similar correlations were less pronounced, suggesting more stable redox conditions.

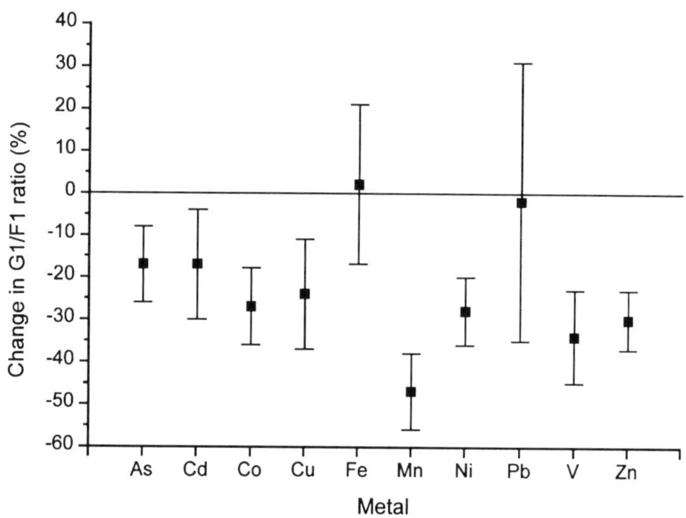


**Figure 11.4.** Relative change in metal concentration ratios calculated by least squares fitting of a straight line to all determined ratios G1/F1, of samples taken at the same time, of each metal. The error bars denote the standard deviation of the slope calculated at the 1$\sigma$ level

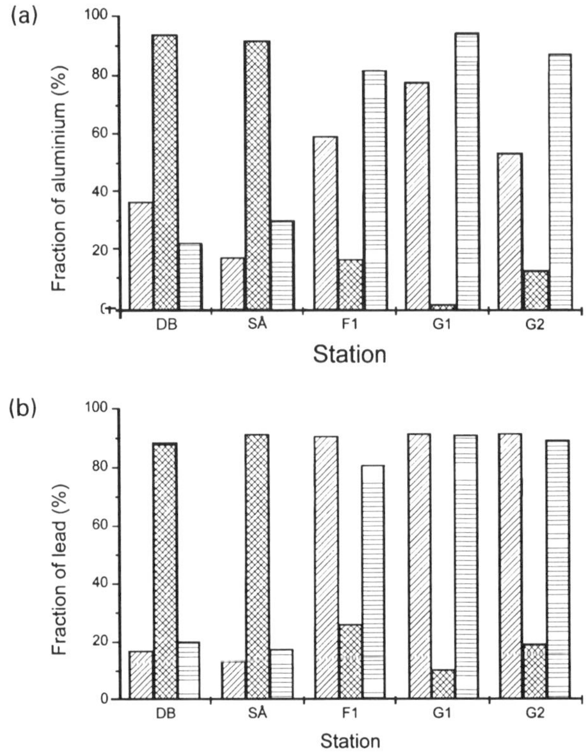


**Figure 11.5.** Results from speciation study of (a) aluminium and (b) lead. ▨ = fraction adsorbed on Chelex-100 (%); ▩ = fraction adsorbed on DEAE (%); ▤ = fraction passing through an ultrafilter with a 10 000 Dalton cut-off (%)

In the roofed catchment, G1, the lowest Fe and Pb mean concentrations and the highest mean concentrations of Cr, Co, Cu, Mn, Zn and Al are found.

## Speciation results

The results from a speciation experiment with aluminium are shown in Figure 11.5a. Total and dissolved concentrations can be found in Table 11.3. The Chelex labile Al fraction is approximately four times as high in the Gårdsjön area compared with the sampling stations in Gothenburg. The large fraction of aluminium bound in low

**Table 11.3.** Total (tot) and filtered (f) concentrations of Al and Pb in the speciated samples

| Station | Al (tot ($\mu$g/l) | Al (f)($\mu$g/l) | Pb (tot)($\mu$g/l) | Pb (f) ($\mu$g/l) |
|---|---|---|---|---|
| SÅ | 158 | 82 | 0.23 | 0.08 |
| DB | 291 | 193 | 0.52 | 0.29 |
| F1 | 695 | 695 | 0.95 | 0.95 |
| G1 | 1181 | 1190 | 0.66 | 0.66 |
| G2 | 1003 | 982 | 1.4 | 1.13 |

molecular weight compounds is likely to be important for the toxicity of aluminium in the Gårdsjön area. The high concentrations of aluminium, in combination with the bioavailability of the forms present, is probably of significance when explaining the toxicity of the water to living organisms such as fish.

The concentration of lead in unfiltered samples from the Gårdsjön area is two to five times higher compared with the samples from Gothenburg (Table 11.3). Furthermore, the low pH in the Gårdsjön area makes the bioavailable fractions significantly higher (Figure 11.1). The fraction that adsorbs to Chelex-100 is less than 20% in the samples from Gothenburg compared with approximately 90% in the Gårdsjön area. The hypothesis that the bioavailable fraction is significantly higher is strengthened by the fact that less than 10% of the lead is bound to colloids. Approximately 80% of the lead in the Gothenburg streams is bound to colloids capable of passing the ultrafilter used.

## CONCLUSIONS

1. The differences in metal concentrations and patterns between the catchments are significant.
2. The concentrations of several heavy metals studied in the runoff water have decreased from March 1991 to October 1995 in the roofed catchment. An example is cadmium in runoff water which has decreased by approximately 50%. This is probably due to the increase in pH.
3. Some metals have decreased in the reference catchment as well as in the roofed catchment, but the decrease was significantly smaller.
4. Most metal concentrations are at their highest in September–October after the summer drought.
5. The speciation studies have shown that the readily bioavailable fraction dominates concentrations of the trace metals in the Gårdsjön area.

## REFERENCES

Haraldsson, C., Lyvén, B., Pollak, M. and Skoog, A. (1993) Multielement speciation of trace metals in fresh water adapted to plasma source mass spectrometry. *Anal. Chim. Acta*, **284**, 327–335.

McKenzie, R. M. (1980) The adsorption of lead and other heavy metals on oxides of manganese and iron. *Aust. J. Soil. Res.*, **18**, 61–73.

Taylor, R. M. and McKenzie, R. M. (1966) The association of trace elements with manganese minerals in Australian soils. *Austr. J. Soil Res.*, **4**, 29–39.

# 12 Cycling of Mercury and Methyl Mercury in the Gårdsjön Catchments

**JOHN MUNTHE, YING-HUA LEE, HANS HULTBERG, ÅKE IVERFELDT, GUNNAR CH. BORG & B. INGVAR ANDERSSON**

A summary of the results of research on the cycling of mercury (Hg) and methyl mercury (MeHg) in the forested catchments at Gårdsjön is given. Previously presented data on Hg and MeHg in precipitation, throughfall and runoff water are evaluated together with more recently measured concentrations. The mechanisms of mobilisation of Hg and MeHg in forest soils are discussed. In the roof-covered catchment, the removal of the throughfall input corresponds to a reduction of the total Hg and MeHg input of about 50%, the remainder being distributed via litterfall. The output measurements performed in the roof-covered catchment initially showed a decrease of MeHg output in comparison to the reference catchment. Including data from a longer time period showed that this effect was not statistically significant due to large inter-annual variations in runoff concentrations during the following years. Continued monitoring of Hg and MeHg in runoff is necessary before the more permanent effects can be assessed.

## INTRODUCTION

Atmospheric deposition and biogeochemical cycling of mercury (Hg) and methyl mercury (MeHg) in forested catchments have been investigated in the Lake Gårdsjön watershed since the mid 1980s. Catchments are valuable in studies of the environmental behaviour of pollutants since they allow both general investigations of element fluxes and large-scale processes, and detailed mechanistic investigations, e.g. of weathering and interactions of air-borne pollutants with the forest. The construction of the roof-covered sub-catchment has also allowed investigations of the responses of the terrestrial ecosystem to decreased atmospheric deposition and of long-term recovery processes.

For Hg and MeHg, the major environmental and health-related issue is the appearance of elevated MeHg concentrations in freshwater fish making them unfit for human consumption. Forests and forest soils are important compartments in the environmental cycling of Hg and MeHg. The forest itself serves as a trap for atmospheric Hg species, increasing the net flux of Hg and MeHg to the terrestrial ecosystem. The pools of Hg and MeHg in forest soils today constitute the largest

*Experimental Reversal of Acid Rain Effects: The Gårdsjön Roof Project.*
Edited by Hans Hultberg and Richard Skeffington. 

potential source of these contaminants to freshwater foodchains. Factors potentially influencing the release of Hg and MeHg from forest soils include temperature, water flowpaths and mineralisation of soil organic matter.

The research on Hg and MeHg in the forested catchments of Lake Gårdsjön has covered a wide variety of processes including some of the first determinations of MeHg in precipitation, surface waters and runoff water (Lee and Hultberg, 1990; Lee and Iverfeldt, 1991). A mass balance approach has also been used to calculate input–output fluxes of Hg and MeHg for a forested catchment calling attention to the importance of atmospheric deposition of MeHg as well as discussing a critical load concept for MeHg (Hultberg *et al.*, 1994). The more mechanistic-orientated investigations include studies of deposition processes of Hg and MeHg (Iverfeldt, 1991b; Munthe *et al.*, 1995a) as well as transport modelling of terrestrial MeHg (Lee *et al.*, 1994a). An analysis of trends in environmental Hg and MeHg fluxes has also recently been made (Munthe *et al.*, 1995b)

The importance of the transfer of Hg and MeHg from forest soils to surface water via runoff has been identified as an important source of the Hg accumulated in the aquatic ecosystem. The role of dissolved organic carbon (DOC), and thus the mineralisation of the soil organic matter, in this transport was also recognised at an early stage (Johansson and Iverfeldt, 1991; Lee and Iverfeldt, 1991).

In this chapter, a brief summary of the results from these investigations is presented along with some more recent results from the evaluation of controlling factors of Hg and MeHg in runoff.

## METHODS

Detailed descriptions of the sampling equipment and procedures for Hg and MeHg analysis have been made elsewhere (Lee and Hultberg, 1990; Iverfeldt, 1991a, b; Lee and Iverfeldt, 1991; Hultberg *et al.*, 1994; Jensen and Iverfeldt, 1994).

Deposition (open field and throughfall) of Hg/MeHg was determined using bulk samplers. Litterfall was sampled in nylon nets mounted on aluminium frames. The collection bottles of the bulk samplers were always pre-acidified with 2.5 mL of HCl (Merck, Suprapur). The bulk sampling technique employed has been shown to be of satisfactory accuracy and precision, in comparison to wet-only and event sampling (Iverfeldt and Munthe, 1993).

Soil profiles of Hg/MeHg were also collected to estimate the total store as well as the amount at different depth (Lee *et al.*, 1994a). Solid samples were frozen immediately after collection and kept in triple plastic bags until time of analysis.

Surface runoff samples were collected at the catchment weir at least once per month, while soil water samples were taken during two seasons. Based on the soil water concentrations of Hg/MeHg, a hydrological model (SOIL) was used to calculate the yearly Hg/MeHg catchment export (Lee *et al.*, 1994a).

Hg/MeHg analysis was performed using techniques described earlier (Lee, 1987; Iverfeldt, 1988, 1991a,b; Lee and Mowrer, 1989; Lee and Hultberg, 1990; Lee *et al.*, 1994b; Padberg *et al.*, 1995). The first MeHg measurements were performed using a capillary column GC technique, which was later used for intercomparison and

**Table 12.1.** Distribution of total Hg and MeHg in the soil profile from two catchments in south-west Sweden

| Soil layer (depth in cm) | Total Hg | | | | MeHg | | | |
|---|---|---|---|---|---|---|---|---|
| | (g $km^{-2}$) | | % | | (g $km^{-2}$) | | % | |
| | F1 | G1 | F1 | G1 | F1 | G1 | F1 | G1 |
| LFE (−9 to 0) | 3200–5100 | | 28 | 18 | 18–25 | | 30 | 11 |
| Ae (0–5) | 900 | | 8 | 3 | 0.5–2.5 | | 3 | 0.4 |
| Bh (3–35) | 5000–13 000 | | 44 | 47 | 44–120 | | 51 | 73 |
| Bf (25–45) | 2300 | | 20 | 8 | 14 | | 17 | 8 |
| BC/C (45–70) | 7000* | | | 24 | 15* | | | 9 |
| Sum | | | 100 | 100 | | | 100 | 100 |

*Data from the soil profile in G1 catchment.

intercalibration exercises (Lee, 1987; Lee and Mowrer, 1989; Lee and Hultberg, 1990; Padberg *et al.*, 1995). After 1991, MeHg was normally analysed using aqueous phase ethylation and GC separation followed by pyrolysis and cold vapour atomic fluorescence spectrometry (CVAFS) detection. Water samples were pre-treated using dichloromethane extraction (Bloom, 1989). Pre-treatment of solid samples before Hg/MeHg analysis was performed through $H_2SO_4/HNO_3$ digestion (total Hg) and distillation (MeHg) (Horvat *et al.*, 1993; Lee *et al.*, 1994b). An extensive summary of our experiences from MeHg analysis in different matrices can be found in Lee *et al.* (1994b). Total Hg analysis was performed after BrCl oxidation followed by $SnCl_2$ reduction and pre-collection of volatilised $Hg^o$ on gold, and CVAFS detection (Bloom and Crecelius, 1983; Bloom and Fitzgerald, 1988; Iverfeldt, 1988, 1991a).

## STORES OF MERCURY AND METHYL MERCURY IN THE GÅRDSJÖN CATCHMENTS

### Hg and MeHg in soil

The major store of Hg and MeHg in the forested catchment is the forest soil. The forest soil in the investigated catchments, G1 (0.0063 $km^2$) and F1 (0.037 $km^2$) is thin with an average depth of 0.6 m (F1) and 0.4 m (G1). The total stores amount to 12–28 kg/km for total Hg and 0.09–0.17 kg $km^{-2}$ for MeHg. The distribution of Hg and MeHg in the different soil horizons at F1 and G1 catchments was surveyed by Lee *et al.* (1994a) and is presented in Table 12.1. The amounts have been recalculated using recently evaluated soil density data (see Chapter 2, this volume).

The concentrations of Hg and MeHg in the different soil layers are presented in Table 12.2. The Hg and MeHg data in the humus layer are mean values from five different areas in each catchment and data from the other layers are from one soil profile in each catchment. Both soil profiles were sampled at the inflow area and near the valley in the catchments.

**Table 12.2.** Total Hg and MeHg concentration in different soil layers of the F1 and G1 catchments

| Soil layer (depth in cm) | MeHg, dw (ng/g (SD)) | *n* | Hg-tot, dw (ng/g (SD)) | *n* |
|---|---|---|---|---|
| Litter (0–6) | 0.97 (0.52) | 12 | 146 (67) | 6 |
| Humus (4–10) | 0.53 (0.18) | 11 | 109 (23) | 5 |
| Bleach (8–13) | 0.035 (0.007) | 5 | 28 (13) | 2 |
| Peat/moss (0–10) | 1.35 (0.23) | 10 | 215 (97) | 3 |
| Mineral, Bhf/Bf (8–45) | 0.30 (0.22) | 10 | 46 (19) | 8 |
| Mineral, BC/C | 0.064 (0.039) | 3 | 22 (3) | 3 |

*n* = Total numbers of samples analysed.

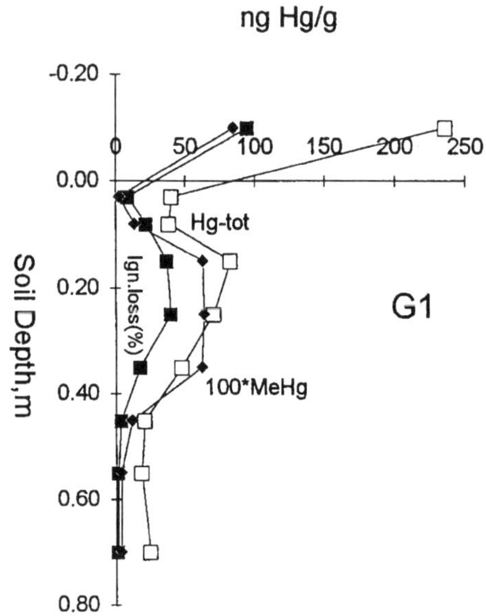


**Figure 12.1.** Concentrations of Hg and MeHg and ignition loss in a soil profile in the G1 catchment

Both Hg and MeHg are accumulated in the organic layers of the soil horizon in the Gårdsjön catchments. The accumulation of Hg and MeHg in the Bh soil horizons, especially in the G1 catchment, is remarkable. The strong association of Hg and MeHg with the soil organic material makes the organic soil horizon an efficient trap for Hg and MeHg deposited from the atmosphere and leads to the accumulation of Hg and MeHg in the Bh/B soil horizons. The similar distribution pattern of MeHg/Hg and ignition loss (%), a measure of the organic content in the soil, clearly indicates the strong association of Hg and MeHg with soil organic material (Figure 12.1).

Recent hydrological investigations performed in the G1 catchment have indicated that dissolved organic substances may be transported laterally, from the shallow soil on

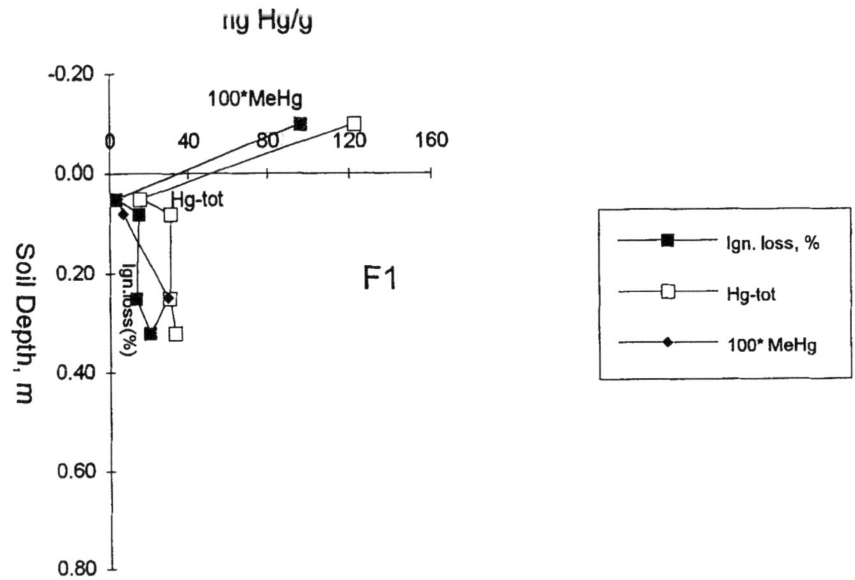


**Figure 12.2.** Concentrations of Hg and MeHg and ignition loss in a soil profile in the F1 catchment

the steep sides of the catchment to the deeper soil layers in the valley, where they can precipitate (Nyberg, 1995). More than 40% of the annual lateral water flow can occur in the upper 30 cm of the soil in the lower part of the G1 catchment. Since Hg and MeHg to a large extent are associated with DOC, this suggests that Hg and MeHg can follow the same lateral flow and thus be transported from the shallow soil in the steeper parts of G1 and retained on the Bh/B soil horizons in the valley of the catchment.

This flow pathway for Hg/MeHg and dissolved organic matter is more important in the G1 area than in the F1 catchment where the soil profiles were sampled. In F1, the accumulation of Hg and MeHg in the Bh horizon is less pronounced in agreement with these observations (see Figure 12.2).

## Hg and MeHg in biomass

Another important store is the growing biomass where Hg is primarily located in the needles and bark of the trees (see Table 12.3). The total Hg store in trees amounts to about 1 kg km$^{-2}$. The older needles (more than 2 years old) contain more than 40% of the total Hg store in plant. The highest biomass concentrations are found in the parts of the tree that have been in contact with air for an extended period, i.e. bark and older needles, which suggests that the origin of the biomass Hg is the atmosphere and not the soil. The relatively large difference between 1-year-old and older needles further supports this observation.

**Table 12.3.** Total Hg in trees (mean values from F1 catchment). All pools are based on limited numbers of samples ($n=2$ to 6)

| Component | Total biomass, dw | | Hg conc. | Hg content | |
|---|---|---|---|---|---|
| | (t/km$^2$) | (t) | (ng/g, dw) | (g) | (g/km$^2$) |
| Wood | 7060 | 261 | 13 | 3.4 | 92 |
| Bark | 1120 | 41 | 200 | 8.2 | 222 |
| Needle, 1 year old | 215 | 8 | 40 | 0.3 | 9 |
| Needle $\geqslant 2$ years old | 675 | 25 | 530 | 13.3 | 358 |
| Root (bark+wood) | 3460 | 128 | 49 | 6.3 | 170 |

## INPUT AND OUTPUT OF MERCURY AND METHYL MERCURY

### Deposition

Hg and MeHg are introduced to the forested catchment via three major deposition pathways; open field precipitation, throughfall and litterfall. The open field deposition consists of wet deposition and is not unique to the forested catchment whereas throughfall and litterfall may contain large contributions from wet as well as dry deposition.

*Hg in open field precipitation, throughfall and litterfall*

The concentration of Hg in open field precipitation (OF) is generally lower than that found in throughfall (TF). This is generally interpreted as the washing-off of Hg previously dry deposited on the needles and branches of the trees. This phenomenon has been described in Iverfeldt (1991b), Hultberg *et al.* (1994) and Munthe *et al.* (1995a). Similar relations between OF and TF Hg have been found in deciduous forests in the USA (Lindberg *et al.*, 1994). In Figure 12.3, an example of concentrations of Hg in OF and TF measured during a period of over 5 years is presented.

Apart from the differences in concentrations, a pronounced seasonal variation can also be seen for Hg in TF with higher concentrations occurring during the summer months and early autumn. The difference between Hg in TF and OF is not as pronounced for deposition fluxes as it is for concentrations owing to the difference in precipitation amounts at the two locations which counterbalances the differences in concentration.

Measured concentrations of Hg in litterfall (LF) vary between 33 and 140 ng g$^{-1}$ (dry weight). The deposition of Hg with LF represents roughly an equal flux of Hg to that depositing via TF. The total deposition in the forest (TF and LF) is approximately three to four times the wet deposition in OF (Iverfeldt, 1991b; Driscoll *et al.*, 1994; Munthe *et al.*, 1995a).

The most likely source of the additional Hg found in TF and LF as compared with OF is dry deposition on the needle surfaces. Gaseous divalent Hg species and

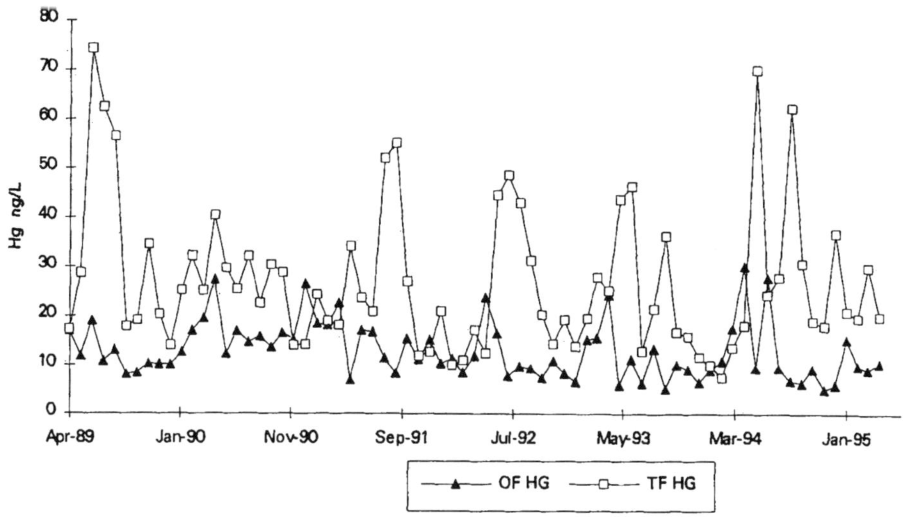


**Figure 12.3.** Concentrations of Hg in throughfall (TF) and open field precipitation (OF) at Gårdsjön, south-west Sweden

particulate Hg are most likely deposited on the surface of the needles whereas elemental mercury vapour can be taken up via the stomata (Hanson *et al.*, 1995). Hg can also be transferred to the canopy via oxidation of $Hg^0$ on the surfaces of the needles. Under humid conditions and when the canopy is wet after fog or rainfall, this may occur in aqueous reactions with ozone.

*MeHg in open field precipitation, throughfall and litterfall*

The differences in concentrations between OF and TF are not as pronounced for MeHg as they are for Hg (Figure 12.4). As for Hg, clear seasonal trends can be found with the highest MeHg concentrations occurring in early spring for both OF and TF (if the two individual peak values are not considered). In contrast to Hg, however, the seasonal variations are slightly less obvious in TF.

Investigations of MeHg in litterfall (LF) have shown that this flux dominates the deposition of MeHg to the forest floor, see Table 12.4 (Hultberg *et al.*, 1994; Munthe *et al.*, 1995a). LF concentrations of MeHg vary from 0.29 to 3.9 ng $g^{-1}$, dry weight. Assuming that the MeHg in LF originates from dry deposition of air-borne MeHg, the dry deposition flux to the canopy, and the subsequent flux to the forest floor, is the overall most important source of MeHg to the ecosystem. Other possible sources of MeHg in litterfall are methylation of inorganic Hg and root uptake from soil water. These sources have yet to be quantified.

## Mercury and methyl mercury in runoff

Hg and MeHg concentrations in runoff from catchment F1 have been monitored since 1986/87. In the roofed catchment (G1) sampling started in 1989.

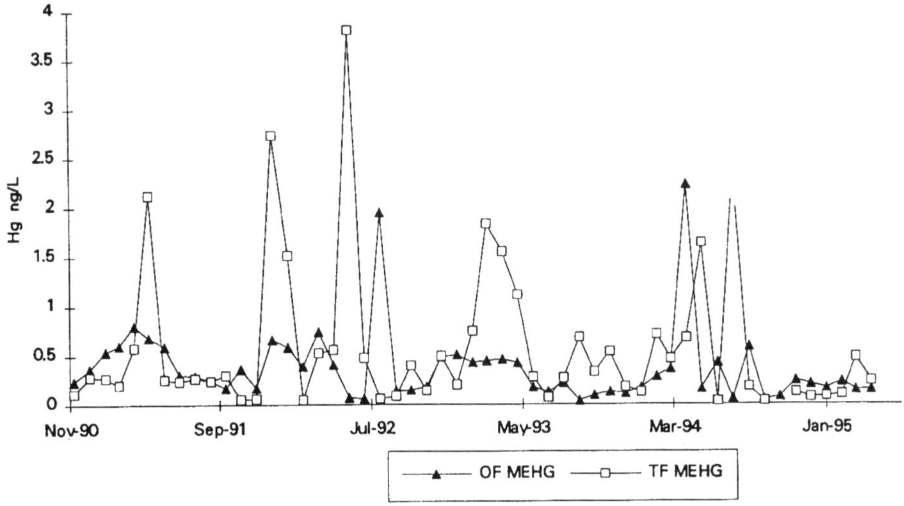


**Figure 12.4.** Concentrations of MeHg in throughfall (TF) and open field precipitation (OF)

**Table 12.4.** Average annual input/output fluxes in g/km$^2$ for Hg and MeHg in the control (F1) and roof (G1) catchments in the Gårdsjön watershed, 1990 to 1995

| Open field | | Throughfall | | Litterfall | | Runoff F1 | | Runoff G1 | |
|---|---|---|---|---|---|---|---|---|---|
| Hg | MeHg | Hg | MeHg | Hg | MeHg | Hg | MeHg | Hg | MeHg |
| 11.2 | 0.34 | 16.0 | 0.20 | 23.0 | 0.55 | 2.2 | 0.10 | 1.0 | 0.06 |

Data from Munthe *et al.* (1995a) and Hultberg *et al.* (1995).

Hg concentrations typically range between 2 and 6 ng L$^{-1}$ in both catchments whereas the MeHg concentrations generally are lower than 1 ng L$^{-1}$, and during the last 3 years lower than 0.5 ng L$^{-1}$.

The concentrations of MeHg and total Hg are not directly correlated and different mechanisms are probably responsible for the mobilisation of these two species. The atmospheric input plays an important role for both species (Hultberg *et al.*, 1994; Munthe *et al.*, 1995a), but, given the large stores of Hg in the soils, other processes resulting in the release of Hg/MeHg or methylation/demethylation will also be of importance for the overall biogeochemical turnover. Degradation of soil organic matter to soluble forms, with capacity to transport Hg and MeHg out of the soil, as well as subsequent chemical interactions (including methylation/demethylation) are here of prime importance.

### *Factors affecting the concentrations and fluxes of Hg and MeHg in runoff*

The runoff data from 1993 to 1995 can be subjected to a close examination with emphasis on how hydrological factors, including water flow and pathways, influence the levels of Hg and MeHg in runoff. Both water discharge rate and soil temperature influence MeHg and Hg concentrations but in different ways. In general, when high water flow occurs during the winter/spring snowmelt period, there will be a dilution of the MeHg concentration whereas the concentration of Hg increases (e.g. during winter 1993 and 1994), see Figure 12.5.

During the period of high water flow in the autumn of 1994, following a dry summer, relatively high concentrations of total Hg were observed whereas the MeHg concentrations decreased compared with the previous year. During the dry summer period in 1994, high concentrations of MeHg in F1 runoff were found. When a period of low water flow occurred during the beginning of autumn 1993, following a wet July and August, the MeHg concentrations increased somewhat whereas total Hg declined.

The independent behaviour of MeHg and Hg concentrations suggests that different factors may be involved. The concentration changes during high flow situations may be explained by a direct influence of precipitation (or snowmelt) on the runoff water. On average, runoff contains lower concentrations of Hg than precipitation which indicates that a direct mixing of precipitation and soil water can lead to the observed increase in runoff Hg. Another possible explanation is that the principal water flow pathways during high discharge are superficial (Seibert, 1993) through soil layers rich in Hg, which may result in increased Hg concentrations compared with the flow during low discharge when the water comes from deeper soil layers with lower Hg concentrations (Lee *et al.*, 1995).

Possible biogeochemical mechanisms for controlling MeHg output include transferring MeHg from soil to soil water, via degradation of the soil organic matter to DOC, and methylation of inorganic Hg in the soil pool to MeHg. Both processes make MeHg available for hydrological transport out of the soil. The results from the investigation in the Gårdsjön catchments suggests the release of MeHg from the forest soil can be promoted by conditions of high temperature and high soil humidity. Higher MeHg concentrations and MeHg output would thus be expected during the wet summer 1993 in comparison to 1994, see Figures 12.5 and 12.6. Similar observations of the effects of seasonal variations in temperature on MeHg output have also been made in catchment studies in other parts of Sweden and North America (Lee and Hultberg, 1990; Lee and Iverfeldt, 1991; Krabbenhoft *et al.*, 1995; Lee *et al.*, 1995).

The pattern of runoff concentrations of MeHg and Hg in the G1 catchment was similar to F1, although much lower output levels of MeHg and Hg were observed. The lower runoff amounts in G1 compared with F1 are caused by a higher evaporation from a larger tree biomass (cf. Chapter 2, this volume) which also results in lower overall MeHg and Hg fluxes from G1, see Figures 12.7 and 12.8.

A major factor influencing the generation of Hg/MeHg from the soil matrix in these Gårdsjön catchments seems to be preferential water flowpaths (and residence times). In general, during stormflow events a substantial part of the precipitation water only infiltrates into upper soil layers as the water table of the saturated zone rises. Then, in the G1 catchment, water reaches the outlet as lateral subsurface flow (at depths of 20–30 cm) in a short time (Lange *et al.*, 1996; Nyberg, 1995). In F1 the lateral flow quickly becomes a subsurface flow as it is discharged to the streamlet in the middle of the valley. Therefore the predominant flowpaths are more superficial

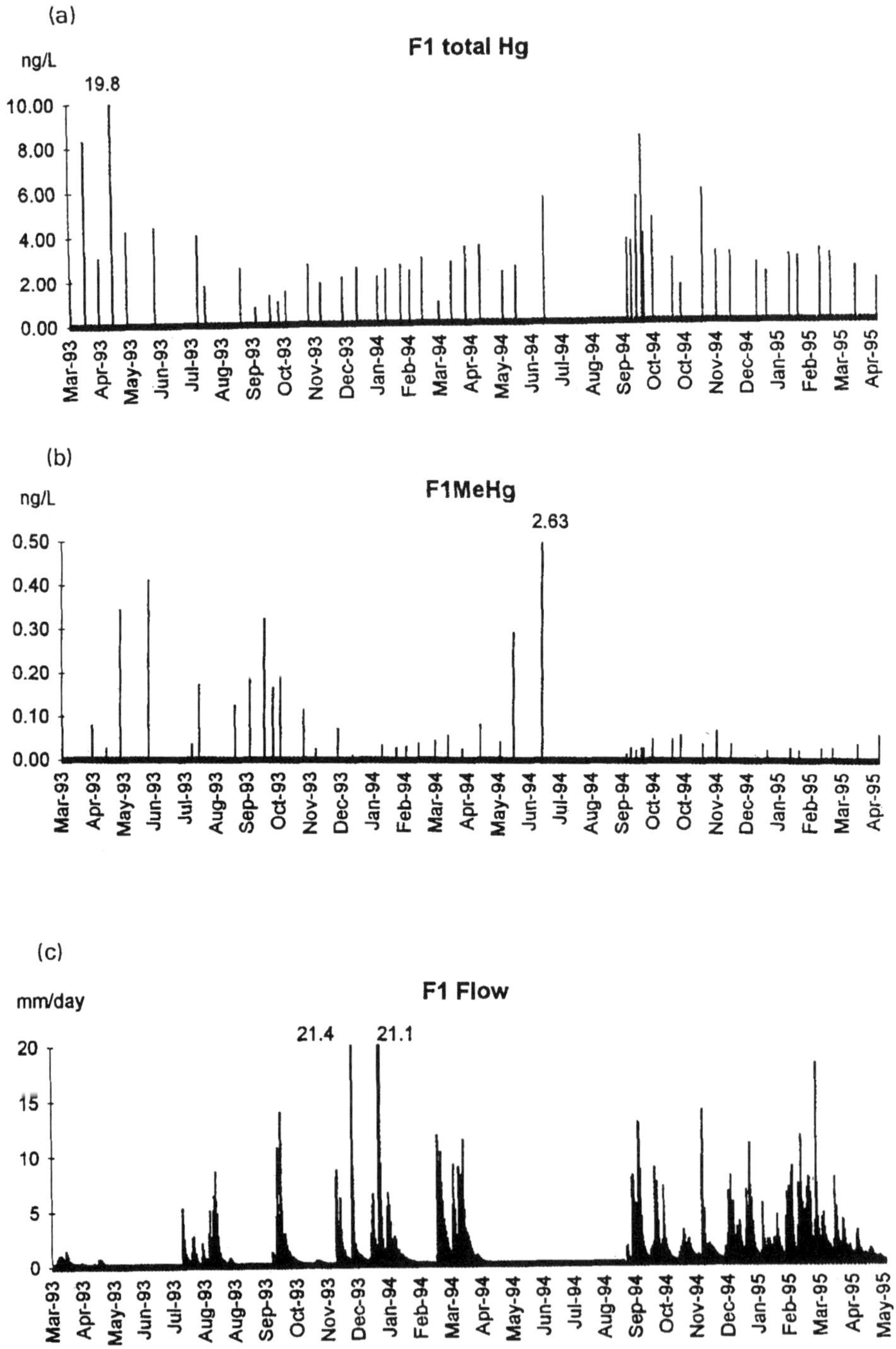


**Figure 12.5.** Concentrations of Hg and MeHg and runoff amounts in the F1 catchment, March 1993 to May 1995

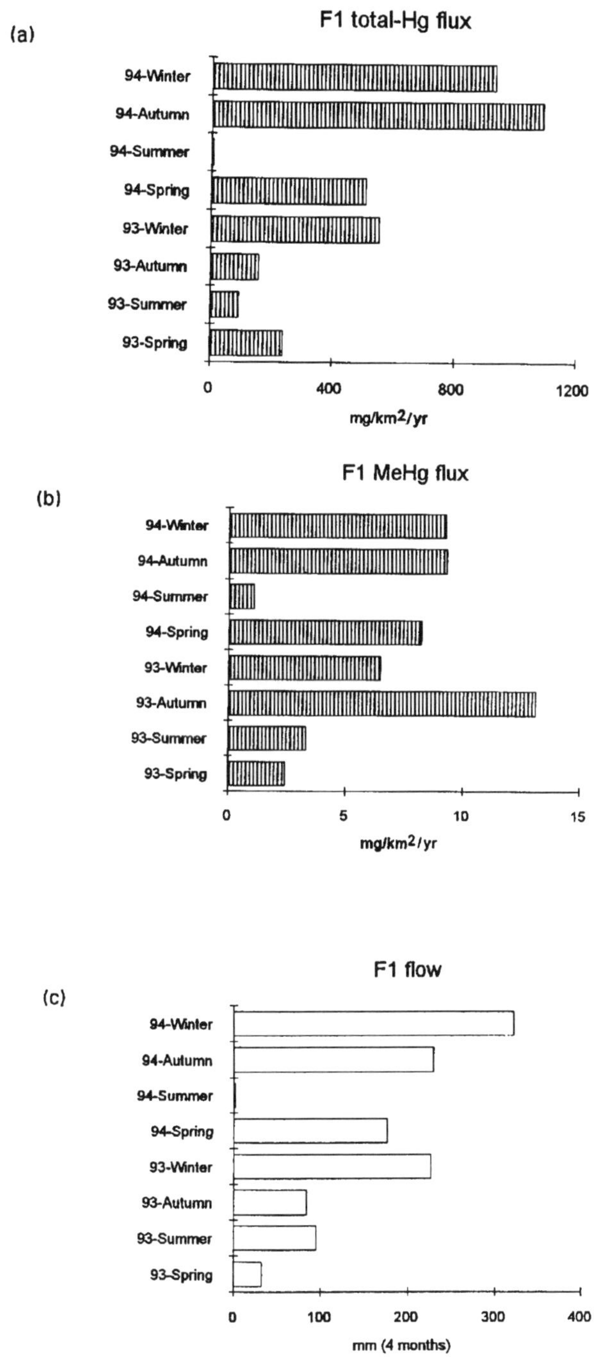


**Figure 12.6.** Seasonal variations in fluxes of Hg, MeHg and runoff amounts in the F1 catchment

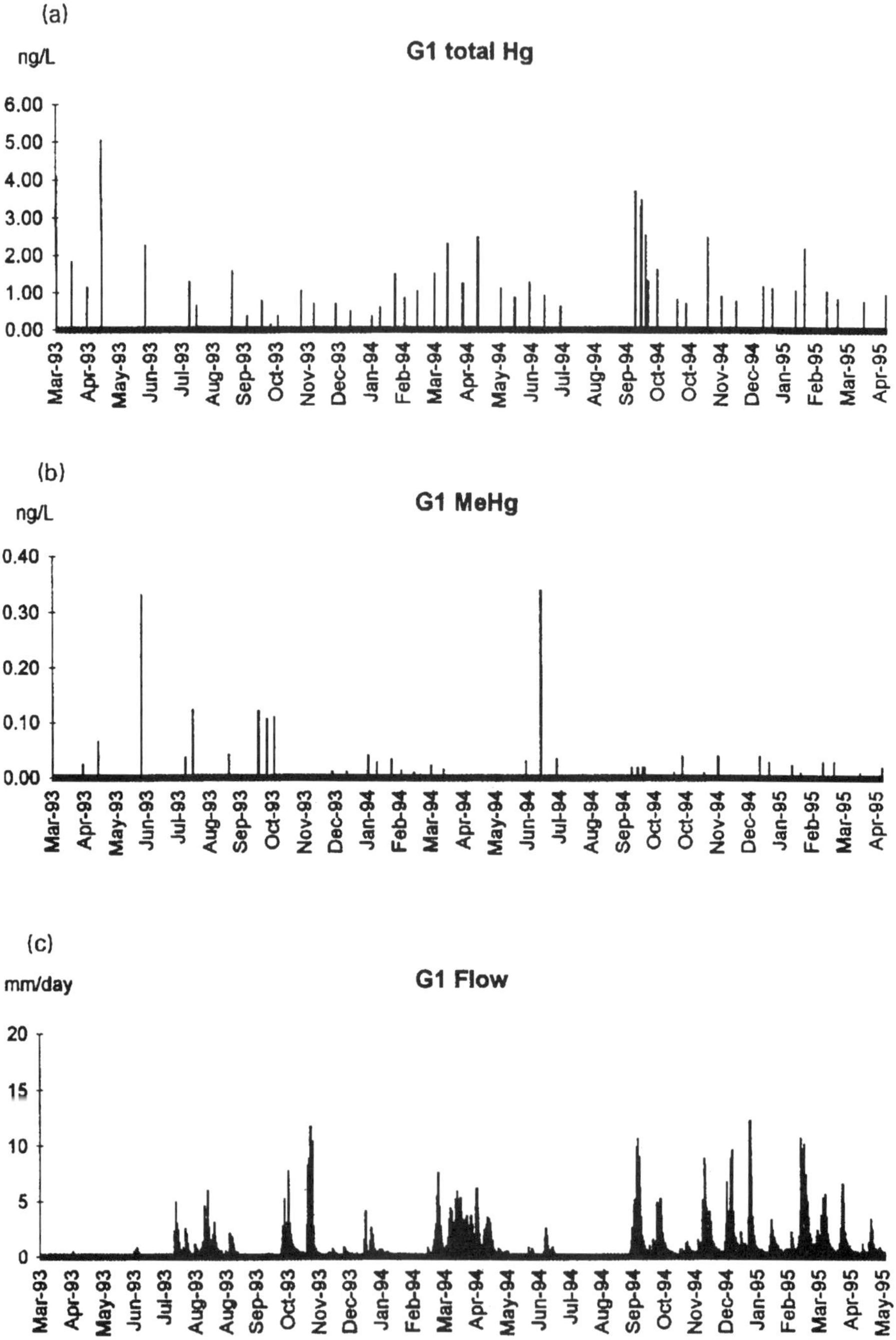


**Figure 12.7.** Concentrations of Hg and MeHg and runoff amounts in the G1 catchment, March 1993 to April 1995

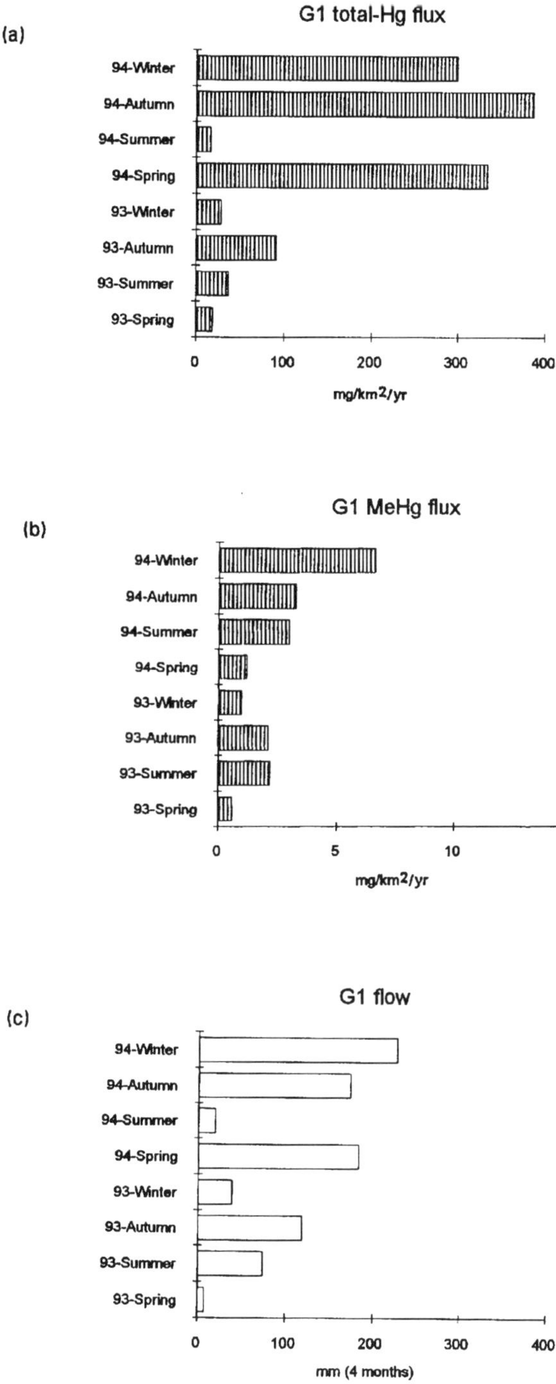


**Figure 12.8.** Seasonal variations in fluxes of Hg and MeHg and runoff amounts in the G1 catchment

in catchment F1 and only a small portion of the yearly runoff in G2 occurs in the corresponding superficial soil layers which are rich in organic matter and Hg.

There is a small area of wetland in the lower part of F1 which may have influenced the higher MeHg concentrations found in F1 runoff than in G1 runoff during the wet summer of 1993. The conditions of high temperatures and high soil humidity during this wet summer makes the wetland an important source area for making MeHg available for hydrological transport out from soil, either via methylation of soil Hg or via degradation of soil organic matter containing MeHg of atmospheric origin.

In summary, the difference in annual water fluxes from G1 and from F1 will result in lower annual MeHg/Hg output fluxes from G1. The existence of a small area of wetland and the more superficial flowpaths in F1 are other factors influencing the MeHg transfer from soil to runoff water.

## EFFECTS OF A DECREASED ATMOSPHERIC INPUT

The yearly input–output measurements showed a sharp decrease in output by runoff for both Hg and MeHg during the third year of the experiment (Hultberg *et al.*, 1994). The decrease, however, was large in both the G1 and F1 catchments, making an evaluation of experimental versus natural climatic effects difficult to interpret. The decreased atmospheric input of Hg and MeHg to the roof catchments resulted in a faster and larger decrease in runoff of both Hg and MeHg than was expected before the start of the experiment in Gårdsjön. Because of this, an additional evaluation of the data from the period after the roof construction was made using a slightly different approach. In the earlier study (Hultberg *et al.*, 1994) measured and calculated output fluxes from the G1 catchment were compared and a decrease of around 30% was found for both Hg and MeHg. Using a different statistical approach, a "randomised intervention analysis" (Carpenter *et al.*, 1989) where the difference in Hg and MeHg concentrations in runoff from F1 and G1 before and after the roof construction are compared, somewhat different results are obtained. The observed decrease of MeHg output in the roofed catchment is confirmed ($p < 0.05$) for the first 3 years after the construction whereas the observed decrease in Hg was more uncertain. If 4 years of data are included in the statistical treatment, i.e. including the most recent period (March 1994 to April 1995), no statistically significant decrease is found for either Hg or MeHg. These results indicate that between-year variations in Hg and MeHg output are important and that the large stores of Hg and MeHg in forest soils must also be taken into account when predicting loads on aquatic ecosystems in the future. In order to be able to evaluate the long-term effects of a decreased atmospheric deposition, further mechanistic studies of Hg and MeHg mobilisation in forest soils as well as a continued monitoring of output levels in the covered catchment are needed.

## CONCLUSIONS

The critical load concept presented in Hultberg *et al.* (1994) still represents the only attempt to quantify the influence of atmospheric MeHg sources on the aquatic accumulation of MeHg. The continued research in the Gårdsjön catchments has shown that the basis for the critical load concept is still relevant but that further research is needed before a more complete description of the mechanisms of MeHg output from forest soils can be given.

The measurements in the covered catchment have shown that between-year variations in catchment output of Hg and MeHg will influence the evaluation of the effects of a reduced input. The initial results suggested a rapid response in the catchment output of MeHg but the subsequent measurements have shown that the time-scale of the catchment response is somewhat uncertain. Further measurements during the next years will be needed before a final answer can be given.

## REFERENCES

Bloom, N. S. (1989) Determination of picogram levels of methylmercury by aqueous phase ethylation, followed by cryogenic gas chromatography with atomic fluorescence detection. *Can. J. Fish. Aquat. Sci.*, **46**, 131.

Bloom, N. S. and Crecelius, E. A. (1983) Determination of mercury in seawater at sub-nanogram per liter levels. *Mar. Chem.*, **14**, 49.

Bloom, N. S. and Fitzgerald, W. F. (1988) Determination of volatile mercury species at the picogram level by low temperature gas chromatography with cold-vapor atomic fluorescence detection. *Anal. Chim. Acta*, **208**, 151–161.

Carpenter, S. R., Frost, T. M., Heisey, D. and Kratz, T. K. (1989) Randomized intervention analysis and the interpretation of whole ecosystem experiments. *Ecology*, **70**, 1142–1152.

Driscoll, C. T., Otton, J. K. and Iverfeldt, Å. (1994) In: B. Moldran and J. Cerny (eds), *The Biogeochemistry of Small Catchments — A Tool for Environmental Research. SCOPE 51.* John Wiley & Sons Ltd, Chichester, 299–322.

Hanson, P. J., Lindberg, S. E., Tabberer, T. A., Owens, J. G. and Kim, K.-H. (1995) Foliar exchange of mercury vapour: Evidence for compensation point. *Water, Air, Soil Pollut.*, **80**, 3373–3382.

Horvat, M., Bloom, N. S. and Liang, L. (1993) Comparison of distillation with other current isolation methods for the determination of methyl mercury compounds in low level environmental samples. Part II Water. *Anal. Chim Acta*, **282**, 153–168.

Hultberg, H., Iverfeldt, Å. and Lee, Y. H. (1994) Methylmercury input/output and accumulation in forested catchments and critical loads for lakes in southwestern Sweden. In: C. J. Watras and J. W. Huckabee (eds), *Mercury Pollution: Integration and Synthesis.* Lewis Publishers, Chelsea, MI, pp. 313–322.

Hultberg, H., Munthe, J. and Iverfeldt, Å. (1995) Cycling of methylmercury and mercury — responses in the forest roof catchment to three years of decreased atmospheric deposition. *Water, Air, Soil Pollut.*, **80**, 415–424.

Iverfeldt, Å. (1988) Mercury in the Norwegian fjord Framvaren. *Marine Chemistry*, **23**, 441–456.

Iverfeldt, Å (1991a) Mercury in canopy throughfall water and its relation to atmospheric deposition. *Water, Air, Soil Pollut.*, **56**, 553–564.

Iverfeldt, Å (1991b) Occurrence and turnover of atmospheric mercury over the Nordic countries. *Water, Air, Soil Pollut.*, **56**, 251–265.

Iverfeldt, Å. and Munthe, J. (1993) Determining the wet deposition of mercury — A comparison of weekly, biweekly and monthly collection of precipitation samples. In: Proceedings from the EPA/A&WMA Symposium. *Measurement of Toxic and Related Air Pollutants*, Durham, NC, 3–7 May.

Jensen, A. and Iverfeldt, Å. (1994) Atmospheric bulk deposition of mercury to the southern Baltic sea area. In: C. J. Watras and J. W. Huckabee (eds), *Mercury Pollution: Integration and Synthesis.* Lewis Publishers, Chelsea, MI, Chapter II 4.

Johansson, K. and Iverfeldt, Å. (1991) Factors influencing the runoff of mercury from small watersheds in Sweden. *Verh. Internat. Verein. Limnol.*, **24**, 2200–2204.

Krabbenhoft, D. P., Benoit, J. M., Babiarz, C. L., Hurley, J. P. and Andren, A. W. (1995) Mercury cycling in the Allequash Creek watershed, northern Wisconsin. *Water, Air, Soil Pollut.*, **80**, 425–433.

Lange, H., Lischeid, G., Hoch, R. and Haus, M. (1995) Water flowpaths and residence times in a small headwater catchment at Gårdsjön (Sweden) during steady state stormflow conditions. *Water Resour. Res.*, **32**, 1689–1698.

Lee, Y.-H. (1987) Determination of methyl- and ethylmercury in natural waters of sub-nanogram per liter level using SCF-adsorbent preconcentration procedure. *Internat. J. Environ. Anal. Chem.*, **29**, 263–276.

Lee, Y.-H. and Hultberg, H. (1990) Methylmercury in some Swedish surface waters. *Envir. Toxic. Chem.*, **9**, 833.

Lee, Y.-H. and Iverfeldt, Å. (1991) Measurements of methylmercury and mercury in run-off, lake and rain water. *Water, Air, Soil Pollut.*, **56**, 309–321.

Lee, Y.-H. and Mowrer, J. (1989) Determination of methylmercury in natural waters at the sub-nanograms per liter level by capillary gas chromatography after adsorbent preconcentration. *Anal. Chim. Acta*, **221**, 259.

Lee, Y.-H., Borg, G., Iverfeldt, Å. and Hultberg, H. (1994a) In: C. J. Watras and J. W. Huckabee (eds), *Mercury Pollution: Integration and Synthesis*. Lewis Publishers, Chelsea, MI, Chapter III 5.

Lee, Y.-H., Munthe, J. and Iverfeldt, Å. (1994b) Experiences with the analytical procedures for the determination of methylmercury in environmental samples. *Appl. Organomet. Chem.*, **8**, 659–664.

Lee, Y.-H., Bishop, K., Petterson, C., Iverfeldt, Å. and Allard, B. (1995) Subcatchment output of mercury and methylmercury at Svartberget in northern Sweden. *Water, Air, Soil Pollut.*, **80**, 455–465.

Lindberg, S. E., Owens, J. G. and Stratton, W. (1994) Application of throughfall methods to estimate dry deposition of mercury. In: C. J. Watras and J. W. Huckabee (eds), *Mercury Pollution: Integration and Synthesis*. Lewis Publishers, Chelsea, MI, 261–271.

Munthe, J., Hultberg, H. and Iverfeldt, Å. (1995a) Mechanisms of deposition of methylmercury and mercury to coniferous forests. *Water, Air, Soil Pollut.*, **80**, 363–371.

Munthe, J., Hultberg, H., Lee, Y.-H., Parkman, H., Iverfeldt, Å. and Renberg, I. (1995b) Trends of mercury and methylmercury in deposition, run-off water and sediments in relation to experimental manipulations and acidification. *Water, Air, Soil Pollut.*, **85**, 743–748.

Nyberg, L. (1995) Water flow path, storage and residence time in a Swedish till catchment. Doctoral Dissertation, Uppsala University.

Padberg, S., Iverfeldt, Å., Lee, Y.-L., Baldi, F., Filippelli, M., May, K. and Stoeppler, M. (1995) Determination of low-level methylmercury concentrations in water and complex matrices by different analytical methods: A methodological intercomparison. In: C. J. Watras and J. W. Huckabee (eds), *Mercury Pollution: Integration and Synthesis*. Lewis Publishers, Chelsea, MI, 567–580.

Seikert, J. (1993) *Water Storage and Flux in a Micro-catchment at Gårdsjön, Sweden*. Uppsala University, Institution of Geosciences, Hydrology Section Report, 46 pp.

# 13 Atmospheric Fluxes, Circulation and Accumulation of Persistent Organic Compounds in the Gårdsjön Area

E. BRORSTRÖM-LUNDÉN & C. LÖFGREN

Measurements of persistent organic pollutants (POPs), such as polycyclic aromatic hydrocarbons (PAHs) and polychlorinated biphenyls (PCBs), have been carried out at the Gårdsjön area. The study included parallel sampling of POPs in air, deposition and spruce needles, which were undertaken in campaigns during different seasons. Recent measurements also included sampling of throughfall and soil.

The aim of the study was to identify the importance of the atmospheric fluxes for the occurrence of POPs in a forest at the Swedish west coast and to get an idea of the major pathways of circulation and accumulation of POPs within a terrestrial ecosystem. The studies of accumulation of POPs in spruce needles were favoured by the opportunity to collect needles from twigs both outside and under the roof.

The measurements showed that atmospheric fluxes are important for the occurrence of POPs in vegetation and soil in a terrestrial ecosystem. The amounts of POPs deposited to the Gårdsjön area (open field) were similar to those obtained during measurements at coastal and sea-based sites at the Swedish west coast. Spruce needles may serve as an important compartment in which the organic compounds are deposited and then released back to the atmosphere again. PAHs and PCBs are accumulated in the soil which will act as a reservoir for POPs. However, there is a decrease of more volatile POPs due to volatilisation processes.

## INTRODUCTION

The atmosphere is a major pathway for transport and deposition of persistent organic pollutants (POPs) to aquatic and terrestrial environments both far from and near to source areas (Oehme, 1991; Barrie *et al.*, 1992; Leister and Baker, 1994). Atmospheric fluxes of POPs, such as polychlorinated biphenyls (PCBs), hexachlorocyclohexanes (HCHs) and polyaromatic hydrocarbons (PAHs), have been shown to be quantitatively important for the occurrence of these contaminants in different ecological systems all over the world (Eisenreich *et al.*, 1981; Matzner, 1983, Calamari *et al.*, 1991; Iwata *et al.*, 1993). The presence of POPs is of great ecological

*Experimental Reversal of Acid Rain Effects: The Gårdsjön Roof Project.*
Edited by Hans Hultberg and Richard Skeffington. 

concern due to their toxic properties, the ability for bioaccumulation in the biota, and biomagnification in food chains.

Many POPs are considered as semivolatile organic compounds, which means that they are transported in the atmosphere in both the gas and particle phases (Pankow, 1987; Bidleman, 1988). The distribution of the compounds between the phases, which depends on factors such as the vapour pressure of the compound, the ambient temperature and the particle concentration, is of importance for the deposition processes.

The deposition of semivolatile organic compounds takes place either as wet or dry deposition and includes scavenging of compounds in both the gas and particle phases (Ligocki *et al.*, 1985a, b). Unlike wet and dry deposition via particles, gas exchange takes place in two directions. Deposition fluxes are controlled by the chemical composition of the compound, the particle concentration in the air and the type of deposition surface, etc. (Bidleman, 1988). The main deposition processes of POPs are illustrated in Figure 13.1.

In terrestrial ecosystems, vegetation will act as an important compartment for deposition and exchange of POPs (Riederer, 1990; Tolls and McLachan, 1994). Plant surfaces with a lipophilic epicuticular wax layer have frequently been discussed as a medium which favours accumulation of organic compounds from the gas phase (Erikson *et al.*, 1989; Paterson *et al.*, 1991; Strachan *et al.*, 1994).

POPs will also be transported to the terrestrial environment via deposition to the soil, either as direct deposition to an open field or in a forested area via throughfall or deposited litterfall. Lipophilic organic compounds like POPs will be likely to be accumulated in the organic-rich humus layer in the soil, which will act as a reservoir for these contaminants (Harrad *et al.*, 1994). A relationship between the organic content of the soil and increased concentrations of POPs has been proved (Jones *et al.*, 1989). The content of POPs in the soil will decrease due to chemical or microbial degradation, or it may be delivered to aquatic environments via runoff water. However, some POPs will also be slowly released back to the atmosphere (Harner *et al.*, 1995).

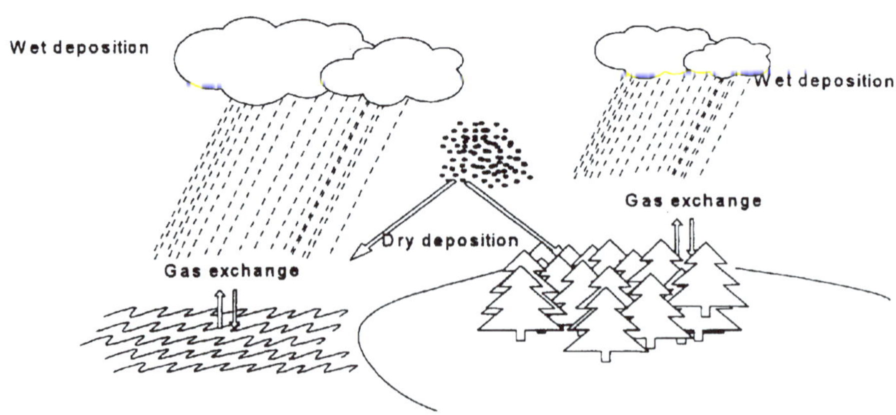


**Figure 13.1.** Deposition processes for persistent semivolatile organic compounds

Most research concerning atmospheric fluxes, circulation and accumulation as well as biological effects of POPs has been focused on the aquatic environment while the occurrence and distribution of POPs in the terrestrial environment is more rarely reported in the literature, especially for the Nordic countries.

The overall aim of this ongoing study is to identify the importance of the atmospheric fluxes for the occurrence of POPs in a forest at the Swedish west coast, the Gårdsjön area. An additional aim is to measure the major pathways of circulation and accumulation of these organic contaminants within a terrestrial ecosystem.

The Gårdsjön project, where the recovery from acid deposition as well as catchment cycling of various pollutants such as mercury is investigated, provides valuable background data on ambient conditions, etc., which supports similar studies of POPs in this area (Bishop and Hultberg, 1995; Chapter 12, this volume). Furthermore, measurements of POPs in air and deposition at coastal and sea-based sites at the Swedish west coast have shown that this area receives a significant amount of these compounds as a result of long-range transport in the atmosphere as well as re-emission from the sea (Brorström-Lundén *et al.*, 1994; Brorström-Lundén, 1996).

## EXPERIMENTAL

### Measurements

Measurements of POPs in air, deposition (open field) and spruce needles have been carried out at Gårdsjön during different seasons between 1991 and 1994, when the uptake of POPs in spruce needles was studied in relation to concentrations in air and amounts in deposition as measured by an artificial surface. The sampling periods were chosen to represent different meteorological conditions and air pollution situations. In order to investigate the occurrence and distribution of POPs in different compartments in the Gårdsjön area, the measurements undertaken in 1994 also included sampling of throughfall and soil.

### Methods

POPs in air, both the gas and particle phases, were collected using a high volume sampler (HVS) equipped with a glassfibre filter and polyurethane foam plugs (PUF). Both wet and dry deposition of POPs (open field) were collected using an open sampler (bulk sampler) placed 1.5 m above the ground level by Lake Gårdsjön. The sampler consisted of a 1 $m^2$ Teflon-coated surface with 10 cm high edges. The bottom declined slightly to a central opening where a cassette with an adsorbent of PUF was attached. The deposition samples included both compounds in the precipitation and compounds deposited to the collection surface of the sampler.

Spruce needles, first-year shoots from Norway spruce (*Picea abies* (L.) Karst), were collected about 1–2 m above the ground level. The uptake of POPs in the needles was studied by comparing the content of POPs in shoots from twigs collected outside and under the roof. In order to study the importance of accumulation in the needles, second-year shoots were also collected.

Throughfall was measured using four stainless-steel buckets placed below spruce trees in the catchment area F1. Soil samples, of the forest floor and the humus layer, were collected in May 1994 at 25 sites in catchment area F1. The samples were pooled in order to represent an average for the entire area of F1.

Descriptions of sampling techniques (air and deposition) as well as the analytical procedures can be found in Brorström-Lundén *et al.* (1994) and Brorström-Lundén (1995).

### Compounds in the study

The compounds investigated in this study are POPs frequently present in the atmosphere at the Swedish west coast; polychlorinated biphenyls (PCBs), hexachlorocyclohexanes (HCHs) and polyaromatic hydrocarbons (PAHs). These compounds, which are emitted from different sources, are widespread in the environment. They are lipophilic with the ability to bioaccumulate, but have somewhat different physical and chemical properties. The differences in physical and chemical properties of the compounds have been used in the evaluation of different deposition processes.

The PAH determination included 11 components, representing PAHs with different volatility and reactivity. Concentrations of seven individual PCB congeners of different degrees of chlorination (IUPAC nos. 28, 52, 101, 118, 153, 138 and 180), which also represent PCBs with different volatility, were determined. These PCB congeners are usually quantitatively important in environmental samples and contribute to about 20% of the total PCBs. The analyses of HCHs included the three isomers alpha, beta and gamma. These isomers have similar boiling points but differ in polarity.

## RESULTS AND DISCUSSION

The average concentrations of PAHs, PCBs and HCHs obtained in different compartments in the Gårdsjön area are summarised in Table 13.1. A yearly deposition flux (open field) of these compounds, which has been estimated from the average values, is also given in Table 13.1. The average concentrations of individual PAHs, PCBs and HCHs found in air, deposition and spruce needles, as well as the variation among the different sampling occasions, are given in Tables 13.2–4. Throughfall has up to now only been determined for one period, 2 weeks in March 1994. The concentrations of PCBs and PAHs in this throughfall sample are given in Figures 13.4 and 13.5, where the distribution of individual PAHs and PCBs is demonstrated for the other media as well.

### Air

Simultaneous air sampling at Lake Gårdsjön and coastal and sea-based stations at the Swedish west coast indicated that the atmospheric concentrations of both PCBs and HCHs were similar at each site, while the concentrations of PAHs were

**Table 13.1.** The average concentrations of PAHs, PCBs and HCHs obtained in different compartments. A yearly deposition flux (open field), which has been estimated from the average values, is also given

| | PAH (sum 11) | PCB (sum 7) | PCB (total)* | HCH (sum 3) |
|---|---|---|---|---|
| Air (ng/m$^3$) | 2.5 | 0.014 | 0.06–0.01 | 0.096 |
| Deposition (open field) | | | | |
| Average flux (ng/m$^2$/day) | 510 | 3.2 | 13–21 | 10 |
| Yearly deposition ($\mu$g/m$^2$/year) | 190 | 1.2 | 5.8 | 3.6 |
| Yearly deposition, Swedish west coast† ($\mu$g/m$^2$/year) | 100–400 | 0.7–1 | 3–5 | 2–5 |
| Vegetation (ng/g dw spruce needles) | 65 | 2.1 | 8–14 | 2.6 |
| Soil‡ (ng/g dw) | | | | |
| Forest floor | 320 | 10 | 40–70 | not determined |
| Humus layer | 2500 | 18 | 70–120 | not determined |

*Estimated from the sum of the 7 PCBs which consists of 15–25% of total PCB.
†Brorström-Lundén (1996).
‡One pooled sample from 25 sites, F1 control catchment.

somewhat lower at Gårdsjön especially in connection with soot episodes (Brorström-Lundén *et al.*, 1994; Brorström-Lundén, 1996). As air-borne PAHs bound to particles to a greater extent than PCBs and HCHs, these results suggest that lower concentrations of compounds in the particle phase will be present in the atmosphere of a forest area compared to coastal and sea areas. A decrease of the suspended particle concentration in the atmosphere in a forest area will probably take place due to deposition of particles to vegetation surfaces.

A seasonal variation in the atmospheric concentrations of POPs was found. However, the highest concentration of PAHs was found during periods in the winter, especially in connection with long-range transport, while the highest concentrations of HCHs and PCBs occurred during spring and summer (Figure 13.2). There was a good correlation between the atmospheric concentrations of PCBs and the ambient air temperature, indicating the importance of a temperature-controlled re-emission of PCBs.

### Deposition

The atmospheric deposition of PCBs, HCHs and PAHs at Lake Gårdsjön appears to be of the same order of magnitude as the fluxes measured at the coastal and sea-based stations at the Swedish west coast, which is also similar to those reported for the Great Lakes region in the USA (Eisenreich and Stachan, 1992). The deposition samples at Lake Gårdsjön were characterised by a greater portion of less volatile PAHs and PCBs, compared with the air samples (Table 13.3), similar to what was found at the coastal and sea-based stations. This also indicates the importance of deposition of particle-bound organic compounds to a forest area. Events with

**Table 13.2.** The average concentrations of individual PAHs, PCBs and HCHs found in air and the variation among the different sampling occasions

| Air<br>$n = 12$ | Average | Median | Min. | Max. |
|---|---|---|---|---|
| **PAH (ng/m$^3$)** | | | | |
| Phenanthrene | 1.1 | 0.98 | 0.65 | 2.0 |
| Anthracene | 0.03 | 0.03 | 0.01 | 0.06 |
| Fluoranthene | 0.52 | 0.45 | 0.20 | 1.1 |
| Pyrene | 0.27 | 0.30 | 0.08 | 0.58 |
| Benzo(a)anthracene | 0.07 | 0.07 | 0.01 | 0.15 |
| Chrysene | 0.16 | 0.15 | 0.03 | 0.41 |
| Benzo(b)fluoranthene | 0.21 | 0.20 | 0.02 | 0.54 |
| Benzo(k)fluoranthene | 0.08 | 0.07 | 0.01 | 0.20 |
| Benzo(a)pyrene | 0.09 | 0.09 | 0.01 | 0.27 |
| Benzo(ghi)perylene | 0.15 | 0.11 | 0.01 | 0.52 |
| Indeno(cd)pyrene | 0.18 | 0.15 | 0.02 | 0.55 |
| Sum | 2.5 | 2.9 | 1.1 | 5.9 |
| **PCB (pg/m$^3$)** | | | | |
| I28 | 5.2 | 4.9 | 3.2 | 4.7 |
| I52 | 3.8 | 3.8 | 2.2 | 4.7 |
| I101 | 2.8 | 2.8 | 1.4 | 3.6 |
| I118 | 1.0 | 0.88 | 0.38 | 1.8 |
| I153 | 1.7 | 1.8 | 0.70 | 2.1 |
| I138 | 1.5 | 1.6 | 0.64 | 1.8 |
| I180 | 0.59 | 0.65 | 0.20 | 0.84 |
| Sum | 14 | 15 | 11 | 18 |
| **HCH (pg/m$^3$)** | | | | |
| alpha | 51 | 42 | 24 | 106 |
| beta | 2.4 | 1.7 | $<0.1$ | 4.9 |
| gamma | 59 | 34 | 10 | 220 |
| Sum | 96 | 68 | 44 | 270 |

long-range air transport to the Swedish west coast, characterised by elevated concentrations of $SO_2$, $NO_2$ and soot, did not significantly increase the atmospheric concentrations of PCBs. However, due to an increased particle concentration in the air, such episodes affected the deposition fluxes of PCBs. The occurrence of precipitation significantly affected the deposition fluxes of the measured POPs when PAHs and PCBs mainly were washed out with particles while HCHs were dissolved in the rainwater.

## Vegetation

The concentrations of the chlorinated compounds found in spruce needles at Gårdsjön were similar to those reported in vegetation from other sites in Scandinavia (Calamari, 1991; Kylin, 1994; Sinkkonen *et al.*, 1995).

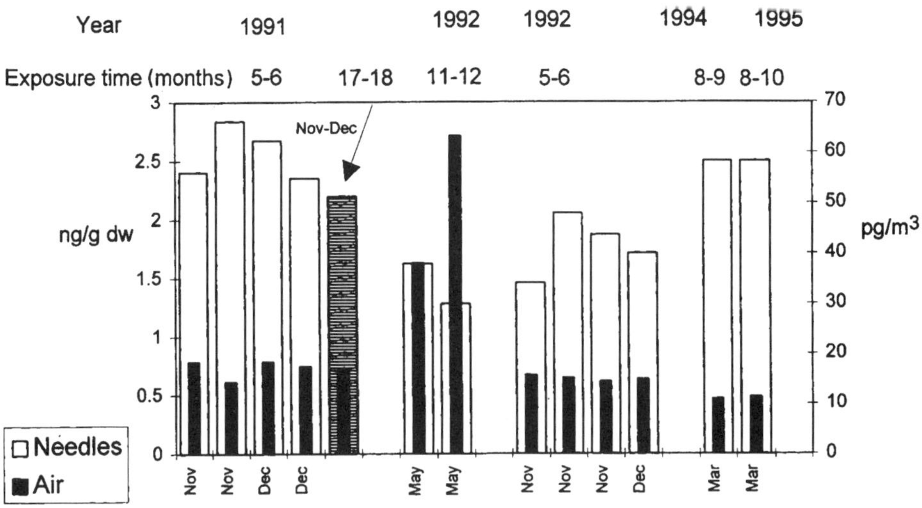


**Figure 13.2.** The variation of the PCB concentrations in air and spruce needles in relation to season and exposure time (age of the needles)

Only small variations in the concentrations of PCBs occurred in the spruce needles (first-year shoots) among different sampling occasions in the autumn and winter periods and no increased concentration of PCBs in relation to the exposure time or age of the needles was found (Figure 13.2). Furthermore, the analyses of PCBs in second-year shoots confirmed that increased accumulation in relation to age of the needles will not take place in the needles.

The lowest PCB concentration was found in needles, collected during a period in spring (May 1991). During this sampling occasion, the highest PCB concentrations in air occurred together with the highest ambient air temperature, which indicates that there is a temperature-controlled re-emission of PCBs.

The PCB concentrations in the needles collected outside and under the roof appeared to be different in the samples from the first winter period 1991 when the concentrations in the needles outside the roof varied between 2.3 and 2.6 ng/g dw (sum of seven PCBs) and under the roof between 1.3 and 1.7 ng/g dw.

The average distribution among the determined PCBs during the two winter periods is shown in Figure 13.3 where the PCB congeners are arranged in order of decreasing volatility. It was especially the more highly chlorinated and non-volatile PCB congeners which tended to be bound on particles that were found in lower concentrations in the needles collected under the roof. This period, November–December 1991, as mentioned above, was characterised by several soot episodes.

No significant differences in the PCB contents were found in the needles collected outside compared with under the roof during the sampling period November–December 1992. This period was characterised by heavy precipitation and low concentrations of soot and other air pollutants at the Swedish west coast indicating the predominance of relatively clean air.

**Table 13.3.** The average concentrations of individual PAHs, PCBs and HCHs found in deposition and the variation among the different sampling occasions

| Deposition $n = 12$ | Average | Median | Min. | Max. |
|---|---|---|---|---|
| **PAH ($\mu$g/m$^2$ day)** | | | | |
| Phenanthrene | 0.049 | 0.041 | 0.007 | 0.13 |
| Anthracene | 0.004 | 0.003 | 0.000 | 0.009 |
| Fluoranthene | 0.091 | 0.083 | 0.008 | 0.23 |
| Pyrene | 0.049 | 0.048 | 0.004 | 0.13 |
| Benzo(a)anthracene | 0.022 | 0.021 | 0.001 | 0.068 |
| Chrysene | 0.069 | 0.064 | 0.003 | 0.23 |
| Benzo(b)fluoranthene | 0.074 | 0.044 | 0.003 | 0.24 |
| Benzo(k)fluoranthene | 0.027 | 0.019 | 0.001 | 0.090 |
| Benzo(a)pyrene | 0.026 | 0.019 | 0.001 | 0.099 |
| Benzo(ghi)perylene | 0.040 | 0.026 | 0.002 | 0.16 |
| Indeno(cd)pyrene | 0.062 | 0.042 | 0.003 | 0.21 |
| Sum | 0.51 | 0.42 | 0.03 | 1.6 |
| **PCB (ng/m$^2$/day)** | | | | |
| I28 | 0.21 | 0.14 | 0.10 | 0.38 |
| I52 | 0.29 | 0.25 | <0.1 | 0.40 |
| I101 | 0.41 | 0.28 | 0.06 | 1.0 |
| I118 | 0.29 | 0.13 | 0.06 | 0.76 |
| I153 | 0.74 | 0.54 | 0.14 | 1.9 |
| I138 | 0.80 | 0.66 | 0.20 | 1.5 |
| I180 | 0.52 | 0.38 | 0.10 | 1.1 |
| Sum | 3.2 | 2.7 | 1.3 | 6.4 |
| **HCH (ng/m$^2$/day)** | | | | |
| alpha | 2.1 | 1.3 | <0.05 | 7.2 |
| beta | 0.30 | 0.21 | <0.1 | 0.34 |
| gamma | 8.3 | 5.4 | <0.05 | 22 |
| Sum | 9.9 | 7.6 | <0.5 | 30 |

The comparison of the PCB content in the needles collected both outside and under the roof with the PCB content in air showed that the uptake of PCBs takes place both from the gas and particle phases, which is demonstrated in Figure 13.5. However, the comparison of the PCBs in the needles collected outside and under the roof indicates that the uptake of PCBs in the particle phase seemed to be important especially during periods with increased particle concentration in the atmosphere. Furthermore, the analyses of PCBs in the needles, collected under the roof, showed that precipitation did not affect the uptake or release in the needles as no significant differences in the PCB content outside and under the roof could be found in relation to the amounts of precipitation.

The PAH concentrations in the needles varied more among different sampling occasions compared with PCBs, but no relation was found between the concentrations and the age of the needles. The concentration of PAHs in the needles increased

**Table 13.4.** The average concentrations of individual PAHs, PCBs and HCHs found in spruce needles and the variation among the different sampling occasions

| Spruce needles<br>$n = 4$ (PAH) $n = 11$ (PCB)<br>$n = 10$ (HCH) | Average | Median | Min. | Max. |
|---|---|---|---|---|
| **PAH (ng/g DW)** | | | | |
| Phenanthrene | 6.1 | 5.3 | 2.9 | 11 |
| Anthracene | 0.86 | 0.77 | 0.16 | 1.7 |
| Fluoranthene | 12 | 9.8 | 4.3 | 24 |
| Pyrene | 8.8 | 6.6 | 4.1 | 18 |
| Benzo(a)anthracene | 3.4 | 1.4 | 0.69 | 10 |
| Chrysene | 9.2 | 5.0 | 2.8 | 24 |
| Benzo(b)fluoranthene | 8.8 | 4.1 | 2.2 | 25 |
| Benzo(k)fluoranthene | 4.3 | 1.6 | 0.84 | 13 |
| Benzo(a)pyrene | 4.8 | 1.6 | 1.0 | 15 |
| Benzo(ghi)perylene | 2.9 | 2.4 | 1.5 | 5.4 |
| Indeno(cd)pyrene | 3.4 | 3.3 | 1.4 | 5.4 |
| Sum | 65 | 50 | 22 | 137 |
| **PCB (ng/g DW)** | | | | |
| I28 | 0.14 | 0.12 | 0.04 | 0.23 |
| I52 | | | | |
| I101 | 0.46 | 0.44 | 0.26 | 0.67 |
| I118 | 0.16 | 0.16 | 0.07 | 0.22 |
| I153 | 0.55 | 0.55 | 0.35 | 0.77 |
| I138 | 0.53 | 0.59 | 0.31 | 0.73 |
| I180 | 0.24 | 0.25 | 0.15 | 0.30 |
| Sum | 2.1 | 2.1 | 1.3 | 2.8 |
| **HCH (ng/g DW)** | | | | |
| alpha | 0.84 | 0.85 | 0.45 | 1.2 |
| beta | 0.12 | 0.11 | 0.05 | 0.17 |
| gamma | 1.1 | 1.1 | 0.67 | 1.5 |
| Sum | 2.0 | 2.2 | 1.3 | 2.6 |

significantly in connection with soot episodes. Consequently the particle phase appeared to be the most important process governing the uptake of PAHs in the needles. Like PCBs there was both an uptake and release of PAHs in the needles, thus the content of PAHs in the needles decreased some time after a soot episode. The decrease of non-volatile PAHs in the needles supports the idea of removal of fine particle matter from plant surfaces due to wind shear and ageing. However, precipitation did not seem to affect the concentrations of PAHs in the needles.

## Soil

The total PCB concentrations found in the soil at Gårdsjön were about 100 ng/g dw in the humus layer and 50 ng/g dw in the forest floor (Table 13.1). Harrad *et al.*

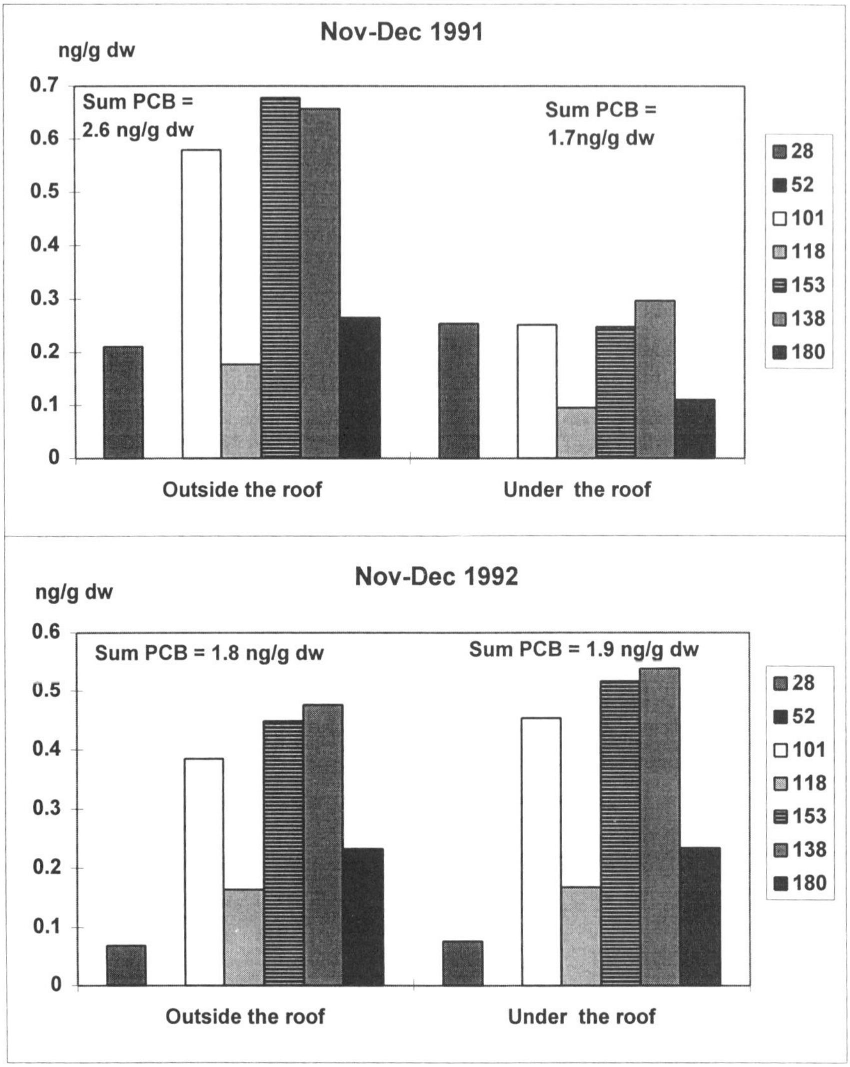


**Figure 13.3.** The distribution among the analysed PCBs in spruce needles collected outside and under the roof. The profiles represent average values for the periods. The PCBs are arranged in order of decreasing volatility

(1994) reported a total PCB concentration of 30 ng/g dw as an average for different types of soil in Great Britain and Alcock *et al.* (1993) reported a variation in the PCB concentrations between 10 and 54 ng/g dw in soils from rural to urban sites. Thus, there seem to be somewhat increased concentrations of PCBs in the Gårdsjön area compared with different sites in England. However, the organic content in the soil will affect the accumulation of organic contaminants and a high organic content in the humus layer at sampling site F1 has been reported (Lee *et al.*, 1994).

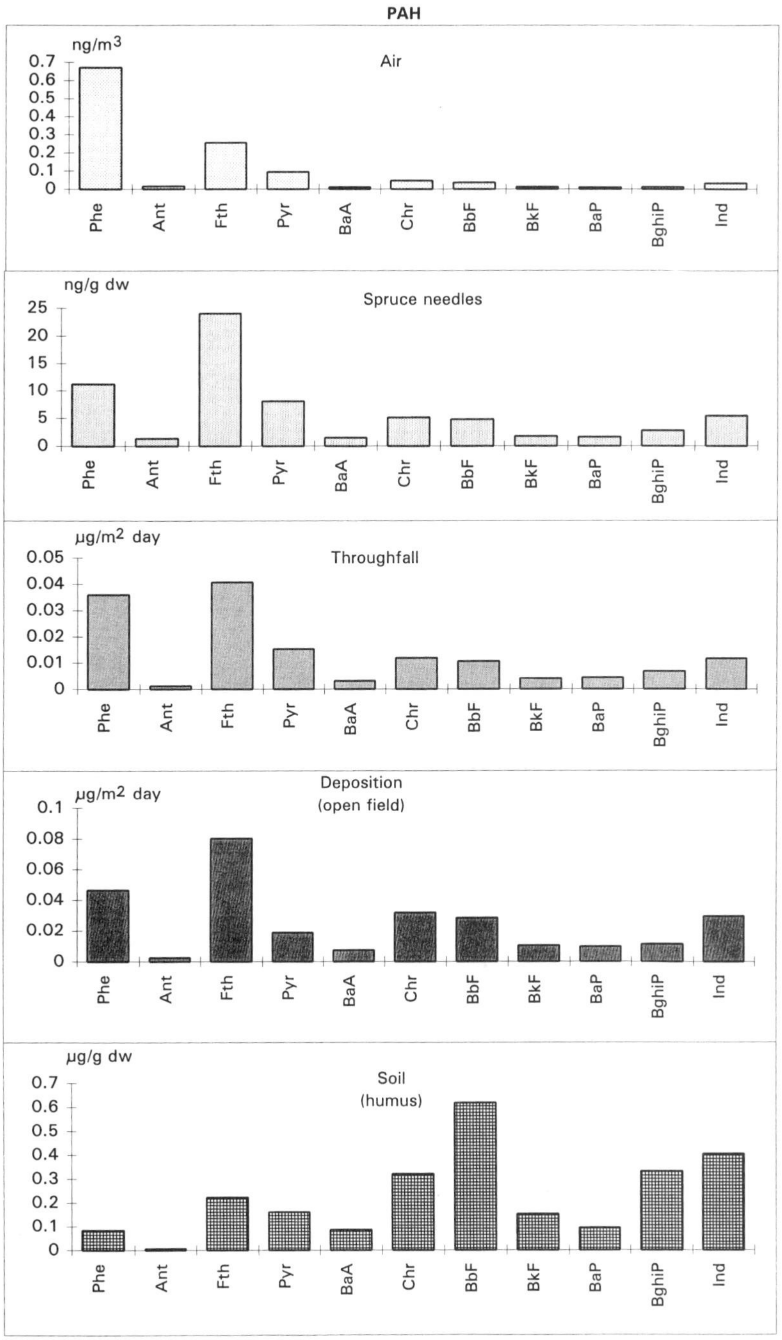


**Figure 13.4.** The PAH profiles, the distribution among individual PAHs, in different sample matrixes from 2 weeks in March 1994. Compound abbreviations represent the compounds in Tables 13.2–13.4. The PAHs are arranged in order of decreasing volatility

As with PCBs, the humus sample contained a higher concentration of PAHs, 2.5 $\mu$g/g dw, compared with the sample from the forest floor, 0.32 $\mu$g/g dw (Table 13.1). Also the content of PAHs was higher in the soil at Gårdsjön compared with Great Britain where a PAH concentration lower than 0.6 $\mu$g/g dw was reported for rural areas (Jones *et al.*, 1989).

### Distribution of POPs

In order to get an idea about circulation and accumulation of POPs in different compartments, the content of PAHs and PCBs in the different sample matrices have been qualitatively compared, which is demonstrated in Figures 13.4 and 13.5. The PAH and PCB profiles, i.e. the distribution among the individual compounds, arranged after decreasing volatility, represent the average values of measurements during 2 weeks in March 1994.

### PAHs

The air sample contained, in agreement with other measurements at Gårdsjön, more volatile PAH compounds compared with the deposition sample, showing the dominance of gas phase PAHs in the air and that particle deposition is the major deposition process for PAHs. A similar distribution among the individual PAHs was found in the spruce needles and throughfall as well as in the deposition (open field) samples. Consequently there is an important influence of particle-bound PAHs in all three media.

The accumulation of less volatile PAHs in the humus layer of soil at the Gårdsjön area is supported by the dominance of these PAH compounds in the soil sample. The relative distribution of PAHs in the soil differed from that in the deposition samples, indicating that different processes, for example evaporation or degradation, may affect the PAH content in the soil.

### PCBs

The PCB profiles showed, in agreement with PAHs, that the air sample contained relatively more volatile compounds compared with the other sample matrices (Figure 13.5). The distribution among PCBs was similar in the throughfall and deposition samples. In contrast to PAHs, however, the PCB profiles in the spruce needles differed somewhat from those in the throughfall and deposition samples indicating the importance of an uptake of PCBs in the gas phase. Except for a relative increase of less volatile congeners, indicating an evaporation of these PCBs from the soil, the distribution among PCBs obtained in the soil agreed well with the PCB profile found in the deposition and throughfall. This supports the hypothesis that atmospheric deposition is the main source for PCBs in this area and suggests that PCBs in soil are persistent.

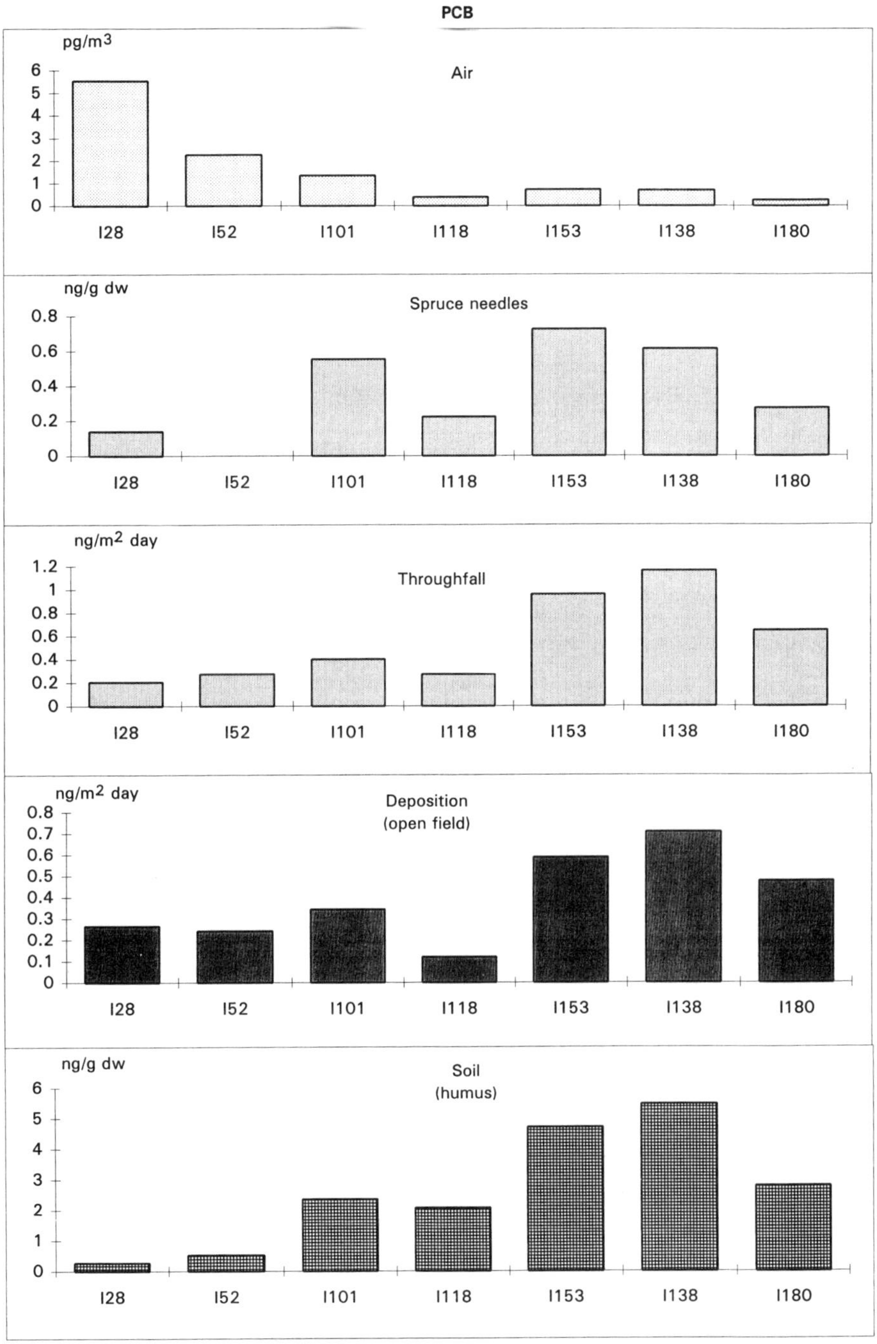


**Figure 13.5.** The distribution among the analysed PCBs in different sample matrixes from 2 weeks in March 1994. "I" values are IUPAC congener numbers. The PCBs are arranged in order of decreasing volatility.

## CONCLUSIONS AND FUTURE RESEARCH

The atmospheric fluxes are important for the occurrence of POPs in a terrestrial ecosystem at the Swedish west coast.

There is both an uptake and release of POPs in spruce needles. Thus, the vegetation may serve as an important compartment in which these organic compounds are deposited and then released back to the atmosphere.

There is an atmospheric deposition of both PCBs and PAHs in soil where the lipophilic compounds accumulate, especially in the humus layer which will act as a reservoir for POPs. However, there is also a decrease of the more volatile POPs, which may be due to volatilisation processes.

This is an ongoing study and the aim of the future research is to gain better understanding of circulation and accumulation of POPs in the terrestrial environment.

## REFERENCES

Alcock, R. E., Johnston, A. E., McGrath, S. P., Berrow, M. L. and Jones, K. C. (1993) Long-term changes in the polychlorinated biphenyl content of United Kingdom soils. *Environ. Sci. Technol.*, **27**, 1918–1923.

Barrie, L. A., Gregor, D., Lake, R., Muir, D., Shearer, R., Tracey, B. and Bidleman, T. (1992) Arctic contaminants: sources, occurrence and pathways. *Sci. Tot. Environ.*, **122**, 1–74.

Bidleman, T. F. (1988) Atmospheric processes. *Environ. Sci. Technol.*, **22**, 361–367.

Bishop, K. H. and Hultberg, H. (1995) Reversing acidification in a forest ecosystem: The Gårdsjön Covered Catchment. *Ambio*, **24**, 84–91.

Brorström-Lundén, E. (1995) Measurements of semivolatile organic compounds in air and deposition. PhD thesis, Department of Analytical and Marine Chemistry University of Göteborg, Sweden.

Brorström-Lundén, E. (1996) Atmospheric transport and deposition of persistent organic compounds to the sea surface. *Netherlands Journal of Sea Research*, **35**, 81–90.

Brorström-Lundén, E., Lindskog, A. and Mowrer, J. (1994) Concentrations and fluxes of organic compounds in the atmosphere of the Swedish west coast. *Atmos. Environ.*, **28**, 3605–3615.

Calamari, D., Bacci, E., Focardi, S., Gaggi, C., Morosini, M. and Vighi M. (1991) Role of plant biomass in the global environment partitioning of chlorinated hydrocarbons. *Environ. Sci. Technol.*, **25**, 1489–1495.

Eisenreich, S. and Strachan, W. (1992) Estimating atmospheric deposition of toxic substances to the Great Lakes — An update — *Proceedings from a Workshop held at the Canada Centre for Inland Waters, Burlington, Ontario, Canada 1992.*

Eisenreich, S. J., Looney, B. B. and Thornton, J. (1981) Airborne organic contaminants in the Great Lakes ecosystem. *Environ. Sci. Technol.*, **15**, 30–38.

Erikson, G., Jensen, S., Kylin, H. and Strachan, W. (1989) The pine needles as monitors of atmospheric pollutants. *Nature*, **341**, 42–44.

Harner, T., Mackay, D. and Jones, K. C. (1995) Model of the long-term exchange of PCBs between soil and the atmosphere in the southern U.K. *Environ. Sci. Technol.*, **29**, 1200–1209.

Harrad, S. J., Sewart, A. P., Alcock, R., Boumphrey, R., Burnett, V., Duarte-Davidson, R., Halsall, C., Sanders, G., Waterhouse, K., Wild, S. R. and Jones, K. C. (1994) Polychlorinated biphenyls (PCBs) in the British environment: sinks, sources and temporal trends. *Environ. Pollut.*, **85**, 131–146.

Iwata, H., Tanabe, S., Sakai, N. and Tatsukawa, R. (1993) Distribution of persistent organochlorines in the oceanic air and surface seawater and the role of oceans on their global transport and fate. *Environ. Sci. Technol.*, **27**, 1080–1098.

Jones, K. C., Stratford, J. A., Waterhouse, K. S. and Vogt, N. B. (1989) Organic contaminants in Welsh soils: polynuclear aromatic hydrocarbons. *Environ. Sci. Technol.*, **23**, 540–550.

Kylin, H. (1994) Airborne lipophilic pollutants in pine needles. PhD thesis, Stockholm University.

Lee, Y.-H., Borg, G., Iverfeldt, Å. and Hultberg, H. (1994) Fluxes and turnover of methylmercury: Mercury pools in forest soils. In: C. J. Watras and J. W. Huckabee (eds), *Mercury Pollution: Integration and Synthesis.* Lewis Publishers, Chelsea, MI, chapter III 5.

Leister, D. and Baker, J. (1994) Atmospheric deposition of organic contaminants to the Chesapeake Bay. *Atmos. Environ.*, **28**, 1499–1520.

Ligocki, M., Leuenberger, C. and Pankow, J. F. (1985a) Trace organic compounds in rain. II Gas scavenging of neutral organic compounds. *Atmos. Environ.*, **19**, 1609–1617.

Ligocki, M., Leuenberger, C. and Pankow, J. F. (1985b) Trace organic compounds in rain. III Particle scavenging of neutral organic compounds. *Atmos. Environ.*, **19**, 1619–1626.

Matzner, E. (1983) Annual rates of deposition of polycyclic aromatic hydrocarbons in different forest ecosystems. *Water, Air, Soil Pollut.*, **21**, 425–434.

Oehme, M. (1991) Further evidence for long-range air transport of polychlorinated aromates and pesticides: North America and Eurasia to the Arctic. *Ambio*, **20**, 293–297.

Pankow, J. F. (1987) Review and comparative analyses of the theories on partitioning between the gas and aerosol particulate phase in the atmosphere. *Atmos. Environ.*, **21**, 2275–2283.

Paterson, S., Mackay, D., Bacci, E. and Calamari, D. (1991) Correlation of the equilibrium and kinetics of leaf–air exchange of hydrophobic organic chemicals. *Environ. Sci. Technol.*, **25**, 866–870.

Riederer, M. (1990) Estimating partitioning and transport of organic chemicals in the foliage/atmospheric system: Discussion of a fugacity-based model. *Environ. Sci. Technol.*, **24**, 829–837.

Sinkkonen, S., Raitio, H., Paasivirta, J., Rantio, T., Lahtiperä, M. and Mäkelä, R. (1995) Concentrations of persistent organochlorine compounds in spruce needles from Western Finland. *Chemosphere*, **30**, 1415–1422.

Strachan, W., Eriksson, G., Kylin, H. and Jensen, S. (1994) Organochlorine compounds in pine needles: Methods and trends. *Environ. Tox. and Chem.*, **13**.

Tolls, J. and McLachlan, M. (1994) Partitioning of semivolatile organic compounds between air and *Lolium multiflorum* (Welsh ray grass). *Environ. Sci. Technol.*, **28**, 159–166.

# 14 Decomposition and Microflora Changes in Spruce-Forested Catchments at Gårdsjön Following Reduced Acidic Deposition

**R. F. HOGERVORST, H. R. ZOOMER & H. A. VERHOEF**

A comparative field study was conducted in Gårdsjön, south-west Sweden to investigate the effects of reduced acidic deposition (by means of a roof beneath the canopy) on decomposition and microflora in a Norway spruce-forested catchment. Stratified litterbag sets, consisting of a litter (Aoo) and fragmentation (Ao) layer, were placed in two plots; one plot was situated in the G1 ROOF catchment which receives clean water and one plot was situated in the F1 CONTROL catchment which receives ambient throughfall water. Litterbag sets were exposed for 2, 4, 6, 8, 10 and 12 months and analysed for dry weight loss, DTAF-active bacteria and FITC-active fungi.

After 1 year no difference in decomposition rate of either L- or F-layer was observed, when acidic deposition was reduced. Dry weight loss of L- and F-material after 1 year in the F1 CONTROL catchment was 17.63 and 7.93% respectively and for the G1 ROOF catchment 18.95 and 6.93% respectively.

The amount of DTAF-active bacteria and FITC-active fungi were significantly higher in the F-layer relative to the L-layer, and showed in both layers a clear pattern of seasonality with peaks in winter. Fungal hyphae length and bacterial numbers showed a significant reduction in the G1 ROOF catchment. Mean hyphal length and mean number of bacteria recorded in the L-layer for the F1 CONTROL catchment was 41.9 m $g^{-1}$ and $6.22 \times 10^9 g^{-1}$ dry weight organic material respectively, and 21.5 m $g^{-1}$ and $5.30 \times 10^9 g^{-1}$ dry weight organic material for the G1 ROOF catchment respectively. Mean hyphal length and mean number of bacteria recorded in the F-layer for the F1 CONTROL catchment was 159.4 m $g^{-1}$ and $1.17 \times 10^{10} g^{-1}$ dry weight organic material respectively, and 120.4 m $g^{-1}$ and $9.71 \times 10^9 g^{-1}$ dry weight organic material for the G1 ROOF catchment respectively. The reduction of nitrogen deposition is discussed as a probable cause for the observed results.

## INTRODUCTION

Decomposition of soil organic matter plays a key role in recycling nutrients into available forms for uptake of higher plants. The process of decomposition is complex

*Experimental Reversal of Acid Rain Effects: The Gårdsjön Roof Project.*
Edited by Hans Hultberg and Richard Skeffington. 

and dependent on both abiotic and biotic factors (Swift *et al.*, 1979). Quality of the substrate and environment factors primarily control the rate of breakdown of soil organic matter (Berg and Staaf, 1980; Meentemeyer and Berg, 1986; Berg *et al.*, 1993).

Chemical composition of the organic matter and climatic factors govern decomposition mainly by influencing soil organisms. Soil organisms immobilize a considerable part of the nutrients, making these nutrients unavailable for plant uptake. Mortality and decomposition enable plant roots and living soil organisms to utilize the released nutrients again. Two main groups of soil organisms co-operate in the decomposition processes: soil animals and soil microflora. These groups, and the interactions between them, play an important role in the process of decomposition and in the availability of nutrients (Persson *et al.*, 1980; Lussenhop, 1992). Fungi and bacteria have the greatest biomass of organisms in temperate coniferous forest soils and generally more than 90% of the energy flow passes through microbial decomposers (Swift *et al.*, 1979), making microbial activity a good measure of organic matter turnover. Next to plant growth and soil fertility, microbial activity is also important for soil structure and the destruction of toxic organic substances. A change or stress in the environment that affects microbial processes may cause a large ecological effect.

Atmospheric deposition is a very important stress factor which threatens the environment. Acidic precipitation plays a part in forest decline (Pitelka and Raynal, 1989), while soil acidification induced by excessive nitrogen inputs is considered an important factor. Soil acidification by atmospheric deposition of nitrogen has been frequently discussed (Van Breemen and Van Dijk, 1988; Schulze, 1989; Schaefer, 1991) and it has been suggested that excessive N inputs induce changes in the soil, resulting in forest decline (Nihlgård, 1985). Studies showed that in the Netherlands enhanced nitrogen deposition (mainly $NH_4^+$) on the forest soil increases the availability of $NH_4^+$, followed by an increase of soil acidification, weathering and leaching of base cations (Van Breemen and Van Dijk, 1988). The excess nitrogen uptake and reduced cation uptake cause nutritional imbalances within trees, reflected in reduced tree growth and impaired tree vitality (Boxman and Roelofs, 1988).

Excessive nitrogen deposition has been found to affect decomposition of organic material and nutrient flow in forest ecosystems (Fog, 1988; Kuyper, 1989). Decomposition of organic matter may be divided into two phases (Berg, 1986b). The mass-loss rate of soluble substances and non-lignified carbohydrates such as celluloses and hemicelluloses in the first phase is regulated by the concentrations of nutrients in the organic matter, while in the second phase the turnover of the remaining lignin and lignified celluloses is controlled by the decomposition rate of lignin (McClaugherty and Berg, 1987). During the initial stages of decomposition nitrogen addition has a positive effect on mass-loss rate (Berg and Staaf, 1980). The late stages of decomposition are energy regulated and negatively influenced by nitrogen (Berg *et al.*, 1987). Furthermore, nitrogen negatively affects activity and composition of micro-organism populations (Fog, 1988; Wardle, 1992). Studies on the effects of deposition of air pollutants on decomposition processes and soil organisms are predominantly based upon experiments adding artificial acid rain. It is difficult to relate the findings of these studies to natural conditions, because most treatments were applied for a short period or at high doses of the pollutant leading to shock effects. In natural situations lower doses of acids and other pollutants are added over a longer period of time, which could have little or no effect on decomposition rate (Fritze, 1988).

In southern Sweden, resampling of forest soils after 50 years revealed a general decrease in soil pH (Hallbäcken and Tamm, 1986), a decrease in exchangeable base cations and vegetational changes (Falkengren-Grerup, 1987). Nitrogen deposition has been ascribed as one of the causes of these changes. Little is known about the reversibility of soil acidification processes induced by nitrogen deposition in terrestrial systems. This study is part of a research project in which the effects of reduced nitrogen deposition on soil biota (soil fauna and microflora) and subsequently on decomposition are investigated in a non-nitrogen-saturated coniferous forest. The research was carried out within the framework of the NITREX programme, which aims to assess among other things the reversibility of nitrogen saturation.

## MATERIALS AND METHODS

### Study site

The research was conducted in a Norway spruce (*Picea abies* L.) stand in the Lake Gårdsjön area, located about 50 km north of Gothenburg near the south-west coast of Sweden. The study site is dominated by >100-year-old Norway spruce (*c.* 460 stems $ha^{-1}$, average height 18 m), combined with some Scots pine (*Pinus sylvestris* L.) in dryer areas. The site ranges in altitude from 113 to 170 m above sea level. The topography is rough, with steep valleys and frequent bedrock outcrops. Soil is predominantly acidic silty and sandy loam, drier in the upper catchment and more peaty in the lower parts, and mainly covered with bilberry (*Vaccinium myrtillus* (L.)), wavy hair grass (*Deschampsia flexuosa* (L.) Trin.) and several species of mosses (mainly *Dicranum* sp., *Leucobryum glaucum* (Hedw.), *Plagiothecium undulatum* and *Sphagnum* sp.).

In April 1991 a manipulation study was started. The manipulated G1 ROOF catchment was compared with an untreated reference F1 CONTROL catchment (3.6 ha). A detailed description of the site is given by Stuanes *et al.* (1992a), Hultberg *et al.* (1993) and in Chapter 2, this volume. Litter was collected from the roof and spread evenly underneath. The present study was conducted from May 1993 till May 1994. During this period the microclimatical conditions of the site could be characterized as initially mild, followed by a period with snow, gradually becoming more wet and ending in a period of drought again.

### Organic material

In April 1993, needle material from the litter (L; Aoo) and fragmentation layer (F; Ao) was collected outside the plots. The material of the L-layer was obtained by collecting intact needles from the surface (0–1 cm depth) of the organic layer. Any green needles were sorted out by hand, while fine material was removed by sieving over a 1 mm sieve. Needle material of the F-layer was collected from 1–3.5 cm depth of the organic layer and sieved until 2–5 mm in length. The needle material was air-dried at 20°C during 3–5 days to about 4–7% moisture and stored at 20°C until it was required for processing of litterbags and placement in the field.

Before introduction to the field, the water contents of litter and fragmentation material were adjusted to 30 and 50% wet weight respectively. An exactly weighed amount (18 g; 30% wet) of litter material was enclosed in 10×10 cm nylon bags with a mesh size of 2 mm, resulting in a 1 cm thickness of the L-layer. For the F-layer, 55 g (50% wet) fragmentation material was enclosed in similar bags, corresponding to a thickness of 3.5 cm.

### Experimental design

L- and F-bags were placed as stratified litterbag sets in the organic layer in the F1 CONTROL and the G1 ROOF catchments. They were fastened to the ground by 15-cm-long wooden pins. The litterbag sets in each catchment were distributed side by side in rows. These rows were placed in an area of 1.5×2 m and had in between a 10 cm strip of the original organic layer.

The experiment started in May 1993 and lasted 1 year. The litterbag sets were randomly sequentially removed after 2, 4, 6, 8, 10 and 12 months. Each sampling occasion had 12 replicates of litterbag sets which were all analysed for weight loss. Subsamples of four replicates were used for determining microflora.

### Weight loss

The collected litterbags were immediately placed in sealed plastic jars for transportation to the laboratory. After microflora determination, roots and soil remnants were removed. The litterbags were then dried in an oven at 50°C for 2–3 days and weighed individually. Using initial and final dry weight of the needle material in the litterbags the dry weight loss was calculated.

### Microflora

Fluorescein-stainable mycelia were determined according to Faber *et al.* (1992), using fluorescein 5-isothiocyanate (FITC) (Sigma Chemical Co., St Louis, Missouri), with some minor modifications; each sample of 2 g fresh weight was a mixture of subsamples of two randomly chosen litterbags, each contributing 1 g; the pH of the McIlvaine buffer was adjusted to natural conditions being 4.82 for the L-layer and 4.27 for the F-layer. One preparation was made for each homogenate and examined microscopically at a magnification of 200×, using a normal light microscope equipped with an epifluorescence attachment (Axioskope, Zeiss, Oberkochen, Germany; main wavelength 495 nm). The total fluorescent fungal hyphae length was determined in 15 random fields using the intersection method (Olson, 1950) and expressed as $m\,g^{-1}$ dry weight of the needle material.

Fluorescein-stainable bacteria were determined using 4-(4,6-dichloro-s-triazin-2-ylamino) fluorescein-dihydrochloride (DTAF) (Fluka Chemie AG, CH-9470 Buchs,

Switzerland) as a protein stain (Sherr *et al.*, 1987; Bloem *et al.*, 1989). Staining solution was prepared by adding 4 mg DTAF to 20 ml 0.05 M $Na_2HPO_4$ buffer adjusted to pH 9 containing 0.85% NaCl and was kept (dark) at 5°C. Staining procedure, rinsing and slide storage were done in darkness.

Samples of 3 g fresh weight were homogenized in a kitchen blender at 16 500 rpm in 57 ml deionized water for 1 minute; 3 ml 37% formaldehyde was added for fixation. Each sample of 3 g fresh weight was a mixture of subsamples of two litterbags (the same pairs as used for mycelia) each contributing 1.5 g. Glass slides were used which had been cleaned by ethanol and sealed with tape containing two holes of 12 mm diameter. In each of the holes 10 $\mu$l of the sample was evenly smeared. After air drying the tape was peeled off. The staining procedure started by covering the glass slides with the staining solution and staining the bacteria for 30 minutes at 20°C. Next the slides were rinsed three times for 20 minutes with the buffer. After air drying the slides were observed at 1000× magnification by epifluorescence microscopy. The total number of fluorescent bacteria was determined in 10 random fields and expressed as numbers per gram dry weight of the needle material.

### Statistical analyses

The variables dry weight loss, fungal hyphae length and bacteria numbers were tested for homogeneity of variances using Bartlett's test (Sokal and Rohlf, 1981). Friedman's test was used to test for differences in decomposition of litter and fragmentation during the year (factors plot and time). Log transformed data on fungal hyphae length and bacteria numbers were subjected to a three-way analysis of variance with the independent factors time, plot and layer. All calculations were performed using SYSTAT version 5.0 (Wilkinson, 1989).

## RESULTS

### Decomposition

Decomposition of spruce needle material of both L- and F-layer, expressed as % dry weight loss, showed a similar pattern and rate in both F1 CONTROL and G1 ROOF catchments (Figure 14.1). Initially decomposition rate was rapid and it stabilized later on. After 12 months average (±s.d.) decomposition of L- and F-material of the F1 CONTROL catchment was 17.63 (±1.72) and 7.93% (±3.36) respectively, versus 18.95 (±1.58) and 6.93% (±1.05) respectively for the G1 ROOF catchment. Friedman's test showed no significant interactions of the factors plot and time for L- and F-layer, indicating no differences in decomposition rates of both organic layers between the F1 CONTROL and the G1 ROOF catchment.

### Fungi

Fragmentation material in both plots contained significantly ($p \leqslant 0.001$) more FITC-stained fungal hyphae than litter material (Figure 14.2); mean fungal hyphae length

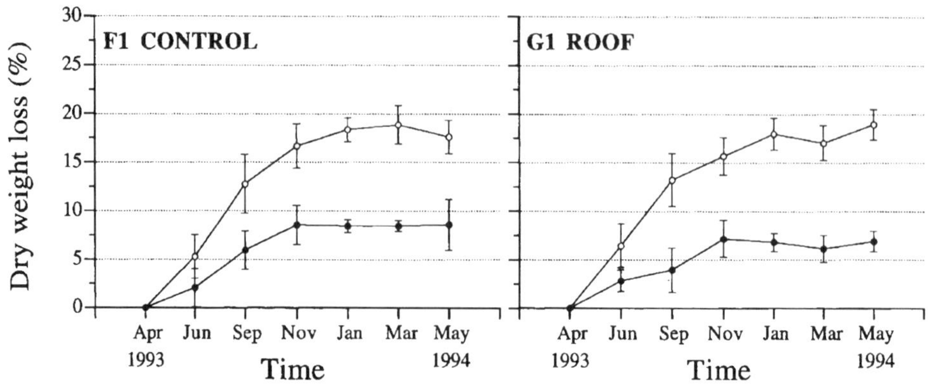


**Figure 14.1.** Bi-monthly dry weight loss (%) of Norway spruce needles for the L (-○-) and F (-●-) layer during one year for the F1 CONTROL and G1 ROOF catchments. For plot treatment see text. Bars indicate standard deviations ($n = 12$)

for the F1 CONTROL and G1 ROOF catchments in the F-layer was 159.4 m g$^{-1}$ and 120.4 m g$^{-1}$ dry weight organic matter respectively, and for the L-layer 41.9 m g$^{-1}$ and 21.5 m g$^{-1}$ dry weight organic matter respectively. Seasonality was present with low amounts of fungal hyphae during the dry summer month June and high amounts during the winter period November till March for both layers in both plots. The highest amounts were found in January. A three-way analysis of variance showed a significant reduction of the fungal hyphae length in the G1 ROOF catchment compared with the F1 CONTROL catchment for both layers ($p \leqslant 0.001$).

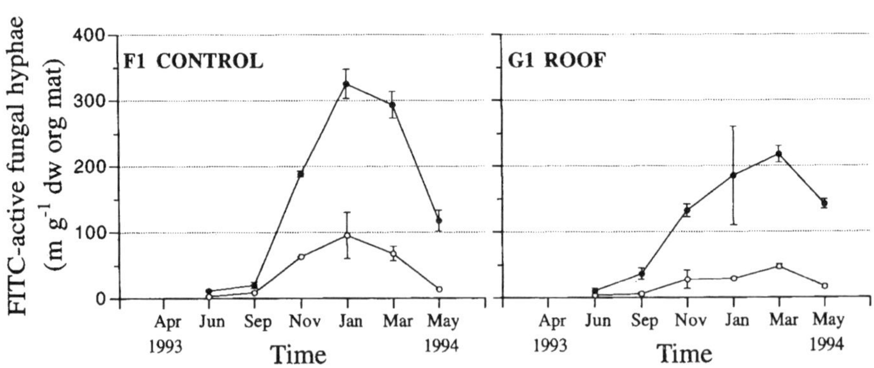


**Figure 14.2.** Bi-monthly FITC-active fungal hyphae length (m g$^{-1}$ dry weight organic material) in the L (-○-) and F (-●-) layer of Norway spruce needles during one year for the F1 CONTROL and G1 ROOF catchments. For plot treatment see text. Bars indicate standard deviations ($n = 2$)

In the L-layer fungal hyphae length fluctuated from 2.7 m $g^{-1}$ in June to 119.6 m $g^{-1}$ dry weight organic matter in January for the F1 CONTROL catchment and from 3.7 m $g^{-1}$ in June to 49.6 m $g^{-1}$ dry weight organic matter in March for the G1 ROOF catchment. The higher amounts of fungal hyphae in the F-layer fluctuated from 10.2 m $g^{-1}$ in June to 342.2 m $g^{-1}$ dry weight organic matter in January for the F1 CONTROL catchment, and 7.2 m $g^{-1}$ in June to 238.1 m $g^{-1}$ dry weight organic matter in January for the G1 ROOF catchment.

## Bacteria

Numbers of DTAF-stained bacteria were also low in summer and highest in the winter period November till March (Figure 14.3), though the numbers were less dynamic than the fungi through the year for both plots and both layers. Furthermore, the peak of the bacterial numbers was in November. Amounts of bacteria were significantly higher in the F-layer in both catchments ($p \leqslant 0.001$); mean bacteria numbers for the F1 CONTROL and G1 ROOF catchments in the F-layer were $1.17 \times 10^{10}$ $g^{-1}$ and $9.71 \times 10^9$ $g^{-1}$ dry weight organic matter respectively, and for the L-layer $6.22 \times 10^9$ $g^{-1}$ and $5.30 \times 10^9$ $g^{-1}$ dry weight organic matter respectively. A three-way analysis of variance showed also a significant reduction of bacterial numbers in the G1 ROOF catchment compared to the F1 CONTROL catchment for both layers ($p = 0.025$).

In the L-layer, bacterial numbers ranged from $2.30 \times 10^9$ $g^{-1}$ in May to $1.04 \times 10^{10}$ $g^{-1}$ dry weight organic matter in March for the F1 CONTROL catchment. Bacterial numbers in the G1 ROOF catchment ranged from $2.86 \times 10^9$ $g^{-1}$ in June to $9.22 \times 10^9$ $g^{-1}$ dry weight organic matter in November. The F-layer had higher numbers of bacteria and ranged from $5.10 \times 10^9$ $g^{-1}$ in June to $2.12 \times 10^{10}$ $g^{-1}$ dry weight organic matter in November for the F1 CONTROL catchment, and $4.28 \times 10^9$ $g^{-1}$ in June to $1.91 \times 10^{10}$ $g^{-1}$ dry weight organic matter in November for the G1 ROOF catchment.

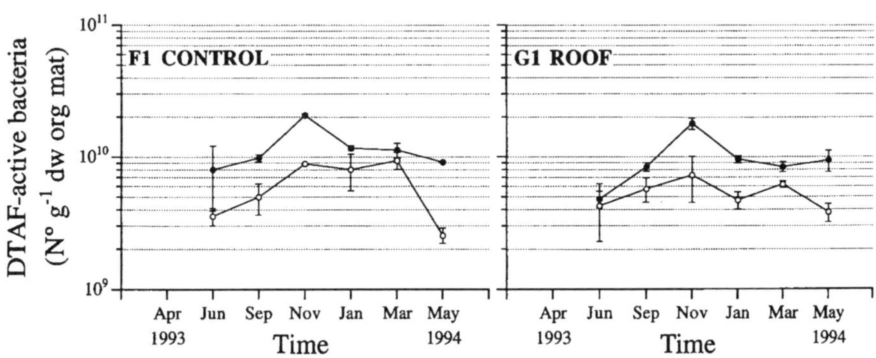


**Figure 14.3.** Bi-monthly numbers of DTAF-active bacteria (number per gram dry weight organic material) in the L (-○-) and F (-●-) layer of Norway spruce needles during one year for the F1 CONTROL and G1 ROOF catchments. For plot treatment see text. Bars indicate standard deviaitions ($n = 2$)

# DISCUSSION

## Decomposition

The 18% dry mass loss of the spruce needles in the F1 CONTROL catchment receiving ambient throughfall water after 1 year of field incubation is low compared with other studies on *Picea abies* litter. Berg (1986a) recorded a mass-loss of spruce needles of 25.2% per annum, while Fritze (1988) found a decomposition rate of 31.6% and Smolander (1996) about 30%. Although the process of decomposition is related to soil moisture and soil temperature (Wilhelmi and Rothe, 1990), the low mass-loss rate in this study is probably not due to climatic influences. Higher mass-loss rates are found far further north than Gårdsjön in south-west Sweden. Hågvar (1991) found after 1 year of litterbag study in a site 40 km north of Oslo a mass loss of spruce needles of about 28%. In central Sweden, a mass loss of spruce needles of 33.3% was observed after 1 year (Berg and Tamm, 1991). In these studies needles were removed from the trees, while the material used in our study is gathered from the soil surface. The age of the litter could thus be a good explanation of the low decomposition rate. The litter needles are probably at least 6 months old and the easily degradable components are already decomposed, while secondary plant metabolites, like phenols, are leached out. The remaining material contains more lignin and lignified substances which have a slower decomposition rate than the freshly fallen green needles (Berg and Ekbohm, 1991). The same argument about ageing holds for the clear differences in decomposition rates between the L- and F-layer.

In south-west Sweden, the main problem for soil acidification is the deposition of nitrogen. Atmospheric input of nitrogen has increased in the area this century (Skeffington and Wilson, 1988) to 20 kg $ha^{-1}$ $year^{-1}$ (Grennfelt and Hultberg, 1986). The reduction of the atmospheric deposition of nitrogen and sulphur by more than 90% in the present research has shown no significant effect on the weight loss of spruce needles after 1 year of incubation in the field. Recovery from acidification has been demonstrated (Stuanes *et al.*, 1992b) and the importance of nitrogen emphasized (Wright and Hauhs, 1991). Reduction of atmospheric deposition of sulphur and nitrogen in the Netherlands showed a quick response of the chemical composition of the soil solution in a Scots pine stand, but the first small biological changes occurred after just 4 years (Van Dijk *et al.*, 1992; Boxman *et al.*, 1995). Litter decomposition of Scots pine was reduced when nitrogen and sulphur deposition was reduced (Boxman *et al.*, 1995), indicating nutrient (N or S) dependent decomposition rate. However, in a succeeding year no such differences between the two treatments were found (Hogervorst, unpublished). A Douglas fir stand in the Netherlands did not show any biological changes after nitrogen and sulphur reduction (Boxman *et al.*, 1995). These two Dutch stands are however highly polluted and obviously nitrogen saturated. The effects of reduction of nitrogen (and sulphur) probably depend on the initial state of the stand, causing nitrogen-limited stands like those in Sweden to react differently. Ecosystem responses following chronic nitrogen deposition can be separated into four stages (Aber *et al.*, 1989). The two Dutch sites are showing effects of stage 2 in the development of nitrogen saturation, such as nitrate leaching, elevated nitrogen concentration in foliage and elevated amino acid content (Boxman *et al.*, 1995), while the Gårdsjön stand still show signs of stage 1. As in Scandinavia most of the forest ecosystems are (still) nitrogen limited, nitrogen inputs in acidic deposition are likely to be taken up and stored in biomass and thereby retained in terrestrial ecosystems. It is questioned though whether nitrogen is still the growth-limiting nutrient in Gårdsjön because at ambient nitrogen deposition of 12 kg N $ha^{-1}$ $year^{-1}$ the ecosystem is near the threshold of nitrogen saturation (Grennfelt and Hultberg, 1986; Dise and Wright, 1995), at which additional N inputs cause an increased nitrate leaching. Nitrate is not yet leached, but after 1 year adding 50 kg N $ha^{-1}$ $year^{-1}$ to the Gårdsjön area, a significant leaching of nitrate in runoff water occurred (Wright *et al.*, 1995). If Gårdsjön is close to saturation, nitrogen is still stored in the soil organic layer of the 3-year-old G1 ROOF catchment, causing the decomposition rate not to decrease. A lack of response in decomposition rate due to the N status of the soil organic layer was also observed in the Netherlands (Boxman *et al.*, 1995, 1997) and Denmark (Hansen *et al.*, 1995; Boxman *et al.*, 1997).

**Microflora**

The average FITC-active hyphae lengths in the different layers are in agreement with data obtained from other *Picea abies* forests in Sweden. For a stand in Kongalunda (south Sweden), the total hyphal length in the litter layer was on average 1155 m $g^{-1}$ dry weight soil (410–1900) and on average 2950 (2300–3600) m $g^{-1}$ dry weight soil in the F-layer. It was estimated that 10% was metabolically alive (Kjøller and Struwe, 1982), resulting in 115 m $g^{-1}$ and 295 m $g^{-1}$ dry weight soil in the L- and F-layer respectively. In Nickobacken, central Sweden, higher fungal hyphae lengths were observed (total hyphae length 9700 and 66 900 m $g^{-1}$ dry weight soil in the L- and F-layer respectively) by Bååth and Söderström (1979). According to Söderström (1979) this showed that approximately 1–5% of the mycelium in different horizons of a Swedish pine forest soil was alive by staining with fluorescein diacetate (FDA). Applying this to the former *Picea* stand resulted in amounts comparable with the length of fungal hyphae of the spruce forest in this study. Not much is known about numbers of bacteria in spruce forests. Data on abundances and activity of bacteria in coniferous forests are mainly obtained from pine forests. In a Scots pine forest in Jädraås, central Sweden, the total organic layer contained a mean value of $1.40\times10^{10}$ total bacteria $g^{-1}$ dw soil (Clarholm and Rosswall, 1980), while in Kroksbo, central Sweden, the organic layer in a Scots pine stand contained an average of $8.4\times10^{9}$ metabolically active (FDA-stained) bacteria $g^{-1}$ dw soil (Bååth *et al.*, 1981). In another site in central Sweden, Ivantjärnsheden, total number of bacteria in organic soil ranged from 5.8 to $63.3\times10^{9}$ $g^{-1}$ dw soil, while FDA-stained bacteria ranged from 2.6 to $20.0\times10^{9}$ $g^{-1}$ dw soil (Lundgren and Söderström, 1983), indicating that, compared with fungi, a larger percentage of the bacterial population is metabolically active. The numbers observed in these studies are comparable with the bacterial numbers recorded in our study on a spruce forest. A unique sampling of the organic soil of a spruce forest recorded $2.28\times10^{9}$ FITC-stained bacteria $g^{-1}$ dw soil, which was in the same range of pine forests (Lundgren, 1981).

A strong seasonal pattern of metabolically active microflora was found with a peak in winter. Soil moisture and soil temperature are factors changing with season.

Studies indicate that both bacterial and fungal numbers in forest soils are correlated with soil moisture (Soderström, 1979; Clarholm and Rosswall, 1980; Kauri, 1982). The influence of soil temperature on soil micro-organisms is not clear. The highest amount of bacteria and fungi was usually recorded in spring or autumn (Söderström, 1979; Clarholm and Rosswall, 1980; Bååth and Söderström, 1982; Diaz-Raviña *et al.*, 1993), when optimal water and temperature conditions are present. In one pine forest, however, the amount of active hyphae increased when the temperature was below 0°C (Söderström, 1979), suggesting that fungi are able to grow under freezing conditions. Studies showing peaks of microbial activity in winter are found for deciduous forests (Gray *et al.*, 1974; Kauri, 1982), but for coniferous forests it was recorded only once (Berg *et al.*, 1997). The density of soil fauna also changes seasonally. Soil fauna strongly interact with microbial populations (Persson *et al.*, 1980; Lussenhop, 1992) and changes in soil fauna abundances or shifts in species compositions could have a strong effect on microbial populations. This is currently being analysed, using the same litterbags.

Hyphal length was higher in the F-layer than in the L-layer, which is usually the case (Kjøller and Struwe, 1982). The length of hyphae decreases going deeper into the soil (Kjøller and Struwe, 1982). The same holds for bacterial numbers which are greater in the F-layer than in the L-layer. Bacterial numbers are known to decrease in mineral soils compared with the organic layer (Clarholm and Rosswall, 1980; Kauri, 1982). The high abundance of micro-organisms in the F-layer could indicate that substrate quality was higher in the F-layer than in the L-layer. However, the soil layers differ in soil fauna (abundances and composition) and environmental conditions (humidity and temperature). Thus, influences of soil fauna and climate must not be ruled out.

Reduction in atmospheric deposition of nitrogen and sulphur in the present study has demonstrated significant negative effects on the abundance of microflora. General responses occurring after acidification are a decrease of total microbial biomass, mainly fungal biomass and activity while bacteria are less influenced, and a change in microbial composition (Fog, 1988; Schaefer, 1991; Wardle, 1992). Both decomposer hyphae and mycorrhizal hyphae decreased after nitrogen application (Bååth *et al.*, 1981; Nohrstedt *et al.*, 1989; Arnebrant *et al.*, 1990; Van Dijk *et al.*, 1990) indicating that nitrogen deposition most likely changes the composition of fungi while bacteria are little influenced. Nitrogen might cause a shift in dominance, from a fungi to a bacteria dominated system (Verhoef and Brussaard, 1990), which could affect soil animals and in turn decomposition and nutrient cycling. A reduction of N input could favour both bacteria and fungi. However, according to Badalucco *et al.* (1992) bacteria became predominant over fungi when soil acidity decreased after liming. In this study, neither bacteria nor fungi was favoured by the reduction of nitrogen. The reduced metabolically active microflora might be explained by a change in soil fauna. Diversity of soil fauna has been reported to increase when nitrogen deposition is reduced in a heavily N-polluted stand (Boxman *et al.*, 1995). This diversity increase occurred in winter, just when we observed the main differences in microflora. Changes in abundances and composition of soil fauna species may reduce grazing pressure resulting in higher amounts of total microflora but less activity.

**General**

The differences in abundance of micro-organisms are probably not caused by the intrinsic differences of the catchments, since the catchments are highly similar in characteristics. The observed results in this study are linked to a reduction in nitrogen deposition, but the findings may coincide with effects caused by sulphur, acidity and the roof itself. Sulphur addition ($H_2SO_4$) is known to reduce decomposition rates (Bååth *et al.*, 1980; Berg, 1986c) and decomposer organisms and decomposition rates are influenced by soil acidity (Bååth *et al.*, 1980; Berg, 1986c). The roof might have had an effect. The organic layers appeared to be a little more humid in the F1 CONTROL catchment compared with the G1 ROOF catchment. Furthermore, during the winter period snow was only present on the F1 CONTROL catchment, causing the soil to be less frozen. However, the observed differences in climatic conditions were small and probably had a slight effect on both decomposition and microflora. The roof may also reduce the amount of ultraviolet B radiation reaching the soil surface. It is suggested that ultraviolet B radiation may affect litter decay and alter the species composition of fungal communities (Moorhead and Callaghan, 1994). In the Netherlands, decomposition was found to be influenced by the roof (Boxman *et al.*, 1995). This was however probably caused by the water regime and not by the roof itself, because in a succeeding year, when the water application changed to automatic, no roof effect was found. At Gårdsjön it is difficult to assess a roof effect properly by lack of a plot underneath the roof receiving ambient throughfall.

Concluding, we can say that chemical responses occur almost immediately after reduction of nitrogen and sulphur deposition but biological response such as nutrients in foliage, tree growth and decomposition rate lag behind several years (Aber *et al.*, 1989; Wright *et al.*, 1995). Soil micro-organisms seem to react faster than other biological factors such as decomposition rate. With nitrogen being limited for most temperate forest ecosystems and the increasing deposition of nitrogen in south-west Sweden, the observed effects are probably caused by the reduction in nitrogen deposition although effects of sulphur reduction or a roof effect may not be excluded.

## ACKNOWLEDGEMENTS

We thank M. Schalkoort, M. Kwadijk, C. Hofman, M. Dijkhuis, L. Lenoir and P. Kniese for their indispensable field assistance. M. Berg is thanked for teaching the technique for counting bacteria and critical reading of the manuscript. H. Hultberg (IVL) is gratefully acknowledged for his help and cheerfulness at the field station. Financial support was provided by the Commission of the European Communities (STEP-CV-90-0056).

## REFERENCES

Aber, J. D., Nadelhoffer, K. J., Steudler, P. and Melillo, J. M. (1989) Nitrogen saturation in northern forest ecosystems. *BioScience*, **39**, 378–386.

Arnebrant, K., Bååth, E. and Söderström, B. (1990) Changes in microfungal community structure after fertilization of Scots pine forest soil with ammonium nitrate or urea. *Soil Biology and Biochemistry*, **22**, 309–312.

Bååth, E. and Söderström, B. (1979) Fungal biomass and fungal immobilization of plant nutrients in Swedish coniferous forest soils. *Revue d'Écologie et de Biologie du Sol*, **16**, 477–489.

Bååth, E. and Söderström, B. (1982) Seasonal and spatial variation in fungal biomass in a forest soil. *Soil Biology and Biochemistry*, **14**, 353–358.

Bååth, E., Berg, B., Lohm, U. *et al.* (1980) Effects of experimental acidification and liming on soil organisms and decomposition in a Scots pine forest. *Pedobiologia*, **20**, 85–100.

Bååth, E., Lundgren, B. and Söderström, B. (1981) Effects of nitrogen fertilization on the activity and biomass of fungi and bacteria in a podzolic soil. *Zentralblatt für Bakteriologie und Hygiene, 1. Abt. Orig. C*, **2**, 90–98.

Badalucco, L., Grego, S., Dell'Orco, S. and Nannipieri, P. (1992) Effect of liming on some chemical, biochemical, and microbiological properties of acid soils under spruce (*Picea abies* L.). *Biology and Fertility of Soils*, **14**, 76–83.

Berg, B. (1986a) The influence of experimental acidification on needle litter decomposition in a *Picea abies* L. forest. *Scandinavian Journal of Forest Research*, **1**, 317–322.

Berg, B. (1986b) Nutrient release from litter and humus in coniferous forest soils—A mini review. *Scandinavian Journal of Forest Research*, **1**, 359–369.

Berg, B. (1986c) The influence of experimental acidification on the decomposition of organic material and nutrient release in the forest floor. *Forest Ecology and Management*, **15**, 195–213.

Berg, B. and Ekbohm, G. (1991) Litter mass-loss rates and decomposition patterns in some needle and leaf litter types. Long-term decomposition in a Scots pine forest. VII. *Canadian Journal of Botany*, **69**, 1449–1456.

Berg, B. and Staaf, H. (1980) Decomposition rate and chemical changes of Scots pine needle litter. II. Influence of chemical composition. *Ecological Bulletins (Stockholm)*, **32**, 373–390.

Berg, B. and Tamm, C. O. (1991) Decomposition and nutrient dynamics of litter in long-term optimum nutrition experiments. I. Organic matter decomposition in *Picea abies* needle litter. *Scandinavian Journal of Forest Research*, **6**, 305–321.

Berg, B., Staaf, H. and Wessén, B. (1987) Decomposition of Scots pine needle litter of differing nutrients concentrations. *Scandinavian Journal of Forest Research*, **2**, 399–415.

Berg, B., Berg, M. P., Bottner, P. *et al.* (1993) Litter mass loss rates in pine forests of Europe and Eastern United States: some relationships with climate and litter quality. *Biogeochemistry* **20**, 127–159.

Berg, M. P., Kniese, P. and Verhoef, H. A. (1997) Wekpad-project II. Dynamics and stratification of bacteria and fungi in the organic layer of a Scots pine forest. *Biology and Fertility of Soils* (in press).

Bloem, J., Ellenbroek, F. M., Bär-Gilissen, M. J. B. and Cappenberg, Th.E. (1989) Protozoan grazing and bacterial production in stratified Lake Vechten estimated with fluorescently labeled bacteria and by thymidine incorporation. *Applied and Environmental Microbiology*, **55**, 1787–1795.

Boxman, A. W. and Roelofs, J. G. M. (1988) Some effects of nitrate versus ammonium nutrition on the nutrient fluxes in *Pinus sylvestris* seedlings. Effects of mycorrhizal infection. *Canadian Journal of Botany*, **66**, 1091–1097.

Boxman, A. W., Van Dam, D., Van Dijk, H. F. G., Hogervorst, R. F. and Koopmans, C. J. (1995) Ecosystem responses to reduced nitrogen and sulphur inputs into two coniferous forest stands in the Netherlands. *Forest Ecology and Management*, **71**, 7–29.

Boxman, A. W., Blanck, K., Brandrud, T.-E., Emmett, B. A., Gundersen, P., Hogervorst, R. F., Kjonaas, O. J., Persson, H. and Stuanes, A. O. (1997) Cross-site comparison of vegetation response as a result of nitrogen input manipulations in coniferous forest ecosystems. *Forest Ecology and Management* (in press).

Clarholm, M. and Rosswall, T. (1980) Biomass and turnover of bacteria in a forest soil and a peat. *Soil Biology and Biochemistry*, **12**, 49–57.

Diaz-Raviña, M., Acea, M. J. and Carballas, T. (1993) Seasonal fluctuations in microbial populations and available nutrients in forest soils. *Biology and Fertility of Soils*, **16**, 205–210.

Dise, N. B. and Wright, R. F. (1995) Nitrogen leaching from European forests in relation to nitrogen deposition. *Forest Ecology and Management*, **71**, 153–161.
Faber, J. H., Teuben, A., Berg, M. P. and Doelman, P. (1992) Microbial biomass and activity in pine litter in the presence of *Tomocerus minor* (Insecta, Collembola). *Biology and Fertility of Soils*, **12**, 233–240.
Falkengren-Grerup, U. (1987) Long term changes in pH of forest soils in southern Sweden. *Environmental Pollution*, **43**, 79–90.
Fog, K. (1988) The effect of added nitrogen on the rate of decomposition of organic matter. *Biological Review*, **63**, 433–462.
Fritze, H. (1988) Influence of urban air pollution on needle litter decomposition and nutrient release. *Scandinavian Journal of Forest Research*, **3**, 291–297.
Gray, T. R. G., Hissett, R. and Duxbury, T. (1974) Bacterial populations of litter and soil in a deciduous woodland II. Numbers, biomass and growth rates. *Revue d'Écologie et de Biologie du Sol*, **11**, 15–26.
Grennfelt, P. and Hultberg, H. (1986) Effects of nitrogen deposition on the acidification of terrestrial and aquatic ecosystems. *Water, Air, and Soil Pollution*, **30**, 945–963.
Hågvar, S. (1991) Succession of microarthropods in decomposing spruce needles: A litter bag study over ten years. In: G. K. Veeresh, D. Rajagopal and C. A. Viraktamath (Eds), *Advances in Management and Conservation of Soil Fauna*. Oxford & IBH Publishing Co. Pvt. Ltd, New Delhi, pp. 485–489.
Hallbäcken, L. and Tamm, C. O. (1986) Changes in soil acidity from 1927 to 1982–1984 in a forest area of southwest Sweden. *Scandinavian Journal of Forest Research*, **1**, 219–232.
Hansen, K., Beier, C., Gundersen, P. and Rasmussen, L. (1995) Experimental manipulations of water and nutrient input to a Norway spruce plantation at Klosterhede, Denmark: 3) Effects on throughfall, soil water chemistry and decomposition. *Plant and Soil*, **168–169**, 623–632.
Hultberg, H., Anderson, B. I. and Moldan, F. (1993) The covered catchment—an experimental approach to reversal of acidification in a forest ecosystem. In: L. Rasmussen, T. Brydges and P. Mathy (eds), *Experimental Manipulations of Biota and Biogeochemical Cycling in Ecosystems*. Ecosystems Research Report 4, Commission of European Communities, Brussels, pp. 46–55.
Kauri, T. (1982) Seasonal fluctuations in numbers of aerobic bacteria and their spores in four horizons of a beech forest soil. *Soil Biology and Biochemistry*, **14**, 185–190.
Kjøller, A. and Struwe, S. (1982) Microfungi in ecosystems: fungal occurrence and activity in litter and soil. *Oikos*, **39**, 389–422.
Kuyper, T. W. (1989) Effects of acid deposition on decomposition processes in forest ecosystems. Internal report, Biological Station Wijster, The Netherlands.
Lundgren, B. (1981) Fluorescein diacetate as a stain of metabolically active bacteria in soil. *Oikos*, **36**, 17–22.
Lundgren, B. and Söderström, B. (1983) Bacterial numbers in a pine forest soil in relation to environmental factors. *Soil Biology and Biochemistry*, **15**, 625–630.
Lussenhop, J. (1992) Mechanisms of microarthropod–microbial interactions in soil. *Advances in Ecological Research*, **23**, 1–33.
McClaugherty, C. and Berg, B. (1987) Cellulose, lignin and nitrogen levels as rate regulating factors in forest litter decomposition. *Pedobiologia*, **30**, 101–112.
Meentemeyer, V. and Berg, B. (1986) Regional variation in rate of mass-loss of Scots pine needle litter in Swedish pine forests as influenced by climate and litter quality. *Scandinavian Journal of Forest Research*, **1**, 167–180.
Moorhead, D. L. and Callaghan, T. (1994) Effects of increasing ultraviolet B radiation on decomposition and soil organic matter dynamics: a synthesis and modelling study. *Biology and Fertility of Soils*, **18**, 19–26.
Nihlgård, B. (1985) The ammonium hypothesis: An additional explanation to the forest dieback in Europe. *Ambio*, **14**, 2–8.
Nohrstedt, H.-Ö., Arnebrant, K., Bååth, E. and Söderström, B. (1989) Changes in carbon content, respiration rate, ATP content, and microbial biomass in nitrogen-fertilized pine forest soils in Sweden. *Canadian Journal of Forest Research*, **19**, 323–328.

Olson, F. C. W. (1950) Quantitative estimates of filamentous algae. *Transactions of the American Microscopical Society*, **69**, 272–279.

Persson, T., Bååth, E., Clarholm, M., Lundkvist, H., Söderström, B. and Sohlenius, B. (1980) Trophic structure, biomass, dynamics and carbon metabolism of soil organisms in a Scots pine forest. *Ecological Bulletins (Stockholm)*, **32**, 419–462.

Pitelka, L. F. and Raynal, D. J. (1989) Forest decline and acidic deposition. *Ecology*, **70**, 2–10.

Schaefer, M. (1991) Effect of acid deposition on soil animals and microorganisms: influence on structures and processes. In: B. Ulrich (ed.), *International Congress on Forest Decline Research: State of Knowledge and Perspectives*, Lake Constance, Germany, 2–6 October 1989. Kernforschungszentrum Karlsruhe GmbH, pp. 415–430.

Schulze, E.-D. (1989) Air pollution and forest decline in a spruce (*Picea abies*) forest. *Science*, **244**, 776–783.

Sherr, B. F., Sherr, E. B. and Fallon, R. D. (1987) Use of monodispersed, fluorescently labeled bacteria to estimate in situ protozoan bactivory. *Applied and Environmental Microbiology*, **53**, 958–965.

Skeffington, R. A. and Wilson, E. J. (1988) Excess nitrogen deposition: issues for consideration. *Environmental Pollution*, **54**, 159–184.

Smolander, A., Kitunen, V., Paavolainen, P. and Malkonen, E. (1996) Decomposition of Norway spruce and Scots pine needles: Effects of liming. *Plant and Soil*, **179**, 1–7.

Söderström, B. (1979) Seasonal fluctuations of active fungal biomass in horizons of a podzolized pine forest soil in central Sweden. *Soil Biology and Biochemistry*, **11**, 149–154.

Sokal, R. R. and Rohlf, F. J. (1981) *Biometry*, 2nd edition. Freeman & Co, San Francisco.

Stuanes, A. O., Andersson, I., Dise, N. *et al.* (1992a) Gårdsjön, Sweden. In: N. B. Dise and R. F. Wright (eds), *The NITREX Project (Nitrogen Saturation Experiments)*. Ecosystems Research Report 2, Commission of European Communities, Brussels, pp. 24–34.

Stuanes, A. O., Van Miegroet, H., Cole, D. W. and Abrahamsen, G. (1992b) Recovery from acidification. In: D. W. Johnson and S. E. Lindberg (eds), *Atmospheric Deposition and Forest Nutrient Cycling*. Ecological Studies, vol. 91. Springer-Verlag, New York, pp. 467–494.

Swift, M. J., Heal, O. W. and Anderson, J. M. (1979) *Decomposition in Terrestrial Ecosystems*. Blackwell Scientific Publications, Oxford.

Van Breemen, N. and Van Dijk, H. F. G. (1988) Ecosystem effects of atmospheric deposition of nitrogen in The Netherlands. *Environmental Pollution*, **54**, 249–274.

Van Dijk, H. F. G., De Louw, M. H. J., Roelofs, J. G. M. and Verburgh, J. J. (1990) Impact of artificial, ammonium-enriched rainwater on soils and young coniferous trees in a greenhouse. Part II—Effects on the trees. *Environmental Pollution*, **63**, 41–59.

Van Dijk, H. F. G., Boxman, A. W. and Roelofs, J. G. M. (1992) Effects of a decrease in atmospheric deposition of nitrogen and sulphur on the mineral balance and vitality of a Scots pine stand in the Netherlands. *Forest Ecology and Management*, **51**, 207–215.

Verhoef, H. A. and Brussaard, L. (1990) Decomposition and nitrogen mineralization in natural and agro-ecosystems: the contribution of soil animals. *Biogeochemistry*, **11**, 175–211.

Wardle, D. A. (1992) A comparative assessment of factors which influence microbial biomass carbon and nitrogen levels in soil. *Biological Review*, **67**, 321–358.

Wilhelmi, V. and Rothe, G. M. (1990) The effect of acid rain, soil temperature and humidity on C-mineralization rates in organic soil layers under spruce. *Plant and Soil*, **121**, 197–202.

Wilkinson, L. (1989) *SYSTAT: The System for Statistics*. SYSTAT, Inc., Evanston, IL.

Wright, R. F. and Hauhs, M. (1991) Reversibility of acidification: soils and surface waters. *Proceedings of the Royal Society of Edinburgh*, **97B**, 169–191.

Wright, R. F., Brandrud, T.-E., Clemensson-Lindell, A. *et al.* (1995) NITREX project: ecosystem response to chronic additions of nitrogen to a spruce-forested catchment at Gårdsjön, Sweden. *Ecological Bulletins (Stockholm)*, **44**, 322–334.

# 15 Ectomycorrhizal Fungus Flora in a Roof-Covered Catchment at Gårdsjön: Changes Following the Removal of Acid Rain

**TOR ERIK BRANDRUD**

The changes in the ectomycorrhizal fungus flora in a small, roof-covered, spruce forest catchment, experimentally manipulated by cleaning the rain, were studied and compared with a control catchment. A substantial increase in the fruit body production was observed in the roof catchment from 1992 (just after the start of treatment) to 1994. This increase was seen in spite of the fairly poor fungal season in south-west Sweden in 1994, and the fruit body production in the roof catchment was six times higher than that of the control. However, the species diversity in the two catchments was of the same magnitude, as was the mycorrhizal density and vitality in the topsoil. The increased production could potentially be explained by changes in the microclimate due to the sheltering effect of the roof. However, no difference in the air humidity was recorded between the roof and the control catchment, and a high fruit body density was seen up to the margins of the roof, where the microclimate should be very similar to the surrounding forest. The increased vitality of the ectomycorrhizal fungi thus seems to be due to the cleaning of the water added. Other evidence suggests that the removal of N from deposition under the roof may be the principal cause of this change. These changes in the fruit body production are one of the first and most pronounced biological responses to the reversed acidification treatment. However, the productivity and vitality is still below that of non-polluted, Scandinavian spruce forests.

## INTRODUCTION

Serious decline, and even an almost total disappearance of the ectomycorrhizal fungus flora has been documented in Central European long-term polluted forests (Derbsch and Schmitt, 1987; Arnolds, 1991; Fellner, 1993). In field experiments, a negative effect has been observed both in the case of sulphuric acid addition (Høiland, 1986; Jansen and Dighton, 1990) and — sometimes very pronounced — in the case of nitrogen addition (Ritter and Tölle, 1978; Rühling and Tyler, 1991; Brandrud, 1995). There are indications that changes in the ectomycorrhizal fungus flora, and especially in the fruit body production, are one of the first observable

*Experimental Reversal of Acid Rain Effects: The Gårdsjön Roof Project.*
Edited by Hans Hultberg and Richard Skeffington. 

biological effects of a high pollution load (Gulden and Høiland, 1985; Fellner, 1993; Brandrud, 1995), and these organisms therefore stand out as suitable bioindicators of forest disturbances (Fellner, 1989). However, the effect of *reversed acidification* has been very little studied (but cf. Boxman *et al.*, 1995).

A large-scale field experiment with reversed acidification in a roof-covered catchment is being carried out at Gårdsjön, south-west Sweden, as part of a series of experimentally manipulated, small, forested catchments. In the roof catchment, the ambient, polluted precipitation is removed, and is substituted by clean water (Chapter 4, this volume). The present study is a part of the NITREX programme at Gårdsjön, primarily focusing on the effects on the mycorrhizal flora of nitrogen addition to the so-called NITREX catchment (Brandrud, 1993, 1995), and with the roof catchment included as a kind of unpolluted control site. The results from the nitrogen addition experiment are not included here.

## MATERIAL AND METHODS

The ectomycorrhizal flora was studied at two small catchments at Gårdsjön (Bohuslän county, south-west Sweden), a roof-covered catchment and a control, during the period 1992–94. The roof catchment (G1 site) receives clean, deionized water, only with the addition of natural sea salts, and this is added underneath the roof through a sprinkling system (Chapter 4, this volume). The roof was constructed and the clean rain sprinkling started in 1991. A control catchment (F1 site), receiving ambient load of pollutants (cf. Chapter 4, this volume), was also included in the study.

The catchments consist of natural, mature, oligotrophic Norway spruce forest of raw humus type and with dominance of *Vaccinium myrtillus* and *Dicranum majus* (*Eu-Piceetum myrtilletosum*). Patches of swampy, *Sphagnum*-dominated spruce forest occur, and some Scots pines are found on drier, shallow-soil patches, see Chapter 2, this volume.

The above-ground production of fruit bodies of ectomycorrhizal fungi has been recorded three to four times during the fungal season (August to October) in three transects per site. The three adjacent transects represent a cross-section of the catchment. These are divided into 52 $5 \times 5$ m permanent plots/relevees per catchment, making up a plot size of 1300 m$^2$ per site.

The taxa treated as ectomycorrhizal fungi follow Gulden *et al.* (1992), except for the inclusion of *Entoloma*. The nomenclature follows Knudsen and Hansen (1992), with a few exceptions in the genus *Cortinarius*, following Brandrud *et al.* (1991–95).

## RESULTS

The fruit body production of ectomycorrhizal fungi in the roof catchment (G1) increased substantially during the period 1992–94, after the construction of the roof. In 1992, 1008 fruit bodies were recorded in the 1300 m$^2$ transect plot area, whilst in 1994 the number was increased to 4003 (Figure 15.1). The production in 1993 was intermediate. This increase after treatment was seen in most genera, but was most pronounced in the dominating genus *Cortinarius* (Figures 15.2 and 15.3, Table 15.1),

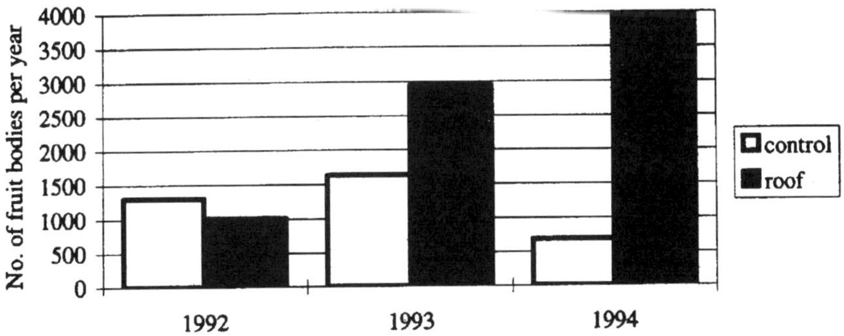


**Figure 15.1.** Fruit body production of ectomycorrhizal fungi in the control and roof catchments at Gårdsjön 1992–94. Transect plot area = 1300 $m^2$

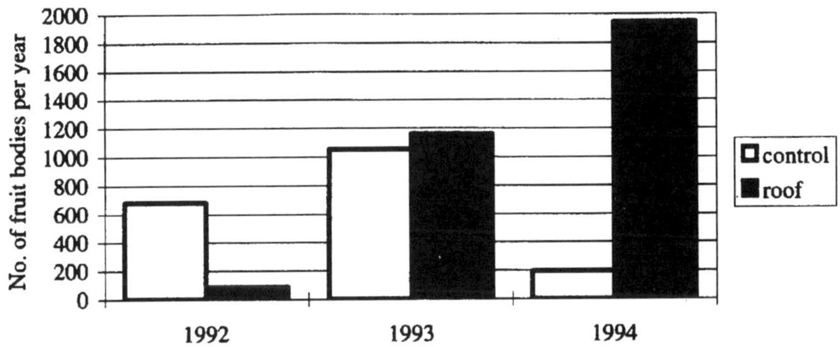


**Figure 15.2.** Fruit body production of the dominant genus *Cortinarius* in the control and roof catchments at Gårdsjön 1992–94. Transect plot area = 1300 $m^2$

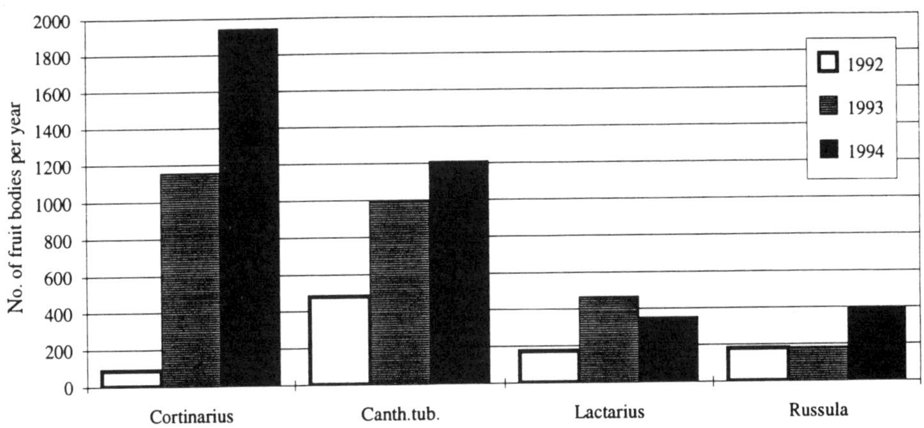


**Figure 15.3.** Fruit body production of the dominating genera of ectomycorrhizal fungi in the roof catchment (1300 $m^2$ plot area) at Gårdsjön 1992–94. (The genus *Cantharellus* was represented in the plots only by the dominant species *Ca. tubaeformis*.)

**Table 15.1.** Numbers of fruit bodies of ectomycorrhizal fungi recorded in the control catchment and the roof catchment at Gårdsjön, 1992–94. Dominant taxa are indicated in bold letters. Taxa occurring in less than four of the columns, are listed below the table (C = control, R = roof)

| | Control | | | Roof | | |
|---|---|---|---|---|---|---|
| | 1992 | 1993 | 1994 | 1992 | 1993 | 1994 |
| *Amanita porphyria* | 1 | 2 | 2 | — | 7 | 5 |
| *Boletus badius* | 6 | 4 | 9 | 56 | 47 | 20 |
| *Boletus subtomentosus* | 1 | 2 | 6 | — | 1 | 1 |
| ***Cantharellus tubaeformis*** | **231** | **257** | **294** | **480** | **996** | **1210** |
| *Cortinarius albovariegatus* | 25 | 38 | 3 | 8 | 16 | 1 |
| *Cort. biformis* | 31 | 37 | 23 | 5 | 14 | — |
| *Cort. evernius* | 7 | 25 | 1 | — | — | 1 |
| ***Cort. flos-paludis*** | **1** | **18** | — | — | **81** | **184** |
| *Cort. lividoochraceus* (= *C. elatior*) | 15 | — | 15 | 2 | — | 5 |
| *Cort. mucifluus* | — | 2 | 1 | 8 | 8 | — |
| ***Cort. obtusus*** var. ***obtusus*** | **428** | **574** | **162** | **42** | **406** | **790** |
| ***Cort. obtusus*** var. ***acutus*** | **74** | **157** | **62** | **9** | **242** | **693** |
| ***Cort. paleaceus*** var. ***paleaceus*** | **25** | **80** | **4** | **2** | **141** | **24** |
| *Cort. paleaceus* var. *paleifer* | 7 | 15 | — | — | 16 | 14 |
| *Cort. pluviorum* | 14 | 3 | 2 | — | 1 | 4 |
| *Cort. sanguineus* var. *sanguineus* | 6 | 2 | — | 1 | 1 | 1 |
| *Cort. sanguineus* var. *vitiosus* | — | 23 | 8 | — | 82 | 28 |
| *Cort. scaurus* var. *scaurus* | 3 | — | 2 | 4 | 4 | 10 |
| *Cort. scaurus* var. *sphagnophilus* | 2 | — | 1 | — | 2 | 3 |
| ***Cort. strobilaceus*** | — | **17** | — | **5** | **123** | **59** |
| *Cort. tortuosus* | 1 | 3 | 7 | — | 7 | 6 |
| *Elaphomyces* sp. | 22 | 19 | — | — | — | — |
| *Entoloma nitidum* | 26 | — | 2 | 10 | — | 3 |
| *Entoloma turbidum* | 10 | — | 5 | 9 | — | 16 |
| *Hebeloma longicaudum* | — | 13 | 2 | — | 2 | 10 |
| *Hygrophorus olivaceoalbus* | 23 | 9 | 2 | — | 4 | — |
| *Inocybe lanuginosa* coll. | — | 5 | 2 | — | 9 | 11 |
| *Inocybe relicina* | 9 | 3 | — | 1 | 1 | 2 |
| *Laccaria laccata* coll. | 2 | 2 | — | 2 | 1 | 4 |
| *Lactarius camphoratus* | 16 | 88 | — | — | 17 | 1 |
| *Lactarius helvus* | 5 | — | — | 10 | 13 | 7 |
| ***Lactarius necator*** | **4** | **10** | — | **59** | **117** | **43** |
| *Lactarius rufus* | 8 | 1 | — | 35 | 10 | 12 |
| *Lactarius sphagneti* | 11 | — | 5 | — | — | — |
| ***Lactarius theiogalus*** | **40** | **23** | **1** | **67** | **425** | **287** |
| *Paxillus involutus* | 1 | 1 | — | 8 | 34 | 3 |
| *Russula consobrina* | 3 | 1 | — | — | 4 | 1 |
| *Russula atrorubens* | — | 4 | 2 | — | 6 | 2 |
| *Russula decolorans* | — | — | — | — | — | 2 |
| *Russula emetica* | 8 | 20 | 6 | — | 31 | 25 |

*(continued)*

**Table 15.1.** (*continued*)

| | | | | | | |
|---|---|---|---|---|---|---|
| *Russula fragilis* | 1 | 2 | — | — | 18 | 46 |
| ***Russula ochroleuca*** | **122** | **33** | **3** | **163** | **88** | **251** |
| *Russula puellaris* | 5 | 1 | 2 | 15 | 17 | 17 |
| *Russula rhodopoda* | 3 | 10 | 1 | — | 14 | 312 |

*Amanita regalis* (C:1,—,—,R:—,—,—), *A. rubescens* (C:—,—,1,R:—,1,2), *Boletus edulis* (C:—,—,4,R:—,—,1). *Cantharellus cibarius* (C:—,8,—,R:—,6,—), *Chalciporus piperatus* (C:8,3,6,R:—,—,—), *Chroogomphus rutilans* (C:—,—,2,R:—,—,5), *Cortinarius anomalus* (C:1,1,—,R:—,—,1), *C. armeniacus* (C:7,—,1,R:—,—,—), *C. bataillei* (C:—,—,1,R:—,—,—), *C. bolaris* (C:1,—,—,R:—,—,—), *C. brunneus* var. *brunneus* (C:15,7,1,R:–,–,–), *C. brunneus* var. *glandicolor* (C:–,–,–,R:–,1,–), *C. camphoratus* (C:–,1,–,R:–,–,–),*C. collinitus* (C:–,–,–,R:–,2,–), *C. comptulus* (C:13,–,–,R:–,4,–), *C. croceus* (C:–,2,–,R:–,–,4), *C.* cf. *eburneus* (C:–,–,1,R:–,–,5), *C. erythrinus* (C:–,–,1,R:–,–,–), *C. gentilis* (C:–,–,–,R:–,–,3). *C. laniger* (C:1,3,–,R:–,–,–), *C. limonius* (C:–,1,11,R:–,–,–), *C. ochrophyllus* (C:–,–,–,R:–,4,–), *C. rubellus* (= *C. speciosissimus*) (C:8,1,–,R:–,–,1), *C. rubicundulus* (C:–,1,—,R:–,–,–), *C. semisanguineus* (C:1,1,–,R:–,–,1), *C. stillatitius* (C:2,2,–,R:–,–,1), *C. subsertipes* (C:–,1,–,R:–,1,3), *C. subtortus* (C:–,16,–,R:–,–,–), *C. traganus* (C:–,1,–,R:–,–,–), *C. turmalis* (C:–,–,–,R:–,–,1), *Entoloma conferendum* (C:–,1,–,R:–,–,–), *Entoloma nitidum* (C:26,–,2,R:10,–,3), *Gomphidius glutinosus* (C:–,–,–,R:2,3,3), *G. roseus* (C:4,–,–,R:–,–,–), *Hydnum repandum* (C:6,–,–,R:–,–,–), *Hydnum rufescens* (C:–,5,–,R:–,–,–), *Hygrophorus camarophyllus* (C:4,–,–,R:–,–,–), *Inocybe boltonii* (C:1,–,–,R:2,–,–), *I. napipes* (C:5,1,–,R:–,1,–), *Lactarius glyciosmus* (C:1,–,1,R:–,2,–), *Lact. hysginus* (C:2,–,–,R:–,–,–), *Lact. lignyotus* (C:–,4,–,R:–,–,–), *Lact. trivialis* (C:5,6,1,R:–,–,–), *Lact. vietus* (C:1,1,–,R:–,–,–), *Lact.* sp. (C:–,3.–,R:–,–,–), *Leccinum variecolor* (C:–,–,2,R:–,–,–), *Piloderma croceum* (C:–,–,100,R:–,–,–), *Russula aquosa* (C:–,–,–,R:–,–,14), *R. badia* (C:–,–,1,R:–,–,20), *R. consobrina* (C:3,1,–,R:–,4,1), *R. nigricans* (C:10,–,–,R:–,–,–), *R. nitida* (C:–,–,–,R:–,–,2),*R. versicolor* (C:3,6,–,R:–,3,–), *R. vesca* (C:–,–,4,R:–,4,1), *R. xerampelina* (C:–,–,–,R:–,1,–), *Suillus bovinus* (C:15,8,1,R:–,–,–), *Su. variegatus* (C:–,–,18,R:–,–,–), *Tricholoma saponaceum* (C:–,1,–,R:–,–,–).

which in 1994 was responsible for almost half of the production in the roof catchment. The two most dominating species in the plots were *Cortinarius obtusus sensu latu* and *Cantharellus tubaeformis*, with more than 1000 fruit bodies each produced in the roof transects in 1994 (Table 15.1).

In 1992 the fruit body production in the roof catchment was somewhat lower than that of the adjacent control site (F1), but in 1994 it had increased to almost six times the production in the control (Figure 15.1). The production of *Corinarius* was approximately 10 times higher in the roof catchment than in the control in 1994 (Figure 15.2).

The changes in α diversity (species richness) were less pronounced, but the number of species showed an increase in the roof catchment from 27 taxa recorded in 1992, via 52 in 1993, to 56 in 1994. In the control catchment the species number varied between 49 and 69 taxa per year, and the α diversity was only slightly lower than that of the roof site in 1994 (Figure 15.4, Table 15.1), even though the productivity was much lower (cf. Figure 15.1). The species richness recorded over the entire investigation period 1992–94 was considerably higher in the control than the roof catchment (93 versus 73 taxa). The different genera showed slightly different patterns with respect to α diversity. The genera *Cortinarius* and *Russula* had an increase in diversity (1992–94) in the roof plots, whilst *Lactarius* showed little change during the investigation period (Figure 15.5).

## DISCUSSION

The increase of the ectomycorrhizal fruit body production in the roof catchment 1992–94 is remarkable since the 1994 fungal season was less productive than normal outside the roof — both in the Gårdsjön area and south-west Sweden as a whole.

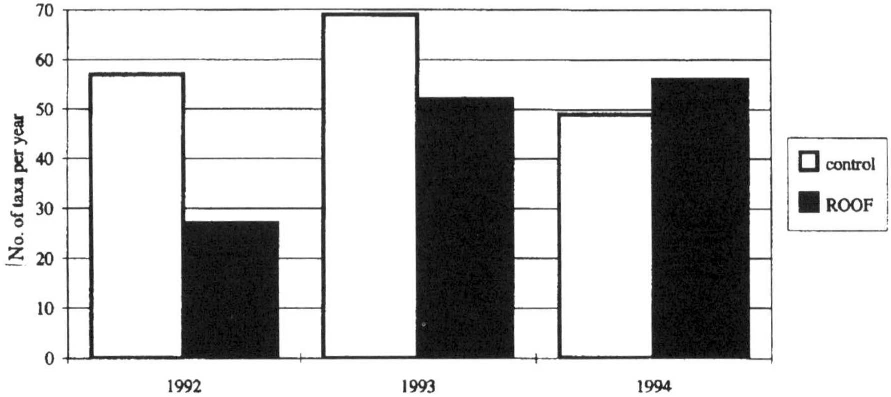


**Figure 15.4.** Species diversity of ectomycorrhizal fungi in the control and roof catchments at Gårdsjön 1992–94. Total number of taxa (species and varieties) recorded per year in the transect plot area of 1300 $m^2$

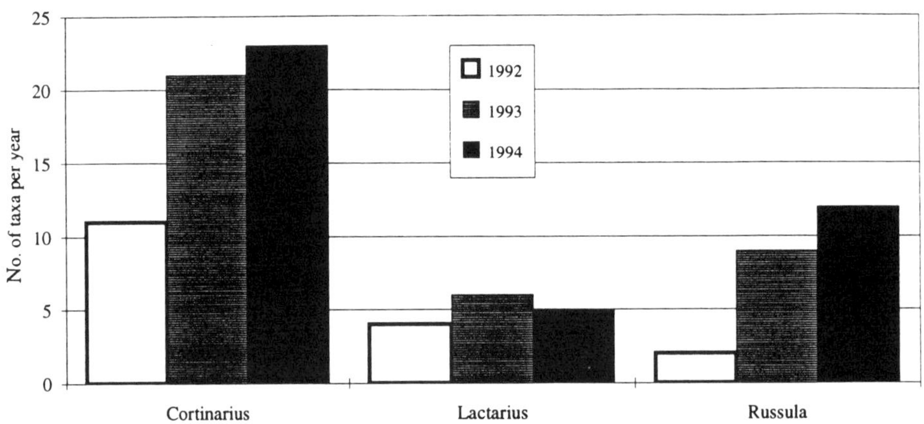


**Figure 15.5.** Species diversity of the dominating genera of ectomycorrhizal fungi in the roof catchment (1300 $m^2$ plot area) at Gårdsjön 1992–94. Total number of taxa (species and varieties) recorded per year

This is seen from the comparatively low 1994 production in the control site (less than a half of the 1993 production, cf. Figure 15.1), and the situation is confirmed also by mycologists conducting regional macrofungi monitoring in western Sweden (A. and L. Stridvall and S. Jacobsson, pers. comm.). The contrast between the fungus-rich roof site, and the fungus-poor surroundings, was easily observable when walking along the margins of the covered catchment in September–October 1994.

With this background, it seems that the steadily increased fungal production since 1992 must be an effect of the treatment in the roof catchment, related to (i) the cleaning of the rain, or (ii) some changes in the microclimate due to the sheltering effect of the extensive cover. In 1992, just after roof construction, some patches of mould development were observed upon the forest floor, indicating a patchwise high air and soil surface moisture in that period. A slightly increased air humidity underneath the roof was also recorded for that period (Moldan, pers. comm.). Consequently, favourable, humid conditions under the roof were suggested as the explanation of the increased production in 1993 by Brandrud (1995). However, a compilation of the 1993–94 humidity data shows that the seasonal average air moisture is not higher within the covered catchment than in the control, in spite of the potential wind and evaporation reduction under the roof (cf. Chapter 4, this volume). It should be added that the mould observed on 1992 soon disappeared after some adjustments in the sprinkling routines. Furthermore, if the increased production is due to increased humidity, the effect would probably decline towards the margins of the covered catchment, where the air is easily mixed with the surroundings. However, the high production in 1994 was observed almost to the very edge of the roof. The roof catchment has a somewhat delayed and reduced frost development in the autumn (pers. obs.), a fact that could influence the fruit body production. However, the fruit bodies recorded after the first periods with frost in 1994 counted for only about 10% of the total production, and therefore cannot explain the differences towards the control site, which also had some (although lower) production after frost periods.

In conclusion, the cleaning of the rain seems to be the most probable explanation of the increase in the ectomycorrhizal fungus production under the roof. Up to 1994, these changes are one of the first and most substantial biological responses to the treatment. This is also in accordance with the results from the neighbouring NITREX treament, where ammonium nitrate has been experimentally added since 1991. Here, like a mirror image of the roof treatment, a *decrease* in fruit body production of ecotmycorrhizal fungi has been one of the earliest and most pronounced effects of the treatment (Brandrud, 1995), indicating that removal of nitrogen might be the key factor to the success of the mycorrhizal fungi under the roof.

In 1992, just after construction of the roof, the productivity and especially the diversity of the ectomycorrhizal fungi in the roof catchment was *lower* than that of the control catchment. Notes on the occurrence and abundance of the fungus flora in the roof catchment in 1991 moreover indicate that the production decreased somewhat from the pre-treament situation. This could be an indication of a temporary stress induced by the construction of the roof, including much trampling on the forest floor in the preceding growth season (cf. Brandrud, 1995). In an ongoing study of Swiss coniferous forests it has been documented that trampling has a temporary, negative effect on the fruit body production of mycorrhizal species (Egli *et al.*, 1990; S. Egli, pers. comm.). An over-representation of some stress-tolerant species such as *Laccaria laccata* and *Paxillus involutus* (cf. Brandrud, 1995) in the roof catchment also indicates a "roof effect". These species were sometimes observed in connection with permanent constructions in the ground (attached to pillars, etc.), and have also been observed in connection with similar constructions in the roof site at the experiment area at Solling, central Germany (pers. obs.). When building a roof, it seems impossible to avoid inducing some changes in the ectomycorrhizal fungus flora due to disturbance from trampling and from the ground-attached constructions themselves.

The Gårdsjön area receives moderately high acid deposition (21 kg S/ha/yr and 13 kg N/ha/yr; 1989–91 mean), and the forest ecosystem shows few signs of reduced vitality (Dise and Wright 1992). On this background the increase in the fungal production has been more pronounced and more sudden than expected. The production achieved in the roof catchment in a non-optimal fungal season in 1994 (3987 fruit bodies per 1300 $m^2$) is almost double the maximum production achieved under non-experimental conditions at Gårdsjön during the period 1990–94 (2279 fruit bodies per 1300 $m^2$ in the NITREX site in 1990 before nitrogen treatment, cf. Brandrud, 1995). When compared with regional studies in oligotrophic *Picea abies* forests, the non-manipulated sites at Gårdsjön have a fairly low maximum production, considerably lower than that of slightly less polluted forests of central Sweden and south Norway (Dahlberg, 1991; Gulden *et al.*, 1992). The high production of the roof 1994 is comparable to that of central Swedish and north Norwegian forests, but still much lower than the peak production recorded in non-polluted forests in central Norway (approximately 12 000 fruit bodies per 1300 $m^2$ per year, cf. Gulden *et al.*, 1992). The short-term increase in production under the roof is probably limited by the growth potential of available fungal populations in the soil, and an establishment of new populations and an increase in species diversity will probably take more time.

In the experimental manipulation with a roof-covered, Douglas fir plantation at Speuld, in the Netherlands, the trend points in the same direction as in Gårdsjön, with the highest production in the plot receiving clean rain (Boxman *et al.*, 1995). An intermediate productivity was seen in a control plot with a roof, indicating that some of the increase could be due to a "roof effect". However, this productivity gradient was established immediately after the construction of the roof (when the first fungal records were obtained), and it is possible that the gradient could be due to original differences between the three sites. It should be added that the $\alpha$ diversity at Speuld was extremely low, with only seven ectomycorrhizal fungi recorded (Boxman *et al.*, 1995). At another roof experimental site in the Netherlands (Ysselsteyn; Scots pine) not a single mycorrhizal fungus was recorded, not even after 4 years of clean rain treatment. At a similar experimental site at Solling, central Germany, the highest fruit body production in 1994 was recorded in the roof plot receiving clean water (pers. obs.), but the production was due mainly to typical stress-tolerant species (*Laccaria laccata* and *Paxillus involutus*), appearing mainly in disturbed soil along the margins of the roof. No vitalization of the original ectomycorrhizal fungus flora could be observed in the clean-water plot at Solling, and a pronounced vitalization is not yet seen in the Dutch sites either. It seems probable that the depauperated fungus flora of the Dutch and German sites will need a long time to re-establish normal fruit body production.

The changes on the below-ground ectomycorrhiza production and vitality seem to be much less pronounced than the above-ground changes at Gårdsjön, or at least it has a lag time in response. The below-ground production was investigated in 1992–93, and the density of mycorrhizal root tips was unchanged in 1993 (150 mycorrhizas per $cm^3$ in the surface root mat), and did not differ much from the mycorrhizal density and vitality in the control catchment (Timmermann, 1994). This also corresponds with the results from the nitrogen addition experiment at

Gårdsjön; the fruit body production responded rapidly, whilst the mycorrhizal root density had changed little 2–3 years after start of treatment (Brandrud, 1993, 1995; Timmermann, 1994). On the other hand, some increase in the fine-root density and vitality has been observed in the roof catchment (cf. Chapter 16, this volume).

## ACKNOWLEDGEMENTS

V. Timmermann, University of Oslo, Dr A. Dahlberg, L. Jonsson and T. Snäll, Agricultural University, Uppsala, and Dr H. Wallander, University of Lund, are thanked for assistance during field work. Financial support for the mycorrhiza project has been provided by the Norwegian Research Council (NFR) and internal funds from the Norwegian Institute for Water Research (NIVA), and, further, from the Swedish Environmental Protection Agency (SNV) and the Commission of the European Communities (CEC) as part of the NITREX–Gårdsjön research programme.

## REFERENCES

Arnolds, E. (1991) Decline of ectomycorrhizal fungi in Europe. *Agric. Ecosyst. Environ.*, **35**, 209–244.

Boxman, A. W., Van Dam, D., Van Dijk, H. F. G., Hogervorst, R. F. and Koopmans, C. J. (1995) Ecosystem responses to reduced nitrogen and sulphur inputs into two coniferous forest stands in the Netherlands. *For. Ecol. Managem.*, **71**, 7–29.

Brandrud, T. E. (1993) The ectomycorrhizal fungi. In: R. F. Wright (ed.), *Gårdsjön. Status Report for 1991–1992: The First Year of Treatment.* NITREX Gårdsjön, Rept 3/93. Norwegian Institute for Water Research Rept 2865.

Brandrud, T. E. (1995) The effects of experimental nitrogen addition on the ectomycorrhizal fungus flora in an oligotrophic spruce forest at Gårdsjön, Sweden. *For. Ecol. Managem.*, **71**, 111–122.

Brandrud, T. E., Lindström, H., Marklund, H., Melot, J. and Muskos, S. (1991–95) *Cortinarius, Flora Photographica.* Vol. 1–3 (English version). Härnösand.

Dahlberg, A. (1991) Ectomycorrhiza in coniferous forest: structure and dynamics of populations and communities. Thesis, Swedish University of Agricultural Sciences, Uppsala.

Derbsch, H. and Schmitt, J. A. (1987) *Atlas der Pilze des Saarlandes. Teil 2: Nachweise, Ökologie, Vorkommen und Beschreibungen. Aus Natur und Landschaft im Saarland.* Sonderb. 3, 816 pp.

Dise, N. B. and Wright, R. F. (eds) (1992) *The NITREX Project (Nitrogen Saturation Experiments).* CEC Ecosystems Research Report 2.

Egli, S., Ayer, F. and Chatelain, F. (1990) Der Einfluss des Pilzsammelns auf die Pilzflora. *Mycol. Helvetica*, **3**, 417–428.

Fellner, R. (1989) Myccorrhizae-forming fungi as bioindicator of air pollution. *Agric. Ecosyst. Environ*, **28**, 115–120.

Fellner, R. (1993) Air pollution and mycorrhizal fungi in Central Europe. In: D. N. Pegler, L. Boddy, B. Ing and P. M. Kirk (eds), *Fungi of Europe: Investigation, Recording and Conservation.* Royal Botanic Gardens, Kew, pp. 239–250.

Gulden, G. and Høiland, K. (1985) The role of ectomycorrhiza in a situation of air pollution and forest death. *Agarica*, **12**, 341–357.

Gulden, G., Høiland, K., Bendiksen, K., Brandrud, T. E., Foss, B. S., Jenssen, H. B. and Laber, D. (1992) Macromycetes and air pollution. Mycocoenological studies in three oligotrophic spruce forests in Europe. *Bibl. Mycol.*, **144**, 1–81.

Høiland, K. (1986) *Storsoppfloraens Reaksjon overfor Forsuring, med Spesiell vekt på Mykorrhizasoppene.* Undersøkelse foretatt i Norsk institutt for skogforsknings forsøksfelter i Åmli, Aust-Agder. Ministry of Environment, Rep. T-671, 62 pp.

Jansen, A. E. and Dighton, J. (1990) *Effects of Air Pollutants on Ectomycorrhizas. A Review.* CEC Air Pollution Research Report 30, 58 pp.

Knudsen, H. and Hansen, L. (1992) *Nordic Macromycetes. Vol. 2. Polyporales, Boletales, Agaricales, Russulales.* Copenhagen.

Ritter, G. and Tölle, H. (1978) Stickstoffdüngung in Kiefernbeständen und ihre Wirkung auf Mykorrhizabildung und Fructification der Symbiosepilze. *Beitr. Forstwirtschaft*, **4**, 162–166.

Rühling, Å. and Tyler, G. (1991) Effects of stimulated nitrogen deposition to the forest floor on the macrofungal flora of a beech forest. *Ambio*, **20**, 261–263.

Timmermann, V. (1994) The mycorrhizal status of Norway spruce under different nitrogen loads. A comparative study of three oligotrophic forest stands in Northern Europe along a regional and an experimentally induced local nitrogen gradient. Thesis, University of Oslo.

# 16 Fine-Root Response to Nitrogen Removal and Addition

A. CLEMENSSON-LINDELL & H. PERSSON

Investigations were carried out on the experimental impact on Norway spruce root standing crop and distribution of fine roots, in three catchment areas subjected to decreased nitrogen and sulphur deposition (G1 ROOF), increased nitrogen deposition (G2 NITREX) and ambient levels of nitrogen and sulphur deposition (F1 CONTROL) respectively, within the Lake Gårdsjön basin, south-west Sweden. Root sampling was performed using soil cores and ingrowth cores. The fine roots were separated into three vitality classes, based on morphological characteristics.

Estimated root variables displayed a high variability between samples from the same area, indicating a high heterogeneity within the catchments. Year-to-year changes in the amount of fine roots of different vitality classes which occurred in all catchment areas were probably related to natural variations in climatic conditions. The bulk of the fine roots (living + dead roots < 1 mm in diameter) in the investigated catchment areas were found in the FH layer (45–82%). Signs of an increased percentage of living fine roots in the organic soil layer in the G1 ROOF catchment suggest that growth conditions for fine roots were improved by reducing N and S deposition. Results from the G2 NITREX catchment suggest that increased nitrogen deposition may reduce the total amount of fine roots.

## INTRODUCTION

Anthropogenic inputs of pollutants have increased during the last century and are thought to be a contributing factor to forest damage in Central Europe (Tomlinson, 1983; Schulze, 1989; Ulrich, 1989; Marschner, 1991). The air pollutants are characterised by a combination of both acidifying and fertilising substances (Marschner *et al.*, 1991). Atmospheric pollutants may influence the growth and anatomy of tree roots directly by changing soil chemistry and indirectly by injuring needles (McQuattie and Schier, 1992). Studies in damaged forests have exhibited a number of abnormalities in fine-root development, which have been related to nutrient deficiencies and metal toxicities (e.g. Hütterman, 1982; Murach, 1984; Meyer, 1985; Persson *et al.*, 1995).

Different approaches have been applied in order to describe and quantify the vitality of fine-root systems. Damage to fine roots and mycorrhiza has been reported throughout a wide geographic range of forest decline (Puhe *et al.*, 1986; Glenn *et al.*,

*Experimental Reversal of Acid Rain Effects: The Gårdsjön Roof Project.*
Edited by Hans Hultberg and Richard Skeffington. 

1991; Schlegel *et al.*, 1992). Several studies have examined the status of fine roots and mycorrhiza as a correlate of tree health, but there are no generally acceptable characteristics of damage signs (Shortle and Smith, 1988; Schlegel *et al.*, 1992). Counting root tips, with or without quantifying the degree of mycorrhizal infection, has been performed in several studies (Blaschke, 1981; Ahlström *et al.*, 1988; Eichhorn *et al.*, 1988). The number of root tips per unit volume of soil is an indicator of the influence of soil properties on forest trees, and it gives an indication of the absorptive capacity of root systems, since the water and mineral nutrients are absorbed most intensively by the fine-root tips and the mycorrhiza.

Puhe *et al.* (1986) found a close relationship between needle loss in Norway spruce (*Picea abies* (L.) Karst) and fine-root loss. Persson (1978) investigated five different parameters related to living fine roots in *Pinus sylvestris* (L.), i.e. weight, length, surface area, volume and the number of root tips, in order to characterise the growth dynamics of fine roots. Variation of all these parameters was closely correlated, since the mean diameter of roots and branching pattern from one sampling occasion to the next will vary very little, when compared with root length. Thus, any one of these parameters may be used for the measurements of the growth dynamics and vitality of fine roots.

Investigations have often been carried out in order to assess the influence of, for example, increased nitrogen levels (Persson, 1980; Ahlström *et al.*, 1988; Gijsman, 1991; Olsthoorn *et al.*, 1991a,b), liming (Persson and Ahlström, 1991; Clemensson-Lindell and Persson, 1993, 1995; Clemensson-Lindell, 1994) and soil acidification and aluminium toxicity (Schlegel *et al.*, 1992; Majdi and Persson, 1993). So far, only limited data are available from field investigations on fine-root development and biomass in relation to reduced or increased nitrogen or sulphur deposition (cf., for example, Aber *et al.*, 1989).

Here we report results of the distribution and development of fine roots from 1990 to 1992 at three catchments in the Lake Gårdsjön basin, south-west Sweden. Catchment G1 ROOF receives experimentally decreased nitrogen deposition, catchment G2 NITREX increased nitrogen and sulphur deposition and catchment F1 CONTROL receives ambient nitrogen and sulphur deposition. Root sampling was carried out using soil core sampling and ingrowth cores (cf. Vogt and Persson, 1991). The fine roots were separated into three classes based on morphological characteristics. The investigation on the catchment G2 NITREX was a part of a nitrogen manipulation programme at seven sites across Europe (Wright and Van Breemen, 1995).

## MATERIALS AND METHODS

### Site description

The experimental site Gårdsjön is described elsewhere in this book (Chapter 2). Fine-root responses have been described in three subcatchments within the Lake Gårdsjön basin. Two vegetation types, *Vaccinium*-dominated and moss-dominated respectively, were distinguished for the sampling programme within each catchment. Treatments at the G1 ROOF and F1 CONTROL catchments have been described elsewhere. At the G2 NITREX site, 35 kg N $ha^{-1}$ $yr^{-1}$ was added as ammonium nitrate to augment the natural N throughfall load (Dise and Wright, 1992). The total nitrogen deposition is within the range received by many forest ecosystems in Central Europe. The nitrogen was distributed weekly by means of a sprinkler system in proportion to the amount of natural precipitation during the previous 7 days. The volume of additional water was about 5% of the natural precipitation. Treatment began in April, 1991.

### Root sampling

Root sampling was performed by the use of two methods: (i) soil cores taken with a 4.5 cm cylindrical steel corer; and (ii) ingrowth cores installed in the holes left by soil coring (Vogt and Persson, 1991). Sets of seven to ten soil cores (depth about 30 cm) were taken in the beginning of October 1990, 1991 and 1992 in each of two vegetation types (*Vaccinium*- and moss-dominated) within each catchment. In G1 ROOF and at G2 NITREX, these represent pre-treatment, 6 months' or 18 months' treatment respectively. Investigation of the distribution pattern of fine roots indicated no dependency on the distance to the nearest tree. The soil cores were divided into humus layer and mineral soil, with the humus layer in turn divided into 5 cm layers. The mineral soil was divided into 5 + 5 + 10 cm layers from the top downwards. The samples were stored in a deep-freeze at −4°C until sorting could take place.

Immediately after thawing, roots were picked out from each soil layer and sorted into diameter classes: less than 1, 1–2 and 2–5 mm. The fine roots less than 1 mm were separated into three classes, based on the following morphological characteristics:

Class 1: Roots more or less brownish/suberised and well branched, with the main part of the root tips light and turgid. Stele white and elastic.
Class 2: Roots darkened. White root tips few or lacking. Stele still elastic and light to slightly brown.
Class 3: Dead roots. Stele brownish and easily broken. No elasticity remaining.

Small-diameter roots (1–2 and 2–5 mm in diameter) were sorted into a living and dead root fraction based on the colour and strength of the stele (Vogt and Persson, 1991). Ingrowth cores of a nylon net filled with perlite were placed into the holes produced by the core sampling each year. After two growing seasons they were resampled and stored in a deep freeze at −4°C. After thawing, each core was divided into an upper 10 cm part, and a lower 20 cm part. The roots were separated into diameter classes less than 1, 1–2 and 2–5 mm, without any further classification. Finally, the roots were dried at 70°C for 48 hours and weighed to the nearest milligram.

### Statistics

Owing to the lack of replicated treatment areas and since the sampling was carried out only once a year, changes were difficult to evaluate statistically. Proper statistical analysis requires longer data series. Only significant differences between different years within each catchment and vegetation area were used. The SAS analysis of variance (SAS GLM) and Student's t-test were used (Ray, 1982).

## RESULTS

### Core sampling

Large variations in the different root fractions were found in the core samples from all catchment areas between different years (Table 16.1). Few significant changes were observed in diameter classes 1–2 and 2–5 mm. The majority of changes in fine roots were observed in the humus layer (FH 0–5 and 5–10 cm).

In the catchment F1 CONTROL, significantly high amounts of dead fine roots (class 3) were found in 1991 in the FH 5–10 cm in the moss-dominated subarea. The percentage of living fine roots in the soil profile was low in 1991 (Table 16.2). Similar tendencies were found in the FH layer of the drier *Vaccinium*-dominated subarea with a decreased percentage of living fine roots in 1991.

In the catchment G1 ROOF there was a gradual increase in the FH layer in the amount of class 2 fine roots (Table 16.1) and in the percentage of living fine roots (Table 16.2), which were most extensive in the moss-dominated area (26–53 and 60–83% in FH 0–5 and 5–10 cm). There was also a significant decrease in the amount of class 3 fine roots in FH 5–10 cm during the same period. In the *Vaccinium*-dominated subarea of G1, no general trend of changes in the FH layer was found (Table 16.2).

In the G2 NITREX catchment there were no distinct year to year effects in the organic layer in any of the subareas (Table 16.1. However, there was a decrease in the amount of fine roots, especially in the upper part of the soil profile in the *Vaccinium*-dominated subarea (the FH layer). Furthermore, an increase in the percentage of living fine roots was observed (10–30 and 44–75% in FH 0–5 and 5–10 cm). This increase was evidently related to a decrease in the amount of dead class 3 fine roots.

The bulk of the fine roots (living + dead roots less than 1 mm in diameter) was found in the FH horizon (45–82%, cf. Table 16.3), which is important to know when evaluating the more substantial changes taking place in this horizon. The percentage share of fine roots in the FH layer was extremely high (82%) in the moss-dominated subarea of F1 CONTROL in 1992. In the *Vaccinium*-dominated part of G1 ROOF, it was very low in 1991 and 1992 (53 and 45% respectively). No significant changes occurred in the percentage of fine roots in the FH layer in G2 NITREX.

### Ingrowth cores

The variation in ingrowth from year to year showed a high variation and few significant changes were observed (Table 16.4). Thus, there was a significantly higher ingrowth in the moss-dominated subarea in the upper part of the ingrowth core in 1991 compared with 1992 of F1 CONTROL. In the moss-dominated part of the G2 NITREX catchment the amount of fine roots was higher in 1992 in the lower part of the ingrowth core. The latter tendency towards a higher ingrowth in the moss-

**Table 16.1.** The distribution of fine roots (<1 mm in diameter) in soil cores taken at catchments F1 CONTROL, G1 ROOF and G2 NITREX. Values with different letters differ significantly (Student's t-test, $p < 0.05$) between the years for each area. Mean values ± SD. Number of samples = 7–10. CONT = CONTROL. Vacc. = *Vaccinium*-dominated vegetation; Moss = moss-dominated vegetation. FH = humus layer; M = mineral soil. (From Clemensson-Lindell and Persson, 1995b, reproduced by permission of Elsevier Science B.V.)

| Catchment/ subarea | Soil level (cm) | Year | Class 1 | Class 2 | Class 3 | Totals |
|---|---|---|---|---|---|---|
| | | | (g dm$^{-3}$ DW) | | | |
| F1 CONT/Moss | FH 5–10 | 1990 | 0.9±0.9ab | 2.5±1.4 | 1.2±0.6a | 4.6±2.2 |
| | | 1991 | 0.2±0.2a | 2.2±1.8 | 2.6±2.0b | 5.0±3.3 |
| | | 1992 | 1.3±1.1b | 3.2±1.9 | 0.9±0.4a | 5.4±2.6 |
| | FH 0–5 | 1990 | 0 | 0.5±0.3ab | 1.1±0.7ab | 1.6±0.9 |
| | | 1991 | 0 | 0.2±0.3a | 1.5±0.7a | 1.8±0.7 |
| | | 1992 | 0.0±0.1 | 0.8±0.5b | 0.5±0.3b | 1.3±0.8 |
| | M 0–5 | 1990 | 0 | 0.4±0.2 | 0.6±0.7 | 1.0±0.9 |
| | | 1991 | 0 | 0.2±0.2 | 0.5±0.4 | 0.7±0.4 |
| | | 1992 | 0.0±0.1 | 0.3±0.4 | 0.4±0.3 | 0.8±0.6 |
| | M 5–10 | 1990 | 0 | 0.1±0.1ab | 0.2±0.2 | 0.3±0.3 |
| | | 1991 | 0 | 0.0±0.1a | 0.1±0.1 | 0.2±0.2 |
| | | 1992 | 0 | 0.2±0.3b | 0.2±0.2 | 0.4±0.4 |
| | M 10–20 | 1990 | 0 | 0.2±0.2 | 0.3±0.2 | 0.5±0.4 |
| | | 1991 | 0 | 0.0±0.1 | 0.2±0.2 | 0.3±0.2 |
| | | 1992 | 0 | 0.1±0.2 | 0.3±0.2 | 0.4±0.4 |
| F1 CONT/*Vacc.* | FH 5-10 | 1990 | 0.6±0.5 | 2.2±1.9 | 2.4±1.8 | 5.2±3.5 |
| | | 1991 | 0.2±0.2 | 1.2±0.8 | 2.0±1.1 | 3.4±1.2 |
| | | 1992 | 0.4±0.4 | 1.8±0.9 | 1.2±0.5 | 3.4±1.1 |
| | FH 0–5 | 1990 | 0.1±0.2 | 0.5±0.4 | 1.2±0.5a | 1.8±0.7 |
| | | 1991 | 0 | 0.1±0.1 | 1.2±0.5a | 1.4±0.4 |
| | | 1992 | 0 | 0.6±0.8 | 0.7±0.3b | 1.3±0.7 |
| | M 0–5 | 1990 | 0 | 0.2±0.2 | 0.6±0.5 | 0.8±0.6 |
| | | 1991 | 0 | 0.0±0.1 | 0.6±0.4 | 0.6±0.4 |
| | | 1992 | 0 | 0.1±0.2 | 0.6±0.3 | 0.8±0.3 |
| | M 5–10 | 1990 | 0 | 0.2±0.4 | 0.6±0.7 | 0.8±0.9 |
| | | 1991 | 0 | 0.0±0.1 | 0.6±0.4 | 0.6±0.4 |
| | | 1992 | 0 | 0.0±0.1 | 0.4±0.4 | 0.4±0.4 |
| | M10–20 | 1990 | 0 | 0.1±0.1 | 0.7±0.5 | 0.8±0.5 |
| | | 1991 | 0 | 0.2±0.2 | 0.5±0.3 | 0.7±0.4 |
| | | 1992 | 0 | 0.1±0.1 | 0.4±0.4 | 0.5±0.4 |
| G1 ROOF/moss | FH 5–10 | 1990 | 1.1±1.5 | 1.6±0.8a | 1.7±0.7a | 4.3±1.1 |
| | | 1991 | 0.5±0.4 | 1.5±0.5a | 1.3±0.7ab | 3.3±1.0 |
| | | 1992 | 1.0±1.0 | 3.1±2.0b | 0.6±0.2b | 4.8±2.7 |
| | FH 0–5 | 1990 | 0.1±0.2 | 0.2±0.1a | 0.9±0.5 | 1.2±0.5a |
| | | 1991 | 0 | 0.4±0.4a | 1.3±0.8 | 1.8±1.1ab |
| | | 1992 | 0.2±0.3 | 1.0±0.7b | 1.0±0.5 | 2.2±0.9b |
| | M 0–5 | 1990 | 0 | 0.1±0.1 | 0.6±0.3 | 0.7±0.3 |
| | | 1991 | 0 | 0.2±0.2 | 0.9±0.8 | 1.0±0.8 |
| | | 1992 | 0 | 0.6±1.6 | 0.5±0.4 | 1.1±1.9 |
| | M5–10 | 1990 | 0 | 0.2±0.3 | 0.6±0.4 | 0.8±0.5 |
| | | 1991 | 0 | 0.1±0.1 | 0.8±1.3 | 0.8±1.3 |
| | | 1992 | 0 | 0.1±0.2 | 0.2±0.2 | 0.3±0.4 |

*(continued) overleaf*

**Table 16.1.** *(Continued)*

| Catchment/ subarea | Soil level (cm) | Year | Class 1 | Class 2 | Class 3 | Totals |
|---|---|---|---|---|---|---|
| | | | (g dm$^{-3}$ DW) | | | |
| | M 10–20 | 1990 | 0 | 0.3±0.4 | 1.0±0.7a | 1.3±0.7a |
| | | 1991 | 0 | 0 | 0.5±0.3ab | 0.5±0.3b |
| | | 1992 | 0 | 0.1±0.2 | 0.2±0.3b | 0.3±0.4b |
| G1 ROOF/*Vacc.* | FH 5–10 | 1990 | 0.7±0.5ab | 1.2±0.7 | 1.3±1.2a | 3.2±1.8 |
| | | 1991 | 0.1±0.1a | 2.1±1.4 | 2.6±1.5b | 4.8±2.3 |
| | | 1992 | 1.4±1.6b | 2.3±1.3 | 1.1±0.7a | 4.8±2.4 |
| | FH 0–5 | 1990 | 0.1±0.2 | 0.3±0.3 | 1.8±1.4a | 2.2±1.5 |
| | | 1991 | 0 | 0.5±1.1 | 1.3±0.7ab | 1.9±1.6 |
| | | 1992 | 0.1±0.2 | 0.4±0.3 | 0.8±0.4b | 1.3±0.4 |
| | M 0–5 | 1990 | 0 | 0.1±0.1a | 0.4±0.4a | 0.5±0.5a |
| | | 1991 | 0 | 0.2±0.3ab | 1.0±0.5b | 1.2±0.7b |
| | | 1992 | 0 | 0.5±0.4b | 0.9±0.3b | 1.4±0.6b |
| | M 5–10 | 1990 | 0 | 0.1±0.1 | 0.4±0.4a | 0.5±0.5a |
| | | 1991 | 0 | 0.1±0.2 | 1.3±0.7b | 1.5±0.8b |
| | | 1992 | 0 | 0.2±0.3 | 0.8±0.5a | 1.0±0.8ab |
| | M10–20 | 1990 | 0 | 0a | 0.2±0.3a | 0.2±0.3a |
| | | 1991 | 0 | 0.0±0.1a | 1.0±0.4b | 1.0±0.3b |
| | | 1992 | 0 | 0.3±0.2b | 1.1±0.3c | 1.4±0.4c |
| G2 NITREX/ Moss | FH 5–10 | 1990 | 0.8±0.8 | 3.3±3.7 | 1.4±0.8ab | 5.5±3.9 |
| | | 1991 | 0.4±0.5 | 2.0±1.6 | 2.2±0.7a | 4.5±2.2 |
| | | 1992 | 0.6±0.9 | 2.0±0.6 | 1.2±0.9b | 3.9±1.3 |
| | FH 0–5 | 1990 | 0.3±0.3a | 0.6±0.4 | 1.3±0.5ab | 2.2±0.8a |
| | | 1991 | 0b | 0.4±0.5 | 1.6±1.4a | 2.0±1.8ab |
| | | 1992 | 0b | 0.4±0.4 | 0.6±0.4b | 1.0±0.3b |
| | M 0–5 | 1990 | 0.1±0.2 | 0.2±0.2 | 0.6±0.3ab | 0.9±0.3 |
| | | 1991 | 0 | 0.1±0.1 | 0.8±0.6a | 0.8±0.7 |
| | | 1992 | 0 | 0.2±0.2 | 0.5±0.2b | 0.7±0.2 |
| | M 5–10 | 1990 | 0 | 0.3±0.3 | 1.0±0.8 | 1.3±0.8 |
| | | 1991 | 0 | 0.1±0.1 | 0.8±0.8 | 0.9±0.8 |
| | | 1992 | 0 | 0.3±0.4 | 0.4±0.4 | 0.6±0.6 |
| | M 10–20 | 1990 | 0 | 0.2±0.1a | 0.8±0.4 | 1.0±0.4 |
| | | 1991 | 0 | 0.2±0.3ab | 0.7±0.5 | 0.9±0.5 |
| | | 1992 | 0 | 0.1±0.2b | 0.5±0.3 | 0.6±0.4 |
| G2 NITREX/ *Vacc.* | FH 5-10 | 1990 | 0.7±0.8 | 3.0±2.2 | 4.3±2.3a | 8.0±2.8a |
| | | 1991 | 0.5±0.5 | 1.5±1.1 | 1.2±0.5b | 3.2±1.1b |
| | | 1992 | 1.2±1.6 | 3.0±1.8 | 1.2±1.0b | 5.4±2.9ab |
| | FH 0–5 | 1990 | 0 | 0.2±0.2 | 2.2±0.7a | 2.5±0.7a |
| | | 1991 | 0.1±0.2 | 0.4±0.6 | 0.9±0.4b | 1.4±1.1b |
| | | 1992 | 0 | 0.4±0.3 | 0.9±0.6b | 1.3±0.7b |
| | M 0–5 | 1990 | 0 | 0.1±0.2 | 0.5±0.4 | 0.6±0.4 |
| | | 1991 | 0 | 0.1±0.2 | 0.8±0.5 | 0.9±0.6 |
| | | 1992 | 0 | 0.2±0.2 | 0.4±0.5 | 0.6±0.6 |
| | M 5–10 | 1990 | 0 | 0.0±0.1 | 0.6±0.5 | 0.7±0.5 |
| | | 1991 | 0 | 0.1±0.1 | 0.5±0.6 | 0.6±0.7 |
| | | 1992 | 0 | 0.1±0.1 | 0.4±0.3 | 0.6±0.2 |
| | M 10–20 | 1990 | 0 | 0 | 0.6±0.5 | 0.6±0.5 |
| | | 1991 | 0 | 0 | 0.4±0.2 | 0.4±0.2 |
| | | 1992 | 0 | 0 | 0.3±0.2 | 0.3±0.2 |

**Table 16.2.** Percentage living fine roots (< 1 mm in diameter) in the organic soil layer, from soil cores taken at catchments F1 CONTROL, G1 ROOF and G2 NITREX. Values with different letters differ significantly (Student's t-test, $p < 0.05$) between the years for each area. Mean values ± SD. Number of samples = 7–10. FH = organic soil layer; M = mineral soil. (From Clemensson-Lindell and Persson, 1995b, reproduced by permission of Elsevier Science B.V.)

| Vegetation/soil level (cm) | Year | F1 CONTROL | G1 ROOF | G2 NITREX |
|---|---|---|---|---|
| Moss-dominated area | | | | |
| FH 5–10 | 1990 | 68 ± 27ab | 60 ± 22a | 72 ± 22 |
| | 1991 | 49 ± 16 | 63 ± 14a | 45 ± 22 |
| | 1992 | 80 ± 14 | 83 ± 8b | 71 ± 17 |
| FH 0–5 | 1990 | 39 ± 16a | 26 ± 26a | 40 ± 13 |
| | 1991 | 16 ± 18 | 25 ± 10a | 29 ± 32 |
| | 1992 | 57 ± 22a | 53 ± 19b | 38 ± 32 |
| *Vaccinium*-dominated area | | | | |
| FH 5–10 | 1990 | 52 ± 20 | 60 ± 21ab | 44 ± 29a |
| | 1991 | 40 ± 24 | 46 ± 25a | 58 ± 21ab |
| | 1992 | 60 ± 22 | 76 ± 11b | 75 ± 15b |
| FH 0–5 | 1990 | 34 ± 25ab | 17 ± 20 | 10 ± 7a |
| | 1991 | 12 ± 12a | 27 ± 26 | 19 ± 25ab |
| | 1992 | 40 ± 32 | 38 ± 27 | 30 ± 18b |

dominated area of the G2 NITREX catchment was not reflected in the amount of living fine roots in the soil cores (Table 16.2).

It is impossible at the present state of the investigation to conclude anything about treatment effects from the ingrowth core data. Since the amount of fine roots had a tendency to be higher in the upper part of the ingrowth core in the F1 and G1 catchment areas in 1991 compared with 1992 and the opposite was observed in G2, one may conclude that root growth is affected in the latter treatment by the additions of ammonium nitrate (50 kg N $ha^{-1}$ $yr^{-1}$). The amount of fine roots in the ingrowth cores in all catchment areas was roughly a quarter of the total amount of living fine roots in the soil profile (Table 16.1), thus indicating a low turnover rate in fine roots.

**Table 16.3.** Fine-root distribution in the soil profile, expressed as percentage of total amount of fine roots (< 1 mm in diameter) in the organic soil layer for different years. Values with different letters differ significantly between different years for each sampling area. Vacc. = *Vaccinium*-dominated vegetation; Moss = moss-dominated vegetation; n.d. = not determined. Mean values ± SD, each based on 7–10 samples. (From Clemensson-Lindell and Persson, 1995b, reproduced by permission of Elsevier Science B.V.)

| Catchment | Vegetation type | 1990 | 1991 | 1992 |
|---|---|---|---|---|
| F1 CONTROL | Vacc. | 61 ± 19a | 64 ± 14a | 64 ± 39a |
| | Moss | 68 ± 16a | 82 ± 8b | 76 ± 16a |
| G1 ROOF | Vacc. | 79 ± 18a | 53 ± 11b | 45 ± 18b |
| | Moss | 55 ± 15a | 64 ± 23a | 70 ± 24a |
| G2 NITREX | Vacc. | 73 ± 9a | 71 ± 16a | 72 ± 11a |
| | Moss | 52 ± 20a | 61 ± 15a | 65 ± 10a |

**Table 16.4.** Fine-root biomass (g $dm^{-3}$ DW) in ingrowth cores taken at catchments F1 CONTROL, G1 ROOF and G2 NITREX. Values with different letters differ significantly (Student's t-test, $p<0.05$) between the years for each area. Mean values ± SD. $n$ = number of samples. Vacc. = *Vaccinium*-dominated vegetation; Moss = moss-dominated vegetation

| Catchment | Subarea | Year | *n* | Upper 10 cm | Lower 20 cm |
|---|---|---|---|---|---|
| F1 CONTROL | Moss. | 1991 | 7 | 2.2 ± 1.0a | 0.7 ± 0.5 |
| | | 1992 | 10 | 1.3 ± 0.4b | 0.4 ± 0.2 |
| F1 CONTROL | Vacc. | 1991 | 0 | n.d. | n.d. |
| | | 1992 | 10 | 1.0 ± 0 | 0.6 ± 0.4 |
| G1 ROOF | Moss. | 1991 | 0 | n.d. | n.d. |
| | | 1992 | 9 | 1.1 ± 0.5 | 0.5 ± 0.6 |
| G1 ROOF | Vacc. | 1991 | 10 | 0.8 ± 0.5 | 0.2 ± 0.1 |
| | | 1992 | 10 | 0.6 ± 0.4 | 0.4 ± 0.4 |
| G2 NITREX | Moss. | 1991 | 9 | 1.7 ± 0.6 | 0.4 ± 0.1a |
| | | 1992 | 10 | 1.8 ± 0.8 | 0.8 ± 0.4b |
| G2 NITREX | Vacc. | 1991 | 14 | 1.5 ± 0.8 | 0.3 ± 0.1 |
| | | 1992 | 10 | 2.0 ± 1.1 | 0.4 ± 0.3 |

## DISCUSSION

Our results from Gårdsjön so far cover only 3 years. This is too short a time to distinguish treatment effects from year to year changes induced by alterations in climatic conditions. Biotic factors, such as the microflora of the rhizosphere, mycorrhizal components, etc., may furthermore influence the year to year changes. Estimated root variables displayed a large variability between samples from the same area and year. This reveals the high degree of heterogeneity within the catchments, and also within distances of a few metres. The difficulties in sampling and processing an appropriate number of samples under these circumstances are obvious.

The signs of increased percentage of living fine roots within the G1 ROOF catchment suggest improved growing conditions for fine roots and reversal of the negative effects caused by high N and S deposition. Results from the G2 NITREX catchment suggest that increased nitrogen deposition may reduce the total amount of fine roots. These results are consistent with those of similar experiments: negative effects of ammonium nitrate application on the development of fine roots in similar Norway spruce stands in south-west Sweden have been reported by Persson and Ahlström (1990/91) and of ammonium sulphate by Clemensson-Lindell and Persson (1995a,b) and Persson *et al.* (1995).

The amount of fine roots in the ingrowth cores provides an estimate of fine-root reproduction and therefore offers a direct test of root vitality (Vogt and Persson, 1991; Persson, 1993). The ingrowth cores were initially root-free; all roots present in the cores were therefore from the same original population and of similar chronological age classes. From a morphological point of view, the fine roots from soil cores and ingrowth cores may differ considerably. Therefore, roots from ingrowth cores were not separated according to morphological classification.

However, since the amount of roots in the ingrowth cores displayed high variability between cores within the same area and year, the root data from the ingrowth cores do not so far suggest any effect of treatment.

## ACKNOWLEDGEMENTS

This work was supported financially by the Swedish Environmental Protection Board (SNV) and the Commission of European Communities (STEP-CV90-0056).

## REFERENCES

Aber, J., Nadelhoffer, K. J., Steudler, P. and Melillo, J. M. (1989) Nitrogen saturation in northern forest ecosystems. *Bioscience*, **39**, 378–386.

Ahlström, K., Persson, H. and Börjesson, I. (1988) Fertilization in a mature Scots pine (*Pinus sylvestris* L.) stand—effects on fine roots. *Plant and Soil*, **106**, 179–190.

Andersson, F. and Olsson, B. (eds) (1985) *Lake Gårdsjön, An Acid Forest Lake and its Catchment. Ecological Bulletin (Stockholm)*, **37**.

Blaschke, H. (1981) Veränderungen bei der Feinwurzelentwicklung in Weisstannenbeständen. *Forstwissenschafltliche Centralblatt*, **199**, 190–195.

Clemensson-Lindell, A. (1994) *Norway Spruce Fine-root Morphology and Function Influenced by Nutrient Applications in Forests.* Dissertation. Department Ecology and Environmental Research, Swedish University Agricultural Science, Report **72**, 68 pp.

Clemensson-Lindell, A. and Persson, H. (1993) Long-term effects of liming on the fine-root standing crop of *Picea abies* and *Pinus sylvestris* in relation to chemical changes in the soil. *Scandinavian Journal Forest Research*, **8**, 384–394.

Clemensson-Lindell, A. and Persson, H. (1995a) Fine-root vitality in a 30-year-old Norway spruce stand subjected to various nutrient supplies. *Plant and Soil*, **168–169**, 167–171.

Clemensson-Lindell, A. and Persson, H. (1995b) The effects of nitrogen addition and removal on Norway spruce fine root vitality and distribution in three catchment areas at Gårdsjön. *Forest Ecology and Management*, **71**, 123–131.

Dise, N. B. and Wright, R. F. (1992) *The NITREX Project (Nitrogen Saturation Experiments).* Ecosystems Research Report 2, Commission of the European Communities, 101 pp.

Eichhorn, von J., Gärtner, E. and Hüttermann, A. (1988) Wurzelspitzenhäufigkeit von Feinwurzeln als Funktionsparameter der Schädigung von Altfichten. *Allgemeine Forst- und Jagdzeitung*, **159**, 37–42.

Gijsman, A. (1991) Soil water content as a key factor determining the source of nitrogen ($NH_{4+}$ or $NO_{3-}$) absorbed by Douglas-fir (*Pseudotsuga menziesii*) and the pattern of rhizosphere pH along its roots. *Canadian Journal Forest Research*, **21**, 616–625.

Glenn, M. G., Wagner, W. S. and Webb, S. L. (1991) Mycorrhizal status of mature and red spruce (*Picea rubens*) in mesic and wetland sites of northwestern New Jersey. *Canadian Journal Forest Research*, **21**, 741–747.

Hütterman, A. (1982) Früdiagnose von Immissionsschäden im Wurzelbereich von Waldbäumen. In: *Sonderheft Mitteilungen der Löflling Recklinghausen*, pp. 26–31.

Majdi, H. and Persson, H. (1993) Spatial distribution of fine roots, rhizosphere and bulk-soil chemistry in an acidified *Picea abies* stand. *Scandinavian Journal Forest Research*, **8**, 147–155.

Marschner, H. (1991) Mechanisms of adaptation of plants to acid soils. *Plant and Soil*, **134**, 1–20.

Marschner, H., Häussling, M. and Eckhard, G. (1991) Ammonium and nitrate uptake rates and rhizosphere pH in non-mycorrhizal roots of Norway spruce (*Picea abies*) (L.) Karst.). *Trees*, **5**, 14–21.

McQuattie, C. J. and Schier, G. A. (1992) Effects of ozone and aluminium on pich pine (*Pinus rigida*) seedlings: anatomy of mycorrhiza. *Canadian Journal Forest Research*, **22**, 1901–1916.

Meyer, F. H. (1985) Einfluss des Stickstoff-Faktors auf den Mykorrhizabesatz von Fichtensämingen im Humus einer Waldschädenfläche, *Allgemeine Forst Zeitschrift*, **40**, 208–219.

Murach, D. (1984) Die Reaktion der Feinwurzeln von Fichten (*Picea abies* (L.) Karst.) auf zunehmende Bodenversäurung. *Göttinger Bodenkundliche Berichte*, **77**, 126 pp.

Olsthoorn, A. F. M., Keltjens, W. G., van Baren, B. and Hopman, M. C. G. (1991) Influence of ammonium on fine root development and rhizosphere pH of Douglas-fir seedlings in sand. *Plant and Soil*, **133**, 75–81.

Persson, H. (1978) Root dynamics in a young Scots pine stand in central Sweden. *Oikos*, **30**, 508–519.

Persson, H. (1980) Fine root dynamics in a Scots pine stand with and without near optimum nutrient and water regimes. *Acta Phytogeographica Suecica*, **68**, 101–110.

Persson, H. (1993) Factors affecting fine root dynamics of trees. *Suo*, **43**, 163–172.

Persson, H. and Ahlström, K. (1991) The effects of liming and fertilization on fine root growth. *Water, Air and Soil Pollution*, **54**, 365–379.

Persson, H., Sun, Y., Majdi, H. and Nilsson, L. O. (1995) Root distribution in a Norway spruce (*Picea abies* (L.) Karst.) stand subjected to drought and ammonium-sulphate application. *Plant and Soil*, **168–169**, 161–165.

Puhe, J., Persson, H. and Börjesson, I. (1986) Wurzelwachtum und Wurzelshäden in Skandinavischen Nadelwäldern. *Allgemeine Forst Zeitschrift*, **20**, 488–492.

Ray, A. A. (1982) *SAS User's Guide: Statistics, Vers. 5*. SAS Institute Inc., Cary, NC, 956 pp.

Schlegel, H., Amundson, R. G. and Hüttermann, A. (1992) Element distribution in red spruce (*Picea rubens*) fine roots; evidence for aluminium toxicity at Whiteface Mountain. *Canadian Journal Forest Research*, **22**, 1132–1138.

Schulze, E. D. (1989) Air pollution and forest decline in a spruce (*Picea abies*) forest. *Science*, **244**, 776–783.

Shortle, W. C. and Smith, K. T. (1988) Aluminium-induced calcium deficiency syndrome in declining red spruce. *Science (Washington D.C.)*, **240**, 1017–1018.

Tomlinson, G. H. (1983) Air pollutants and forest decline. *Environmental Science and Technology*, **17**, 246–256.

Ulrich, B. (1989) Effects of acid precipitation on forest ecosystems in Europe. In: D. C. Adriano and H. H. Johansson (eds), Advances in Environmental Sciences, *Acidic Precipitation Vol. 2*. Springer, New York, pp. 189–272.

Vogt, K. A. and Persson, H. (1991) Measuring growth and development of roots. In: J. P. Lassoie and T. M. Hincley (eds), *Techniques and Approaches in Forest Tree Ecophysiology*. CRC Press, Boca Raton, pp. 477–501.

Wright, R. F. and van Bremen, N. (1995) The NITREX project: an introduction. *Forest Ecology and Management*, **71**, 1–5.

# 17 Forest Health Responses in the Gårdsjön Catchments

**ARNE O. STUANES, MAGNE HUSE, O. JANNE KJØNAAS & PER HOLM NYGAARD**

A forested catchment in Gårdsjön has been sheltered from ambient precipitation by a continuous roof built under the canopy. Since 1991 the catchment has been subjected to precipitation of pre-industrial quality. The roof catchment has been compared with a control catchment without a roof.

During the period from 1990 to 1994, the vitality of the forest expressed as crown density and discoloration has changed in the direction of lower crown density and stronger discoloration. A lower level of discoloration in 1994 in the roof catchment may be due to a slightly delayed watering caused by frost during winter that partly reduced the effect of the subsequent dry summer. The nutrient content of the needles mirrors the crown discoloration pattern in both control and roof catchments during the experimental period. The radial growth of the trees has been low. Some disturbance in the ground vegetation has been caused by the litter redistribution in the roof catchment.

The reported changes seem not to be linked to the treatment, since the responses are similar in both the roof and the control catchment.

## INTRODUCTION

In 1971, Norwegian scientists hypothesized that the deposition of sulphuric acid would reduce forest production (Dahl and Skre, 1971). They calculated that forest growth would decrease by 1–1.5% annually in southernmost Norway due to the deposition of acid from the atmosphere. According to such calculation a similar growth reduction should also be expected in south-western Sweden due to a similar deposition load as in southern Norway. However, no such reduction in forest production has been reported either in Norway or in Sweden. Rather, the forest production in Europe has increased in the period 1971 to 1990 (Kauppi *et al.*, 1992).

In the beginning of the 1980s, extensive forest damage, claimed to be caused by acid rain, was reported from Germany (Ulrich *et al.*, 1980; Wentzel, 1982). This stimulated new interest, and funding, in studying the relationship between forest production and vitality and pollution load. Except for direct effects close to

*Experimental Reversal of Acid Rain Effects: The Gårdsjön Roof Project.*
Edited by Hans Hultberg and Richard Skeffington. 

pollution sources, however, clear regional links between pollution load and forest vitality were not established.

Correlation analyses between emissions of S and N in 1988 in the different European countries, and the corresponding percentage of trees with more than 25% defoliation revealed no significant correlation (Abrahamsen, 1990). For south Sweden no correlations were found between forest growth and acidity, aluminium and base cation concentrations, and the ratio between base cations and aluminium of the soil solution (Örlander *et al.*, 1994). However, in experiments with artificial rain, a rain acidity of pH 3 and below increased the base cation leaching significantly and reduced the forest growth (Abrahamsen *et al.*, 1994).

The Gårdsjön roof project is one among other roof projects with the aim of studying the effects of reduced pollution load on runoff, soils and vegetation. Here we report on forest vitality, forest growth, needle nutrient content, litterfall and ground vegetation.

## METHODS

### Forest vitality

As a measure of forest vitality both crown density and discoloration have been recorded since 1990 for the same 100 trees each in G1 (ROOF) and G2 (NITREX), and the same 10 trees in F1 (CONTROL). The methods used are those recommended by the International Co-ordinated Programme on Assessment and Monitoring of Air Pollution Effects on Forests (ICP-Forests) (Aamlid *et al.*, 1990, 1992).

### Forest growth

Tree height and stem diameter at breast height (dbh) of all trees in G1 (ROOF) and G2 (NITREX) and of six selected trees in F1 (CONTROL) were measured in 1990 and the diameter was re-measured in the autumn of 1994. Based on these measurements, total standing biomass for 1990 was calculated using the Marklund biomass functions relevant for Swedish forests (Marklund, 1998). The standing biomass data are given in Chapter 2.

### Needle nutrient content

Six spruce trees from each of the three major soil moisture regimes and ground vegetation subtypes in G1 (ROOF) were sampled together with six from one vegetation type (*Vaccinium*) in the more intensively studied area of F1 (CONTROL). Foliage was collected once a year from the 7th whorl from the top of the tree. When it was impossible to collect foliage in the autumn due to the weather conditions the sampling was done early next spring. The needles from each branch were sorted by age (current year, previous year, and remainder) and analysed for total elements by a simultaneous ICP technique (Ogner *et al.*, 1991).

The first needle samples were taken in the spring of 1990 followed by new samplings in the autumn of 1991, spring of 1993, autumn of 1993 and autumn of 1994.

### Litterfall

Prior to the construction of the roof the litterfall in G1 (ROOF) was collected using litterfall traps consisting of metal rings fitted with nylon bags. The traps were systematically placed along lines covering the catchment.

### Ground vegetation

Vegetation analysis has been carried out on 10 permanent 1 $m^2$ plots spanning the natural moisture gradient. Percentage cover was recorded for all species in the plots. Each sample plot was divided into 25 subplots, and within these 250 subplots presence or absence of all species was recorded. The data were analysed by the technique of detrended correspondence analysis (DCA) (Hill and Gaush, 1980; ter Braak, 1987).

As a basis for the vegetation maps (Chapter 2) percentage cover was recorded for all species in a 5×5 m grid in the G1 (ROOF) and G2 (NITREX) catchments. This recording has not been repeated.

### Statistical analysis

The roof experiment in the G1 (ROOF) catchment and also the nitrogen addition experiment in the G2 (NITREX) catchment are conducted on a catchment level without any replication of treated or control sites. In spite of large heterogeneity in the catchments, strict statistical criteria of spatial independence are not met. This limits the conclusions drawn from the results. For the three plots with different vegetation subtypes within the catchments and for the F1 (CONTROL), a two-way ANOVA analysis has been used to test differences between years and plots, and the interaction between year and plot as indicative of treatment effects. In addition, a non-parametric Mann–Whitney U-test has been used to compare medians of results from years and plots in order to omit the effect of possible outliers. The data on crown density was arc sin transformed prior to statistical analysis. For further description see Kjønaas *et al.* (1997).

## RESULTS AND DISCUSSION

### Forest vitality

There has almost been a linear decrease in crown density in all three catchments from 1990 to 1993 with no change from 1993 to 1994 (Figure 17.1). The crown densities in G1 (ROOF) and G2 (NITREX) are similar and have decreased from slightly to moderately defoliated to clearly moderately defoliated. The recorded trees in the F1 (CONTROL) catchment have been in the slightly defoliated group except for the two latest years. Since all three catchments have reacted in parallel there are no treatment effects on crown density. The decrease in crown density observed between 1990 and 1994 was significant ($p<0.01$) for G1 (ROOF), G2 (NITREX) and F1 (CONTROL).

Large differences in crown discoloration between the catchments were not recorded until 1994 (Figure 17.2). Both 1992 and 1994 were years with extremely dry

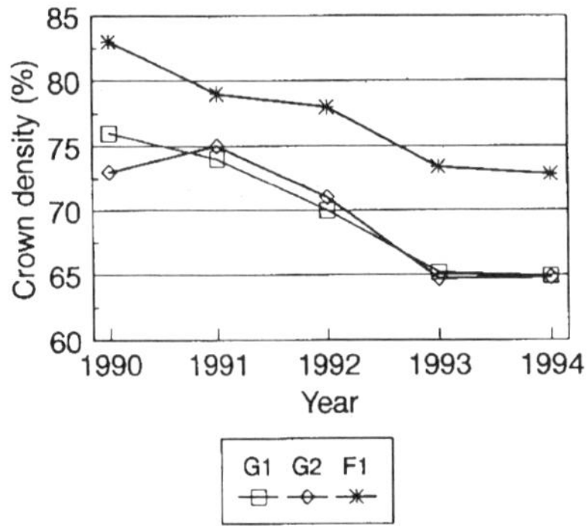


**Figure 17.1.** Mean crown density in the G1 (ROOF), G2 (NITREX) and F1 (CONTROL) catchments. Data are produced by the Norwegian Institute for Land Inventory

summers. In 1994 the discoloration increased in both G2 (NITREX) and F1 (CONTROL) catchments while almost no change was recorded in G1 (ROOF). This may, at least partly, be explained by a delayed watering of the G1 catchment. During the winter a snowpack accumulated in the G2 (NITREX) and F1 (CONTROL) catchments and at the same time it was too cold to spray water under the roof. To simulate snowmelt and keep an annual water balance similar to the other catchments, the water distribution curve for G1 (ROOF) was slightly different with more added water in late spring. This might have given less water stress for the trees in G1 and thereby less discoloration.

## Forest growth

The radial growth during the 4 years between the measurements has been low and of the same extent in all catchments (Table 17.1). This slow growth is of course due to

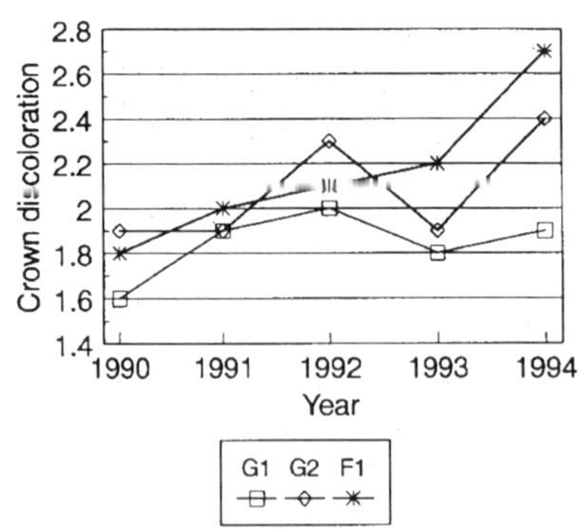


**Figure 17.2.** Mean crown discoloration in the G1 (ROOF), G2 (NITREX) and F1 (CONTROL) catchments. 1 = 0–10% of the tree crown is discoloured, 2 = some needles are clearly discoloured but less than 10%, 3 = 11–25% of the tree crown is discoloured. Data are produced by the Norwegian Institute for Land Inventory

**Table 17.1.** Mean and range of diameter at breast height for spruce and pine in G1 (ROOF) and G2 (NITREX) and for spruce in F1 (CONTROL) in 1990 and 1994

| Year | | G1 (ROOF) | G2 (NITREX) | F1 (CONTROL) |
|---|---|---|---|---|
| **Spruce** | | | | |
| 1990 | Mean (cm) | 24.7 | 22.6 | 31.6 |
| | Range (cm) | 8.4–51.1 | 7.7–48.7 | 26.8–35.0 |
| 1994 | Mean (cm) | 25.4 | 23.1 | 32.6 |
| | Range (cm) | 9.1–52.5 | 7.9–49.6 | 27.5–36.0 |
| **Pine** | | | | |
| 1990 | Mean (cm) | 25.4 | 26.2 | |
| | Range (cm) | 11.9–39.7 | 11.3–34.9 | |
| 1994 | Mean (cm) | 26.2 | 27.1 | |
| | Range (cm) | 11.9–42.7 | 11.6–38.1 | |

the age and vitality of the stands. The differences in growth reflect the site index, being lowest in G1 (ROOF) and G2 (NITREX) and highest in F1 (CONTROL). The differences in stand age and site index make it difficult to draw any conclusions on the treatment effect, especially since we have only two measurements of radial growth. Treatment effects have also been difficult to detect from the "clean rain" roof plot experiments in Norway spruce at Klosterhede, Denmark, and Solling, Germany (Beier *et al.*, 1997; Bredemeier *et al.*, 1997). At these sites the stands were 74 and 60 years old respectively. Bredemeier *et al.* (1997) conclude that spruce forests exhibit a time-lagged response to a changing chemical environment, a conclusion that can also be supported by acidification experiments (Abrahamsen *et al.*, 1994).

## Needle nutrient content

The nutrient content has generally decreased from spring 1990 to spring 1993 with elevated content in autumn 1993 and a drop in 1994 both in G1 (ROOF) and F1 (CONTROL) (Table 17.2 and Figure 17.3). There was a significant difference in N concentration between different year classes, where the current year and previous year needles had significantly larger N concentration than the older needles (not shown). The year to year variation of N concentrations was large, exemplified by a significant increase ($p<0.05$) of N concentration in needles of all age classes at all plots in 1993 followed by a similar significant decrease in previous year and older needles in 1994. The ratios of the other nutrients to N showed similar year to year variation. So far there are no changes that can be directly related to the treatment. The needle nutrient content seems to be more correlated with the vitality of the forest. The variation in needle nitrogen (Figure 17.3) shows the same pattern as the variation in crown discoloration (Figure 17.2).

The needle nutrient content in the roof experiments at Klosterhede and Solling varied within the experimental period, but the variation was parallel in the "clean rain" and control with no significant differences caused by treatment (Beier *et al.*, 1997; Bredemeier *et al.*, 1997). Results from acidification experiments have shown that improvement of the nutrient status of the trees can take a long time after the stress factor is removed (Abrahamsen *et al.*, 1994).

**Table 17.2.** Needle nutrient contents 1990–94. Foliage was collected from the 7th whorl from the top of the tree. Figures are means for all sampled trees in the G1 (ROOF) and F1 (CONTROL) catchments

| | | N | S | P | K | Mg | Ca | Zn |
|---|---|---|---|---|---|---|---|---|
| | | <- - - - - - - - - - - | | (mmol/kg) | | - - - - - - - - - - -> | | ($\mu$mol/kg) |
| G1 (ROOF) | | | | | | | | |
| Current year | Spring 1990 | 1023 | 32 | 47 | 175 | 43 | 110 | 662 |
| | Autumn 1991 | 907 | 32 | 48 | 199 | 41 | 53 | 499 |
| | Spring 1993 | 770 | 25 | 40 | 150 | 42 | 57 | 408 |
| | Autumn 1993 | 860 | 28 | 43 | 180 | 36 | 77 | 423 |
| | Autumn 1994 | 790 | 25 | 44 | 203 | 38 | 46 | 384 |
| Previous year | Spring 1990 | 963 | 32 | 47 | 153 | 46 | 114 | 613 |
| | Autumn 1991 | 781 | 28 | 33 | 135 | 35 | 93 | 463 |
| | Spring 1993 | 771 | 27 | 35 | 134 | 36 | 80 | 413 |
| | Autumn 1993 | 884 | 29 | 41 | 139 | 33 | 95 | 399 |
| | Autumn 1994 | 696 | 25 | 35 | 149 | 30 | 90 | 378 |
| F1 (CONTROL) | | | | | | | | |
| Current year | Spring 1990 | 947 | 33 | 44 | 163 | 44 | 123 | 685 |
| | Autumn 1991 | 972 | 35 | 47 | 199 | 50 | 66 | 746 |
| | Spring 1993 | 804 | 26 | 38 | 137 | 47 | 82 | 524 |
| | Autumn 1993 | 925 | 30 | 43 | 172 | 38 | 101 | 582 |
| | Autumn 1994 | 832 | 26 | 41 | 183 | 46 | 71 | 519 |
| Previous year | Spring 1990 | 960 | 36 | 46 | 132 | 58 | 140 | 643 |
| | Autumn 1991 | 899 | 32 | 34 | 144 | 45 | 124 | 672 |
| | Spring 1993 | 829 | 28 | 35 | 136 | 39 | 100 | 450 |
| | Autumn 1993 | 993 | 33 | 42 | 132 | 40 | 134 | 493 |
| | Autumn 1994 | 741 | 27 | 31 | 140 | 39 | 139 | 536 |

## Litterfall

The roof prevents the litter falling directly on the forest floor. Based on litter collection in a period before the construction of the roof the litterfall in G1 (ROOF) was calculated to approximately 4 t/ha.yr. This total litterfall can be divided into

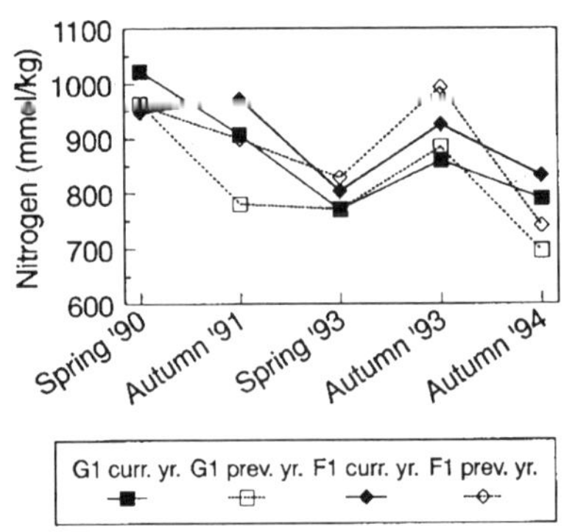


**Figure 17.3.** Needle nitrogen in current and previous year needles for G1 (ROOF) and F1 (CONTROL)

1350 kg needles/ha.yr and 2650 kg of twigs and cones/ha.yr. The total litterfall contributes 46 kg N, 4 kg S, 9 kg K, 3 kg Mg and 13 kg Ca ha.yr.

The roof has acted as a huge littertrap and the litterfall has been collected and redistributed beneath the roof. There are some element losses from the litter before it is removed from the roof, especially of K, Mg and Ca. During the short-term storage in piles before redistribution some decomposition of the litter occurs. It is also very difficult to distribute the litter evenly without extensive use of labour. The litter redistribution seems not to have introduced problems for the vegetation parameters except for the ground vegetation.

### Ground vegetation

The number of species and the percentage cover have not shown any systematic changes influenced by the roof (data not shown). The most pronounced effect is the indirect effect of litter being brought back and hand spread beneath the roof. This has caused a disturbance, especially to some of the bryophytes. Changes in vegetation structure cannot be expected in an experimental period of 4 years.

## CONCLUSION

The most pronounced changes measured during the experimental period were connected to forest vitality and needle nutrient content. A small change in crown discoloration from 1993 to 1994 in the roof catchment compared with the control may be due to delayed watering in the roof catchment in combination with a dry summer. However, none of the measured changes could be linked to the effect of the roof and cleaner precipitation.

## REFERENCES

Aamlid, D., Venn, K. and Stuanes, A. O. (1990) Forest decline in Norway; monitoring results, international links and hypotheses. *Norwegian Journal of Agricultural Sciences Suppl.*, **4**, 27 pp.

Aamlid, D., Venn, K. and Stevens, A. O. (1992) Regional evaluations of acid deposition effects on forests: Europe. In: D. W. Johnson and S. E. Lindberg (eds), *Atmospheric Deposition and Forest Nutrient Cycling*. Ecological Studies 91, Springer-Verlag, New York, pp. 543–570.

Abrahamsen, G. (1990) Hypotheses concerning the "new forest decline". *Report of XIX World Congress of the International Union of Forest Research Organizations*, vol. B, Montreal, Canada, pp. 151–163.

Abrahamsen, G., Stuanes, A. O. and Tveite, B. (eds) (1994) *Long-Term Experiments with Acid Rain in Norwegian Forest Ecosystem.* Ecological Studies 104, Springer-Verlag, New York, 342 pp.

Beier, C., Blanck, K., Bredemeier, M., Lamersdorf, N., Rasmussen, L. and Xu, Y.-J. (1997) Field scale "Clean rain" treatments to two Norway spruce stands within the EXMAN project—effects on soil solution chemistry, foliar nutrition and tree growth. *Forest Ecology and Management* (in press).

Bredemeier, M., Blanck, K., Dohrenbusch, A., Lamersdorf, N., Meyer, A. C., Murach, D., Parth, A. and Xu, Y.-J. (1997) The Solling roof project — site characteristics, experiments and results. *Forest Ecology and Management* (in press).

Dahl, E. and Skre, O. (1971) En undersøkelse over virkningen av sur nedbør på produktiviteten i landbruket. [An investigation of the effect of acid precipitation on land productivity.] In: *Konferens om avsvavling, Publ. 1971 (1)*. Nordforsk, Miljøvårdssekretariatet, Helsingfors, pp. 27–40.

Hill, M. O. and Gaush, H. G. (1980) Detrended correspondence analysis: an improved ordination technique. *Vegetation*, **42**, 47–58.

Kauppi, P. E., Mielikäinen, K. and Kuusela, K. (1992) Biomass and carbon budget of European forests, 1971 to 1990. *Science*, **256**, 70–74.

Kjønaas, O. J., Stuanes, A. O. and Huse, M. (1997) Effects of weekly nitrogen additions on N cycling in a coniferous forest catchment, Gårdsjön, Sweden. *Forest Ecology and Management* (in press).

Marklund, L. G. (1988). *Biomass Functions for Pine, Spruce and Birch in Sweden*. Sveriges Lantbruksuniversitet, Institut för Skogstaxering, **45**, 73 pp.

Ogner, G., Opem, M., Remedios, G., Sjøtveit, G. and Sørlie, B. (1991). *The Chemical Analysis Program of The Norwegian Forest Research Institute, 1991*. Norwegian Forest Research Institute, Ås, Norway.

Örlander, G., Westling, O. and Petersson, P. (1994) *Markvattnets innehåll av baskationer och aluminium och dess påverkan på tillväxt och kådflöde i kraftigt försurad granskog*. [Base cation and aluminium content of the soil solution and the influence on tree growth and resin discharge in spruce forest.] IVL Report B 1155, 18 pp.

ter Braak, C. J. F. (1987) Ordination. In: R. H. G. Jongman, C. J. F. ter Braak and O. F. R. van Tongeren (eds), *Data Analysis in Community and Landscape Ecology*. Pudoc, Wageningen, pp. 91–173.

Ulrich, B., Mayer, R. and Khanna, P. K. (1980) Chemical changes due to acid precipitation in a loess-derived soil in Central Europe. *Soil Science*, **130**, 193–199.

Wentzel, K. F. (1982) Immissionen oder Saurer Regen — wovon sterben Wälder und Seen? *Forst Holz*, **37**, 410–413.

# 18 Fish Toxicity Studies to Evaluate Water Quality Changes in the Gårdsjön Experimental Catchments: Six Experiments Performed During the Initial Phase of Treatment.

**ULF CARLSSON & FRODE KROGLUND**

During the initial period of treatment, six bioassays with brown trout *Salmo trutta* L. were performed on runoff from the G1 ROOF, G2 NITREX and F1 CONTROL catchments. The water from all catchments was rapidly lethal to the fish as expected, but the aim of these initial tests was to establish a baseline and quantify variations in mortality rates due to differences in test conditions. The conditions that mostly affected fish survival were variations in water temperature and size of the fish used for the tests: a higher temperature caused higher mortality rates, and smaller fish, in general, survived longer in the acid, aluminium-rich runoff water from the roofed catchment. The difference in survival between fish of different size was not pronounced in the strongly acid water from the nitrogen-enriched catchment. Thus, future bioassays will have to be carefully controlled for variations in temperature and size of fish.

## INTRODUCTION

In the Gårdsjön area, long-term acidification has brought about extensive damage to fish and other biota (Henriksson and Oscarsson, 1985). One of the notable characteristics of an acidified podzolized soil is the increased leaching of inorganic aluminium and the depletion of base cations such as calcium and magnesium. Runoff from forested catchments is acid with high aluminium concentrations and the response of runoff to different treatments of the catchments at Gårdsjön has been experimentally tested (Hultberg and Grennfelt, 1992; Hultberg *et al.*, 1994).In order to test acidification reversibility one small catchment (G1 ROOF) has been roofed (Chapter 1, this volume) and in another (G2 NITREX) a nitrogen loading experiment has been performed (Hultberg *et al.*, 1994). The runoff waters from the experimental catchments are very acid (pH 3.9–4.5), and have low Ca concentrations (25–100 $\mu$eq/l) and high aluminium (Al) concentrations (56–333 $\mu$eq/l). Al occurs

*Experimental Reversal of Acid Rain Effects: The Gårdsjön Roof Project.*
Edited by Hans Hultberg and Richard Skeffington. 

predominantly in inorganic forms. Inorganic Al is toxic to fish (Driscoll *et al.*, 1980; Muniz and Leivestad, 1980) while Ca has a mitigating effect. Even relatively small differences in Ca concentration of fresh water can be of importance (Brown, 1983; Playle *et al.*, 1989). Fish toxicity experiments were performed in a similar experiment, the RAIN Project, where fish survival in runoff from a roofed catchment was tested (Kroglund and Rosseland, 1992). The results showed that after 7 years of clean precipitation, when sulphate concentrations had decreased by approximately 70% and mean pH had increased by about 0.1, mortality was significantly higher. This was probably due to the reduced concentration of Ca (15 $\mu$eq/l).

In our experimental design, it was hypothesized that acidification reversal in the roofed catchment (G1) would generate improved water quality, thus resulting in decreased fish toxicity as reported from lakes in the Sudbury area (Gunn and Keller, 1990). An alternative hypothesis was that toxicity to freshwater biota would initially increase due to reduced Ca concentrations and increasing Al-toxicity with rising pH. This hypothesis was supported by the results from the RAIN Project (Kroglund and Rosseland, 1992). Reaching the point of nitrogen saturation aimed at in catchment G2 NITREX would increase the acidity in soil and runoff water, and thus fish toxicity should increase there also.

We report here results from an initial phase of the experiments when the catchments G1 and G2 had been treated with artificial clean rain and extra nitrogen for approximately one year. At that time, no significant changes in concentrations of $H^+$, base cations or Al in runoff had been recorded, in spite of a substantial decrease of the sulphate concentrations.

## METHODS

For the toxicity tests, runoff from the G1 ROOF catchment and from the G2 NITREX catchment was collected in two waterflux tanks, and led to storage tanks from which water was pumped to two series of five round, plastic (PVC) barrels (80 l). From the barrels, water runs passively back to the storage tanks (Figure 18.1). The water in the test system is thereby recirculated, but partly exchanged every time the waterflux tanks from the catchments (720 l) empty. The emptying frequency of the waterflux tanks is flow-dependent (1–30 times per day during the toxicity tests). At the reference area (F1), one barrel is installed. Runoff from catchment F1 is also acid with high Al concentrations and low Ca concentrations, as in the G1 and G2 catchments. Water from the limed lake Gårdsjön serves as neutral control. The design is similar to that used for testing of runoff from the RAIN catchment at Risdalsheia, Norway (Kroglund and Rosseland, 1992).

During the experiments, which took place during the period January 1991 to April 1992, water samples were taken daily from all treatments. These samples were analysed for: pH, Ca, total-Al (TAl), total inorganic Al (TiAl), monomeric Al (MAl) and monomeric inorganic Al (MiAl) according to Barnes (1975) and Driscoll (1984).

Brown trout *Salmo trutta* L. of a local strain from Lake Mjörn, which is a circumneutral mesotrophic lake, were used as a test species. This strain has been reared for 25 years at the Anten fish farm, 50 km west of Gothenburg. Fish size varied between 3.0 and 21.5 cm in length, and their age was 0+ to 2.0 years (i.e. those in their first year of life to those just past their second winter).

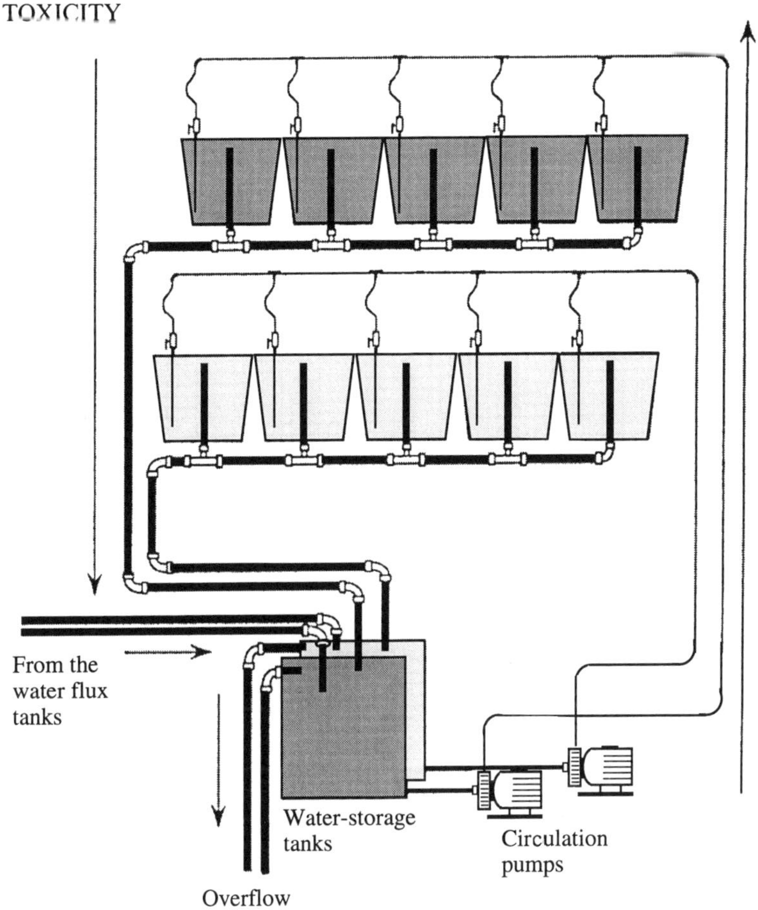


**Figure 18.1.** Schematic diagram of the fish toxicity test setup at Gårdsjön

During travel from the fish farm, the fish were placed in oxygenated plastic bags at approximately ambient temperature. The total time from the fish farm plus handling-time until the start of an experiment was about 5 hours. For the tests, about 20 fish (18–22) were used in each barrel.

Mortality was monitored several times per day. All dead fish, and fish surviving the experimental period, were recorded and length measured. The experiments were terminated when none, or only a few, fish remained alive. At this moment, the time to 50% mortality (LT-50) was well exceeded. Six experiments were performed from January 1991 to April 1992.

## RESULTS

Water chemistry for each of the different tests performed during the initial phase is presented in Table 18.1. The water quality in runoff differs substantially between the three catchments G1 ROOF, G2 NITREX and F1 CONTROL.

**Table 18.1.** Water chemistry values (averages) from fish toxicity tests at the Gårdsjön catchments. TAl = total Al; TiAl = total inorganic Al; MAl = monomeric Al; MiAl = monomeric inorganic Al. Missing values have not been measured

| Catchment | Exp. no. | Temp. | pH | Ca ($\mu$eq/l) | TAl ($\mu$eq/l) | TiAl ($\mu$eq/l) | MAl ($\mu$eq/l) | MiAl ($\mu$eq/l) |
|---|---|---|---|---|---|---|---|---|
| F1 | 1 | | | | | | | |
| F1 | 2 | | | | | | | |
| F1 | 3 | 9.0 | 4.2 | 53.5 | 101.89 | 63.56 | | |
| F1 | 4 | 6.5 | 4.0 | 56.0 | 102.67 | 105.78 | 79.22 | |
| F1 | 5 | 4.8 | 4.1 | 70.0 | 141.11 | 96.67 | 40.00 | 30.00 |
| F1 | 6 | 3.5 | 4.1 | 88.0 | 91.11 | 47.78 | 88.66 | 46.00 |
| G1 | 1 | 3.5 | 4.3 | 65.0 | 242.00 | 209.56 | 21.67 | 186.00 |
| G1 | 2 | 6.0 | 4.5 | 66.0 | 223.33 | 190.33 | | |
| G1 | 3 | 13.8 | 4.5 | 80.5 | 228.67 | 198.00 | | |
| G1 | 4 | 7.8 | 4.3 | 68.5 | 289.56 | 259.33 | 271.00 | |
| G1 | 5 | 6.8 | 4.3 | 65.5 | 318.89 | 265.56 | 156.67 | 137.78 |
| G1 | 6 | 6.8 | 4.4 | 70.5 | 196.67 | 166.67 | 144.44 | 124.44 |
| G2 | 1 | 5.0 | 3.9 | 38.0 | 189.33 | 129.78 | 152.00 | 91.67 |
| G2 | 2 | 8.0 | 3.9 | 32.5 | 176.33 | 126.44 | | |
| G2 | 3 | 11.5 | 3.9 | 23.0 | 206.33 | 160.56 | | |
| G2 | 4 | 7.5 | 3.8 | 23.0 | 180.11 | 105.78 | 78.44 | |
| G2 | 5 | 6.5 | 3.9 | 36.5 | 262,2222 | 212.22 | 111.11 | 94.44 |
| G2 | 6 | 5.6 | 3.9 | 32.0 | 161.11 | 96.67 | 121.11 | 57.78 |

G2 had the most acidic runoff water, with mean pH 3.88 during the six experiments, and had the lowest concentrations of calcium with mean concentrations 31 $\mu$eq/l but intermediate, though still high, concentrations of total inorganic aluminium (TiAl; mean 1.25 mg/l). G1 had the highest pH in all six experiments and the highest Ca (mean 154 $\mu$eq/l) and TiAl values (mean 1.93 mg/l). The reference catchment, F1, was intermediate with respect to pH (mean 4.11) and Ca (mean 67 $\mu$eq/l) levels and had the lowest TiAl concentration (mean 0.71 mg/l) of the three catchments tested. Water from the limed Lake Gårdsjön with the following composition was used as neutral control (pH 6.8, TiAl 0.02–0.05 mg/l, Ca 250–300 $\mu$eq/l). Six experiments were performed with water from G1 and G2, and four in runoff water from catchment F1. The mortality rate is expressed as time to 50% mortality; LT-50. Mortality and mean length of the fish tested are presented in Table 18.2.

No mortality occurred in the neutral control during a timespan of maximum 100 hours for the tests. There were considerable variations in LT-50 values in runoff from the three experimental catchments: from 7 hours as the highest mortality rate, to 51 hours as the lowest. In general, runoff water from G2 was most toxic to fish, with the exception of experiment number 3 when G1 had the highest mortality rate. Coincidentally, during this experimental run, fish in runoff water from F1 responded with a substantially decreased mortality rate. This may be because, on this occasion, F1 water was several degrees colder than water from G1 and G2 (Table 18.1).

**Table 18.2.** LT-50 values and mean length of fish used in toxicity tests at catchments F1, G1 and G2

| Catchment | Exp. no. | LT-50 (hours) | Length (cm) |
|---|---|---|---|
| F1 | 1 | | |
| F1 | 2 | | |
| F1 | 3 | 42 | 14.0 |
| F1 | 4 | 14 | 16.4 |
| F1 | 5 | 34 | 17.0 |
| F1 | 6 | 34 | 17.9 |
| G1 | 1 | 51 | 10.7 |
| G1 | 2 | 35 | 10.3 |
| G1 | 3 | 7 | 14.0 |
| G1 | 4 | 20 | 15.8 |
| G1 | 5 | 45 | 17.7 |
| G1 | 6 | 24 | 19.0 |
| G2 | 1 | 34 | 10.6 |
| G2 | 2 | 31 | 11.0 |
| G2 | 3 | 11 | 14.0 |
| G2 | 4 | 14 | 16.0 |
| G2 | 5 | 25 | 17.4 |
| G2 | 6 | 23 | 19.3 |

The only variable that was significantly correlated to the LT-50 values was temperature. The correlation was negative ($r = -0.63$, $p = 0.009$). Both Ca and TiAl showed a significant positive correlation with pH ($r = 0.82$, $p = 0.0001$; and $r = 0.54$, $p = 0.046$ respectively). The covariance between pH, Ca and Al makes it difficult to interpret how the individual chemical parameters affect fish survival. Mortality data from the three experimental catchments are presented in Figures 18.2a–f, 3a–f and 4a–d.

In experiments 5 and 6, two different size-classes of brown trout, 0-year fish (0+) mean length 3.7 cm (exp. 5) and 5.0 cm (exp. 6); 1-year fish (1+), mean length 17.6 cm and 19.1 cm respectively, were tested in runoff water from G1 and G2 (Table 18.3). In both experiments with G1 runoff water, the larger fish had the highest mortality rate. In G2 mortality of 0+ and 1+ was almost identical (Figures 18.5a–d). The results indicate a size-dependent component in G1, which was not evident in G2.

## DISCUSSION

The fish toxicity tests at Gårdsjön were performed to measure water quality changes during the run of experiments in the covered catchment (G1) and the nitrogen-enriched catchment (G2). Expected changes included an enhanced fish survival in water from G1, though during an initial period toxicity may be even worse, and an increased mortality rate for fish tested in G2 water as runoff became increasingly

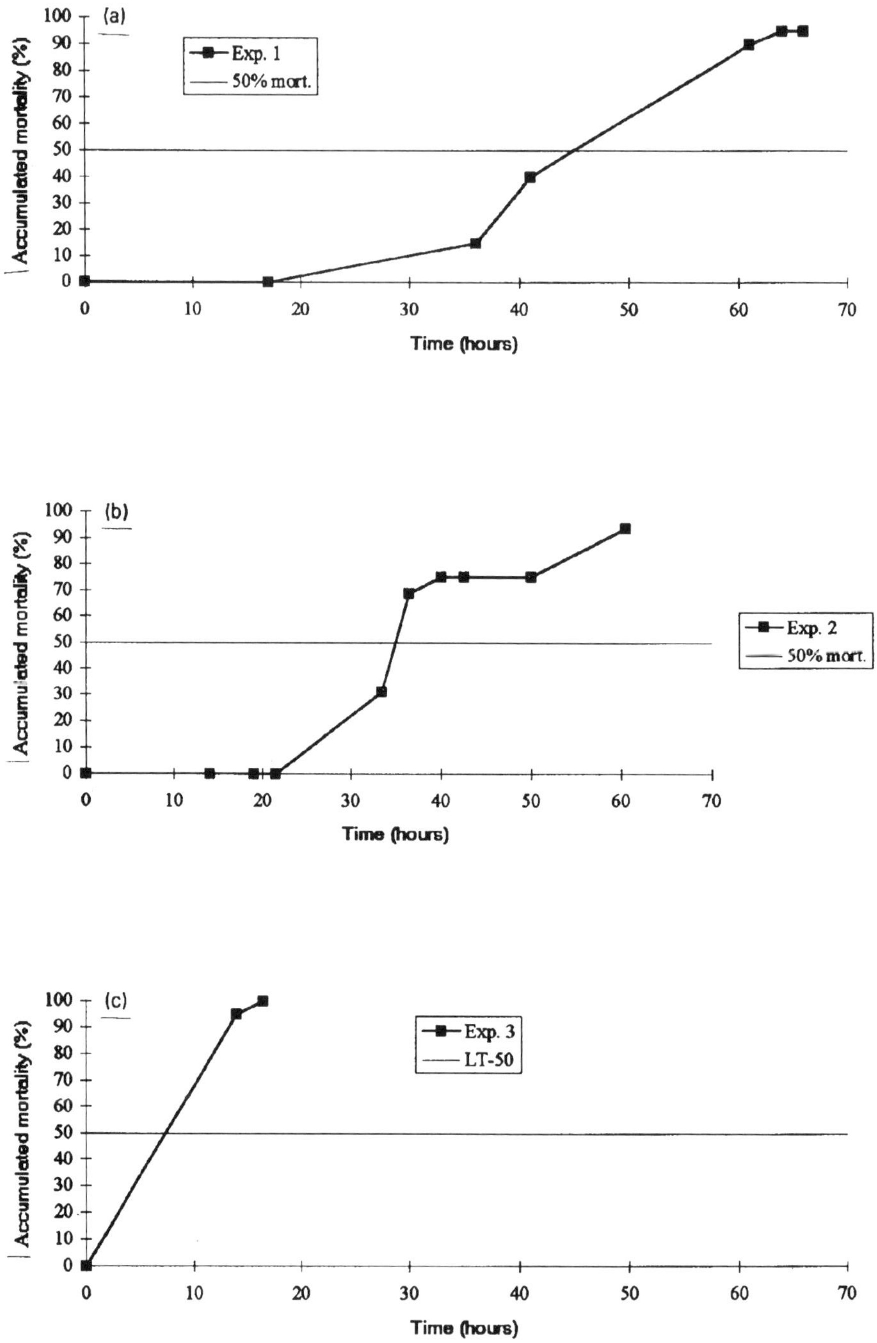


**Figure 18.2.** Mortality rates, as time to 50% mortality (LT-50), in experiments in runoff from the roofed catchment (G1)

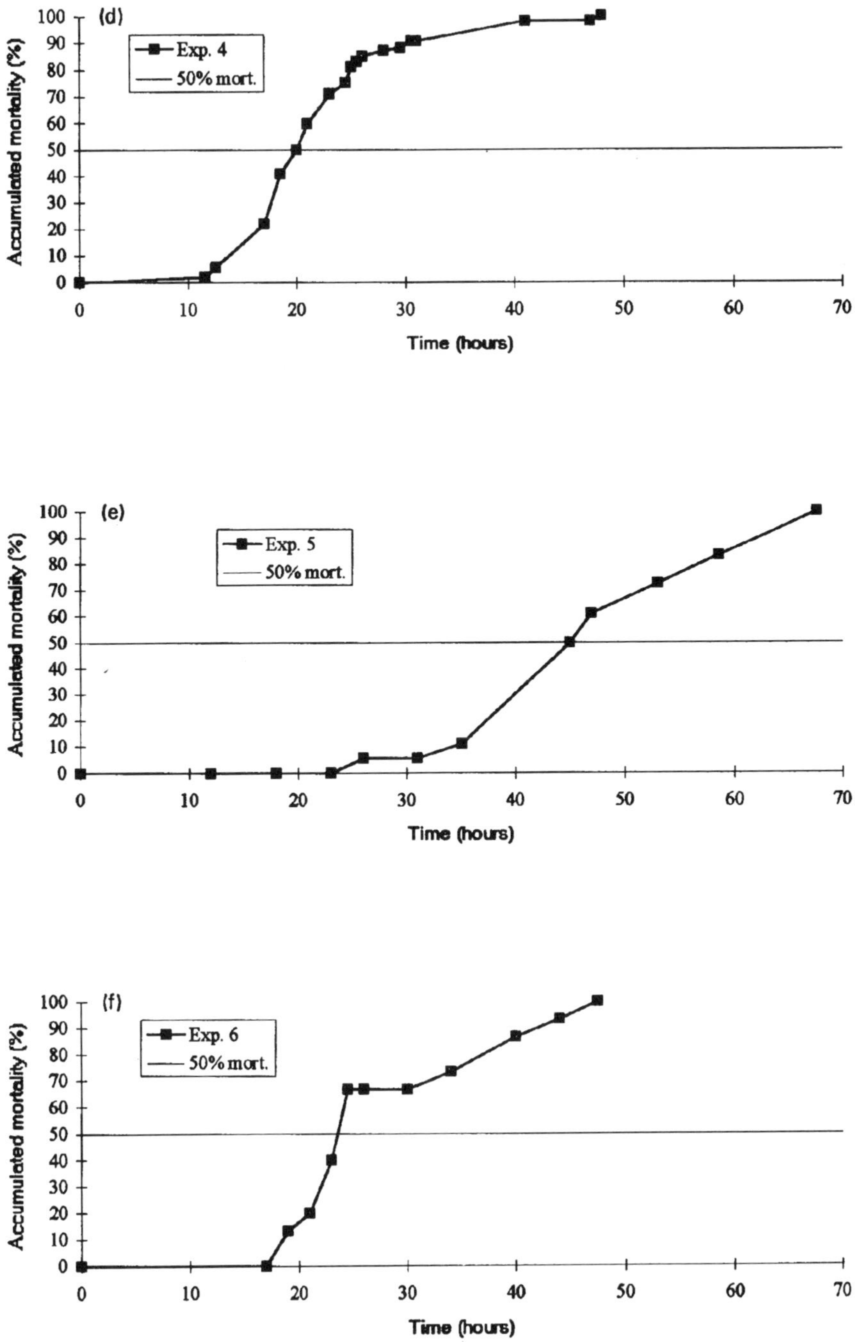


**Figure 18.2.** *continued*

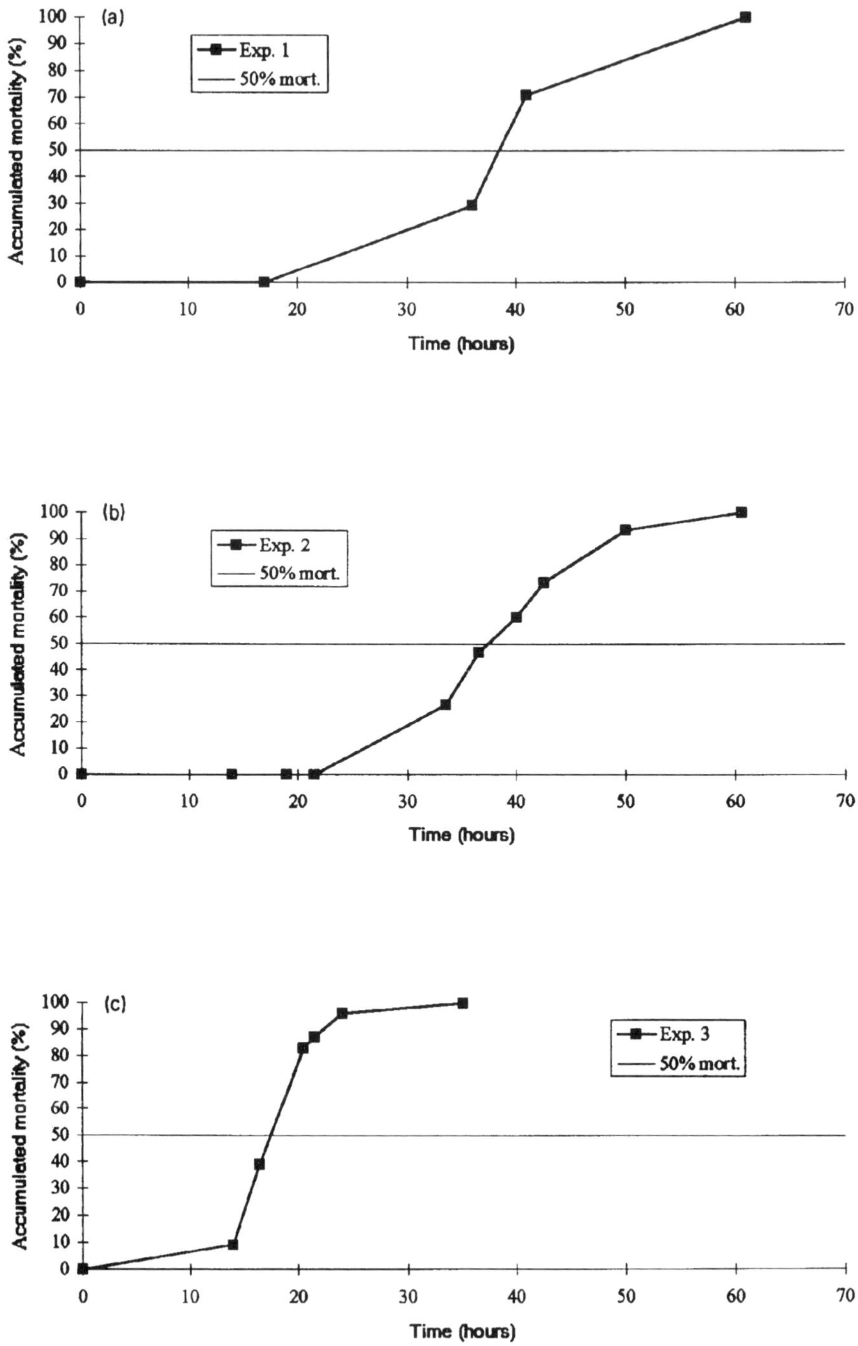


**Figure 18.3.** Mortality rates, as time to 50% mortality (LT-50), in experiments in runoff from the nitrogen-enriched catchment (G2)

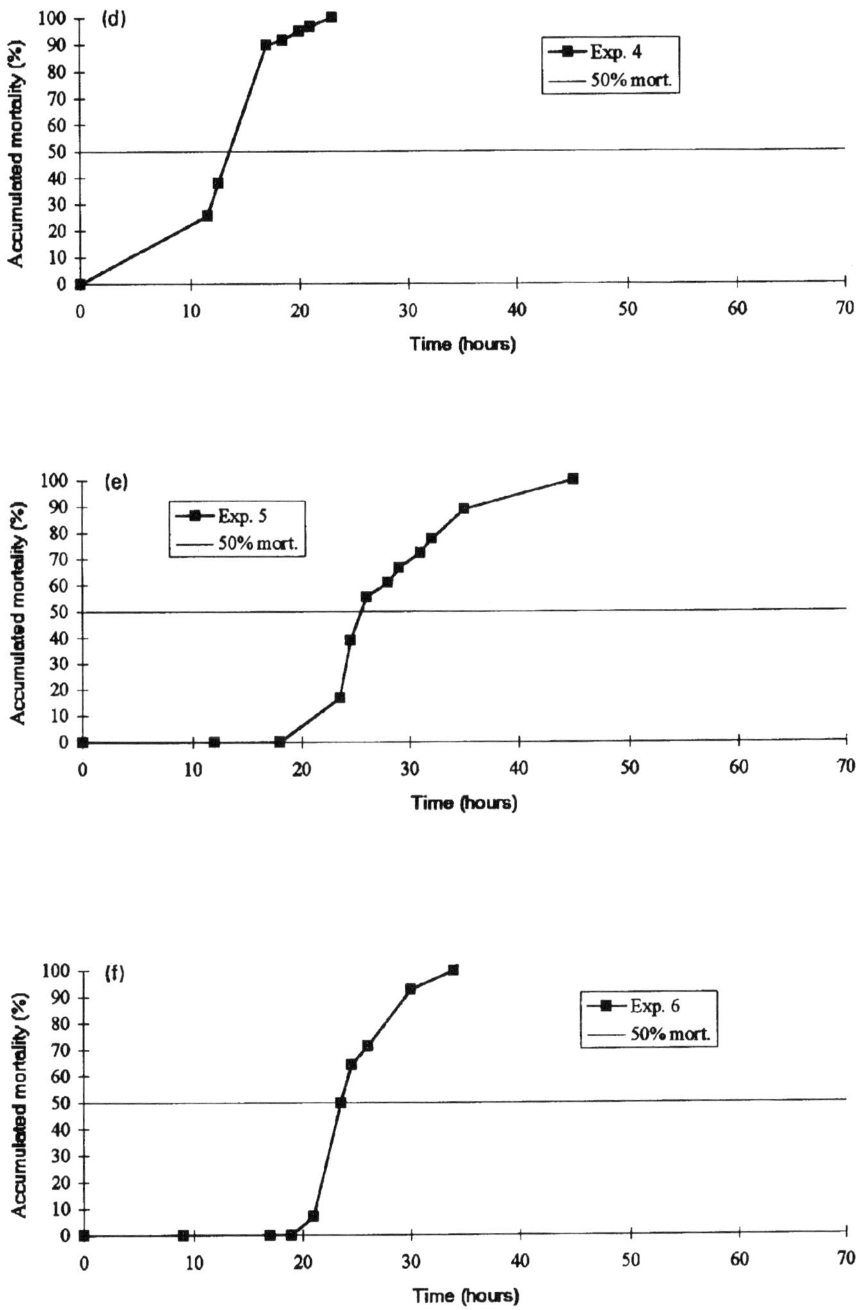


**Figure 18.3.** *continued*

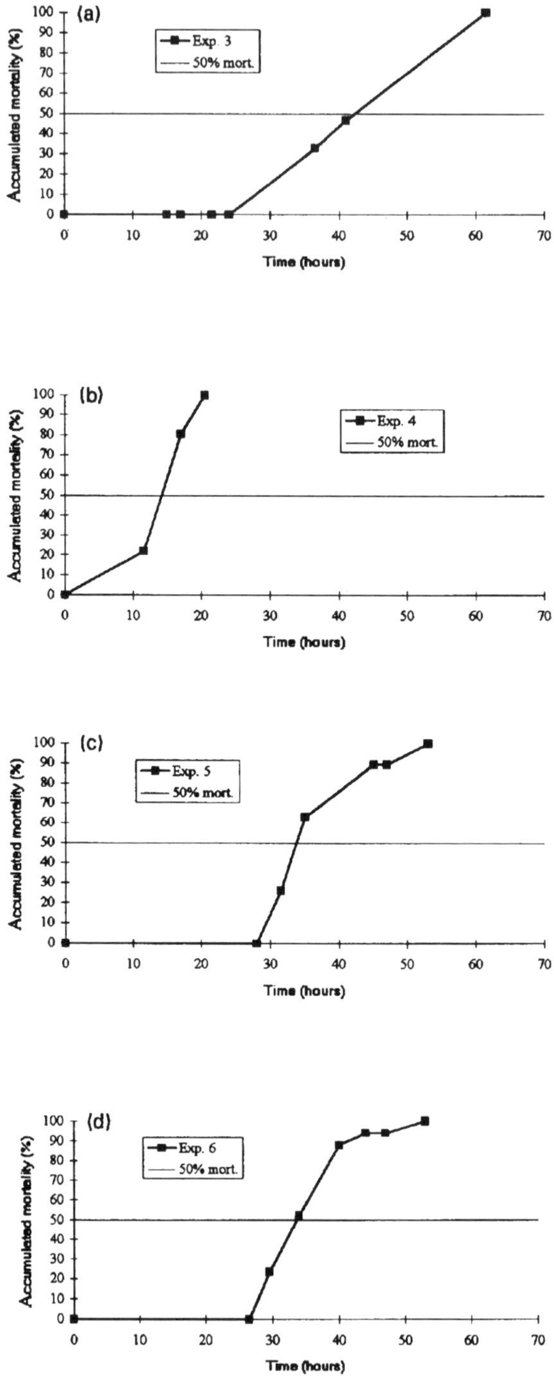


**Figure 18.4.** Mortality rates, as time to 50% mortality (LT-50), in experiments 3, 4, 5 and 6 in runoff from the acid reference catchment (F1).

**Table 18.3.** LT-50 values and mean length of fish used in size-dependent toxicity tests at catchments G1 and G2

| Catchment | Exp. no. | LT-50 (hours) | Length (cm) |
|---|---|---|---|
| G1 | 5 | 45 | 17.7 |
| G1 | 5 | 61 | 3.8 |
| G2 | 5 | 25 | 17.4 |
| G2 | 5 | 41 | 3.7 |
| G1 | 6 | 24 | 19.0 |
| G1 | 6 | 31 | 5.0 |
| G2 | 6 | 23 | 19.3 |
| G2 | 6 | 31 | 5.0 |

acidic. Fish toxicity tests have, so far, only been performed during the initial period of clean water treatment and extra nitrogen addition. At the time of the last experiments (5 and 6), G1 and G2 had been treated with artificial "clean rain" and extra nitrogen for 10 and 11 months respectively. The water chemistry changes show a decreased sulphate concentration (about 20%), no changes in acidity (pH) and only minor changes in base cation concentrations (Chapter 7, this volume). The changes were probably insignificant in relation to fish toxicity. Results from the fish tests support this assumption and the general impression is that water quality changes, during this relatively short initial period, were too minor to affect fish survival. Thus, all six tests may be regarded as a pre-experimental basis for comparisons with future tests.

## pH, calcium and aluminium

$H^+$ stress has proved to be a major cause of death in the pH interval 3.8–4.5 (Baker and Schofield, 1980; Brown, 1983). The combined effects of Al with pH in the range tested in these experiments, have been more diverse. Brown (1981) found that adding 0.25 and 0.5 mg/l Al as $AlCl_3$, ameliorated acid stress at pH 3.5 and 4.0. Muniz and Leivestad (1980) found that addition of 0.38 mg Al/l decreased loss of plasma chloride in brown trout at pH 4.0. Wood *et al.* (1990) described a protective effect for whole body ion content of swim-up fry of brook trout (i.e. fish which have just used their yolk sac) by Al in concentrations in the range 0.012–0.11 mg/l. However, generally negative effects of Al in low pH waters on fish survival have been observed (Baker and Schofield, 1982; Dalziel *et al.*, 1987; Gunn and Noakes, 1987; McDonald *et al.*, 1989). In the runoff waters tested at the Gårdsjön experiments, $H^+$ and Al concentrations were very high (pH: 3.8–4.5, and Al: 0.8–2.9 mg/l), and thus both should act as toxicants.

Calcium has been shown by several authors to have a protective effect against osmotic stress induced by low pH and high Al concentrations (Brown, 1983; Ingersoll *et al.*, 1990).

## Temperature

In a bioassay with brook trout fry *Salvelinus fontinalis* Mitchill., temperature showed a weak but significant negative correlation to LT-50 in the pH range 4.4–5.6

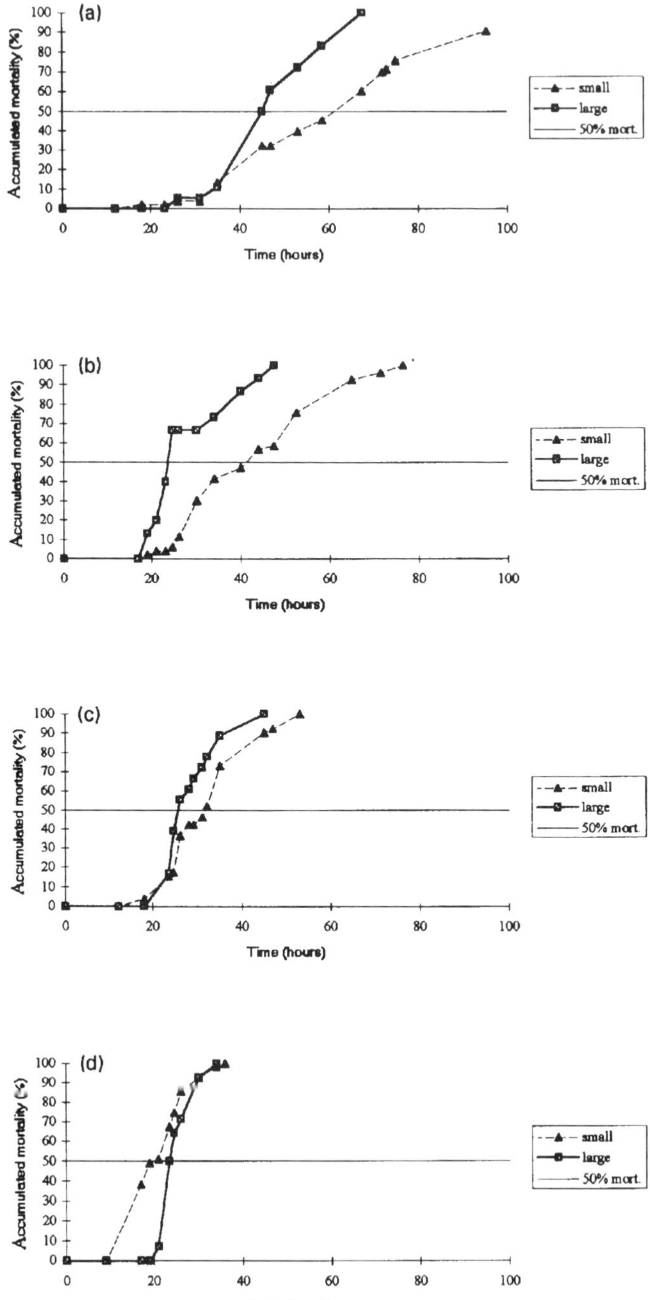


**Figure 18.5.** Mortality rates, expressed as time to 50% mortality (LT-50). Data from Experiments 5 and 6 were from two different size-classes of trout (mean length: 3.7 and 17.6 cm respectively in Exp. 5, and 5.0 and 19.1 cm respectively in Exp. 6). These fish were tested in runoff from the roofed catchment, G1 (Figures 18.5a and b), and the nitrogen-enriched catchment, G2 (Figures 18.5c and d)

(Parkhurst *et al.*, 1990). In an experiment with brown trout, Karlsson-Norrgren *et al.* (1986) found that gill lesions were more severe in a test medium with pH 5.5, Al 0.5 mg/l at 2.5°C than at 15°C. Poléo *et al.* (1991) found that increased temperature increased the mortality rates of fingerlings of Atlantic salmon *Salmo salar* at pH 5.0. In our experiments, temperature was negatively correlated to LT-50 indicating that increased temperature decreased fish survival. Thus, the results from these experiments are consistent with those of Poléo *et al.* (1991). The negative effect of temperature on mortality can probably be explained as an increased polymerization of inorganic Al on fish gills with increasing temperature, making fish suffer from acute oxygen depletion and osmotic stress (Lydersen *et al.*, 1990; Poléo *et al.*, 1991).

### Fish size

In two separate experiments, two different size classes of fish were tested in runoff water from G1 and G2. The results showed a higher mortality rate among the larger-sized, 1-year-old fish compared with smaller, 0-year-old fish; the effect was most pronounced in G1. Several authors have reported differences in survival of different size-classes of salmonids in low pH/high Al waters (Rosseland and Skogheim, 1984; Cleveland *et al.*, 1986; Ingersoll *et al.*, 1990). Small fish are reported to be very sensitive to $H^+$ stress and they become less sensitive as they grow (Baker and Schofield, 1980). Al stress appears to act in the opposite way and the fish become more sensitive with age. The mechanisms for the size-dependent toxicity are not yet clear, but some of the differences can be linked to differences in life history stages (Rosseland *et al.*, 1986). The differences in size-dependent toxicity in this study were most clear in G1, that is, in water with pH values between 4.3 and 4.5, as opposed to G2 with pH 3.8–3.9. In this case, it indicates that Al could be the component of importance, because Al-toxicity is more pronounced in the higher pH interval.

Variations in pH, Ca and Al did not explain the variations in mortality rates in and *between* the catchments. The variation in pH, Ca and Al level *within* the catchments are rather small, and the variables covaried with each other making it hard to discriminate the effects on mortality from the separate chemical parameters. pH, Ca and Al are expected to influence fish survival but the combined effect of these components are simultaneously temperature- and fish-size-dependent. This should be explored in future experiments.

## CONCLUSION AND FUTURE AIMS

The aim of the tests performed in the initial period was to assess and quantify, as far as possible, variations in mortality rates due to differences in test conditions. The most important variables were, as shown above, temperature and size of the fish used. During a later phase of the recovery process at the roof experiment, fish toxicity studies should be repeated. The future changes in water chemistry in runoff from the roofed catchment will affect the water quality experienced by biota, and future fish toxicity tests at Gårdsjön can be used as a tool to evaluate and forecast the biotic response in natural, acidic aluminium-rich waters recovering from acid stress.

## REFERENCES

Baker, J. P. and Schofield, C. L. (1980) Aluminium toxicity to fish as related to acid precipitation and Adirondack surface water quality. In: D. Drablos and A. Tollan (eds), *Ecological Impact of Acid Precipitation*. SNSF-project, Oslo, pp. 292–293.

Baker, J. P. and Schofield, C. L. (1982) Aluminium toxicity to fish in acidic waters. *Water, Air, and Soil Pollution*, **18**, 289–309.

Barnes, R. B. (1975) The determination of specific forms of aluminum in natural water. *Chemical Geology*, **15**, 177–191.

Brown, D. J. A. (1981) The effects of various cations on the survival of brown trout *Salmo trutta* at low pHs. *Journal of Fish Biology*, **18**, 31–40.

Brown, D. J. A. (1983) Effects of calcium and aluminium concentration on survival of brown trout *Salmo trutta* at low pH. *Bulletin of Environmental Contamination and Toxicology*, **30**, 382–387.

Cleveland, L., Little, E. E., Hamilton, S. J., Buckler, D. R. and Hunn, J. B. (1986) Interactive toxicity of aluminum and acidity to early life stages of brook trout *Salvelinus fontinalis* Mitchill. *Transaction of the American Fisheries Society*, **115**, 610–620.

Dalziel, T. R. K., Morris, R. and Brown, D. J. A. (1987) Sodium uptake inhibition in brown trout *Salmo trutta*, exposed to elevated aluminium concentrations at low pH. *Annales de la société royale zoologique de Belgique*, **117**, 421–434.

Driscoll, C. T., Baker, J. P., Bisogni, J. J. and Schofield, C. L. (1980) Effect of aluminum speciation on fish in dilute acidified waters. *Nature*, **284**, 161–164.

Driscoll, C. T. (1984) A procedure for the fractionation of aqueous aluminum speciation in dilute acidic waters. *International Journal of Environmental Analytical Chemistry*, **16**, 267–284.

Gunn, J. M. and Keller, W. (1990) Biological recovery of an acid lake after reduction in industrial emissions of sulphur. *Nature*, **345**, 431–433.

Gunn, J. M. and Noakes, D. G. (1987) Latent effects of pulse exposure to aluminum and low pH on size, ionic composition, and feeding efficiency of lake trout *Salvelinus namaycush* alevins. *Canadian Journal of Fisheries and Aquatic Sciences*, **44**, 1418–1424.

Henriksson, L. and Oscarsson, H. (1985) History of the acidified lake Gårdsjön: the development of chironomides. *Ecological Bulletins (Stockholm)*, **37**, 58–63.

Hultberg, H. and Grennfelt, P.-I. (1992) Sulphur and seasalt deposition as reflected by throughfall and runoff chemistry in forested catchments. *Environmental Pollution*, **75**, 215–222.

Hultberg, H., Dise, N. B., Wright, R. F., Andersson, I. and Nyström, U. (1994) Nitrogen saturation induced during winter by experimental $NH_4NO_3$ addition to a forested catchment. *Environmental Pollution*, **84**, 145–147.

Ingersoll, C. G., Mount, D. R., Gulley, D. D., LaPoint, T. W. and Bergman, H. L. (1990) Effects of pH, aluminum, and calcium on survival and growth of eggs and fry of brook trout *Salvelinus fontinalis*. *Canadian Journal of Fisheries and Aquatic Sciences*, **47**, 1580–1592.

Karlsson-Norrgren, L., Björklund, I., Ljungberg, O. and Runn, P. (1986) Acid water and aluminium exposure: experimentally induced gill lesions in brown trout, *Salmo trutta* L. *Journal of Fish Disease*, **9**, 11–25.

Kroglund, F. and Rosseland, B. O. (1992) *Reversibility of Acidification: Fish Responses in Experiments at Risdalsheia*. NIVA Acid Rain-Report, 27/1992.

Lydersen, E., Poléo, A. B. S., Muniz, I. P., Salbu, B. and Bjornstad, H. E. (1990) The effects of naturally occurring high and low molecular weight inorganic and organic species on the yolk-sac larvae of Atlantic salmon *Salmo salar* L. exposed to acidic aluminium-rich lake water. *Aquatic Toxicology*, **18**, 219–230.

McDonald, D. G., Reader, J. P. and Dalziel, T. R. K. (1989) The combined effects of pH and trace metals on fish ionoregulation. In: R. Morris, D. J. A. Brown, E. W. Taylor and J. A. Brown (eds.), *Acid Toxicity and Aquatic Animals*. Society for Experimental Biology Seminar Series. Cambridge University Press, Cambridge, pp. 221–242.

Muniz, I. P. and Leivestad, H. (1980) Toxic effects of aluminium on the brown trout *Salmo trutta* L. In: D. Drabløs and A. Tollan, (eds.), *Ecological Impact of Acid Precipitation.* SNSF-project, pp. 320–321.

Parkhurst, B. R., Bergman, H. L., Fernandez, J., Gulley, D. D., Hockett, J. R. and Sanches, D. A. (1990) Inorganic monomeric aluminum and pH as predictors of acidic water toxicity to brook trout *(Salvelinus fontinalis). Canadian Journal of Fisheries and Aquatic Sciences,* **47**, 1631–1640.

Playle, R. C., Goss, G. G. and Wood, C. M. (1989) Physiological disturbances in rainbow trout *Salmo gairdneri* during acid and aluminum exposures in soft water of two calcium concentrations. *Canadian Journal of Zoology*, **67**, 314–324.

Poléo, A. B. S., Lydersen, E. and Muniz, I. P. (1991) The influence of temperature on aqueous aluminium chemistry and survival of Atlantic salmon *Salmo salar* L. fingerlings. *Aquatic Toxicology*, **21**, 267–278.

Rosseland, B. O. and Skogheim, O. K. (1984) A comparative study on salmonid fish species in acid aluminium-rich water. II. Physiological stress and mortality of one and two year old fish. *Report of the Institute for Freshwater Research Drottningholm*, **61**, 186–194.

Rosseland, B. O., Skogheim, O. K., Kroglund, F. and Hoell, E. (1986) Mortality and physiological stress of year-classes of landlocked and migratory Atlantic salmon, brown trout and brook trout in acid aluminium-rich soft water. *Water, Air, and Soil Pollution*, **30**, 751–756.

Wood, C. M., McDonald, D. G., Ingersoll, C. G., Mount, D. R., Johansson, O. E., Landsberger, S. and Bergman, H. L. (1990) Whole body ions of brook trout (*Salvelinus fontinalis*) alevins: responses of yolk-sac and swim-up stages to water acidity, calcium, and aluminum, and recovery effects. *Canadian Journal of Fisheries and Aquatic Sciences*, **47**, 1604–1615.

# 19 Simulating the Gårdsjön Covered Catchment Experiment with the MAGIC Model

**FILIP MOLDAN, RICHARD F. WRIGHT, ROBERT C. FERRIER, INGVAR ANDERSSON AND HANS HULTBERG**

The G1 ROOF catchment runoff response to the clean rain treatment was modelled with the MAGIC Model. MAGIC was calibrated to the 2 years prior to the treatment, and then run forward in time to predict runoff chemistry in response to the changed deposition. After 4 years of treatment, measured concentrations of sulphate in runoff declined from 394 to 214 $\mu$eq/l and inorganic Al declined from 277 to 110 $\mu$eq/l. Less dramatic declines were measured in concentrations of several base cations. ANC increased from $-264$ to $-159$ $\mu$eq/l. There was no change in pH. These changes were successfully predicted by the model. The clean rain treatment was projected by the model for 6 more years. At the end of 10 years of modelled treatment, the $SO_4$ concentration was 132 $\mu$eq/l, inorganic Al 88 $\mu$eq/l, ANC $-111$ $\mu$eq/l, and pH slightly increased, from 4.2 to 4.3. The results indicate that the experimental conditions created by the roof artificially reduce base cation (especially Ca) supply to the soil surface. With ambient base cation cycling a more rapid improvement of runoff quality could be expected.

## INTRODUCTION

The purpose of reducing emissions of acidifying compounds is to diminish and reverse the environmental impacts of acidification. The causal relationship between reductions and restoration of acidified soils and fresh waters is well accepted, but the time required to obtain recovery is much more uncertain. Dynamic process-based mathematical models of acidification processes provide a tool by which these time delays can be quantified. The reliability of such models can be established by comparison with relevant experimental data. Large-scale whole-catchment experiments such as the Gårdsjön Covered Catchment Experiment (Chapter 2, this volume) provide just such information.

The Gårdsjön Covered Catchment Experiment is designed to provide experimental data on acidification recovery. The intent of the G1 ROOF catchment manipulation is to approximate a situation where acidic deposition to an acidified

*Experimental Reversal of Acid Rain Effects: The Gårdsjön Roof Project.*
Edited by Hans Hultberg and Richard Skeffington. 

catchment (Andersson and Olsson, 1985) is removed and replaced with a deposition which is not affected by anthropogenic emissions, i.e. with an atmospheric deposition similar to that prior to the industrial revolution. The roof at the Gårdsjön site is constructed beneath the canopy and is the largest such construction to date. The 0.7 ha roof covers an entire well-defined headwater catchment such that the runoff response can be measured directly rather than extrapolated from measurements of soil solution. Also the catchment presumably encompasses a wide range of hydrobiogeochemical interactions that may influence the recovery dynamics.

MAGIC (Model of the Acidification of Groundwater in Catchments) (Cosby *et al.*, 1985a,b) was used to simulate the runoff response to the clean rain treatment. MAGIC is an intermediate complexity, process-orientated model and has been applied to a variety of sites to evaluate the dynamics of soil and runoff response to acidification and land-use change (cf. Wright *et al.*, 1994). MAGIC has also been used to evaluate the rate of reversibility at the Norwegian RAIN roofed catchment experiment (Cosby and Wright, 1995; Beier *et al.*, 1996). Here we calibrate MAGIC to the pre-treatment period, compare the modelled and observed response of the runoff chemistry after the first 4 years of treatment, and use the model to extrapolate into the future.

## METHODS

### Site description

The G1 ROOF catchment is located near Lake Gårdsjön, where acidification research has been conducted for over two decades (Grahn *et al.*, 1974; Andersson and Olsson, 1985; Dickson, 1988). The G1 ROOF catchment has a closed stand of 80–100-year-old Norway Spruce (*Picea abies* (L.) Karst.), with some Scots pine (*Pinus sylvestris* L.) present. The bedrock is siliceous gneiss. Soils are podzols and brown earths developed on silty and sandy glacial moraine. Mean soil depth is 43 cm. A detailed site description is given in Andersson *et al.* (Chapter 2, this volume).

### The experiment

The G1 ROOF catchment was hydrologically and chemically calibrated during a 2-year period (1989–90) before the roof construction. Treatment with clean rain commenced in April 1991. The roof intercepts all throughfall which is then led away from the catchment. Clean rain is applied beneath the roof by means of a sprinkling system. The clean rain produced by ion-exchanging water from the lake and adding natural concentrations of sea salts. The result is a near-100% reduction in deposition of the anthropogenic pollutants $SO_4$, $NO_3$, $NH_4$ and $H^+$. The roof interrupts the natural cycling of base cations. During the first 4 years of treatment the removal of base cations in throughfall has not been fully compensated by additions to the clean rain. Moldan *et al.* (Chapter 4, this volume) give details of the experimental set-up.

## MODEL DESCRIPTION AND CALIBRATION

The major processes in MAGIC are deposition, cation exchange, weathering, plant uptake, inorganic and organic carbon reactions, sulphate sorption, aluminium mineral solubility and reactions of Al-ligands. Time-steps of months or years can be used. The model applies lumped parameters to describe the soil. The model was calibrated using measured input and output fluxes from the two pre-treatment years (Table 19.1).

The model requires a number of fixed parameters to describe the stream and soil in the modelled catchment (Table 19.2). These include mean annual discharge, soil temperature, soil depth, bulk density and cation exchange capacity, which were all obtained from field measurements and the soil survey at the site (Chapter 2, this volume). Sulphate adsorption capacity in the soil was set 15% higher than the $NaHCO_3$ extractable sulphate pool of 704 mmol/m$^2$ measured by Mörth *et al.* (1995). Weathering rates and historical soil base saturation (BS) were calculated using an automated optimisation routine (Cosby *et al.*, 1990). This routine finds an optimal combination of historical BS and weathering rates for base cations such that the model hindcast arrives at the observed stream and soil chemistry in the target year. The model was driven by reconstructed historical rates of acid deposition and forest growth over a 140-year period from 1850 to 1990. Cation

**Table 19.1.** Measured runoff chemistry and soil base saturation used as target values for the MAGIC at the end of historical period 1850–1990

| **G1 runoff** | **mean pre-treatment ($\mu$eq/l)** |
|---|---|
| $H_2O$ (mm) | 326 |
| $H^+$ | 64 |
| Ca | 66 |
| Mg | 183 |
| Na | 646 |
| K | 23 |
| Kex | 9 |
| Al inorg | 227 |
| $NH_4$ | 0 |
| $NO_3$ | 2 |
| $SO_4$ | 394 |
| Cl | 785 |
| A- | 28 |
| SBC | 917 |
| SAA | 1181 |
| ANC | −264 |
| **Soil base saturation (%)** | |
| Ca | 11.5 |
| Mg | 5.9 |
| Na | 2.5 |
| K | 3.3 |
| BS sum | 23.2 |

**Table 19.2.** Fixed parameters used in MAGIC

| | |
|---|---|
| **Stream parameters** | |
| Discharge (mm/yr) | 330 |
| Solubility $Al(OH)_3$ log 10 | 8.4 |
| Temperature °C | 8 |
| $CO_2$ partial pressure (atm) | 0.001 |
| Total organic acid (mmol/m$^3$) | 40 |
| **Soil parameters** | |
| Soil depth (m) | 0.43 |
| Porosity (%) | 50 |
| Bulk density (kg/m$^3$) | 647 |
| CEC (meq/kg) | 93 |
| Temperature °C | 8 |
| $CO_2$ partial pressure (atm) | 0.005 |
| Total organic acid (mmol/m$^3$) | 150 |
| Solubility $Al(OH)_3$ log 10 | 8.4 |
| Sulphate adsorption 1/2 sat. (meq/m$^3$) | 300 |
| Sulphate adsorption max cap. (meq/kg) | 6 |
| Ca saturation historical (1850) (%) | 13.5 |
| Mg saturation historical (1850) (%) | 10.9 |
| Na saturation historical (1850) (%) | 3.1 |
| K saturation historical (1850) (%) | 3.5 |
| BS historical (1850) (%) | 30.9 |
| Weathering Ca (meq/m$^2$/yr) | 10.5 |
| Weathering Mg (meq/m$^2$/yr) | 7.4 |
| Weathering Na (meq/m$^2$/yr) | 4.7 |
| Weathering K (meq/m$^2$/yr) | 2.5 |
| Select. coef. Al–Ca (log) | −2.82 |
| Select. coef. Al–Mg (log) | −0.62 |
| Select. coef. Al–Na (log) | −0.85 |
| Select. coef. Al–K (log) | −5.55 |

exchange selectivity coefficients between each of four individual base cations and aluminium were calculated by the model given the optimised historical base saturation.

The pre-treatment total deposition of each ion was obtained from measurements of bulk precipitation and throughfall. The marine fraction was calculated from the ratios of each ion to chloride in seawater using the assumption that all chloride in deposition is of marine origin. The amounts of base cations in throughfall coming from exchange in the canopy were estimated by Ferm and Hultberg (Chapter 3, this volume). According to these estimates, all Na and Mg in deposition is of marine origin whereas deposition of Ca and K is about twice that of marine origin. Deposition of $NO_3$, $NH_4$, $SO_4$ and Cl was set equal to the throughfall deposition. The net biomass uptake of base cations was estimated from base cation accumulation in spruce forest (Rosén, 1988). Uptake of nitrogen was set proportional to the nitrogen atmospheric input; measured input–output fluxes of nitrogen show that practically all nitrogen is retained in the catchment.

**Hindcast**

The G1 ROOF catchment was modelled as forested with a mixed stand and no significant cutting up to 1904, when the catchment was clear-cut and forested with spruce, canopy closure was assumed at 30 years. Dry deposition and nutrient uptake were scaled with forest history. Precipitation, runoff volume, wet deposition of sea salt and primary weathering were assumed constant. Deposition of $SO_4$ was estimated from data on sulphur emissions in Europe (Mylona, 1993) combined with estimated local emissions from the power plant in Stenungsund (Figure 19.1). Nitrate and ammonium deposition were scaled parallel to sulphate.

**Treatment**

The calibrated model was then driven by the actual deposition under the roof during the first 4 treatment years. The simulation was then run for an additional 6 years. The influence of the roof on the internal cycle of base cations was modelled so that the biomass uptake was increased from net uptake before the roof to the net uptake plus canopy exchange during the treatment, and the deposition of BC· was changed to the total sprinkling input (i.e. sea-salt deposition together with K and Ca additions). It was assumed that the circulation of elements in litterfall was not changed by the roof treatment. Litterfall is periodically swept from the roof and spread beneath it on the soil. Further it was assumed that the treatment did not

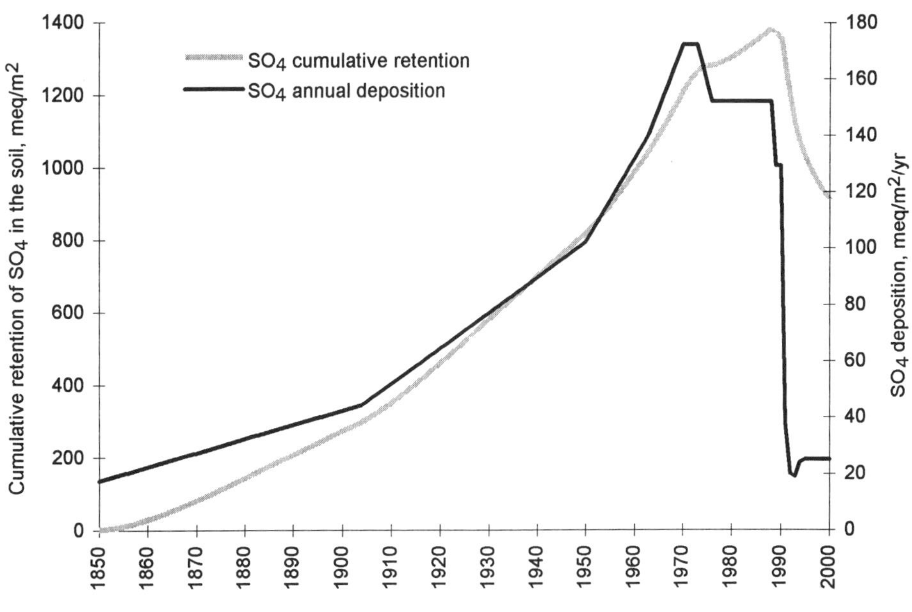


**Figure 19.1.** Estimated (measured 1989–94) S deposition and the cumulative retention (deposition – runoff) of $SO_4$ in catchment G1 ROOF at Gårdsjön. Runoff values are simulated by MAGIC

affect the hydrology of the catchment, and that there was no dry deposition beneath the roof.

## RESULTS AND DISCUSSION

### Acidification history

The MAGIC simulation provides a picture of the 140-year course of acidification at the catchment. Over the 140-year hindcast period the concentration of $SO_4$ in runoff increased as a result of an increased S loading from the atmosphere. The model indicates that a portion of the deposited S accumulated in the soil, and that the catchment reached approximate steady state only after the S deposition peaked and then started to decline around the year 1980 (Figure 19.1).

The base cations were desorbed from the soil and lost from the catchment through runoff (Figure 19.2) due to the increased deposition of $SO_4$, cation exchange and the increased $SO_4$ output. The base cations cumulative loss of 2000 meq/m$^2$ from 1850 to 1990 is equivalent to a 25% decline in the soil base saturation estimated in 1850 to a present value of 23.2% (Tables 19.1 and 19.2). As

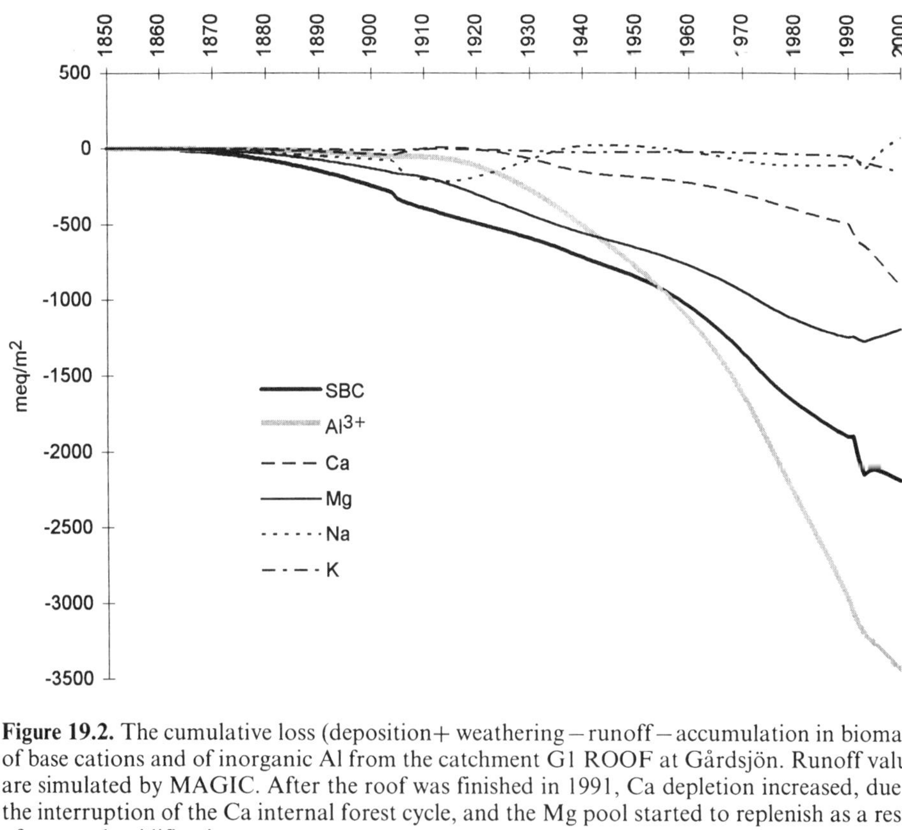


**Figure 19.2.** The cumulative loss (deposition+ weathering – runoff – accumulation in biomass) of base cations and of inorganic Al from the catchment G1 ROOF at Gårdsjön. Runoff values are simulated by MAGIC. After the roof was finished in 1991, Ca depletion increased, due to the interruption of the Ca internal forest cycle, and the Mg pool started to replenish as a result of reversed acidification

acidification of the soil proceeded, the pH of the runoff declined from an initial value of about 5.1 (Figure 19.3). The pH increased temporarily following a clear-cut in 1904, associated with a modelled drop in the total dry deposition. Since then, the runoff pH has steadily decreased up to the present value of around 4.2. Beginning about 1920 inorganic Al (modelled as a $Al^{3+}$) began to act as a major cation and counter ion to the $SO_4$ in runoff.

The long-term decrease in acid neutralising capacity (ANC) defined as the equivalent sum of base cations (SBC) less the equivalent sum of strong acid anions (SAA) (Figure 19.4) is due to the inability of the catchment to supply enough base cations to balance the increasing concentrations of strong acid anions; over time these are increasingly balanced by inorganic Al.

Weathering rates for the four base cations obtained by this calibration of MAGIC indicate a rate of about 25 meq/m$^2$/yr, similar to those obtained by several other methods (Chapter 10, this volume).

The modelled historical change of stream pH at the G1 ROOF catchment indicates that the stream had an initially lower pH and acidified earlier than the Lake Gårdsjön itself. Independent information on the historical pH of the lake comes from analysis of diatom remains in the sediments (Renberg and Wallin, 1985). The G1 catchment and other small catchments at Lake Gårdsjön are inherently more sensitive to acidification and acidified earlier and more severely. Factors contributing to the inherent differences are the more shallow soils with lower weathering rates and thus lower initial pH and ANC in runoff, the increased dry deposition and evapotranspiration due to the presence of the forest. Furthermore in-lake processes such as redox reactions, mixing and buffering by the lake sediments will tend to slow and reduce the acidification of the lake itself.

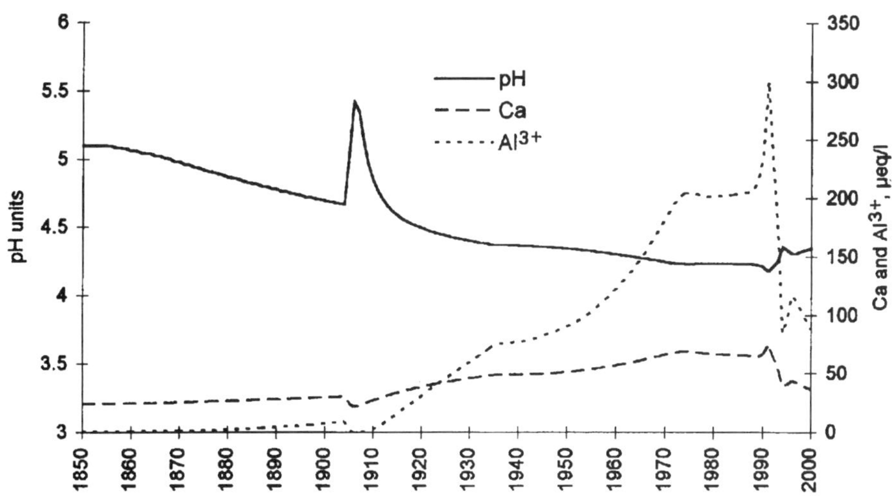


**Figure 19.3.** Modelled runoff acidity and concentration of inorganic Al (modelled as $Al^{3+}$) and Ca for the years 1850 to 2000

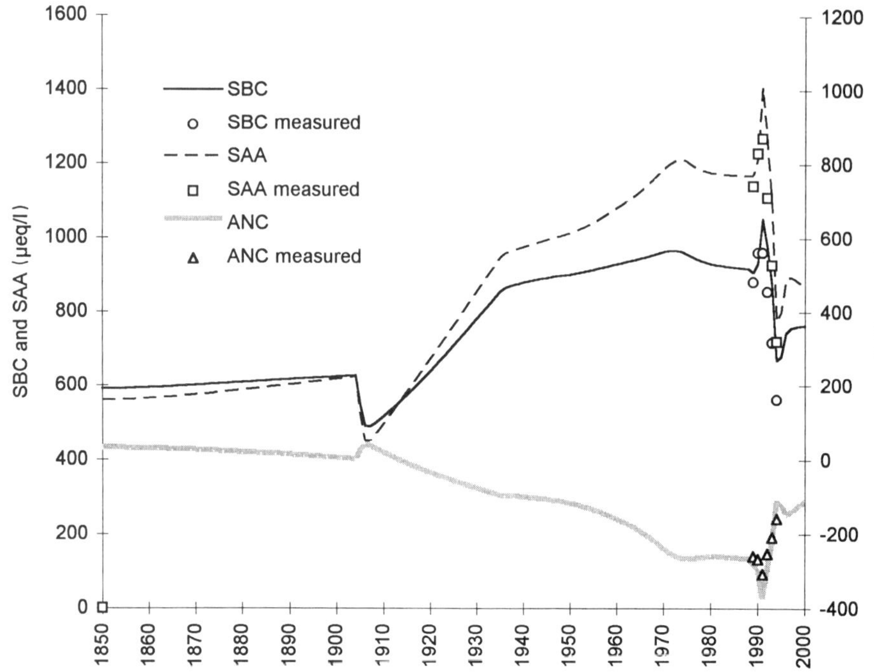


**Figure 19.4.** Runoff concentrations of sum of base cations (SBC) and sum of acid anions (SAA) reconstructed by MAGIC since the year 1850, measured data 2 years prior to the clean rain treatment (1989–90) and 4 years of clean rain treatment (1991–95). ANC is the distance between SBC and SAA lines, positive from the beginning of the simulation till the year 1915, negative for the rest of the modelled period

## Clean rain treatment

The clean rain treatment resulted in declining concentrations of $SO_4$, inorganic Al and Mg, and secondarily also of Ca and K and an increase of ANC in runoff (Moldan *et al.*, 1995). Since the roof was introduced, the output of $SO_4$ in runoff has exceeded the S input. The model simulation indicates that after 10 years of the experimental treatment about one third of S accumulated in the soil over the last 140 years will be leached from the catchment (Figure 19.1).

The roof treatment apparently did not slow the depletion of base cations from the soil (Figure 19.2), because of the less-than-full compensation of the cycled base cations in throughfall (Chapter 4, this volume).

MAGIC reproduced the major trends in runoff chemistry during the 4 years of treatment (Figure 19.5), although the year to year variation was not quite the same modelled and observed. The large differences between the individual years at Gårdsjön area are primarily due to variable deposition of sea salt (Hultberg and Grennfelt, 1992). The chloride output follows the episodic input with a time lag of months to a few years; the processes causing this lag are poorly known. There is no

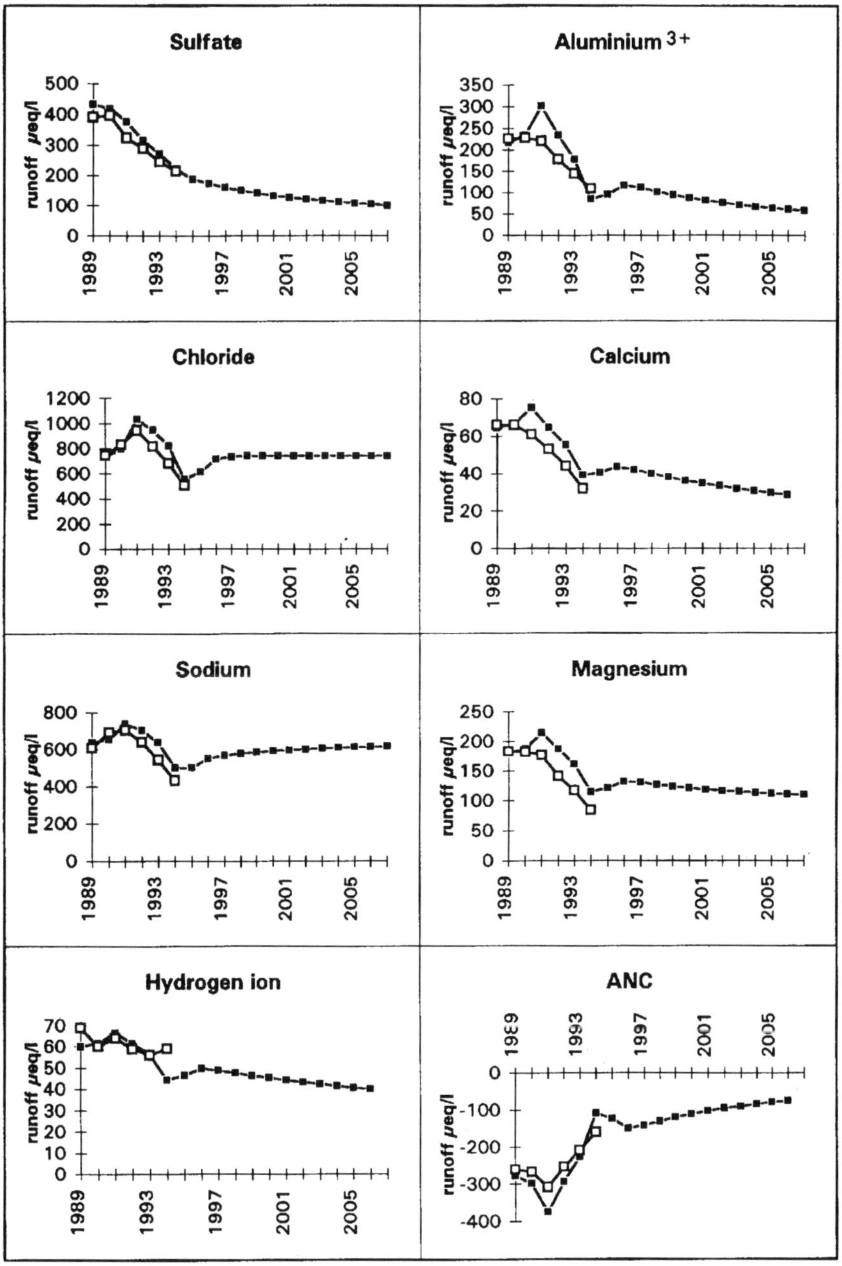


**Figure 19.5.** Measured (open squares) and simulated (solid squares) changes in runoff concentrations of G1 ROOF catchment after the beginning of the clean rain treatment in 1991 and extrapolated to 10 years of treatment

process in MAGIC for retention of large pulses of chloride. During the first year of treatment (1991) the sea-salt input was relatively high. The MAGIC response was larger than observed but the difference smoothed out with additional treatment years. However, even after 4 years of treatment, the measured input and output of chloride did not completely balance. The catchment retained about 70 meq/m$^2$ of Cl (about 6% of input), whereas the model simulated no Cl retention. This explains the somewhat higher modelled chloride concentrations and consequently higher modelled concentrations of base cations.

Model projection of the clean rain treatment for an additional 6 years indicates further declines of all already declining species, and only a very slight decline of $H^+$. According to MAGIC, 10 years of the treatment will result in the pH of runoff changing from 4.20 to 4.34, ANC from $-304$ to $-111$ $\mu$eq/l and inorganic Al from 239 to 88 $\mu$eq/l.

The relatively rapid improvement of runoff water quality is in spite of the continued depletion of soil-exchangeable Ca (Figure 19.2). Without the presence of the roof, about 15 meq/m$^2$/yr of Ca and 35 meq/m$^2$/yr of K (Chapter 3, this volume) is taken up by the trees and leached with precipitation in throughfall. This cycle is interrupted at G1 by the presence of the roof beneath the canopy. The K is re-added to the sprinkling solution in amounts comparable to estimated loss. Ca, however, was insufficiently compensated. From the total deposition of base cations, only the marine part is compensated by addition of sea salt to the sprinkling solution. The Mg soil pool is being replenished from sea salt, but Ca is depleted even faster than during the acidification period. Na and K inputs and outputs are approximately balanced both before and after the roof with some retention or losses following changes in deposition due to the ion exchange processes.

This application of MAGIC to the Covered Catchment Experiment shows that the model satisfactorily simulates the measured response of runoff chemistry to reductions in acid deposition. This corroborates previous MAGIC applications to other similar large-scale experiments including the roof experiments at the Norwegian RAIN project (Cosby and Wright, 1995) and the Danish EXMAN experiment at Klosterhede (Beier *et al.*, 1996). Together these applications indicate that MAGIC describes the acidification process in sufficient detail to be used as a tool for predicting the response of soils and surface waters to future planned and projected changes in acid deposition.

## ACKNOWLEDGEMENTS

This work was supported by the Swedish Environmental Protection Agency, the Swedish Environmental Research Institute, the Norwegian Institute for Water Research and the Macaulay Land Use Research Institute.

## REFERENCES

Andersson, F. and Olsson, B. (eds) (1985) Lake Gårdsjön, An Acid Forest Lake and its Catchment. *Ecological Bulletins (Stockholm)*, **37**.

Beier, C., Hultberg, H., Moldan, F. and Wright, R. F. (1996) MAGIC applied to the roof experiments (Risdalsheia, N; Gårdsjön, S; Klosterhede, DK) to evaluate the rate of reversibility of acidification following experimentally reduced acid deposition. *Water, Air, and Soil Pollution*, **85**, 1745–1751.

Cosby, B. J. and Wright, R. F. (1995) An acidification model (MAGIC) with organic acids evaluated using whole-catchment manipulations in Norway. *Journal of Hydrology*, **170**, 101–122.

Cosby, B. J., Hornberger, G. M., Wright, R. F. and Galloway, J. N. (1985a) Modelling the effects of acid deposition: assessment of a lumped parameter model of soil water and streamwater chemistry. *Water Resources Research*, **21**, 51–63.

Cosby, B. J., Wright, R. F., Hornberger, G. M. and Galloway, J. N. (1985b) Modelling the effects of acid deposition: estimation of long term water quality responses in a small forested catchment. *Water Resources Research*, **21**, 1591–1601.

Cosby, B. J., Jenkins, A., Miller, J. D., Ferrier, R. C. and Walker, T. A. B. (1990) Modelling stream acidification in afforested catchments: long-term reconstruction at two sites in central Scotland. *Journal of Hydrology*, **120**, 143–162.

Dickson, W. (ed.) (1988) *Liming of Lake Gårdsjön*. SNV rept 3426, Solna, 327 pp.

Grahn, O., Hultberg, H. and Landner, L. (1974) Oligotrophication — a self-accelerating process in lakes subjected to excessive supply of acid substances. *Ambio*, **3**, 93–94.

Hultberg, H. and Grennfelt, P. (1992) Sulphur and seasalt deposition as reflected by throughfall and runoff chemistry in forested catchments. *Environmental Polution*, **75**, 215–222.

Moldan, F., Hultberg, H. and Andersson, I. (1995) Covered catchment experiment at Gårdsjön: Changes in runoff chemistry after four years of experimentally reduced acid deposition. *Water, Air, and Soil Polution*, **85**, 1599–1604.

Mörth, C.-M., Torssander, P., Sjöberg, L. and Hultberg, H. (1995) Acidification reversal observed from a clean rain experiment in the roof catchment at Lake Gårdsjön: Implications from modeling of the sulphur dynamics. *Water Resources Research* (submitted). Also in: Mörth, C.-M. (1995) Sulphur isotopes used as a tracer of acidification reversal, dispersion of acid mine drainage and sulphur dynamics in small and large catchments. PhD thesis, Stockholm University.

Mylona, S (1993) Trends of Sulphur Dioxide Emissions, Air Concentrations and Depositions of Sulphur in Europe since 1880. EMEP MSC-W rep. 2/93, Oslo, 96 pp.

Renberg, I. and Wallin, J.-E. (1985) The history of the acidification of lake Gårdsjön as deduced from diatoms and Sphagnum leaves in the sediment. *Ecological Bulletins (Stockholm)*, **37**, 47–52.

Rosén, K. (1988) Effects of biomass accumulation and forestry on nitrogen in forest ecosystems. In: J. Nilsson and P. Grennfelt (eds), *Critical Loads for Sulphur and Nitrogen*, NORD 1988:15, Stockholm, pp. 269–294.

Wright, R. F., Cosby, B. J., Ferrier, R. C., Jenkins, A., Bulger, A. J. and Harriman, R. (1994) Changes in acidification of lochs in Galloway, southwestern Scotland, 1979–1988 — The MAGIC model used to evaluate the role of afforestation, calculate critical loads, and predict fish status. *Journal of Hydrology*, **161**, 257–285.

# 20 Modelling Soil Acidification in a Roofed Catchment: Application of the SAFE Model

HARALD SVERDRUP, PER WARFVINGE, HANS HULTBERG, INGVAR ANDERSSON & FILIP MOLDAN

Application of the SAFE model to a roofed subcatchment at Gårdsjön, Sweden, indicates that the response to a large acidity deposition reduction follows only after many years. The roof was installed in 1991 and the immediate change in pH is predicted to be small. A more significant decrease in both base cations and Al concentration after 1995 is predicted. The roofed catchment was designed as an experiment to investigate how a deposition reduction may lead to recovery, and to quantify the rate and degree of such a recovery. The results indicate that only a partial recovery of stream and soil chemistry can be expected by the year 2000. The reduction in acid input under the roof is not sufficient to cause complete recovery, mostly because of large losses of base cations from the nutrient cycle. The predictions of recovery are limited in accuracy because of uncertainties in the deposition and nutrient circulation inputs.

## INTRODUCTION

Dynamic models of soil acidification are being used to assess the effects of acid deposition on forest soils and aquatic ecosystems. Such assessments, together with mapping of critical loads of acidity, are being used as input to European negotiations on sulphur and nitrogen emission reductions. The objective of this study is to show that acid deposition reductions down to or below the critical load are required to protect forest soil, fish-spawning streams and the lake ecosystem at Gårdsjön. Another objective is to test and evaluate the performance of the dynamic soil chemistry model developed in Sweden for forest soils: SAFE.

The model will be used to estimate the different sources of acidity input to the soil and explain the changes in soil chemistry observed in the experimental roofed catchment at Gårdsjön. The model will be used to assess the reaction time of the G1 catchment to a reduction in deposition.

It is of importance that models used for prediction of the future are capable of reconstructing a reasonable past. A model that cannot reconstruct the past will make

*Experimental Reversal of Acid Rain Effects: The Gårdsjön Roof Project.*
Edited by Hans Hultberg and Richard Skeffington. 

less reliable predictions about the future. Thus the reasonableness of the past reconstructions gives indications about the reliability of a forecast.

## MODEL DESCRIPTION

SAFE has been developed with the objective of studying the effects of acid deposition on soils and water. It calculates the values of different chemical state variables as a function of time. It can therefore be used to study the process of acidification and recovery, as affected by deposition rates, soil parameters and hydrological variations. The model includes mass balances for ANC, base cations (BC), phosphorus, sulphate, chloride, nitrate and ammonium. SAFE contains the following chemical subsystems (Figure 20.1):

- Deposition, leaching and accumulation of dissolved chemical components.
- Chemical weathering reactions of soil minerals with the soil solution.
- Cation exchange reactions.
- The net results of reactions of N compounds; complete nitrification.
- Internal cycling of N in the canopy, such as canopy exchange, litterfall and net mineralization, and biological net uptake of base cations and nitrogen.
- Solution equilibrium reactions involving $CO_2$, Al and organic acids.

The model is structured in different compartments to represent the natural vertical differences in soils (Figure 20.2). To assemble input data for the model, the soil was

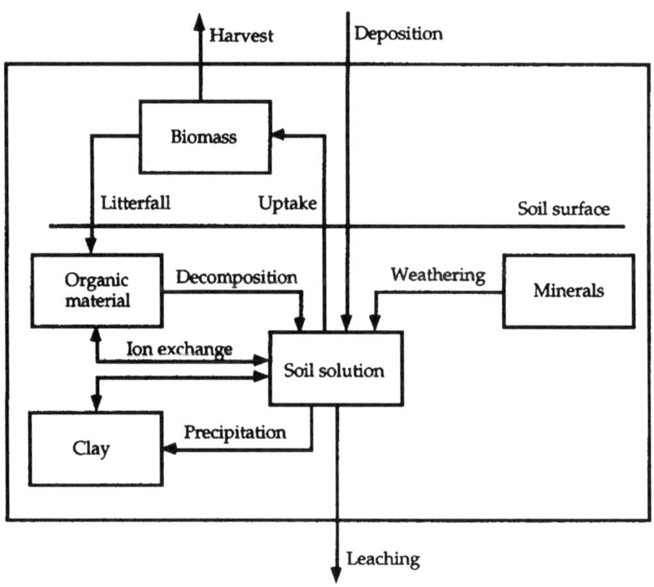


**Figure 20.1.** Coupling of the individual processes in the SAFE model. All processes and equilibria in the model are affected by temperature

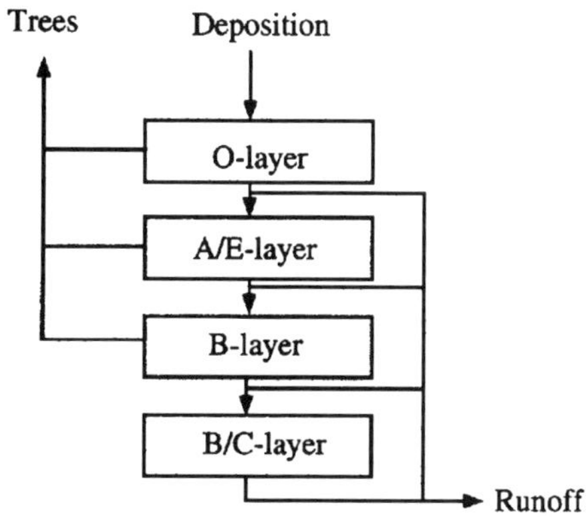


**Figure 20.2.** Flux pathway between the layers in the SAFE model. The scheme of all the processes are repeated within each layer box in the model

divided into different layers. This was done based on available soil chemistry data for different depths in the soil profile and the natural soil stratification.

The biology of the catchment is in most cases very important for the cycling of nutrients, such as base cations, phosphorus and nitrogen. It is also of great importance for the balance of acidity in a catchment; acidity may be a strong driving force in many soil processes. Major biological processes in this context are the growth of plants, decomposition of organic matter, nitrification and denitrification.

The Al concentration in the SAFE model is controlled by the gibbsite forcing function. The $pK_{Gibb}$ used had the values systematically used in mapping critical loads (Sverdrup *et al.*, 1990) 6.5, 7.6, 8.6, 9.2. In this test of the model, the $pK_{Gibb}$'s used were not subject to any adjustment, for the sake of not masking away any shortcomings of the present model formulation. The gibbsite model is obviously an oversimplification of reality, and the authors are working on a model based on weathering, ion exchange, precipitation reactions and complexing with organic acids.

## Weathering rate kinetics

The outstanding difference between SAFE and other comparable soil chemistry models is that the weathering rate is calculated from independent geophysical properties of the soil system. Several chemical reactions between soil minerals and constituents in the soil solution contribute to the base cation release rate, and the total chemical weathering rate will be the sum of the weathering rates of the individual reactions. Several weathering reactions have been identified experimentally:

1. The reaction with $H^+$.
2. The reaction with $H_2O$.
3. The reaction with $CO_2$.
4. Reactions with soil organic acids.

The rate of dissolution is inhibited by the reaction products due to complexing on active dissolution sites. High concentrations of Al and BC will effectively retard the weathering rate (Sverdrup, 1990; Sverdrup and Warfvinge, 1993a).

### Cation exchange kinetics

The exchange reaction between exchangeable calcium and acidity is modelled as a reversible chemical reaction only limited in rate by the transport of base cations between the bulk of the soil solution and active sites on the exchanger surface. The change in exchangeable bases, $\bar{X}_{BC}$ is proportional to the flux of base cations to the exchanger, which is driven by the concentration difference between the surface of the exchange complex and the concentration in the bulk of the soil solution. The rate equation becomes:

$$\frac{d\bar{X}_{BC}}{dt} = k_x([BC^{2+}] - [BC^{2+}]_{ex}) \qquad (20.1)$$

Subscript *ex* indicates that the conditions on the exchanger surface are considered. The above equation thus states that the flux of base cations to the exchange surface is proportional to the concentration gradient between the solution and the surface.

The base cation concentration at the liquid–solid interface $[BC]_{ex}$, is given by an equilibrium model, the Gapon exchange equation:

$$[BC]_{ex} = \frac{[H^+]_{ex}^2 \, \bar{X}_{BC}^2}{K_{H/BC}^2 (1 - \bar{X}_{BC})^2} \qquad (20.2)$$

where $K_{H/BC}^2$ is the Gapon selectivity coefficient for H/BC exchange. Calculating $[BC^{2+}]_{ex}$ will require computation of the pH at the exchanger phase surface, and one equation in addition to an exchange equilibrium equation is needed. The additional relationship required can be derived from the charge balance evaluated at the surface as well as for the soil solution. When this is combined with the Gapon equation, it becomes possible to solve for $[H^+]_{ex}$. This procedure has been described in detail by Warfvinge and Sverdrup (1992).

If exchangeable aluminium is dominating on the exchanger, however, the interface concentrations are calculated from an Al–Ca exchange isotherm. One may combine this expression with the assumption that $[Al^{3+}]$ is given by a simple gibbsite equilibrium. It then becomes evident that it does not matter whether H–Ca or Al–Ca cation exchange is assumed, provided that the Gapon exchange equation is applied.

### Calibration

SAFE is calibrated on initial base saturation alone. As in all dynamic models, SAFE calculations are triggered with a set of steady-state initial values. As defined above, the state variables involved are ANC, base cations, exchangeable bases or acidity, nitrate and ammonia. To make sure that the simulations reflect changes due to changes in external and internal loads, rather than just unstable initial conditions, SAFE is initiated with steady-state initial conditions. These initial values are calculated with a separate routine, PROFILE (Sverdrup and Warfvinge, 1993a), a special version of SAFE that solves the algebraic equations that result from setting all differentials to 0.

### Assessing forest vitality effects

To run the SAFE model, tree uptake and nutrient circulation is specified as input, as SAFE does not yet incorporate a proper growth model. In order to analyse effects of soil acidity on growth, a back-calculation of the growth and defoliation is carried out using a growth modification function based on three components:

$$G = G_{max}\, f(BC/Al)\, f(Nutr)\, f(w) \tag{20.3}$$

where $f(BC/Al)$ expresses the action of soil acidity as expressed by the BC/Al ratio (Sverdrup and Warfvinge, 1993b) with a function value between 0 and 1, $f(Nutr)$ expresses the action of Liebig's Law with a function value between 0 and 1, and $f(w)$ the action of water stress or surplus with a function value between 0 and 2. The effects act independently but additively as described, but by decreasing the general uptake possibility in soil layers, the effect of $f(BC/Al)$ is to inhibit uptake of necessary base cation nutrients. $G_{max}$ is the maximum growth obtainable, not considering any base cation nutrient limitations or any limiting effects caused by soil acidification. The reference level for growth in our calculations is the growth calculated using standard yield tables for growth modified for site characteristics.

In the growth modification function, growth is also subject to nutrient restrictions. Growth is assumed to be proportional to uptake of the limiting nutrient, i.e. according to Liebig's Law: "The nutrient in least supply will limit growth". In the model, nitrogen and base cation uptake is coupled, and either nitrogen or base cation deficit will eventually reduce total growth.

Limits for uptake are determined by the availability of nutrient either by deposition (N) or deposition and weathering (BC, P). Thus "Liebig's Law" is expressed as the ratio between nutrient requirements to sustain growth according to the yield model, and nutrient availability:

$$f(Nutr) = \min_{i = nutrients} \sum_{j=1}^{4\ layers} \frac{W_{j,i} + XXdep_{j,i}}{G_{max}\, x_i} \tag{20.4}$$

where $x_i$ is the content of the nutrient $i$ in the tree, $W$ is weathering and $XX$ is the substance considered (Ca, Mg, K, P, N). If the amount required by the plant is less than the amount available, then the function has the value 1, and this nutrient will not limit growth. The deposition term is zero below the first soil layer. Weathering is zero for N.

It is known from laboratory studies that Al in the soil solution will be able to disturb the growth of trees, and this can be expressed in terms of the (Ca+Mg+K)/Al ratio (Sverdrup and Warfvinge, 1993b). Under severe acidification, growth will be reduced by a low ratio of base cations necessary for growth and Al which can interfere with uptake. If uptake is less, then the base cations will stay in the soil solution, and the concentration will be changed by the fact that the trees have decreased the growth rate or stopped growing. Decomposition and nitrification is coupled to soil chemistry in a similar way.

The expression used for chemical feedback on uptake is (Sverdrup and Warfvinge 1993b):

$$f(BC/Al) = \frac{u}{u_{max}} = \sum_{i=1}^{layers=4} \frac{u_i}{u_{tot}} \frac{(BC/Al)_i}{(BC/Al)_i + K_{BC/Al}} \tag{20.5}$$

where $u_i$ is uptake in soil layer $i$. A value of 0.33 was used for $K_{BC/Al}$, ignoring the effect of tree age. This can be included by letting $K_{BC/Al}$ have a value of 0.15 for younger trees from 5 to 25 years, pass 0.3 at 30 years and stabilize at 0.5 for trees older than 50 years. Such a treatment would allow younger trees to be more robust than older.

Warfvinge and Sverdrup (1992) and later Jönson *et al.* (1994) showed that growth and decomposition are both dependent on soil moisture. The expression was normalized to describe the deviation in effect from the level at a soil moisture saturation of 0.2. The expression then becomes:

$$f(w) = \frac{247\, w^{3.4}}{1 + 9.4\, w^{3.4}} \tag{20.6}$$

where $w$ is soil moisture saturation. Soil moisture was calculated as:

$$w = 0.2 \frac{Q(t)}{Q_{mean}} \tag{20.7}$$

where $Q$ is the amount of total water coming in to the soil. This was used to approximate the effect on growth by drought.

Defoliation ($DF$) is calculated from the uptake reduction (Dr U. Söderberg's empirical equations as described in Sverdrup and Warfvinge, 1993b):

$$DF = 100 - 25.229 \frac{G}{G_{max}} - 63.159 \left(\frac{G}{G_{max}}\right)^2 \tag{20.8}$$

The above equation is a purely empirical relationship derived from the Swedish Forest Inventory database.

## SITE-SPECIFIC INPUT DATA

It has been determined that Lake Gårdsjön acidified in the early 1960s (Renberg *et al.*, 1985). As a part of the lake and soil acidification study, the mineralogy of the soil as well as the bedrock was determined along with parameters concerning the soil texture and chemical status at a number of sites in the watershed (Melkerud, 1983; Olsson *et al.*, 1985). Input data have whenever possible been derived from actual measurements from the G1 catchment. Sometimes data from a subcatchment 150 m to the north-east, F1, were used as a substitute. The data used to run the model are shown in Table 20.1.

**Table 20.1.** Soil model input data for the G1 subcatchment at Gårdsjön. The soil property inputs are based on actual measurements. Initial conditions were calculated using PROFILE

| Parameter | Unit | 1 | 2 | 3 | 4 |
|---|---|---|---|---|---|
| Soil layer type | | O | A/E | B | B/C |
| Soil layer thickness | m | 0.05 | 0.05 | 0.23 | 0.13 |
| Moisture content | $m^3 m^{-3}$ | 0.20 | 0.15 | 0.25 | 0.30 |
| Soil bulk density | $kg m^{-3}$ | 200 | 880 | 610 | 1250 |
| Specific surface area | $m^2 m^{-3} \times 10^{-6}$ | 0.35 | 1.34 | 1.40 | 1.18 |
| CEC | $keq kg^{-1} \times 10^{-6}$ | 300 | 200 | 300 | 350 |
| $CO_2$ pressure | times ambient | 2 | 5 | 20 | 30 |
| Dissolved organic carbon | $mg l^{-1}$ | 35 | 30 | 20 | 5 |
| log Gibbsite constant | $kmol^2 m^{-3}$ | 6.5 | 7.6 | 8.6 | 9.5 |
| Inflow | % of precipitation | 100 | 50 | 46 | 42 |
| Percolation | % of precipitation | 50 | 46 | 42 | 42 |
| Mg+Ca+K uptake | % of total max | 20 | 30 | 30 | 20 |
| N uptake | % of total max | 50 | 40 | 10 | 0 |
| K-feldspar | % weight | 15 | 15 | 18 | 19 |
| Oligoclase | % weight | 14 | 14 | 15 | 16 |
| Hornblende | % weight | 0 | 0.5 | 1.5 | 1.5 |
| Epidote | % weight | 0.1 | 0.5 | 0.5 | 0.5 |
| Garnet | % weight | 0 | 0.1 | 0.1 | 0.1 |
| Biotite | % weight | 0 | 0.5 | 0.5 | 0.5 |
| Chlorite | % weight | 0 | 0.4 | 0.4 | 0.4 |
| Vermiculite | % weight | 0 | 0 | 3 | 0 |
| Apatite | % weight | 0 | 0.1 | 0.2 | 0.3 |
| Kaolinite | % weight | 0 | 0 | 5 | 0 |

The sulphur deposition was studied by long-term monitoring (Hultberg, 1985). The deposition values delivered to the modelling effort correspond to a long-term deposition of 1.41 keq $ha^{-1} yr^{-1}$ of S, 1.05 keq $ha^{-1} yr^{-1}$ of N, 2.68 keq $ha^{-1} yr^{-1}$ of Cl, and 2.22 keq $ha^{-1} yr^{-1}$ of Na. Throughfall was 0.66 keq $ha^{-1} yr^{-1}$, 0.57 keq-$ha^{-1} yr^{-1}$ and 0.45 keq $ha^{-1} yr^{-1}$. Canopy leaching and total deposition amounted to 1.68 keq $ha^{-1} yr^{-1}$ of Ca+Mg+K. This all amounts to a total acidity input of approximately 1.8 keq $ha^{-1} yr^{-1}$.

Historic deposition scenarios were reconstructed using a number of sources, such as historic deposition measurements at Rothamsted (Sverdrup *et al.*, 1995b), and data from EMEP based on historical emission records (Lövblad *et al.*, 1992; Mylona, 1993). It was assumed that the roof would be installed forever to make the effect of the roof clear. The deposition inputs are shown in Figure 20.3.

The catchments have mixed coniferous stands (60 to 110 years old) of Norway spruce. Average soil thickness is 0.43 m. Tree uptake of nutrients is available from 1979 to 1981 in F1. Additional data are available from G1 on growth, but not uptake. The same elemental content as measured in F1 was used to estimate uptake. Growth in 1992, 1993 and 1994 was very small. We assume 20% of the N deposition to be absorbed in the lichens and bacteria in litter which is spread out under the roof. Uptake and cycling fluxes used in the calculation are shown in Figure 20.4a. Precipitation variations are available for the time 1960–91 in the region; these were

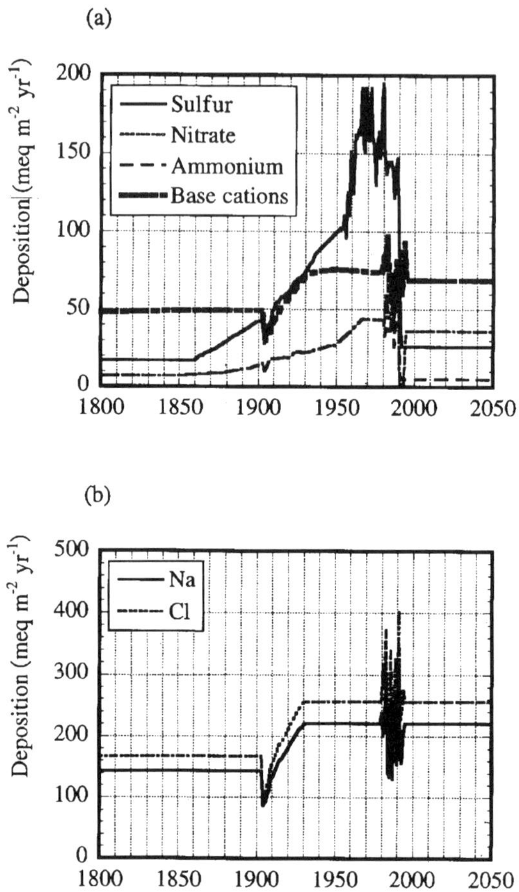


**Figure 20.3.** Deposition of sulphur, nitrogen, chloride and base cations used as inputs to the model calculations

used to generate infiltration rates (Figure 20.4b). Outside this range the series was randomly repeated.

## RESULTS

Outputs from the model included soil solution concentrations of all major ions layer by layer during the period 1800–2050. Base saturation and weathering rate were also produced for every layer as a function of time. In addition, for this run, the value of the BC/Al function in each layer, the nutrient limitation, the value of the water function and the defoliation at the site were considered, and some of the results are shown below.

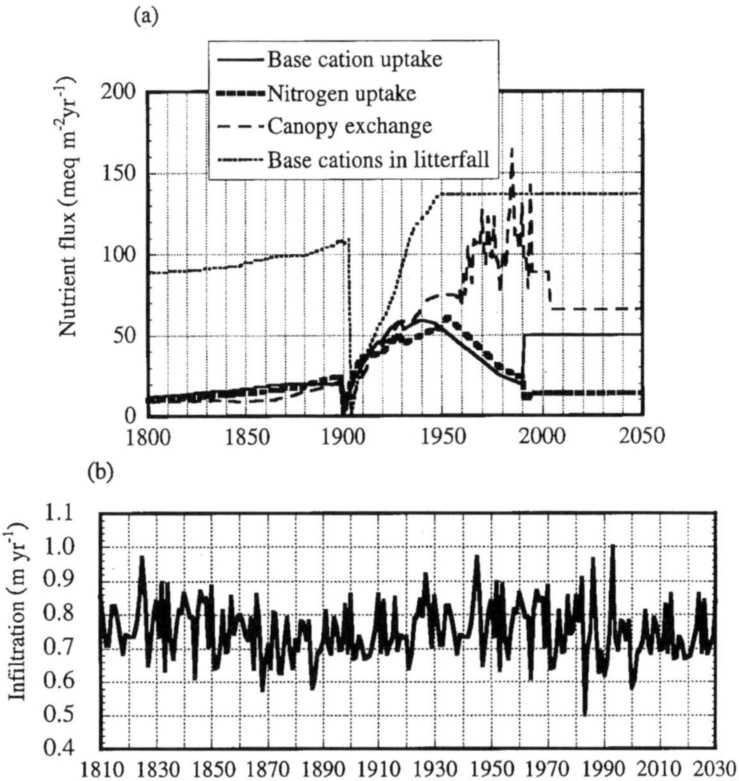


**Figure 20.4.** (a) Uptake and circulation of substances in the SAFE calculations. Uptake was scaled using standard yield curves from forestry handbooks. (b) Annual variations in net infiltration rate since 1810, used in the calculations

## The long perspective on soil chemistry 1800–2050

### *pH and base saturation*

Soil solution pH and base saturation are shown in Figure 20.5. The current runoff chemistry is correctly predicted, but with less variability than observed. The soil acidified slowly starting around the turn of the century. Then also base saturation started to change. The soil at G1 is a little different from the normal forest soil, because of its high CEC and content of organic matter down the whole profile. This makes the changes in base saturation appear low. This probably arises from the simplifications in hydrology, where annual variations are averaged.

There is a slight recovery in pH after installation of the roof, but the residual sulphate and nitrogen returned under the roof is sufficient to keep the pH suppressed to a low level. This suggests that drastic deposition reductions are required to make Lake Gårdsjön recover from acidification. Base saturation in the upper layers also recovers quickly under the roof, but recovery of the deeper layers requires a very long time.

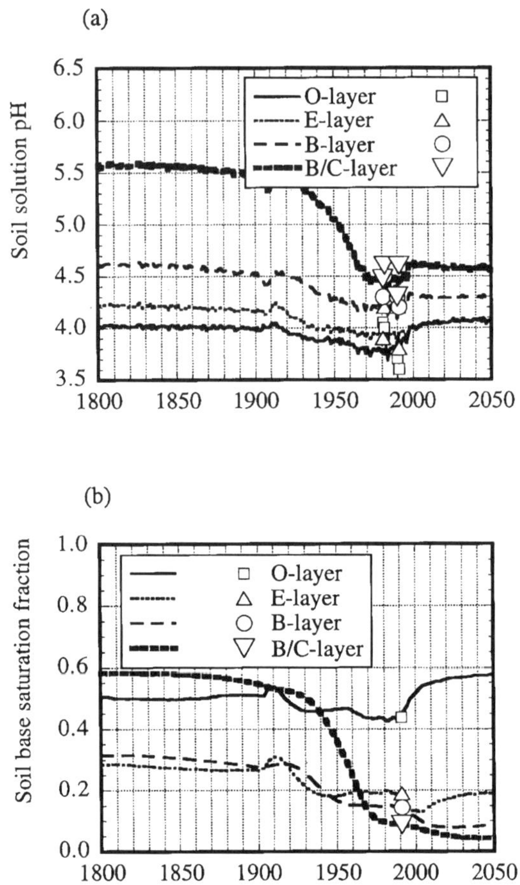


**Figure 20.5.** Calculated change in soil solution pH (a) and base saturation (b) during the period 1800–2050. Model output is shown as a full line, data are represented by points

*BC and BC/Al-ratio*

Figure 20.6 shows calculated base cation concentrations (excluding sodium) and the BC/Al ratio during the period 1800–2050. The increased base cation concentration starting early in the 20th century was the first sign of soil acidification. The increase is supplied by the decrease in base saturation. After installation of the roof, the base cation concentrations decline rapidly in the soil solution and in the stream. This is caused by the lack of acid to drive base cations off the ion exchange matrix at the same time as we allow vegetation uptake to stay high in the calculations. The predicted decline is corroborated by the field observations. The scenario would have been different for base cations if base cation uptake had been coupled with N uptake.

The BC/Al ratio is used as a diagnostic parameter for forest acidification effect. The BC/Al ratio attains values below 1.0 only in the B/C-layer, but this is sensitive to the size of net uptake. The model predicts that the BC/Al-ratio will not recover above 1 before 2050 in the B/C-layer.

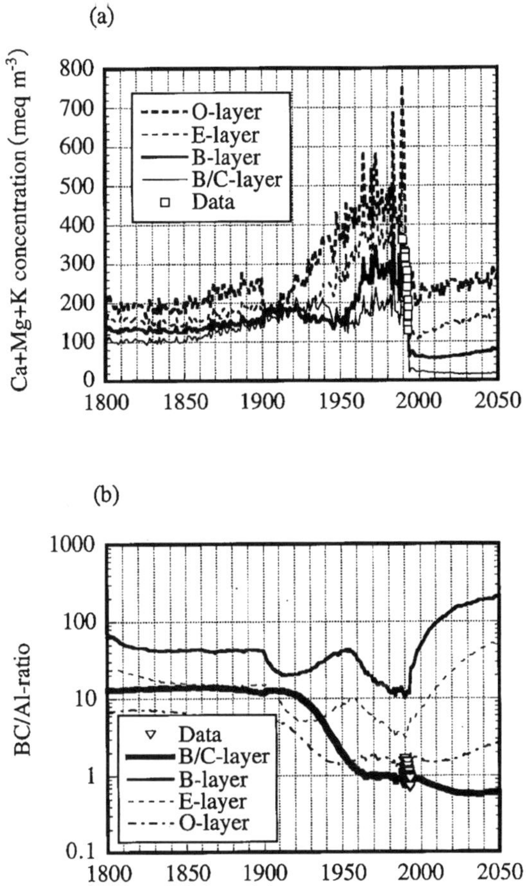


**Figure 20.6.** Calculated change in base cation concentration and the BC/Al ratio during the period 1800–2050. The BC/Al ratio observed in the runoff corresponds to the ratio calculated for the fourth layer

### *Al in soil and runoff*

Figure 20.7 shows the development of soil Al and runoff pH during the period 1800–2050. It can be seen that the soil had acidified, pH had decreased and Al had already increased 60–70 years ago, before the time when the first Swedish soil scientists collected samples and characterized the "natural" chemical state of forest podzols. The calculation for Al shows an increase during the 1940s to significant levels in the 1950s. This suggests that acidification episodes may have occurred in the lake by then. The general concentration of Al in the runoff in modern times is reproduced by the model. The sharp decline in Al predicted by the model may also be seen in the data. Al will have decreased significantly in runoff before 2000 as a result of the roof installation. This suggests that drastic deposition reductions are required to make Lake Gårdsjön recover from acidification.

pH recovers, but not back to the values in 1800. The main reason for this is that throughfall is diverted away from the catchment and only substituted with a clean

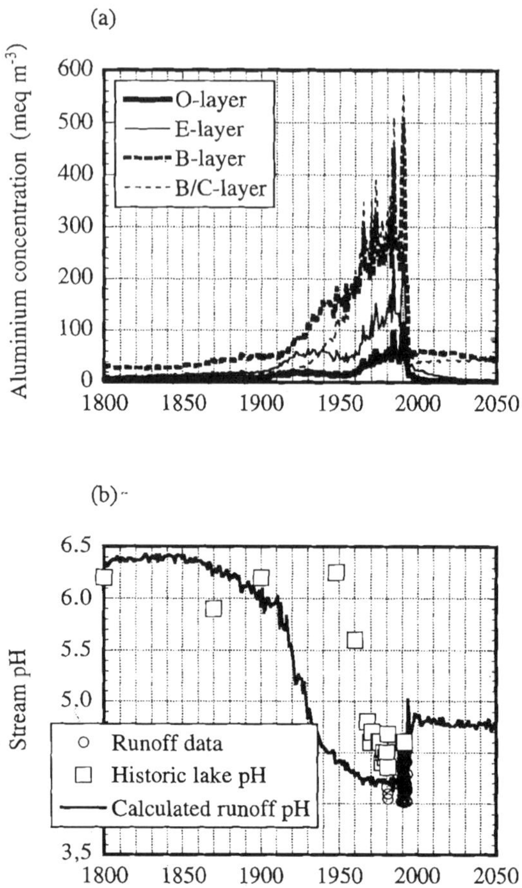


**Figure 20.7.** Calculated change in soil solution total Al and stream pH during the period 1800–2050. After installation of the roof, soil solution and runoff Al are predicted to fall to insignificant levels

rain made of diluted seawater. This short-cuts one important part of the return path of the nutrient cycle. Effective uptake by the trees is now equal to total uptake, which is quite large. Net base cation uptake is thus increased to 2.5–3 times the amount of N uptake, implying a strong biological acidification.

*Lake Gårdsjön pH*

Figure 20.8 shows pH in the lake for the period 1800–2050 as calculated with the model, but using data from subcatchment F1 as proxy for the whole catchment. The properties of F1 are very similar to G1, but soil depth is 0.69 m instead of 0.43 m in G1. F1 is a reference catchment and has no roof. Inputs of acidity and water at F1 may be 10% less. The mineralogy is the same in both catchments. The fit with the lake data is apparently good, suggesting that F1 is representative of the whole catchment.

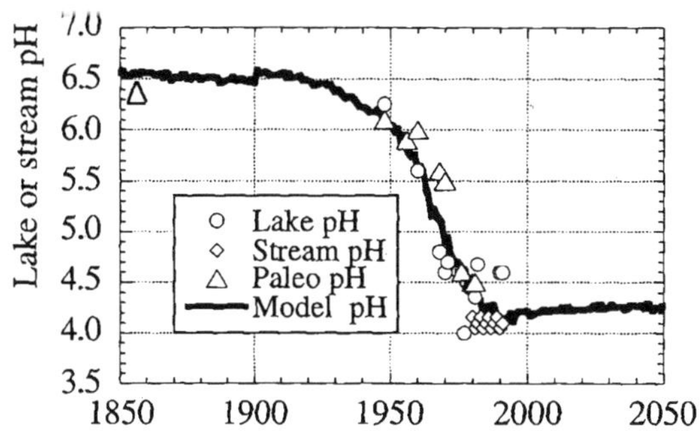


**Figure 20.8.** Lake Gårdsjön pH calculated with the SAFE Model using data from Catchment F1 as proxy for the whole catchment, compared with measured pH values for the lake and the F1 output stream and paleological inferred pH values (the latter from Olsson *et al.*, 1985)

### The short perspective on soil chemistry, 1980–2000

#### *pH and Al*

Figure 20.9 shows pH and Al for the period 1980–2000. The symbols represent *in situ* measurements. The general outlines of the pH values observed in the individual layers are predicted by the model. After installation of the roof in 1991, the pH recovers somewhat erratically, an effect mainly of variations in water flux. The peaks in Al are also caused by variations in hydrology and deposition. The general level of Al is also predicted, but the model cannot predict the short-term dynamics. This is mostly caused by the use of a 1-year step in the resolution of the input data; the model itself could operate on a finer time-scale, provided input with finer resolution could be generated. After the installation of the roof, Al decreases quickly to low levels. This response may indicate that the acid input reduction caused by the roof would be sufficient to re-create viable conditions for fish in the lake and its tributaries.

#### *BC, Cl and $SO_4$*

Figure 20.10 shows the base cation concentrations, and Figure 20.11 the sulphate and chloride concentrations for 1980–2000. The data in the runoff correspond to the concentrations predicted for the third soil layer. The mismatch between the runoff chloride data and the model calculation is diagnostic of uncertainty in the hydrological data or the chloride analysis. Base cation concentrations fall to very low levels soon after the roof was installed; this is also reflected in the measurements. The main reason for this is not returning all the canopy leaching and exchange to the soil. Thus uptake rises from 20 meq $m^{-2}$ $yr^{-1}$ to 50 meq $m^{-2}$ $yr^{-1}$. Since at the same time a pH increase occurs, then additional base cations are consumed by ion exchange to restore base saturation. Little is then left for leaching, and possibly base cation deficiency in the trees will occur.

Calculated change in soil solution chemistry for sulphate during the period 1980–2000 is seen in Figure 20.11. The mismatch between the data and the calculations may explain some of the anomalies in the output. The sulphate signal suggests either of two things: (1) sulphate adsorption may be active in the catchment, delaying the decline in sulphate concentration, or (2), there is an error in hydrology together with

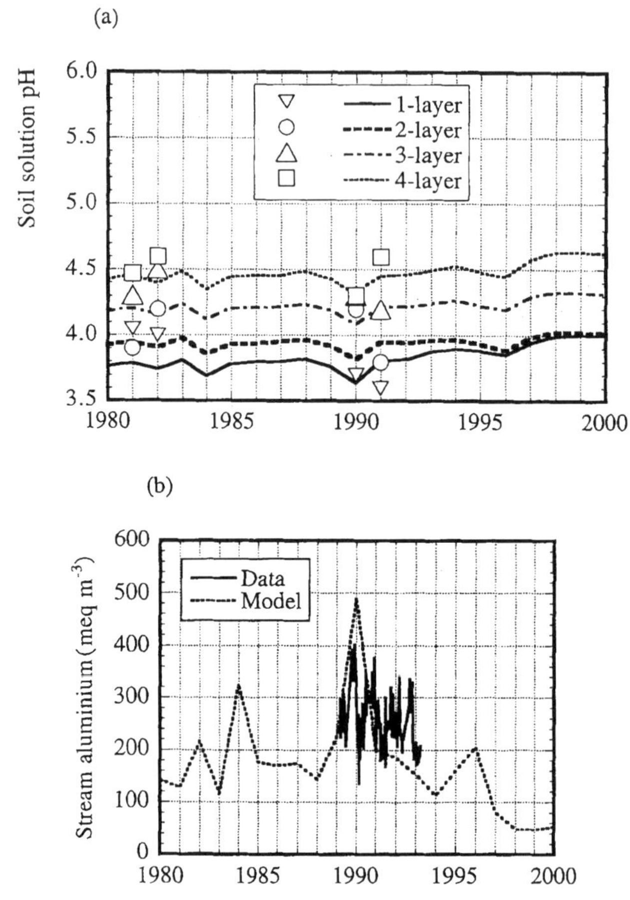


**Figure 20.9.** Calculated change in (a) solution pH and (b) stream aluminium concentrations during the period 1980–2000

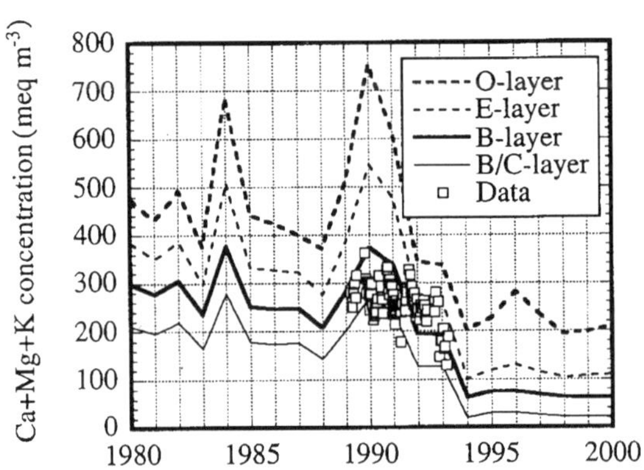


**Figure 20.10.** Calculated change in base cation concentrations during the period 1980–2000

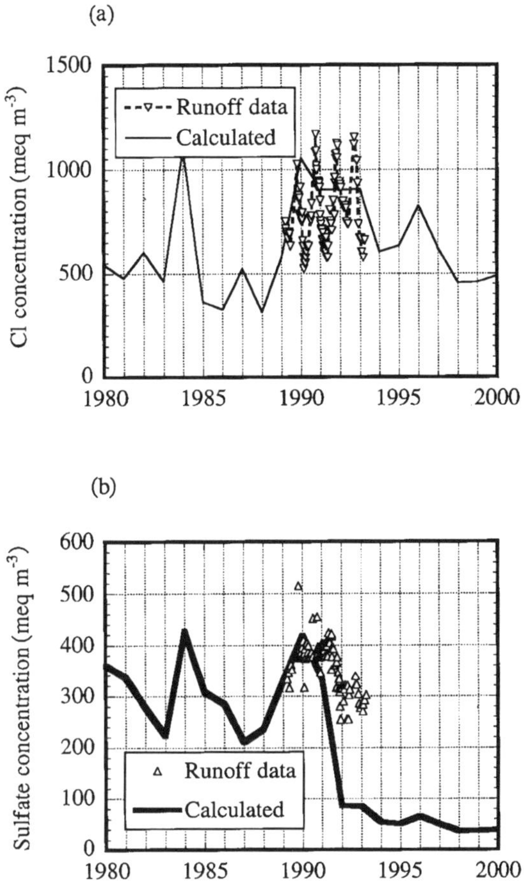


**Figure 20.11.** Calculated change in (a) chloride and (b) sulphate concentrations during the period 1980–2000

sulphate being transported in under the roof with litterfall contaminated by sulphur deposition. The fact that using average hydrology gives a better match with the observed data is a problem to consider when evaluating the results.

In general, the runoff chemistry is reasonably well predicted with respect to absolute levels, but with less variability than observed. This probably comes from the simplifications in hydrology and deposition, where annual variations are averaged. Earlier studies have shown that daily or weekly hydrology inputs together with water availability feedback on uptake are required to provide the proper chemical dynamics (Warfvinge and Sverdrup, 1992).

## Mass balances for base cations and alkalinity

The model outputs were used to make mass balances for base cations and alkalinity. Figure 20.12 shows the mass balance for base cations, where cation exchange was calculated as:

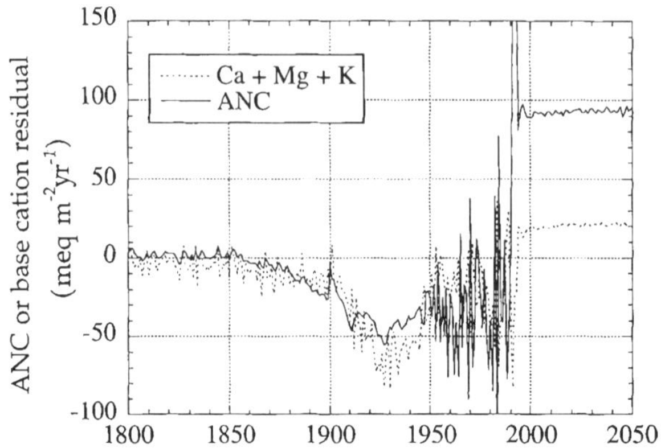


**Figure 20.12.** Mass balance for base cation and ANC at G1 based on the model calculations

$$\Delta = W_{BC} + D_{BC} - [BC^{2+}]Q - U_{BC} \tag{20.9}$$

where $W$ is the weathering rate, $D$ the base cation deposition, and $U$ the uptake. The deficiency in ANC is covered completely by the base cations taken from the soil ion exchange complex. It can be seen that during the first phase of acidification, more and more base cations are released from the exchange complex. But as the reservoir becomes more and more depleted, fewer base cations can be released. After the roof has been installed, leaching decreases strongly and most weathering is used for filling up the ion exchanger. The weathering rate in the roofed catchment is at present 0.30–0.35 keq $ha^{-1}$ $yr^{-1}$, but only 0.28 keq $ha^{-1}$ $yr^{-1}$ is release of Ca, Mg and K. The difference between net ANC and base cations after roof installation helps explain why the water chemistry is recovering faster than the base saturation. The decrease in acidity input does not really manifest itself to any degree in the runoff, since the ion exchange mechanism does not really deliver alkalinity. pH in the runoff will not start to improve before base saturation has improved significantly. The process that delayed the acidification of the runoff water is now delaying recovery by the reverse process.

Figure 20.13 shows the individual terms in the mass balance for base cations at G1. Net uptake increases after the roof has been installed as some of the base cations released through canopy exchange are removed and not returned under the roof.

### Assessing possible forest effects

Defoliation was calculated with the empirical expressions given earlier, based on calculated BC/Al and soil moisture as input. Calculated change in defoliation with time for catchment G1 is shown in Figure 20.14. The defoliation level in the catchment can be approximately reproduced with the model. The figures illustrate the resolution required in biological data to decide the issue of how much was caused by aluminium and how much by drought. Proper observations on defoliation are available for 1990–95, whereas the data for 1982 and 1984 are based on hearsay and old photographs. In the present calculation, the effects on growth by drought and

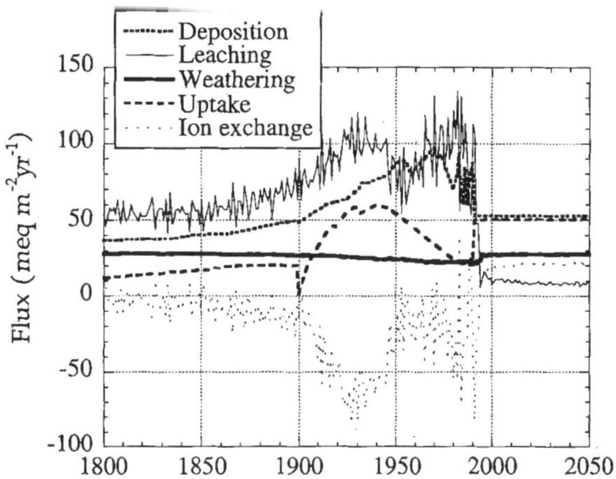


**Figure 20.13.** The individual terms of the mass balance, expressed as ANC

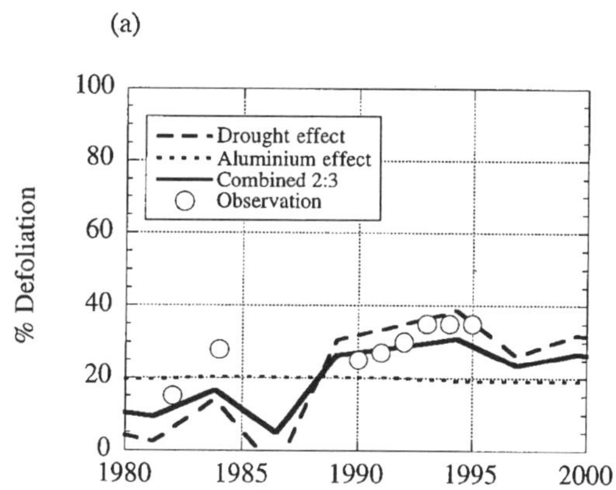


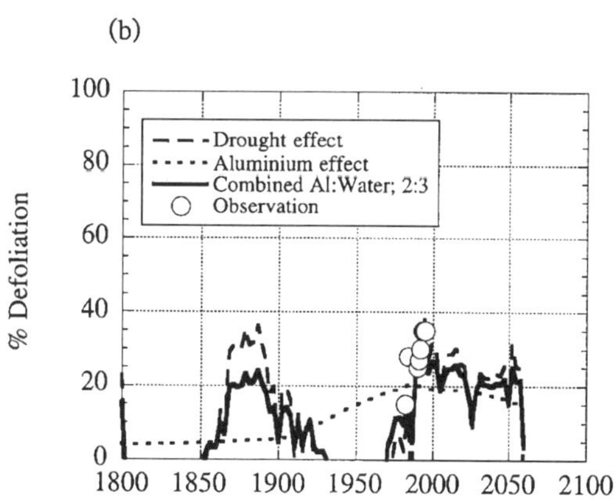


**Figure 20.14.** Calculated change in defoliation with time for catchment G1. The average defoliation in Gårdsjön is obtained by using simple empirical plant-chemistry and soil moisture feedback relationships

aluminium were calculated independently, and the effects were put in linear supposition.

## DISCUSSION

The observed changes in soil chemistry are explained by the model mainly as a result of changes with time in S and N inputs. Thus the accuracy of the input for these variables and the accuracy for the inputs on nutrient cycling largely determines the quality of the output. The estimates for total inputs of N and base cations under the roof are actually quite uncertain. The rate of growth and the N transferred under the roof are very important for the long-term predictions. The time required for recovery of the soil after a reduction in emissions, and the degree of recovery connected with a specific reduction, is of importance for evaluating different reduction strategies.

The model calculations indicate that the soil chemistry will recover, but slowly. Recovery of the soil solution chemistry will take several decades, recovery of the base saturation will require several centuries.

The calculations indicate that the roofed experiment should continue for another 5 years in order to observe the initial stages of chemical recovery of a catchment. More emphasis on soil chemistry is recommended for a longer time but with longer intervals between data collection. Intensified monitoring of biological variables and diagnostic parameters for forest damage in both F1 and the roofed catchment G1 could be very important for answering some of the key questions of exactly what is causing defoliation and forest decline. The impact on forest vitality could then be better assessed as large enough soil chemistry changes will have occurred.

The observed change in soil chemistry is explained by the model as mainly a result of soil acidification due to S and N inputs. The evolving canopy of the forest stand at the site acts as an effective collector of deposition, thus increasing the acidity load to the site by perhaps 50% when the canopy closes in the mature stand. Uptake of base cations for incorporation in biomass contributes to the removal of base cations from the soil.

The time required for recovery of the soil after a reduction in emissions, and the degree of recovery connected with a specific reduction, is of importance for evaluating different reduction strategies. Some of the damage caused to the soil structure at the site by acid deposition may not be reversible. Whilst some of the decrease in base saturation is undoubtedly an effect of pH-dependent surface charge, much is not. The Macaulay Land Use Research Institute was able to detect some measurable changes in clay mineral properties with time in the Geesecroft Wilderness at Rothamsted samples. Increasing amounts of Al-hydroxy interlayer material blocks interlayer spaces, and reduce expansibility, changing CEC. Such alterations are not necessarily reversible, in which case the soil has been permanently changed.

The calculations show that the time required for recovery is very long indeed. It emphasizes the fact that if acidification is allowed to proceed to a point where the ion exchange reservoir is nearly empty or tree damage is evident, then recovery is slow and the time for the system to self-repair back to a state which can be called acceptable in terms of ecological damage will be in the order of centuries. Thus, any policy that envisions waiting until damage is evident is bound to fail if ecosystem protection is an objective.

## CONCLUSIONS

The model calculations indicate that the soil chemistry will recover, but slowly unless acid deposition is very dramatically reduced. The model calculations also indicate that the future development of the nitrogen dynamics in the soil is important for the outcome.

A better fit could have been obtained by manipulating the deposition scenarios, the weathering rate or the fluxes of elements returned under the roof. This was not done at the present time, however, as it conflicts with the input data specified by the monitoring team. Before input data to the model are changed in any significant way we felt it was necessary to assess carefully whether such changes can be justified within the observed range of these parameters. There are some uncertainties as to exactly how large these return fluxes are, and the accuracy of some of the input variables.

The calculations indicate that the experiment should continue for at least another 5 years in order to observe the chemical recovery of the catchment. The next 5 years should also be used to improve the accuracy of some of the element fluxes, especially in the internal cycling of base cations and nitrogen, as well as to continue observations of biological parameters and indicators of tree damage.

## ACKNOWLEDGEMENTS

Several colleagues contributed data to this study, and we would like to thank Professor I. Nilsson, Soil Science, and Dr P. A. Melkerud, Forest Soils, Swedish Agricultural University, Uppsala, and G. Borg, Geology Department, Gothenburg University.

Both the SAFE and PROFILE models are available from the authors at Lund University free of cost, or can be downloaded via the World Wide Web from our FTP server at the address: HTTP://WWW2.CHEMENG.LTH:SE//.

## REFERENCES

Hultberg, H. (1985) Budgets of base cations, chloride, nitrogen and sulphur in the acid Lake Gårdsjön. *Ecological Bulletins (Stockholm)*, **37**, 133–157.

Jönsson, C., Warfvinge, P. and Sverdrup, H. (1994) Uncertainty in predicting weathering rate and environmental stress factors with the PROFILE model. *Water, Air, and Soil Pollution*, **81**, 1–23.

Lövblad, G., Amann, M., Andersen, B., Hovmand, M., Joffre, S. and Pedersen, U. (1992) Deposition of sulfur and nitrogen in the Nordic countries: Present and future. *Ambio*, **21**, 339–347.

Melkerud, P. (1983) *Quarternary Deposits and Bedrock Outcrops in an Area Around Gårdsjön, South-West Sweden, with Physical, Mineralogical and Geochemical Investigations.* Reports on Forest Ecology and Forest Soils No. 40, Swedish University of Agricultural Sciences, Uppsala.

Mylona, S. (1993) *Trends of Sulphur Dioxide Emissions, Air Concentrations and Depositions of Sulphur in Europe since 1880*. EMEP MSC-W rep. 2/93, Oslo, 96 pp.

Olsson, B., Hallbäcken, L., Johansson, S., Melkerud, P.-A., Nilsson, I. and Nilsson, T. (1985) The Lake Gårdsjön area — physiogeographical and biological features. *Ecological Bulletins (Stockholm)*, **37**, 10–28.

Renberg, I., Hellberg, T. and Nilsson, M. (1985) Effects of acidification on diatom communities as revealed by analyses of lake sediments. *Ecological Bulletins (Stockholm)*, **37**, 219–223.

Sverdrup, H. (1990) *The Kinetics of Chemical Weathering*. Lund University Press.

Sverdrup, H. and Warfvinge P. (1993a) Calculating field weathering rates using a mechanistic geochemical model — PROFILE. *Journal of Applied Geochemistry*, **8**, 273–283.

Sverdrup, H. and Warfvinge, P. (1993b) *Soil Acidification Effect on Growth of Trees, Grasses and Herbs, Expressed by the (Ca+Mg+K)/Al Ratio*. Reports in Environmental Engineering and Ecology 1:93, Department of Chemical Engineering II, Lund University, Sweden, 178 pp.

Sverdrup, H., de Vries, W. and Henriksen, A. (1990) *Mapping Critical Loads*. 1990:14. Nordic Council of Ministers, Copenhagen.

Sverdrup, H., Warfvinge, P., Blake, L. and Goulding, K. (1995b) Modeling recent and historic soil data from the Rothamsted Experimental Station, England, using SAFE. *Agriculture, Ecosystems and Pollution*, **52**, 223–259.

Warfvinge, P. and Sverdrup, H. (1992) *Hydrochemical Modelling*. In: P. Sanden and P. Warfvinge (eds), Vol. 2 of SMHI-RH *Modelling Groundwater Response to Acidification*, Norrköping, Sweden, pp. 79–118.

# 21 Modelling the Hydrology of the Covered Catchment

M. HAUHS, H. LANGE, R. CALLENBERG, R. HOCH, F. MOLDAN, A. RODHE, J. SEIBERT & K. H. BISHOP

Four different models were applied to the G1 ROOF dataset. Two of them (a mixing model and CXTFIT) are intimately connected to the residence time approach. The other two models (TOPMODEL and SilVlow) address flowpaths within the G1 catchment. Each application has been successful with respect to a reconstruction of the observed behaviour in runoff. Different attempts to interpret these model results in terms of a more spatially resolved (in the case of residence times) or more time-averaged (in the case of flowpaths) picture did not arrive at consistent representation of the internal dynamics. However, many of the problems encountered correspond to long-standing similar puzzles concerning the hydrology of the Gårdsjön area. The G1 catchment seems to recapitulate in a nutshell some typical unsolved problems in contemporary hydrology. It is suggested that it serves in the future as a primary experimental site where practical or theoretical limitations of prominent methods can be tested in a reproducible way.

## INTRODUCTION

The goal of the Covered Catchment Experiment (CCE) at Gårdsjön G1 is to study the reversibility of soil and streamwater acidification and ultimately the resulting implications for biological recovery (Hultberg and Bishop, 1995). There is widespread empirical evidence that soil chemistry is the crucial link by which acidic deposition causes changes in aquatic and terrestrial biota. Therefore numerous observations and measurement campaigns have been devoted to gaining an adequate understanding of the chemical and physical processes involved in transport of solutes through catchment soils at G1 (Chapter 5, this volume).

In this chapter four different types of models applied to these datasets will be synthesized and discussed in order to specify further the meaning of the term "adequate understanding" of the influence of transport process in controlling runoff chemistry. For this purpose, modelling is entirely restricted to the abiotic variables. Such an understanding of the physical and chemical aspects of the catchment's response to the altered depositional load is a prerequisite for subsequent studies on biological effects, and it is hoped that the population and ecosystem response can

*Experimental Reversal of Acid Rain Effects: The Gårdsjön Roof Project.*
Edited by Hans Hultberg and Richard Skeffington. 

eventually be scaled up from local and (instantaneous) chemical reactions. Therefore we assume as a first approximation that an interaction, e.g. with trees or micro-organisms, can be neglected for this purpose. Henceforth we will use only a "frozen" version of these biological aspects within the models. The implications and limits of this assumption will be revisited in the final discussion.

## THE MODELS

The observational techniques used at G1 correspond to the two complementary concepts of flowpaths and residence times (Chapter 5, this volume). Here we will try to arrive at a consistent and quantitative reconstruction of the G1 dataset with the help of four available models. However, these models employ widely different abstractions and each of them can only be related to separate aspects of the dataset. That is why we will introduce the modelling applications also from the alternative perspectives of residence times (rt) and flowpath analysis (fp).

The models that have so far been applied at G1 are:

| | |
|---|---|
| I—Mixing Model (Rodhe *et al.*, 1996) | rt |
| II—CXTFIT (Parker and van Genuchten, 1984) | rt |
| III—TOPMODEL (Beven *et al.*, 1984) | fp |
| IV—SilVlow (Schmidt *et al.*, 1995) | fp |

The first two models were used to describe the breakthrough of (inert) tracer solutes at G1. The first application (Rodhe *et al.*, 1996) uses the catchment response to the step pulse in $^{18}O$ after the start of irrigation with lake water. In this application the flow is transient and much of the "old" water that was initially contained in the catchment leaves through the "transpiration route". The second model was used to evaluate the steady-state Br tracer experiments in November 1993 and June 1995 (Hoch, 1995; Lange *et al.*, 1996). These models typically contain two to four parameters.

The third model, TOPMODEL, provides an explicit analysis of those flow pathways that are controlled by the catchment topography, i.e. surface runoff along zones of flow convergence (Seibert, 1993). In this model subsurface processes are approximated by a steady-state assumption (instantaneous equilibrium in the vertical direction). TOPMODEL usually includes about 12 different parameters. Difficulties with respect to the identifiability of these parameters through inverse modelling have been studied for other catchments (Beven, 1993, 1996).

The last model, SilVlow, uses an almost complementary approach to TOPMODEL. It assumes instantaneous infiltration into the soil, neglects all surface runoff and calculates the transient flow along a vertical cross-section through the catchment. This last model is fully distributed and represents a "physical" model of the soil water transport processes. Thus in contrast to TOPMODEL it actually solves a transport problem numerically. It contains a potentially unlimited number of parameter functions as the hydraulic characteristics can be varied independently for each point in the cross-section. (As the model uses a finite element scheme to solve for the water potential distribution, these "points" are actually two-dimensional triangles.) For this type of finite element model, which solves the Richards equation with appropriate boundary conditions, the corresponding inverse problem has also been studied (Binley *et al.*, 1991).

In the following sections the key assumptions and main results of these model applications will be summarized. Detailed descriptions of these applications are given elsewhere (Seibert, 1993; Hoch, 1995; Rodhe *et al.*, 1995; Callenberg *et al.*, unpublished).

## RESULTS

### Model I — The Mixing Model

The response of the catchment is described via a simple mixing reservoir into which the irrigated water with enriched $^{18}O$ is added (treatment started in April 1991). Until November 1991, the runoff remained isotopically distinguishable from irrigation water (Chapter 5, this volume). In the model the output from the single reservoir representing all soil water for this time interval occurs via runoff (60 mm) and transpiration (126 mm) (Rodhe *et al.*, 1996). The transient gradual response is clearly influenced and superimposed by rapid changes in groundwater levels and hence runoff during episodes (Figure 21.1). Yet over the 7 months such short-term variations can be neglected and approximated by the simple mixing approach.

This gradual breakthrough of the enriched $^{18}O$ water in runoff was additionally studied in several lysimeters in G1. Location 2B (see map, Figure 5.1, this volume), though only a few metres away from lysimeter nest 2A, nevertheless showed a very rapid breakthrough relative to the latter and to runoff. This variation may be reconstructed in the model by varying the amount of soil water and thus "old water" in the respective subcatchments upstream of these lysimeters. The group 2B is apparently dominated by water from a relatively small slope subcatchment with shallow soils, whereas the subcatchment feeding 2A seems to be similar to the whole of G1.

For any predictive application of such a model two conditions must hold. The total mixing volume of the reservoir needs to be determined independently and the (transient) flow conditions encountered during the tests must be "representative". It remains to be tested whether the results from the breakthrough under transient flow conditions lasting for the 7 months from April until November 1991 can be reconciled with the residence time distribution function derived from other tracer signals. An important question for flowpath modelling is whether the large differences in the respective size of upstream areas and mixing volume could be independently derived.

### Model II — CXTFIT

The second model was applied to all experiments where steady-state flow conditions were maintained between application and breakthrough of a tracer solute. This is a theoretical precondition to derive true travel time distribution functions. Data from three such experiments are available: the first two were conducted in the same

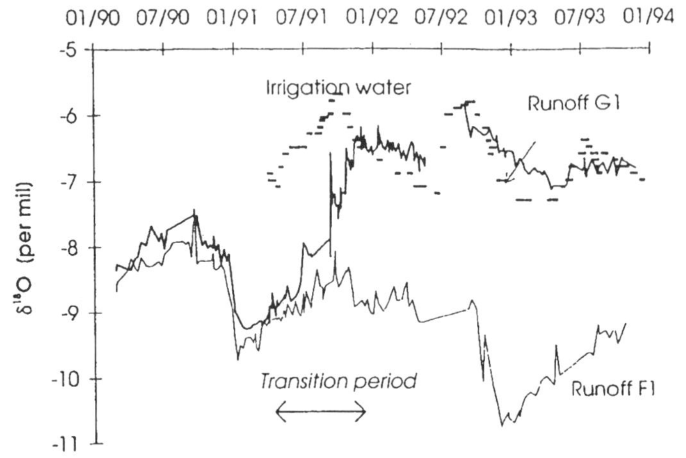


**Figure 21.1.** $\delta^{18}O$ of irrigation and runoff from G1 and of runoff from the uncovered reference catchment F1, February 1990 to December 1993 (from Rodhe *et al.*, 1996)

subcatchment of G1 (Lange *et al.*, 1996; Callenberg *et al.*, unpublished). The third experiment consists of a soil column taken from this subcatchment during steady-state flow, but prior to tracer application. This column was sprinkled in the laboratory at the same rate and an equivalent amount of tracer (same solution concentration) was applied (Seiler, 1995).

Unlike the simple (zero-dimensional) Mixing Model, CXTFIT describes the transport as a one-dimensional convection–dispersion phenomenon. That is why the different length scales in the field and laboratory experiments can be accounted for. Table 21.1 summarizes the characteristics of these three experiments. In the second field experiment, the applied tracer concentration was increased tenfold in response to the low Br recovery of the first experiment (Lange *et al.*, 1996).

CXTFIT was used in the so-called two-region version, modelling transport of an inert tracer in a flow region with mobile and immobile water. This model allows three parameters to vary independently: the Peclet number $P$; the ratio of mobile to immobile water ($\beta$); and a transfer coefficient describing the passage between the two types of soil water (represented here as the dimensionless $\omega$). The model is forced to have an asymptotic recovery rate of 100%. Therefore the missing Br was accounted for by an *a priori* reduction of the effective input concentrations for the various model runs (for a discussion of the Br budget see Lange *et al.*, 1996; Callenberg *et al.*, unpublished).

The best values for the three parameters (according to a least squares fit) were determined by an automatic calibration procedure (Parker and van Genuchten, 1984). Table 21.2 shows that the results reveal large discrepancies among the three datasets (see Figure 21.2 for the corresponding curves).

The expectation for the second experiment was that the conditions of the first one would be reproduced, with the exception of the higher tracer concentration. For an inert tracer, a linear scaling-up of concentrations should be possible after appropriate background subtraction. Due to technical limitations, the irrigation

**Table 21.1.** Main characteristics of the three steady-state experiments

| Experiment | November 1993 | June 1995 | Soil column 1995 |
|---|---|---|---|
| Flow rate (mm/h) | 3.7 | 3.1 | 3.1 |
| Effective length $L$ (m) | 40 | 40 | 0.08 |
| Br input conc. (mg/l) | 44 | 440 | 440 |
| Recovery rate (%) | 13.6 | 14.7 | 89.1 |

**Table 21.2.** Optimal parameter sets for CXTFIT. The values of $v$ and $D$ are "typical" combinations with the correct $P$ as generated by the program for the optimal parameter set

| Experiment | November 1993 | June 1995 | Soil column 1995 |
|---|---|---|---|
| $P(= vL/D)$ | 19.3 | 1.67 | 22.6 |
| $\beta = \theta_m/\theta$ | 0.255 | 0.388 | 0.652 |
| $\omega$ | 4.50 | 4.83 | 0.097 |
| $r^2$ | 0.975 | 0.982 | 0.990 |
| $v$ (m/d) | 36.6 | 10.6 | 0.11 |
| $D$ (m$^2$/d) | 75.7 | 254.6 | 0.00039 |

rate in 1995 was unfortunately about 20% lower than in 1993 (see Table 21.1). However, this relatively small difference does not explain the more than tenfold difference in the optimal Peclet number. Thus the transport parameters for Br were poorly reproduced without any clear hints to the reason for this unexpected variability. It is obvious from Table 21.1 that the assumption of conservative behaviour of Br is a very questionable one at least for the two field experiments. Some speculations about possible reasons for this lack of conservation are to be found in Lange *et al.* (1996). A satisfying explanation of this problem will hopefully emerge from an experiment with a "truly conservative" tracer. This will be studied in future experiments.

Even worse is the comparison between the field and laboratory parameter set. Mobile water becomes the dominant fraction as typical for soil column experiments. The $v$ and $D$ values suggested by the program provide hints towards the principal differences: whereas transport along a distance of about 40 m at the field scale is mainly by lateral saturated flow, this type of flow does not exist in the column, where transport is governed by unsaturated vertical flow. As in the first example comparing and reconciling different tracer experiments needs independent and extremely detailed spatially resolved flowpath information.

## Model III — TOPMODEL

An analysis of the varying water storage in the catchment as monitored by the groundwater level showed a strong correlation for most wells whenever groundwater exceeded a location-specific threshold value (Seibert, 1993; Chapter 5, this volume). This was especially true for groundwater wells located in topographically convergent flow zones, i.e. along the valley bottom. Motivated by this observation the

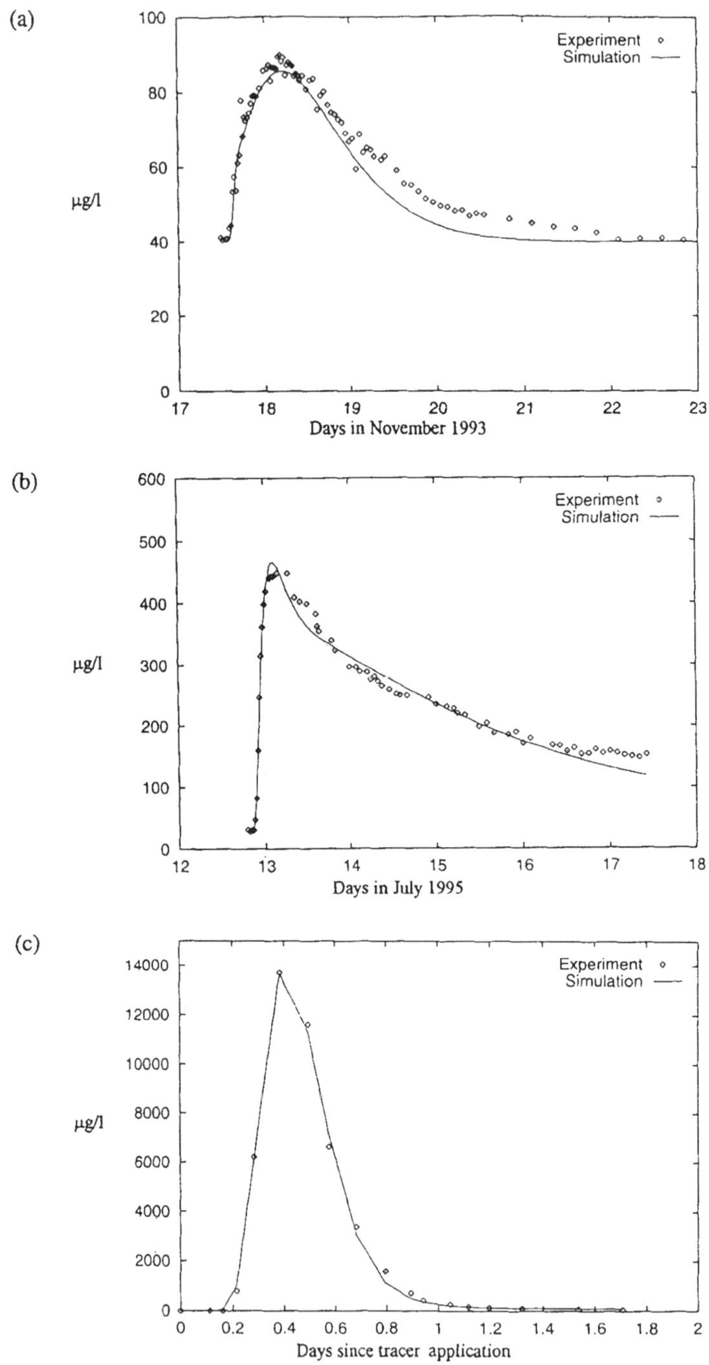


**Figure 21.2.** Bromide concentrations — measurements and fitted curves (based on the Parker–van Genuchten model): (a) 1993, (b) 1995, (c) soil column in laboratory

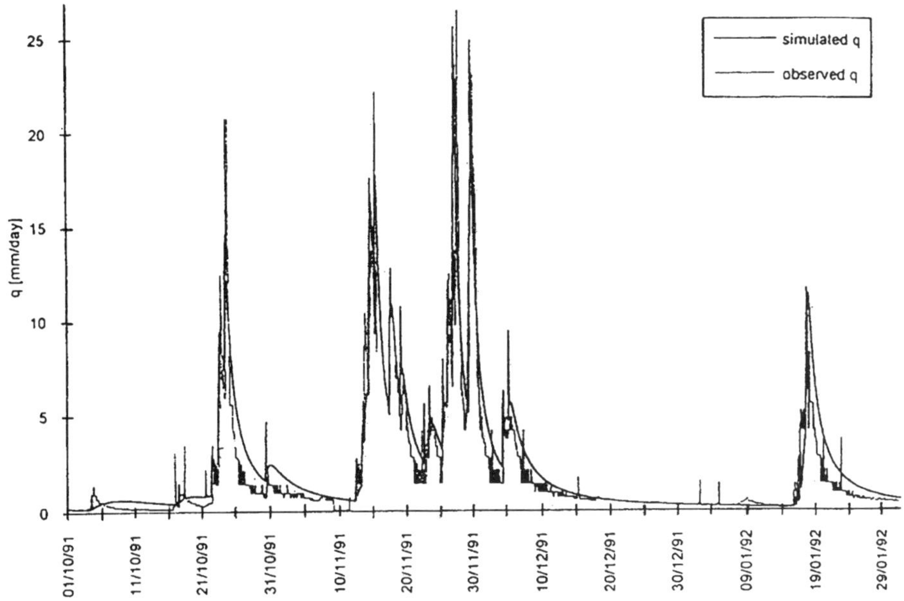


**Figure 21.3.** Runoff simulation from TOPMODEL (October 1991–January 1992). (From Seibert, 1993)

catchment was subdivided into eight topographically similar zones (Chapter 5, this volume). Zone number three of these eight areas is almost identical with the subcatchment irrigated during the two steady-state experiments.

In a quasi-distributed fashion TOPMODEL summarizes the total runoff response of G1 from the mean topographic characteristics of these subareas. However, this model application requires a calibration phase to derive the model parameters that were in this case shown to be most sensitive (soil transmissivity and a scaling factor for the decline of hydraulic conductivity with soil depth; Seibert, 1993).

A comparison of measured and calculated runoff for data outside the calibration period showed a satisfactory agreement (Figure 21.3). However, the subsequent analysis of model sensitivity revealed severe limitations for the interpretation of this result. The approximation of the runoff by TOPMODEL that is based on the stratified topographic index is no better than that of a simple linear reservoir model in which no stratification into subareas was used (Seibert, 1993). In addition any distributed soil variables such as the groundwater levels within the eight subareas that were included into TOPMODEL were only poorly reproduced (Seibert, 1993).

Thus, the explicit use of detailed information on topography and soil depth could not be shown to yield a more powerful reconstruction of the observed dataset than much simpler models. The advantage that TOPMODEL is closer to a physical description of the transport process is fully compensated by the fact that the two most sensitive parameters needed to be specifically calibrated to G1.

## Model IV — SilVlow

The first steady-state tracer experiment in November 1993 represents one of the most extreme runoff events at G1. It combines rapidly changing transient flow situations (due to two breaks in power supply and at the end of the irrigation) prior to and within the steady-state regime (Lange *et al.*, 1996). Therefore it was selected as the most challenging example for the saturated–unsaturated two-dimensional flow modelling.

We performed a simulation of the water status in the catchment G1 based on a two-dimensional finite element Richards equation solver (Schmidt *et al.*, 1995). In order to reduce three dimensions to two, a typical cross-section was assumed. The basis for this was a contour plot of the area. The cross-section has an upper boundary at the water divide of the sprinkled area towards G2 (see map, Figure 5.1, this volume). Then the line followed the steepest descent down to the weir.

The depths of the humus and mineral soil respectively were available for a regular $5 \times 5$ m mesh across G1 (Nyberg *et al.*, 1993). We divided the two major soil types into a thin humus layer with high water conductivity and storage capacity, and a mineral soil layer with a lower conductivity and storage capacity. As a basis for pK and pF functions we used results of detailed measurements carried out in the laboratory (Nyberg, 1995). Those measurements suggested a $K_s$ of $10^{-3}$ m/s and $10^{-5}$ m/s respectively and a $\theta_0$ of 0.97 and 0.7.

The actual numerical solution was performed using a program based on a finite element algorithm (SilVlow). The altitudes and depths of the soil layers at the grid points lying on the path served as the core for the mesh of nodes. Between these points additional nodes were added by a linear interpolation to achieve a finer discretization. This resulted in a mesh of 300 nodes and 472 triangles.

The first tracer experiment took place in winter. Therefore no sink terms accounting for transpiration were used. The precipitation, distributed homogeneously onto the whole area, was the influx for the Neumann condition on the upper boundary. At the outlet, boundary conditions were chosen as $q.\mu^{\perp} = 0$, where $q$ is the output flux, and $\mu^{\perp}$ the vector perpendicular in direction to the lower boundary at the weir. Zero flux conditions held for the rest of the boundary.

Due to the high precipitation applied during the steady-state experiment, large parts of the soil were saturated. This caused severe problems for the convergence behaviour of the algorithm. The only solution we found was to assume much higher saturated conductivities relative to the laboratory measurements ($K_s$ = 0.2 m/s and $9 \times 10^{-3}$ m/s respectively). Then it was possible to achieve a solution including mixed saturated as well as unsaturated parts of the flow region. Just increasing conductivity values over the whole range of water potential leads to solutions with more or less unsaturated conditions throughout the cross-section, in contradiction to the field observations.

Comparing measured and simulated courses of runoff and potentials at several points, a fairly good agreement can be observed (see Figure 21.4). The problems of the simulation are the unrealistic high conductivity in the saturated domain and the fact that it was not possible to achieve acceptable results when restricting the precipitation to the slope, in agreement with the experiment. The runoff reacted

(a)

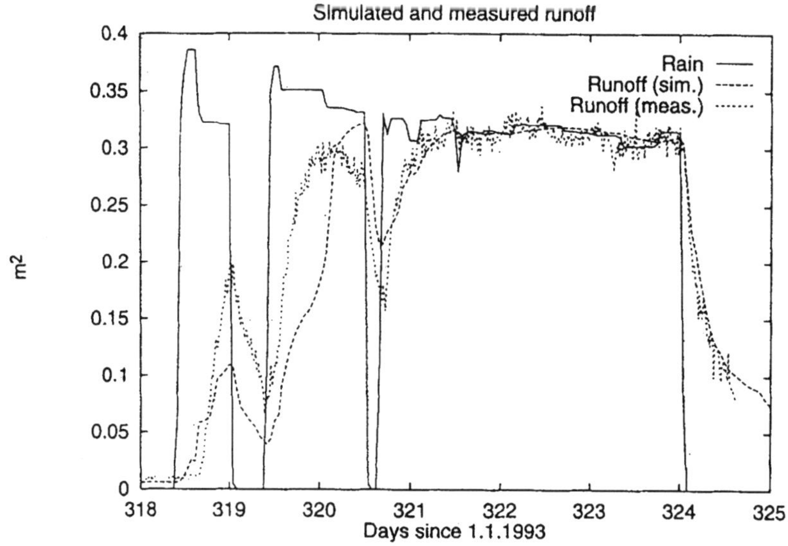


(b)

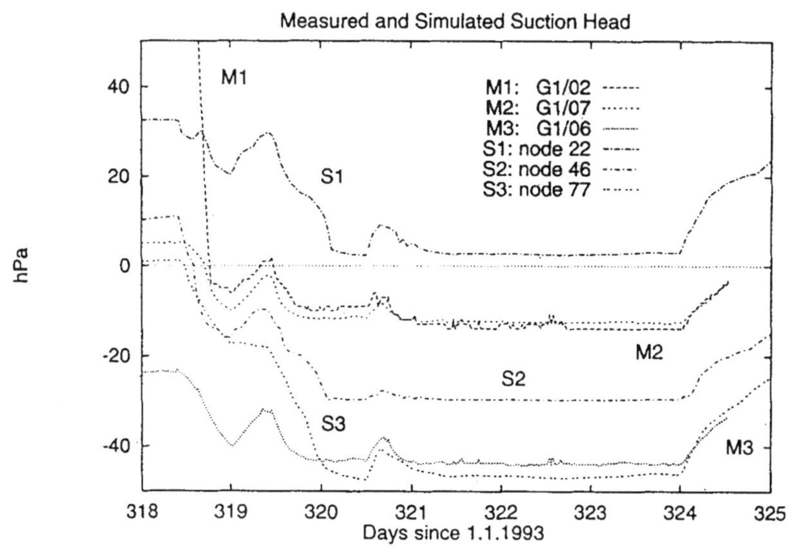


**Figure 21.4.** Results of SilVlow-simulations: (a) runoff, (b) suction head at different points

much too slowly. We suppose this could be explained by the drawback of the two-dimensionality and/or the transfer of hydraulic parameter measurements on soil columns to field conditions.

In order to answer the flowpaths questions posed by the residence times models (see above) a three-dimensional version seems to be indispensable. However, it is already known that (i) the inverse problem is at least practically unsolvable (Binley *et al.*, 1991) and (ii) the measurement effort necessary to derive the full heterogeneity of the parameter functions at the appropriate scale would practically destroy the G1 catchment (Hultberg, pers. comm.).

## DISCUSSION

In this chapter we discussed four different model applications to G1, covering a wide range of approaches to explain the observed hydrology. It is not possible directly to compare the results of (for instance) the "zero-dimensional" mixing model with the distributed two-dimensional model, but even if one restricts the comparison to different applications of the same model, the conclusions remain unclear.

Laboratory experiments do not match with field observations. This has been observed in several attempts to use laboratory conductivity results under field conditions (Lischeid, 1995). Here, however, a deeper inconsistency seems to be encountered: (i) residence time distributions derived under transient flow conditions do not match those derived under steady state (even when related to flow instead of time), and (ii) the reproducibility of steady-state experiments is elusive, e.g. it is highly sensitive to the actual flow rate established or requires completely inert tracers.

On the other hand the total mixing volume for G1 derived by Seibert (1993) (see also Chapter 5, this volume) matches the mean value of about 200 mm given by Rodhe *et al.* (1996) for the $^{18}O$ breakthrough reconstruction. Thus under transient flow situations it appears as if the complete soil water content is involved in the mixing process. This behaviour contrasts with the parameter set describing steady-state experiments (see Table 21.2). In 1993, for example, the tracer concentration already approached background values again after about 25% of total soil water storage was turned over (Lange *et al.*, 1996). These results obtained under steady-state, high-flow conditions were also not unexpected. It is very likely that the pore water velocities are highly stratified under such circumstances: (i) vertically into the most conductive upper layers and (ii) microscopically into the largest pore classes.

Differences in the preceding tracer experiments and catchment manipulations at Gårdsjön can also be understood in the light of these discrepancies. In two $^{3}H$ tracer experiments at F1 and F3 less than 50% was recovered over the following 10 months under transient flow conditions (Nyström, 1985). The difference from the $^{18}O$ step pulse behaviour is probably due to the fact that the $^{3}H$ application occurred within a regular mesh of points ($25 \times 25$ m), each at a soil depth of 40 cm. The effect of direct application within the mineral soil is probably that it takes much longer for the tracer to reach any of the (shallow or large pore) flowpaths that are characterized by short residence times.

A different tracer behaviour more typical of the above tracer experiments with Br was revealed from the application of NaCl and elementary sulphur at the soil surface of L1 and F5 respectively (Hultberg *et al.*, 1990, pers. comm.). After 7 years the sulphate budget showed that the applied sulphur (after microbial oxidation) was almost entirely exported and thus recovered as sulphate.

Residence time distributions are a spatially integrated measure linking observations at the outlet to precipitation or infiltration events over a certain period. The above discussion shows that the generality of the results depends on the details of the two limiting cases: (i) flow conditions are transient but typical (stationary statistics assumed) for the long-term behaviour; or (ii) are artificially controlled. Thus this concept describes a procedure where assumptions with respect to the spatial structure are implicit to the method but dependencies on time are made explicit through the derived distribution of residence times.

Flowpaths can be viewed as a complementary abstraction compared with the residence time concept. Instantaneously equilibrated variables such as soil water potential represent the implicit assumptions with respect to time dependencies. The influence of spatial gradients between these local variables is, however, made explicit by the method (e.g. in form of the gradients used in Richards equation). Flowpaths are a spatially aggregated measure that delineates those parts of a larger flow region that can be characterized by a typical flow response. Examples are the zones of vertical unsaturated infiltration, which is confined at G1 to the few uppermost centimetres of soil, or the zone of saturated lateral groundwater flow near the gauging station of a perennial stream.

The application of two flowpath models shows the limits inherent to this method. A matching pattern with observed runoff values alone (TOPMODEL) or in combination with soil water potentials (SilVlow) does not imply a *unique* interpretation with respect to internal flows. The problem is that a model reconstruction of the runoff and soil water potential dataset is relatively easy, but interpreted flowpaths depend critically on local (calibrated) conductivities used by the model. Without such independently tested internal flow fields, the typology of flowpaths remains *ad hoc*. For short periods the flow can become lateral or saturated even close to the surface of the slope soils, or after extreme droughts the flow during the rewetting can become unsaturated and vertical also in the boggy part of the catchment. Even less can be said quantitatively for the larger part of the G1 catchment where the flow situation may transiently change between these two extreme flow regions.

That is why the optimal outcome of a flowpath analysis in a practical situation such as at G1 is a reproducible potential field for one instant in time, i.e. a clear picture of probable internal flow lines for one specific set of reproducible boundary conditions. The subsequent time-averaging of this spatial pattern may already be limited by the theoretical constraints implicit in the methods (as long as the exact geometry of the pore space in the flow region is not known by construction).

Thus not only the abstraction but also the outcome of these approaches appears complementary. The residence time concept yields at best a detailed model for time dependencies for one specific location defining a "subcatchment" within the flow region. The flowpath concept yields spatial connectivities of internal variables for a well-defined instance in the sequence of infiltration events.

What are the chances that the two methods complement each other in a consistent way? Can the various allowed and conflicting interpretations of the flowpath models be made unique by combining them with the residence times approach, or vice versa? At present the answer to this question seems to be "no". It does not appear feasible in practice to reconstruct the local and transient catchment behaviour from the respective analysis of all potential subcatchments and all potential flow instances. However, *a posteriori* it is easy to select for a given question (e.g. a specific ion) a transport behaviour that can either be approximated by the residence time or flowpath analysis.

A final remark attempts to trace these unexpected difficulties back to our initial assumptions. It is suspected that neglecting the influence of biological organization at the catchment scale may be highly inappropriate. To test this, a formal analysis of various complexity measures and the information content of the respective datasets describing input and output from G1 respectively is suggested. The qualitative impression is that large differences exist for the two types of concepts. It is hypothesized that depending on whether or not the recorded input–output signal is within the realm of "biological sensitivity", the resulting models will be either identifiable but biologically irrelevant (residence times) or relevant but non-unique (flowpaths) (Hauhs and Lange, 1996). G1 appears to be a site which is perfectly suited to testing such a hypothesis by future experiments and model applications, in order either to overcome the encountered inconsistencies or resolve them theoretically.

## ACKNOWLEDGEMENTS

We appreciate the contribution of H. Bürgmann, E. Plötscher, M. Radke, P. Rothenhöfer and P. Thres for finding an optimal parameter set for CXTFIT. Parts of this work were funded by the German Ministry of Education, Research and Technology (BMBF) under grant No. PT BEO 51-0339476B.

## REFERENCES

Beven, K. (1993) Prophecy, reality and uncertainty in distributed hydological modelling. *Advances in Water Resources*, **16**, 41–51.

Beven, K. (1996) The limits of splitting: Hydrology. *Science of the Total Environment*, **183**, 89–97.

Beven, K., Kirkby, M. J., Schofield, N. and Tagg, A. F. (1984) Testing a physically-based flood forecasting model (TOPMODEL) for three U.K. catchments. *Journal of Hydrology*, **69**, 119–143.

Binley, A. M., Beven, K., Calver, A. and Watts, L. G. (1991) Changing responses in hydrology: Assesssing the uncertainty in physically based model predictions. *Water Resources Research*, **27**, 1253–1261.

Callenberg, R., Hauhs, M., Hoch, R., Köstner, B., Lange, H., Lischeid, G., Müller, M., Prechtel, A., Wörlen, C. and Zimmermann, F. (1995) The second tracer experiment at Gårdsjön G1 during steady state stormflow conditions: Preliminary report (unpublished).

Hauhs, M. and Lange, H. (1996) Ecosystem dynamics viewed from an endoperspective. *Science of the Total Environment*, **183**, 125–136.

Hoch, R. (1995) Modellierung von Fließwegen und Verweilzeiten in einem Einzugsgebiet unter stationären Fließbedingungen. Diploma thesis, University of Bayreuth.

Hultberg, H. and Bishop, K. H. (1995) Reversing acidification in a forest ecosystem: The Gårdsjön covered catchment. *Ambio*, **24**, 85–91.

Hultberg, H., Lee, Y.-H., Nyström, U. and Nilsson, I. (1990) Chemical effects on surface-, ground- and soil-water of adding acid and neutral sulphate to catchments in southwest Sweden. In B. J. Mason (ed.), *The Surface Waters Acidification Programme*. Cambridge University Press, Cambridge, pp. 167–182.

Lange, H., Lischeid, G., Hoch, R. and Hauhs, M. (1996) Water flowpaths and residence times in a small headwater catchment at Gårdsjön (Sweden) during steady state stormflow conditions. *Water Resources Research*, **32**, 1689–1698.

Lischeid, G. (1995) Prozeßorientierte hydrologische Untersuchungen am kleinen Gudenberg bei Zierenberg (Nordhessen) in verschiedenen Skalenbereichen. Dissertation, University of Göttingen. *Berichte des Forschungszentrums Waldökosysteme, Reihe A*, Bd. **128**.

Nyberg, L. (1995) Water flow path interactions with soil hydraulic properties in till soil at Gårdsjön, Sweden. *Journal of Hydrology*, **170**, 255–275.

Nyberg, L., Bishop, K. H. and Rodhe, A. (1993) Importance of hydrology in the reversal of acidification in till soils, Gårdsjön, Sweden. *Applied Geochemistry*, Suppl. Issue No. **2**, 61–66.

Nyström, U. (1985) Transit time distributions of water in two small forested catchments. *Ecological Bulletins* (Stockholm), **37**, 97–100.

Parker, J. C. and van Genuchten, M. Th. (1984) Determining transport parameters from laboratory and field tracer experiments. *Virginia Agricultural Experiment Station, Bulletin* **84–3**.

Rodhe, A., Nyberg, L. and Bishop, K. H. (1996) Transit times for water in a small till catchment from a step shift in the oxygen-18 content of the water input. *Water Resources Research*, **32**, 3497–3511.

Schmidt, J. P., Blendinger, C. and Lange, H. (1995) silVlow—eine Modelldokumentation. In: G. Hörmann and J. P. Schmidt (eds), *Dokumentation von Wasserhaushaltsmodellen*. Berichte des Forschungszentrums Waldökosysteme, Reihe B, Bd. **42**, Göttingen, pp. 71–83.

Seibert, J. (1993) Water storage and flux in a micro-catchment at Gårdsjön, Sweden. Master thesis, University of Uppsala.

Seiler, J. (1995) Bericht zum LiBr-Tracer-Versuch von Bodensäulen aus Schweden. BITÖK internal report, Bayreuth.

# 22 European Experience of Manipulation of Forest Ecosystems by Roof Cover: Possibilities and Limitations

C. BEIER, P. GUNDERSEN & L. RASMUSSEN

Manipulation of forest ecosystems by the use of roofs constructed above the stand or the soil allows the effects of changes in the atmospheric load to natural ecosystems to be studied. So far, attention has been focused on the effects of reduced emissions of sulphur and nitrogen and the possibility of reversing soil and streamwater acifidication, but climate change effects, nutrient circulation and heavy metals have also been studied. Results from such roof studies may be especially useful for testing of mathematical models used to predict the effects of various emission abatement strategies. The main results from the European experiments are summarized. Apart from the obvious advantages offered by the roof cover approach, unintended changes caused by the roof construction and the experimental setup are unavoidable. The possible unintended changes caused by the roof approach as well as the possibilities of reducing the problems are discussed with special reference to the roof at Gårdsjön.

## INTRODUCTION

Ecosystem manipulation performed as large field experiments has become an important research tool in the study of environmental impacts on forest ecosystems (Rasmussen *et al.*, 1990; Beier and Rasmussen, 1994a). Within ecosystem manipulation research a "roof project" concept has developed, where roofs have been constructed below the tree canopy with the aim of studying the biological and biogeochemical responses of forest stands or catchments to environmental changes such as reduced atmospheric deposition (clean rain) and drought.

Manipulations by covering catchments with a roof have various possibilities and benefits. They are generally supposed to give valuable information about the possible effects of changes in the environment on forest ecosystems. In particular, roof projects raise the possibility of reducing the input of various elements and thereby simulating the effects of air pollution abatement strategies under field conditions as close as possible to ambient conditions. Furthermore, mathematical models have been used to predict the effects of such strategies, and the data from the roof manipulation projects are useful for evaluating and validating these.

*Experimental Reversal of Acid Rain Effects: The Gårdsjön Roof Project.*
Edited by Hans Hultberg and Richard Skeffington. 

The benefits and obvious possibilities offered by the "roof cover" concept has to be weighed against the possible problems linked to the use of roof covers. The massive experimental setup linked to the roof covers may in itself cause changes in the ecosystem which have to be assessed when the results are evaluated. Consequently, experience from the roof experiments may be useful, not only in existing and future roof projects, but also in other types of manipulation experiments.

The present paper summarizes the main results and experiences from previous and ongoing roof projects and addresses some of the possible problems and unintended changes connected to the roof approach, with special reference to the covered catchment experiment at Gårdsjön.

## Roof projects

Roof projects have involved the use of transparent roofs constructed above the whole ecosystem as in the Norwegian project "Reversing Acidification in Norway", RAIN (Wright *et al*, 1993) or below the canopy 2–6 m above the soil as in the co-ordinated European manipulation network "EXperimental MANipulation of forest ecosystems in Europe", EXMAN (Rasmussen *et al.*, 1990). Roofs have been constructed below the canopy on the following sites: Klosterhede, DK (Beier *et al.*, 1995a; Gundersen *et al.*, 1995; Hansen *et al.*, 1995), Kootwijk, NL (de Visser, 1991), Solling, D (Bredemeier *et al.*, 1993), Ballyhooly, IRL (Ryan *et al.*, 1995), Gårdsjön, S (Hultberg *et al.*, 1993; Bishop and Hultberg, 1995), Skogaby, S (Nilsson and Wiklund, 1992), Speuld and Ysselsteyn, NL (van Dijk *et al.*, 1992; Boxman *et al*, 1995), Rajec, CZ (Cermak *et al.*, 1993) and in Walker Branch, US (Turner *et al.*, 1993). Under the roof the water and nutrient input to the soil/ecosystem can be controlled by addition of water or nutrient solutions by sprinkler systems. Roofs placed below the canopy offer the possibility of distinguishing direct (canopy mediated) and indirect (soil mediated) effects of air pollution.

So far, roof projects have been used to study:

- The effect of removing anthropogenic inputs of N, S and acidity (Rasmussen *et al.*, 1990; van Dijk *et al.*, 1992; Bredemeier *et al.*, 1993, 1995; Hultberg *et al.*, 1993; Wright *et al.*, 1993; Boxman *et al.*, 1994, 1995; Bishop and Hultberg, 1995).
- Combined removal of anthropogenic N, S and acidity and addition of fertilizers (Rasmussen *et al.*, 1990; de Visser *et al.*, 1994).
- Reduced water addition or drought (Rasmussen *et al.*, 1990; Nilsson and Wiklund, 1992; Bredemeier *et al.*, 1993; Cermak *et al.*, 1993; Turner *et al.*, 1993; Ryan *et al.*, 1995).
- Combined drought and fertilization (Nilsson and Wiklund, 1992).
- Addition of extra water (Rasmussen *et al.*, 1990; Turner *et al.*, 1993).

Recently the roofs at Risdalsheia in Norway (the former RAIN Project) have been utilized in a climate change research project, CLIMEX, where $CO_2$ concentrations and temperature are manipulated (Jenkins *et al.*, 1993).

## MANIPULATION EFFECTS

### General effects of manipulations on soil processes

Several roof projects have focused on the possibility of reversing soil acidification and increasing soil pH by manipulating/reducing the input. In general reduced input of acidity or increased input of base cations caused no or only slight changes in the pH in the soil or the soil solution (Wright *et al.*, 1988, 1993; Beier *et al.*, 1995a; Bishop and Hultberg, 1995; Hansen *et al.*, 1995). Therefore, reversibility of acidification has been shown mainly by increased alkalinity and base saturation in runoff water (Wright *et al*, 1993), reduced soil solution concentrations of base cations (Boxman *et al.*, 1995; Hansen *et al.*, 1995) or reduced base cation runoff (Wright *et al.*, 1993) with only small corresponding changes in pH.

The sulphur cycle is sensitive to changes in the input. Reduced input of sulphate has caused almost immediate reductions in the soil solution concentrations as well as percolation from the soil (van Dijk *et al.*, 1992; Blanck *et al.*, 1993; Hultberg *et al.*, 1993; Wright *et al.*, 1993; Boxman *et al.*, 1994, 1995; Beier *et al.*, 1995b; Bishop and Hultberg, 1995; Hansen *et al.*, 1995; Moldan *et al.*, 1995, Chapter 4, this volume). However, differences in response time have been observed depending on, for example, adsorption capacity, organic S retention and Al–S chemistry.

The internal N cycle is quantitatively more important than for S. Thus, the N cycle could be expected to be less sensitive to changes in input. Surprisingly reduced input of $NO_3^-$ and/or $NH_4^+$ has caused immediate reduction of $NO_3^-$ in runoff water from a small catchment (Wright *et al.*, 1993) or in the soil solution concentrations at Speuld and Ysselsteyn (van Dijk *et al.*, 1992; Boxman *et al.*, 1994, 1995) and at Solling (Bredemeier *et al.*, 1995) (Figure 22.1). The rapid response of the $NO_3^-$ in the soil solution and thereby the decrease in $NO_3^-$ leaching in the N saturated systems in Speuld, Ysselsteyn and Solling shows that the excess N is related to the atmospheric input. N saturation seems to be a reversible condition of the ecosystem. Part of the immediate response to the N reduction in these systems may however be related to the unintended effects of the roof manipulation discussed in subsequent sections.

### Effects on biological parameters

It is the general hypothesis that the biological effects from input manipulation will be governed by the changes in the soil/soil solution, which means that if the manipulation is buffered in the soil the corresponding changes in the vegetation will probably be small or absent. The biological effects of the manipulation may further depend on the biological part of the system. Microbial biomass, roots and ground vegetation have a rapid turnover and may therefore be more sensitive to the manipulations than trees. Manipulations of water and nutrients have caused changes in microbial activity (Beier and Rasmussen, 1994b; Hansen *et al.*, 1995) and ground vegetation (de Visser *et al.*, 1994). Changes in microbial biomass and ground vegetation may even buffer manipulations, thereby hindering or delaying effects on other parts of the ecosystem, as has been reported for nitrogen (Feger, 1993). The

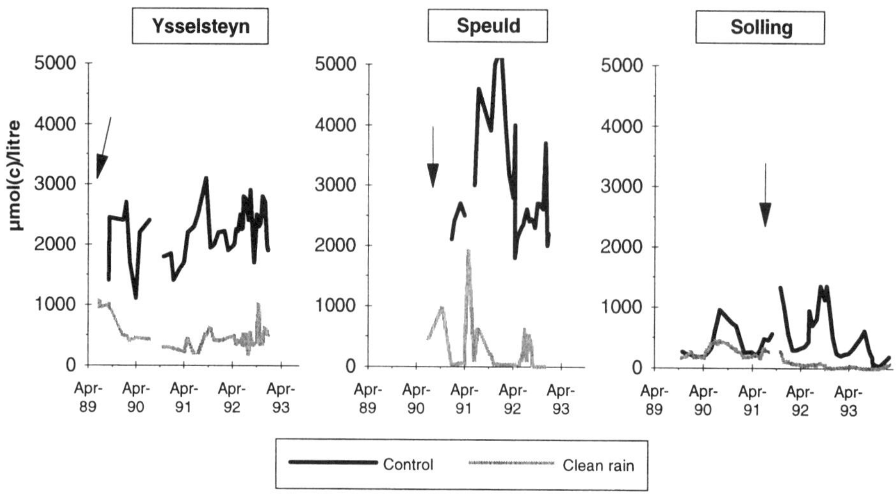


**Figure 22.1.** $NO_3^-$ concentrations in the soil solutions in the control and "clean rain" plots at Speuld, Ysselsteyn (Boxman *et al.*, 1995) and Solling (Bredemeier *et al.*, 1995). Arrows indicate the initiation of the manipulation

influence of such changes may be especially important if the microbial biomass or the ground vegetation constitute a significant part of the ecosystem.

Trees have shown a sensitive and fast response to manipulations if the changes lead to decreased or enhanced availability of growth-limiting substances like water (Nilsson and Wiklund, 1992; Nilsson, 1993; de Visser *et al.*, 1994; Beier *et al.*, 1995a) (Figure 22.2). However, if the manipulations do not affect limiting factors, growth responses are unlikely to occur as illustrated at Klosterhede where addition of nutrients in combination with water did not have an additional growth effect compared with addition of water alone (Rasmussen *et al.*, 1995).

Needle chemistry has become a frequently used indicator for nutritional stress, especially nutrient deficiency caused by acidification and nutrient imbalances caused by high inputs of nitrogen. In both cases roof manipulations serve as a unique method of testing the effect of reduced inputs of air pollution on the nutritional status of the trees. In roof projects differences in needle chemistry between manipulated and control sites can be linked directly to the manipulation, since the air exposure is the same for all plots at a site. This is illustrated by the results from the roof project at Ysselsteyn, NL, where the N content has not changed after 4 years of manipulation, but the K/N relationship has improved (Figure 22.3), probably because the competition between $K^+$ and $NH_4^+$ has changed in favour of $K^+$. Furthermore, the concentration of the N-containing amino acid arginine has decreased significantly after reduction of the N input (Boxman *et al.*, 1995). Since the canopy of the trees is exposed to the same air, the observed changes in the needles must be caused by the soil-mediated changes due to the manipulations.

The needle chemistry can, however, only be expected to change if the soil solution chemistry has changed as described above. In many of the roof projects the ability of

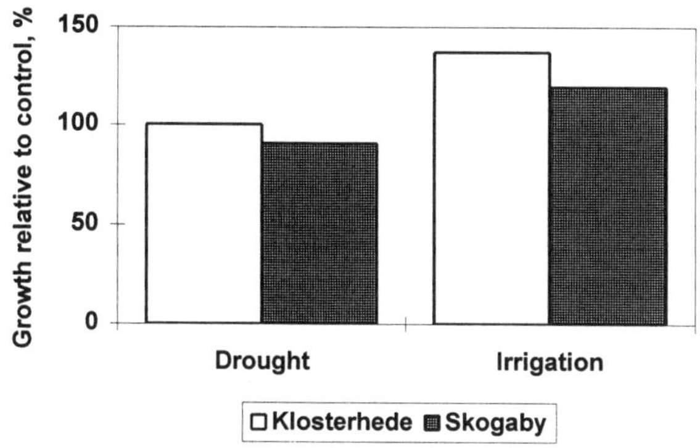


**Figure 22.2.** Growth response to changes in water supply in drought and irrigation plots relative to control at Klosterhede, DK (diameter increment 1988–92) (Rasmussen *et al.*, 1995) and Skogaby, S (total carbon uptake 1987–90) (Nilsson, 1993)

the forest ecosystem to buffer the induced manipulations has been quite extensive and in most cases the nutrients have not changed significantly in the soil solution, and correspondingly not in the needles (Figure 22.3). However, the total pools of nutrients in the canopy may in some cases have changed, but this is still under study at some sites (Rasmussen *et al.*, 1995).

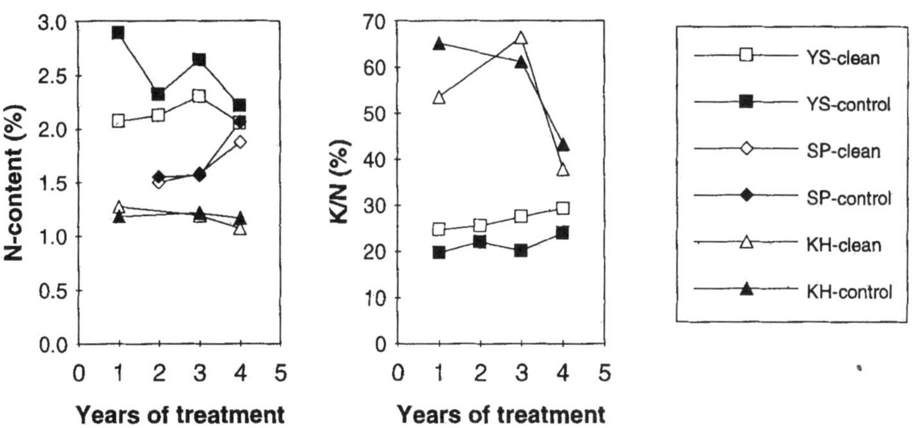


**Figure 22.3.** Needle content of nitrogen and K/N ratio in the control and "clean rain" plots at Ysselsteyn (YS) and Speuld (SP), NL (Boxman *et al.*, 1995) and Klosterhede (KH), DK (Rasmussen *et al.*, 1995)

### Unintended changes in the system caused by manipulations

The roof approach is meant as a tool for manipulation intended to cause effects by the chosen manipulations, and generally it is the intended changes which are examined. But the roof concept will inevitably cause unintended changes as well. Introducing a roof in a forest is a significant manipulation which will interact on a range of parameters that are difficult to control independently. The roof construction and manipulation techniques involve a lot of choices which will influence how well the natural situation is mirrored. In Table 22.1 most of the possible or detected unintended changes are listed, and the extent of the changes observed in the Klosterhede roof project is shown as an illustration (Gundersen, 1995; Gundersen *et al.*, 1995). Unintended changes due to the Gårdsjön Roof are discussed in Chapter 4 of this volume.

### Climatic effects

Construction of a roof, although transparent, will inevitably cause a reduction of the light reaching the forest floor due to reflection from the cover plates and shadow effects from the beam construction. Litter, algal growth, and dust on the cover plates may contribute significantly to the light reduction, although this problem can be reduced by frequent cleaning.

Due to the reduction of irradiation under a roof, a temperature decrease could be expected in the growing season compared with outside the roof, whereas an increase could be expected in the dormant season due to reduced heat radiation. These expectations have been confirmed at the Klosterhede roof with a monthly mean soil temperature increase under the roof during winter at maximum 0.5°C and average 0.2°C. The decrease during summer was maximum 0.6°C and average 0.3°C (Gundersen, 1995).

### Water input

The intensity of natural throughfall is highly variable, and only approaches the intensity from the sprinklers in short intervals of a rain event. In particular, the constant intensity of the sprinklers will cause all the small events to disappear. Therefore the duration of a wet and moist humus layer is most likely shorter compared with the ambient control, even if water is applied in "real time".

In spruce forests the amount of throughfall decreases close to the stems and decreases with tree size due to interception of water from the canopy cover (Beier *et al.*, 1993; Gundersen *et al.*, 1995). This distinct pattern of throughfall water found outside the roof will be changed to a pattern created by the distribution from the sprinklers under the roof. Further, the sprinklers will cause a "catch effect" on the exposed sides of the stems and a "shadow effect" behind stems. The extent of this catch and shadow effect will depend strongly on the site and the sprinkler type, but it can be substantial with up to 7% of the applied water amount being "caught" by the stems (Gundersen, 1995). The amount of stemflow for old Norway spruce will be

**Table 22.1.** Unintended differences between natural and roof covered forest plots. Quantitative differences observed in the roof experiment at Klosterhede are indicated when possible (Gundersen, 1995)

| Factor/process | Change from natural plots → roof plots | Klosterhede |
|---|---|---|
| **Climate** | | |
| Light | Significant reduction | −50% |
| Soil temperature | Slight increase during winter | +0.2°C |
| | Slight decrease during summer | −0.3°C |
| Humidity | Minor increases during sprinkling | |
| **Water input** | | |
| Event frequency | Decreasing, small events disappear | |
| Event size | Increasing | < 3 mm → 15 mm |
| Rain intensity | Variable → constant (high) | → 6 mm/h |
| Droplet size | Variable → constant (sprinkler dependent) | → large |
| Timing | Changed; e.g. rain on warm and sunny days | ±300 mm → ±700 mm |
| Spatial pattern of throughfall amount | Canopy cover related distribution → overlapping sprinkler distributions (increasing variability) | |
| Stemflow | Dramatically increasing (stem "catch") | < 0.5% → *c.* 10% |
| Water amount | Drought and roof control: Loss of throughfall water due to limited storage capacity and overflow in the pipe system in heavy rain storms | −10% |
| **Element input** | | |
| Spatial pattern | | |
| • Concentrations | Related to canopy cover → constant | a factor 2–5 → 5–10 |
| • Amounts | Related to canopy cover → proportional to sprinkling water distribution; large stem flows (increasing variability) | |
| **Element cycling** | | |
| Throughfall | | |
| • Suspended matter | Missing; lost in storage tanks or pipe systems | ? → 0 |
| • Canopy exchange | Drought and roof control: Internal cycling disturbed if water is lost; can be restored by addition | |
| • Dissolved organic matter | Clean rain: Missing | |
| | Drought and roof control: Transformed during storage | |
| Litterfall | | |
| • Timing | Continuous → a limited number of events | → 4 events |
| | Decomposition and loss of nutrients in the collection period | |
| • Spatial pattern | Related to canopy cover → "uniform" | |
| **Derived biological differences at Klosterhede** | | |
| Ground vegetation | Decline of moss cover. Cause: Light reduction and possibly reduced rain event frequency | 70% → 20% cover |
| Litter decomposition | Decreased weight loss. Cause: Reduced rain event frequency, increased event size, increased rain intensity and reduced moss cover, which in combination may have reduced the moisture of litter | 30% reduction |

constant in all sprinkling events whereas natural stemflow only occurs in heavy rainstorms and in total only as a small part of the yearly throughfall amount.

### Element input

Concentrations of all elements in throughfall exhibit a pattern controlled by the canopies, as a consequence of dry deposition, foliar uptake and leaching, and evaporation. The pattern is highly dependent on the tree species. For Norway spruce in Denmark, concentration differences in throughfall of a factor 5 to 10 depending on distance from stems and a factor 2 to 3 depending on tree size have been observed (Beier *et al.*, 1993). Consequently, the element fluxes may still be higher near the stems compared with between the stems by a factor 2 to 3 at Klosterhede (Beier *et al.*, 1993). Such distinct spatial variability in throughfall concentrations will be lost in the manipulated plots where concentrations will be constant in space and normally constant in time as well. The spatial variability in element fluxes under the roof will thus be determined by the distribution of the water flux alone. In the Norway spruce case mentioned above this is the reverse of the natural situation where the highest element fluxes occur with the lowest water fluxes close to the stems. The strong spatial variability in the input has been reported to be reflected in the soil solutions as well (Gundersen, 1993).

### Element cycling

The removal of throughfall by a roof will cut off the internal cycles of nutrients with throughfall, especially the canopy leaching of K, Mg and Ca. This will be the case if water is lost from the storage tanks during drought or heavy rainstorms, or if resupply with artificial throughfall does not include at least the amount of nutrients leached from the canopy.

Apart from the nutrient ions, throughfall also contains suspended matter and dissolved organic compounds, including organic N leached from needles, bark and epiphytic organisms. In general, throughfall sprinkled under the roof will not contain the original organic compounds found in natural throughfall.

Normally, litterfall will be redistributed under the roof to maintain the internal cycling of nutrients with litterfall. However, some readily available nutrients and soluble organic compounds may be leached from the litter by throughfall or released by decomposition while lying on the roof before the collection period. In this way they might be lost from the system in the roof runoff. Further, redistribution of litter will often be done in larger portions compared with the situation with the more continuous supply of litter to the decomposer community on the control plots. Finally, the redistribution of litter will cause a minor change from the natural situation where litterfall amounts are slightly related to the canopy cover.

## DISCUSSION

It is obvious that the quality of roof manipulation projects and the results obtained depends strongly on the quality of the research design and the efforts invested to reduce or to quantify the unintended changes.

One of the most critical components is the light reduction under the roof. Reduced light penetration to the forest floor will affect the ground floor vegetation significantly as has been reported for Klosterhede, DK, which exhibited a strong decrease in moss cover in roof covered plots (Gundersen, 1995). Such changes may also influence the nutrient turnover and the microbial activity. Careful choice of highly light transmitting roof plates is therefore of great importance as well as careful and regular cleaning. Although the roof plates at Gårdsjön have been chosen to transmit maximum amounts of light, UV light is still excluded, and shadow effects from the construction and "dirt" on the roof, as well as some reflection due to the steep slope, cannot be avoided. Compared with the roof at Klosterhede, the Gårdsjön roof has slightly more transparent plates, and no opaque seals around the stems, thus the light reduction may be less than the observed 50% reduction at Klosterhede.

Another critical component in the experimental setup is the sprinkling system and the sprinkling strategy which influences both the water and the nutrient input. Changes in water quality as well as water application characteristics, e.g. size, frequency and intensity of the events and droplet size may lead to changes in the ground flora and to changes in the microbial activity as reported by Hansen *et al.* (1995) and Gundersen (1995). At Gårdsjön this will cause changes, since irrigation is based on deionized lake water with sea water added. Therefore, small organic particles and suspended material as well as the rain events with low intensity are excluded, probably influencing the decomposer activity.

The manipulation will cause changes in chemical balances in cases where water and elements are unintentionally lost or not resupplied in sufficient amounts to compensate for the amounts removed, as illustrated for potassium at Klosterhede, DK (Rasmussen *et al.*, 1995). Furthermore, sprinkling changes the natural variability. The sprinkler system at Gårdsjön consists of a large number of small sprinklers which may reduce the previously described "catch/shadow-effects" by the trees and reduce the variability compared with the finding at Klosterhede (Table 22.1). However, sprinkling will still cause a fundamental change in the chemical input from a situation where the high chemical fluxes are at spots with low water fluxes to a situation where the magnitude of the water and chemical fluxes are positively linked. This change must cause unintended changes in the soil chemical reactions in the soil.

The importance of changes in spatial variability induced by introducing a sprinkling system will depend on the scale and duration of the experiment. Since root turnover in forests is relatively rapid, the root distribution may quickly adjust to the new water distribution from the sprinklers and in a catchment the changes may be buffered during the water and chemical transport through the catchment. Therefore, on the scale of a root system of a tree, a plot or a catchment, the difference from the natural situation may thus be small. But at each sampling point, for example a soil water sampler, the difference will be of great importance, especially since a low number of samplers is usually used to represent a plot.

It is crucial that the nutrients possibly removed by filtering, overflow or disposal of collected throughfall are carefully calculated and resupplied, and that the water and nutrient supply is performed as close to "real time" as possible. For new experiments it is recommended that different sprinkler types be carefully tested, and to test the effects of different positions in relation to stems and different water pressure to optimize the spatial distribution. Finally, since the input differences in the treatment plot will be constant in time, and because these differences will probably be larger than in a natural plot, it is crucial for the interpretation of the results to quantify the differences between the samplers.

The covered catchment at Gårdsjön is subject to almost all the same conceptual problems as described above. However, the Gårdsjön Roof is unique in the sense that it covers a whole catchment of 6300 $m^2$ and the overall effect on the soil chemistry is measured in the streamflow in the bottom of the catchment. This will probably reduce the problem connected to changes in spatial variability, although single soil solution samples obtained by lysimeters inside the catchment will still face the problem.

### Roof control and pre-treatment period

The obvious solution to some of the difficulties discussed above would be to establish a roof control plot/catchment, where collected throughfall is sprinkled under the roof. Ideally, this plot would show the effect of the roof itself, but the problem is more complex. In practice, it is difficult to run a throughfall collection and storage system in parallel to a "clean" water system and at the same time mimic the situation on the ambient plot. The roof control treatment may well be something in between the ambient and the roof manipulation which will still be difficult to interpret. The roof controls at Ysselsteyn and Speuld (Boxman *et al.*, 1995) and at Solling (Bredemeier *et al.*, 1993) have emphasized some of the problems, since the roof controls in all cases show strong differences in soil solution chemistry compared with the unroofed controls (Boxman *et al.*, 1995; Bredemeier *et al.*, 1995). The problem seems to be most strongly connected to the timing in water addition (Boxman *et al.*, 1994) and to the unavoidable sedimentation and filtering of organic particles in the storage and pumping system (Lamersdorf and Blanck, 1995).

In the Gårdsjön case establishment of a control roof would be even more difficult because of the catchment scale. This would require an identical catchment (size, slope, hydrology, direction, etc.) which is obviously not practically and economically feasible.

The change in spatial variability and the possible changes in the transition phase underline the importance of establishing pre-treatment data for at least one year prior to construction of a roof. With the short grants usually obtained to initiate these expensive experiments this is often difficult. However, it should be given high priority, since treatment differences may easily be masked by pre-treatment differences. Comparison of the effect of the change in input above each individual soil solution sampler will give much more information on the transition phase and the treatment effect than the average of the samplers.

## CONCLUSIONS

Roof manipulation studies have been shown to be a valuable experimental tool in forest ecosystem studies despite the difficulties and limitations discussed above. On a

European scale they have all shown very well the direct link between S input and S output from forest ecosystems. They have also shown that the K cycle in forest ecosystems is very tight and small disturbances will lead to detrimental effects on the K supply for the trees. For N the manipulation experiments have demonstrated that large changes in input also imply changes in output in N-saturated systems, whereas the effects of moderate N manipulation may to begin with be hidden by internal accumulation and turnover processes in the ecosystems. For the other nutrient elements the ion balance depends very much on the supply from the soil.

The results obtained from roof studies so far constitute important data for improving and validating mathematical models for forested ecosystems. Even the results from the manipulation control treatments (control roofs), where several small changes are integrated, may comprise a demanding test of the models. Ecosystem manipulations have the advantage over traditional, controlled laboratory experiments, that they involve the complexity of the whole ecosystem, but this inevitably raises difficulties in interpretation. Furthermore, compared with traditional fertilization experiments, the roof experiments allow the possibility of testing moderate and environmentally realistic changes; and of separating the atmospheric and soil-mediated effects. Finally, the unintended manipulations revealed in the roof experiments have created a new understanding of the ecosystems and have facilitated the focus on important processes that were not recognized before.

The experimental design of the roof project at Gårdsjön as well as at other sites has in every case been a compromise between a range of scientific questions to be investigated by the experiments, the practical possibilities at the site, and economic constraints. The choices made, the practical solutions found, and the limitations set by economic resources have caused some of the specific problems reported above, which, looking back, might have been partly eliminated. However, some of these experiences may be generally valid for (forest) ecosystem manipulations. It is important to realize the inherent limitations of this type of experiment and to be able to adapt the research ambitions and the experimental design to these limitations.

On the basis of our existing experience, ecosystem manipulation by roof cover will be well suited for further studies of forest ecosystem function, such as climate change effects on natural ecosystems and studies of nutrient cycling, for example, by experimental exclusion of base cation recycling with throughfall to investigate deficiency levels.

## REFERENCES

Beier, C. and Rasmussen, L. (eds) (1993) EXMAN — Experimental manipulation of forest ecosystems in Europe, Project Period 1988–1991. *Ecosystem Research Report*, **7**, 124 pp.

Beier, C. and Rasmussen, L. (1994a) Effects of whole-ecosystem manipulations on ecosystem internal processes. *TREE*, **9**, 218–223.

Beier, C. and Rasmussen, L. (1994b) Organic matter decomposition in an acidic forest soil in Denmark as measured by the cotton strip assay. *Scandinavian Journal of Forestry Research*, **9**, 106–114.

Beier, C., Hansen, K. and Gundersen, P. (1993) Spatial variability of throughfall fluxes in a spruce forest. *Environmental Pollution*, **81**, 257–267.

Beier, C., Gundersen, P., Hansen, K. and Rasmussen, L. (1995a) Experimental manipulation of water and nutrient input to a Norway spruce plantation at Klosterhede, Denmark: 2) Effects on tree growth, and nutrition. *Plant and Soil*, **168–169**, 613–622.

Beier, C., Blanck, K., Bredemeier, M., Lamersdorf, N. and Rasmussen, L. (1995b) Effects of reduced input of N and acidity to Norway spruce — Results of roof experiments within the EXMAN Project. *Ecosystem Research Report*, **20**, 77–85.

Bishop, K. H. and Hultberg, H. (1995) Reversing acidification in a forest ecosystem: The Gårdsjön covered catchment. *Ambio*, **29**, 85–91.

Blanck, K., Lamersdorf, N. and Bredemeier, M. (1993) Bodenchemie und stofhaushalt auf den dachflächen im Solling. *Forstarchiv*, **64**, 164–172.

Boxman, A. W., van Dijk, H. F. G. and Roelofs, J. G. M. (1994) Soil and vegetation responses to decreased atmospheric nitrogen and sulphur inputs into a Scots pine stand in the Netherlands. *Forest Ecology and Management*, **68**, 39–45.

Boxman, A. W., van Dam, D., van Dijk, H. F. G., Hogervorst, R. F. and Koopmans, C. J. (1995) Ecosystem responses to reduced nitrogen and sulphur inputs into two coniferous forest stands in the Netherlands. *Forest Ecology and Management*, **71**, 7–29.

Bredemeier, M., Blanck, K. and Widey, G. A. (1993) Experimentelle Manipulation des Wasser-und Stoffhaushalts in einem Fichtenwald — das Dachprojekt im Solling. *Forstarchiv*, **64**, 154–158.

Bredemeier, M., Blanck, K., Lamersdorf, N. and Wiedey, G. A. (1995) Response of soil water chemistry to experimental "clean rain" in the NITREX roof experiment at Solling, Germany. *Forest Ecology and Management*, **71**, 31–44.

Cermak, J., Kucera, J. and Janicek, R. (1993) The "roof" project in Rajec. In: L Rasmussen, T. Brydges and P. Mathy (eds), Experimental manipulations of biota and biochemical cycling in ecosystems — Approach, methodologies, findings. *Ecosystem Research Report*, **4**, 165–171.

de Visser P. H. B. (1991) *Manipulation of the Water and Nutrient Supply in Two Forest Ecosystems in the Netherlands that were Exposed to Acid Atmospheric Depositions*. Dutch Priority Programme on Acidification. Rep. 100.1, RIVM, Bilthoven.

de Visser, P. H. B., Beier, C., Rasmussen, L., Kreutzer, K., Steinberg, N., Bredemeier, M., Blanck, K., Farrell, E. P. and Cummins, T. (1994) Biological response of five forest ecosystems in the EXMAN project to input changes of water, nutrients and atmospheric loads. *Forest Ecology and Management*, **68**, 15–29.

Feger, K. H. (1993) The ARINUS project: Effects of whole-watershed nitrogen and sulphur additions on biogeochemical cycling in spruce forests. In: L. Rasmussen, T. Brydges and P. Mathy (eds), Experimental Manipulation of Biota and Biochemical Cycling in Ecosystems — Approach, Methodologies, Findings. *Ecosystem Research Report*, **4**, 12–22.

Gundersen, P. (1993) Spatial variability of soil water chemistry in a spruce forest. In: J. Cerny (ed.), *Abstracts from BIOGEOMON and Workshop on Integrated Monitoring*. Czech Geological Survey, Prague, pp. 108–109.

Gundersen, P. (1995) Unintended differences between natural and manipulated forest plots. European Commission. *Ecosystem Research Report*, **20**, 335–343.

Gundersen, P., Andersen, B. R., Beier, C. and Rasmussen, L. (1995) Experimental manipulation of water and nutrient input to a Norway spruce plantation at Klosterhede, Denmark: 1) Unintended physical and chemical changes by roof experiments. *Plant and Soil*, **168–169**, 601–611.

Hansen, K., Beier, C., Gundersen, P. and Rasmussen, L. (1995) Experimental manipulation of water and nutrient input to a Norway spruce plantation at Klosterhede, Denmark: 3) Effects on throughfall, soil water chemistry and decomposition. *Plant and Soil*, **168–169**, 623–632.

Hultberg, H., Andersson, B. I. and Moldan, F. (1993) The covered catchment — an experimental approach to reversal of acidification in a forest ecosystem. In: L. Rasmussen, T. Brydges and P. Mathy (eds), Experimental manipulations of biota and biochemical cycling in ecosystems — Approach, methodologies, findings, *Ecosystem Research Report*, **4**, 46–54.

Jenkins, A., Wright, R. F., Berendse, F., van Breemen, N., Brussaard, L., Schulze, E. D. and Woodward, F. I. (1993) The "CLIMEX" project — climate change experiment. In: L. Rasmussen, T. Brydges and P. Mathy (eds), Experimental manipulations of biota and biochemical cycling in ecosystems — Approach, methodologies, findings. *Ecosystem Research Report*, **4**, 71–77.

Lamersdorf, N. and Blanck, K. (1995) Evaluation of fine material input with throughfall for a spruce forest in Solling, FRG, by means of a roof construction. *Ecosystem Research Report*, **20**, 168–173.

Moldan, F., Hultberg, H. and Anderson, I. (1995) Gårdsjön covered catchment experiment: changes in runoff after three years of throughfall exclusion. *Ecosystem Research Report*, **20**, 113–118.

Nilsson, L.-O. (1993) Carbon sequestration in Norway spruce in south Sweden as influenced by air pollution, water availability, and fertilization. *Water, Air, and Soil Pollution*, **70**, 177–186.

Nilsson, L.-O. and Wiklund, K. (1992) Influence of nutrient and water stress on Norway spruce production in south Sweden — the role of air pollutants. *Plant and Soil*, **147**, 251–265.

Rasmussen, L., Beier, C., Van Breemen, N., de Visser, P., Kreutzer, K., Schierl, R., Matzner, E. and Farrell, E. P. (1990) *Study on acid deposition effects by manipulating forest ecosystems*. New title: *EXMAN — EXperimental MANipulation of forest ecosystems in Europe*. Commission of the European Communities. Directorate-General for Science, Research and Development, ERP. *Air Pollution Research Report*, **24**, 44 pp.

Rasmussen, L., Beier, C. and Andersen, B. R. (1995) Effects of experimental manipulations with biogeochemical cycles in a Norway spruce plantation in Denmark: the Danish part of the EXMAN project. European Commission. *Ecosystem Research Report*, **20**, 96–103.

Ryan, M., O'Toole, P. and Farrell, E. P. (1995) Nitrogen dynamics in a forest ecosystem influenced by drought at Ballyhooly, Ireland: the EXMAN project. *Ecosystem Research Report*, **20**, 171–173.

Turner, R. S. Hanson, P. J., Huston, M. A., Garten, Jr, C. T. and Mulholland, P. J. (1993) A large-scale throughfall experiment on Walker Branch watershed. In: L. Rasmussen, T. Brydges and P. Mathy (eds), Experimental manipulations of biota and biochemical cycling in ecosystems — Approach, methodologies, findings. *Ecosystem Research Report*, **4**, 96–105.

van Dijk, H. F. G., Boxman, A. W. and Roelofs, J. G. M. (1992) Effects of a decrease in atmospheric deposition of nitrogen and sulphur on the mineral balance and vitality of a Scots pine standing in the Netherlands. *Forest Ecology and Management*, **51**, 207–215.

Wiedey, G. A., Blanck, K., Bredemeier, M. and Lamersdorf, N. (1995) The roof experiment in Solling: a general description of the site and experimental design. *Ecosystem Research Report*, **20**, 153–154.

Wright, R. F., Lotse, E. and Semb, A. (1988) Reversibility of acidification shown by whole-catchment experiments. *Nature*, **334**, 670–675.

Wright, R. F., Lotse, E. and Semb, A. (1993) RAIN project: Results after 8 years of experimentally reduced acid deposition to a whole catchment. *Canadian Journal of Fisheries and Aquatic Sciences*, **50**, 258–268.

# 23 Assessment of Critical Loads for Coniferous Forests and Surface Waters at Lake Gårdsjön in South-west Sweden

B. I. ANDERSSON, R. A. SKEFFINGTON, H. HULTBERG & H. U. SVERDRUP

In this paper, an assessment of critical loads (CLs) for sulphur and nitrogen is conducted for the Gårdsjön site where many of the necessary basic data are available as measured values. CLs were developed as a model to assist in the negotiations on abatement strategies for acidifying substances. Despite general agreement on the applicability and usefulness of the CL concept, some uncertainties still reside within its scientific framework. Intensive research continually advances our understanding of ecosystems and the CL concept, leading to a need for regular testing and updating of the methods in use.

Calculation of CLs has been performed using a number of more-or-less "official" methods. CL values for S fell in the range 0.04–1.25 keq $ha^{-1} yr^{-1}$ (1–20 kg S $ha^{-1} yr^{-1}$), depending on methods and assumptions. Comparison with historical data suggests that the true critical load is between 0.1 and 0.44 keq $ha^{-1} yr^{-1}$. Calculated N critical loads fall in the range 8–15 kg N $ha^{-1} yr^{-1}$ (0.57–1.07 keq $ha^{-1} yr^{-1}$). Currently, there is a lack of good data to validate N critical loads. Both non-marine S and N deposition thus need to be reduced from their present values of around 0.85 keq $ha^{-1} yr^{-1}$ and 16 kg N $ha^{-1} yr^{-1}$ (1.15 keq $ha^{-1} yr^{-1}$) respectively. At present, the G1 ROOF catchment results are difficult to use in validating CLs.

## INTRODUCTION

Extensive and severe acidification of the environment in Europe and North America has attracted much scientific and public concern since around 1970. The understanding of causes and effects, as postulated originally by Odén (1968), has steadily increased, and various actions were quickly implemented in an attempt to remedy the problems. Far-reaching reduction of Swedish sulphur emissions, starting around 1975, have resulted in a total decrease of more than 85% to date, but still a rather restricted improvement of the environmental situation. This is because many air pollutants are transported over long ranges, across borders, meaning that a

*Experimental Reversal of Acid Rain Effects: The Gårdsjön Roof Project.*
Edited by Hans Hultberg and Richard Skeffington. 

substantial part of the acid loading deposited in Scandinavia is due to Central European and UK emissions.

A direct countermeasure is the large-scale operational liming that Sweden started in 1977 (Henrikson and Brodin, 1995) and which up to present includes treatment of more than 6300 lakes and 6000 km of streams. Principally, this has been a successful method to guarantee survival or recolonisation of freshwater biota, especially fish, and save biological and genetic diversity. However, during the time elapsed since the late 1970s, signs of damage to forests over large regions of Europe have occurred. What causes decreased vitality and the observed damage is not yet clear, but involvement of air pollutants has been suggested. The effects of these pollutants may either be direct (cf. Abrahamsen and Tveite, 1983; Ashmore *et al.*, 1985; Ashmore and Wilson, 1994), or indirect via significant acidification effects on the soils (cf. Ulrich, 1983; Cronan, *et al.*, 1989; Gundersen, 1992). In both cases, improvement of the situation would demand substantially decreased anthropogenic emissions over the continental scale. This cleaning-up process started early in certain countries (e.g. Sweden) and was later carried on to the European scale via international negotiations under the auspices of the United Nations Economic Commission for Europe (UN-ECE). It finally reached a state of implementation by the signing of various protocols to the Convention on Long Range Transboundary Air Pollution (LRTAP). The original protocols to the Convention committed parties to flat-rate emission reductions—e.g. 30% of 1980 $SO_2$ emissions for the Sulphur Protocol adopted in Helsinki in 1985. It was recognised, however, that ecosystems differ considerably in sensitivity to acidification, and that it would be more efficient to reduce those acidic emissions which have most impact on the sensitive areas. This led to the adoption of the critical load (CL) concept as the basis for the development of the UN-ECE conventions.

## The critical load concept

The scientific framework for the critical load (CL) approach was founded by a Nordic working group (Nilsson, 1986). Further development has come through a series of workshops (Nilsson and Grennfelt, 1988; Grennfelt and Thörnelöf, 1992; Hornung *et al.*, 1995) where methods and data have been scrutinised and revised. In parallel, a large-scale application of practical techniques for mapping CLs (see e.g. Henriksen *et al.*, 1990; Sverdrup *et al.*, 1990; Hettelingh *et al.*, 1991; Downing *et al.*, 1993) has been operative and continually evaluated within the UN-ECE Task Force connected to the Convention on LRTAP. Hitherto, all of this work has been undertaken based on the definition of the critical load concept as it was stated in the proceedings of the workshop held at Skokloster in 1988 (Nilsson and Grennfelt, 1988):

> A quantitative estimate of an exposure to one or more pollutants below which significant harmful effects on specified sensitive elements of the environment do not occur according to present knowledge.

A few aspects of this concept may be questioned though, at the same time, it is versatile and as a parallel to the eco-toxicological "threshold" it could with due adaptation be applied to a number of problems connected to a wide range of different pollutants. Concurrently, the CL approach has effectively disclosed gaps in our scientific knowledge and understanding of ecosystem processes and by that triggered research. Some of these gaps are not only crucial for calculations of the CL but probably also important for our wider understanding of how different ecosystems function. Most of the shortcomings and a sometimes annoying lack of data have been overcome by the use of distribution functions and a low resolution on the European scale. For future regional and even local work there is a need for improvement in those respects. Three examples will illustrate this. The accuracy of deposition data over Europe as estimated by EMEP has been questioned, on both local and regional scales, i.e. what is the actual input to local areas or even single landscape elements (Chapter 3, this volume). The availability and applicability of regional soil data needs to be increased, and a more accurate knowledge of weathering rates on the same scale is necessary to evaluate CLs further. Finally, the alkaline input (base cation deposition) is poorly quantified, and usually relates to specific sites, or is derived from late short-term high quality networks (cf. Hedin *et al.*, 1994; Chapter 3, this volume).

The total impact of different management practices on different vegetation/forest systems is another factor of importance especially with regard to the CL for nitrogen. This has partially focused critical load work on relations that are independent of land use and deposition, which has not been easily achieved. So far, primary mineral weathering of undisturbed natural ecosystems (far away from human impact) may be a single unique candidate that fulfils this requirement. Hence the value of such "theoretically" defined factors may only find restricted use in a world so widely impacted by human management and misuse.

Whatever criticism that has been stated, the process of practical implementation of the CL concept has forced the scientific community to concentrate on an urgent issue. Focus has been on the development of valid equations that are rigid enough to be used for calculations and mapping of CLs and exceedances. Critical load exceedances were used as the basis for the second Sulphur Protocol, adopted in Oslo in 1994. Hence, it is now even more important to evaluate and refine the procedures for calculation of CLs. The aim should be to validate existing methods or to propose improved methods, this process of scientific assessment and validation being an integral part of the framework to promote progress.

There are only a restricted number of sites where enough work on the ecosystem level has been performed to enable such detailed testing of the CL approach, among them the small catchments surrounding Lake Gårdsjön. The assessment and calculations presented in this chapter give some perspectives on the critical load concept based on a well-known acid-sensitive research site with a multitude of available data.

## MATERIALS AND METHODS

### Critical load methods

A number of different equations for calculations of critical loads have been suggested. This reflects both the increased understanding of the basics of the concept, a demand to be able to describe specific systems and also the need to be able to assess the CL using a variety of already available data resulting in different levels of accuracy. In this paper, CL calculations have been performed with eight methods varying much in complexity and in need for basic data. The equations are all more or less "officially" endorsed and in use by national working groups or they have been used in published work. The methods used are listed below with a reference to either original work or the current version in use as described in different "mapping manuals" or proceedings from workshops on CLs.

1. CL for acidity on soil — Level Zero Method (Nilsson and Grennfelt, 1988).
2. CL for acidity on soil — Simple Mass Balance Method (Sverdrup and de Vries, 1994).
3. CL for acidity on water — Steady-State Water Chemistry Method (Sverdrup *et al.*, 1990; Henriksen *et al.*, 1992; Kämäri *et al.*, 1993; Posch *et al.*, 1993).
4. CL for acidity on water — "Diatom Model" (Hornung and Skeffington, 1993).
5. CL for nitrogen and sulfur on soil — Mass Balance Method (cf. Grennfelt and Thörnelöf, 1992) including derivation for nitrogen on soil — Eutrophication Method (Hettelingh *et al.*, 1992).
6. CL for nitrogen and sulphur on water — First-order Acidity Balance (FAB) Method and determination of a response surface for Lake Gårdsjön (Henriksen *et al.*, 1993; Posch *et al.*, 1993).
7. CL for sulphur using the methods 1–4 and a "sulphur fraction" approach (Sverdrup *et al.*, 1990; Hettelingh *et al.*, 1992; Posch *et al.*, 1993).
8. CL for acidity on soil — Steady-State Mass Balance Method, PROFILE variant (Sverdrup *et al.*, 1990).

**Lake Gårdsjön and the database**

The Lake Gårdsjön catchment comprises headwater parts of the major branch of the Anråseå River, near the Swedish west coast. The catchment and experimental small forest catchments are described in Andersson *et al.* (Chapter 2, this volume).

In this paper, some of the data used were compiled during the earlier study of the Lake Gårdsjön area (cf. Andersson and Olsson, 1985), but most of the values used are recently updated or new data from the current studies within the covered catchment (see elsewhere in this volume) and NITREX (see Dise and Wright, 1992; Wright and van Breemen, 1995) projects. With regard to water chemistry, elemental fluxes, etc., data are updated as close as possible to present and, when convenient, we have made use of long-term measured averages. Whenever possible or feasible, data for different catchments are contrasted, which often gives a range for the critical load calculations as there are some variations like soil thickness and standing forest biomass between otherwise comparatively uniform catchments. The data used in our calculations are shown in Table 23.1, referenced with some brief notes. Further comments on some values introduced and used in the CL equations will be found when the results of each method are being described.

**Table 23.1.** Compilation of input parameter values used in the calculations of critical loads for the Gårdsjön site

| Parameter | Value | | Remarks |
|---|---|---|---|
| Annual precipitation | 1100 mm | | For more details see Chapter 2, this volume |
| Annual runoff | 550 mm | | |
| Annual mean temperature | 6.3°C | | |
| Lake Gårdsjön, volume (total lake volume) | $1.500 \times 10^6$ ($2.161 \times 10^6$) $m^3$ | | Data from Olsson *et al.*, 1985 |
| Lake Gårdsjön, area (total lake area) | $31.2 \times$ (45.8) ha | | |
| Lake Gårdsjön total catchment area | 211.4 ha | | |
| Forested part of the catchment | ~71.6% | | |
| Lake Gårdsjön water retention time (all lakes) | 1.3 yr (1.9 yr) | | |
| Weathering. F1 | 0.56 keq $ha^{-1}$ $yr^{-1}$ | | Chapter 10, this volume. Soil depth 0.63 m |
| Weathering. G1 | 0.37 keq $ha^{-1}$ $yr^{-1}$ | | Chapter 10, this volume. Soil depth 0.43 m |
| Vegetation uptake, base cations. F1 | 0.64 keq $ha^{-1}$ $yr^{-1}$ | | Data from Stuanes *et al.* unpubl. and Hallbäcken unpubl. compiled in Chapter 2, this volume. |
| Vegetation uptake, base cations. G1 | 1.07 keq $ha^{-1}$ $yr^{-1}$ | | |
| Vegetation uptake, nitrogen. F1 | 0.50 keq $ha^{-1}$ $yr^{-1}$ | 7 kg $ha^{-1}$ yr | |
| Vegetation uptake, nitrogen. G1 | 0.84 keq $ha^{-1}$ $yr^{-1}$ | 11.8 kg $ha^{-1}$ $yr^{-1}$ | |
| Average wet + dry deposition, sulphur 1983/86 | 1.58 keq $ha^{-1}$ $yr^{-1}$ | 25.2 kg $ha^{-1}$ $yr^{-1}$ | Data from Hultberg and Grennfelt, 1992; Moldan *et al.*, 1995; Chapter 3, this volume |
| : Excess sulphur fraction | 1.34 keq $ha^{-1}$ $yr^{-1}$ | 21.4 kg $ha^{-1}$ $yr^{-1}$ | |
| Average wet + dry deposition, sulphur 1991/93 | 1.16 keq $ha^{-1}$ $yr^{-1}$ | 18.5 kg $ha^{-1}$ $yr^{-1}$ | |
| : Excess sulphur fraction | 0.86 keq $ha^{-1}$ $yr^{-1}$ | 13.8 kg $ha^{-1}$ $yr^{-1}$ | |
| Average wet + dry deposition, total nitrogen | 1.14 keq $ha^{-1}$ $yr^{-1}$ | 16 kg $ha^{-1}$ $yr^{-1}$ | |
| : Fraction as nitrate | 0.64 keq $ha^{-1}$ $yr^{-1}$ | 9 kg $ha^{-1}$ $yr^{-1}$ | |
| Average wet + dry deposition, base cations | 0.57 keq $ha^{-1}$ $yr^{-1}$ | 10.1 kg $ha^{-1}$ $yr^{-1}$ | Sum of Ca, Mg, and K, Chapter 3, this volume |
| : Non-marine fraction | 0.13 keq $ha^{-1}$ $yr^{-1}$ | 3.4 kg $ha^{-1}$ $yr^{-1}$ | |
| Average wet + dry deposition, chloride 1983/86 | 2.34 keq $ha^{-1}$ $yr^{-1}$ | 83 kg $ha^{-1}$ $yr^{-1}$ | |
| Average wet + dry deposition, chloride 1991/93 | 2.87 keq $ha^{-1}$ $yr^{-1}$ | 102 kg $ha^{-1}$ $yr^{-1}$ | |
| Average wet + dry deposition, sodium 1983/86 | 1.85 keq $ha^{-1}$ $yr^{-1}$ | 42.5 kg $ha^{-1}$ $yr^{-1}$ | |
| Average wet + dry deposition, sodium 1991/93 | 2.46 keq $ha^{-1}$ $yr^{-1}$ | 56.5 kg $ha^{-1}$ $yr^{-1}$ | |
| Average long-term annual immobilisation of nitrogen (store divided by 12 000 years) | 0.06 keq $ha^{-1}$ $yr^{-1}$ | 0.8 kg $ha^{-1}$ $yr^{-1}$ | Variation based on the F1 and G1 catchments: 0.6 to 0.9 kg $ha^{-1}$ $yr^{-1}$ |

## RESULTS AND DISCUSSION

### Calculation of critical loads

*Method 1. Critical load for acidity on soil—"Level Zero" Method*

The dominant bedrock elements around Lake Gårdsjön are mainly granite and gneissic granodiorite of older and younger series, and their mineral composition is stated to be of intermediate or acid composition. As the predominant part of the parent till material is derived locally, the weathering rates are characterised by this mineralogical composition. Furthermore, the total capacity of the thin soil cover becomes important as is obvious in the weathering assessments made by Sverdrup *et al.* (Chapter 10, this volume). Earlier estimates by Sverdrup and Warfvinge (1988a) give for soils with a component of the fast and intermediate weathering silicates (amphiboles, nesosilicates and pyroxenes), the following:

If mineral component is

| | |
|---|---|
| <1% | Critical load is <0.20 keq ha$^{-1}$ yr$^{-1}$ |
| <3% | Critical load is 0.20–0.50 keq ha$^{-1}$ yr$^{-1}$ |
| <8% | Critical load is 0.50–1.00 keq ha$^{-1}$ yr$^{-1}$ |

These critical load values could be related to weathering rates for the type of soil occurring at Gårdsjön where the most important minerals for the total weathering rate were plagioclase, K-feldspar, hornblende and epidote amounting to approximately 15, 15, 0.5–1 and 0.1–0.5% respectively (Melkerud, 1983; Sverdrup and Warfvinge, 1988a,b; Sverdrup *et al.*, Chapter 10, this volume). Of these minerals, only epidote and hornblende relate to the above scheme and applying this mineralogy would suggest a critical load in the range 0.2 to 0.5 keq ha$^{-1}$ yr$^{-1}$.

Using the table presented by Sverdrup *et al.* (1990, p. 102) and roughly summing up by including plagioclase and K-feldspar weathering (slow to very slow rate group) results in a weathering rate of approximately 0.5 keq ha$^{-1}$ yr$^{-1}$ for soil of 1 m depth. However, accounting for mean soil depths of approximately 0.4 m (G1 and G2) or 0.6–0.7 m (F1), a value of total weathering of 0.2 to 0.35 keq ha$^{-1}$ yr$^{-1}$ is calculated for the catchments at Gårdsjön. These values are on the low side of both earlier estimates (cf. Sverdrup and Warfvinge, 1988a, b) and the recent estimates made by Sverdrup *et al.* (Chapter 10, this volume). The latter calculated, with various approaches, average weathering rates of 0.38 keq ha$^{-1}$ yr$^{-1}$ and 0.56 keq ha$^{-1}$ yr$^{-1}$ for the catchments G1 and F1 respectively.

*Method 2. Critical load for acidity on soil—Simple Mass Balance (steady state)*

This method was originally formulated by Sverdrup and Warfvinge (1988a) and de Vries (1988), and further discussed in Sverdrup *et al.* (1990) where *ANC* replaces *BC* in the formula below:

$$CL(acidity) = BC_{weathering} - BC_{uptake} - AC_{nitrogen} - ANC_{leaching} \qquad (23.1)$$

The authors argue (with Gårdsjön as example) that the net acidity efflux from soils could equal an $ANC_L$ of $-1.35$ keq ha$^{-1}$ yr$^{-1}$ as critical limit value with no effects on forest growth. However, leaching losses from soils of this size that substantially exceed the weathering rate would decrease available base cation storage and severely acidify the forest soil. In their calculation example, the result was a critical load of 0.95 $\mathrm{keq\,ha^{-1}\,yr^{-1}}$. Updated figures for tree uptake, a deposition of $NH_4 = 7\,\mathrm{kg\,ha^{-1}\,yr^{-1}}$ (Table 23.1) and $ANC_L$ values of −1.07 (F1) and −1.35 $\mathrm{keq\,ha^{-1}\,yr^{-1}}$ (G1) would result in $CL$ values of 0.49 to 0.77 $\mathrm{keq\,ha^{-1}\,yr^{-1}}$ for G1 and F1 respectively. The range between catchments gives a perspective on total small-scale variation in this type of calculation especially where there are differences in forest stand uptake and weathering.

The equation has been reformulated basing the criterion $ANC_L$ on a permissible leaching of $H^+$ and aluminium (Sverdrup and de Vries, 1994), where the critical aluminium leaching is related to the need of having the molar ratio (Ca+Mg+K)/Al adapted for target species.

$$CL(acidity) = ANC_{weathering} - ANC_{leaching} \tag{23.2}$$

This equation resides on several assumptions simplifying the system to a degree that can be questioned. Two other conditions were introduced, taking into account the fact that some of the available base cations escape uptake for physiological reasons (including also the winter dormancy) and that uptake cannot exceed the available amount. The relation between $H^+$ and aluminium is controlled by the gibbsite equilibrium and usually observed values are used. With $W = ANC_{weathering}$, and the limiting $ANC_{leaching}$ substituted by expressions for $H^+$ and Al, the resultant expression to solve is:

$$\begin{aligned} CL(acidity) = W + \left(1.5\,\frac{(0.7\,BC_W + BC_D - BC_U)}{((Ca + Mg + K)/Al)_{crit} \cdot K_{gibb}}\right)^{1/3} Q^{2/3} \\ + 1.5\left(\frac{(0.7\,BC_W + BC_D - BC_U)}{((Ca + Mg + K)/Al)_{crit}}\right) \end{aligned} \tag{23.3}$$

where $BC_W$ is weathering, $BC_D$ is deposition, $BC_U$ is base cation uptake (vegetation) and $K_{gibb}$ is the equilibrium constant for gibbsite dissolution.

The following values for input parameters were used for the calculations:

- Weathering; according to Table 23.1 but reducing the rate by 30% due to the default sodium fraction.
- Estimated net base cation uptakes of 0.64 and 1.07 $\mathrm{keq\,ha^{-1}\,yr^{-1}}$ (Table 23.1) violate the condition set by Sverdrup and de Vries (1994, equation 18) that uptake cannot exceed deposition + weathering. Assuming a centred value for the physiological uptake limit (2 to 15 $\mu\mathrm{eq\,L^{-1}}$) the minimum leaching terms were 0.04 and 0.05 $\mathrm{keq\,ha^{-1}\,yr^{-1}}$ resulting in adjusted uptake values of 0.47 and 0.35 $\mathrm{keq\,ha^{-1}\,yr^{-1}}$ for the F1 and G1 catchments respectively.
    - $Q$ was set equal to runoff values; 580 mm for F1 and 450 mm for G1.
    - The critical ratio of BC/Al was set to 1.0 as used by Sverdrup and de Vries (1994), although this is a much debated and questioned ratio.
    - For the gibbsite equilibrium a $pK_{gibb} = 8.5$ was used.

- For base cation (Ca + Mg + K) deposition, which is usually rather difficult to estimate accurately, values according to Table 23.1 were used.

The results of calculations in accordance with the assumptions above were 0.40 and 0.60 keq $ha^{-1}$ $yr^{-1}$ for G1 and F1 respectively. The current large biomass uptake of base cations was a problem in the calculations and was solved via applications of the conditions as above. It is obvious that the amount of base cations deposited with sea salt must have a substantial impact, as the very low non-marine input plus low weathering rate can not sustain present forest growth rate. Adaptation of lower growth rates seems reasonable as a long-term growth increment (Sverdrup and de Vries, 1994) when import of nitrogen to the system is less. Other shortcomings of the method have called for an updating of the procedure (Sverdrup *et al.*, 1995) so that total base cation deposition is included in the calculations and the physiological limit for uptake is set to $<2$ $\mu$eq $L^{-1}$. Applying these new conditions would allow a substantial increase of base cation uptake values, to the true value in the case of catchment F1. However, results of these calculations are very close to the earlier computations 0.39 keq $ha^{-1}$ $yr^{-1}$ in G1 and 0.64 keq $ha^{-1}$ $yr^{-1}$ in F1.

*Method 3. Critical load for acidity on water — Steady-State Water Chemistry (SSWC) according to Henriksen*

This is an empirical method which has been developed through several steps (Henriksen and Brakke, 1988; Henriksen *et al.*, 1990). A calculation example is given in Sverdrup *et al.* (1990) and followed here. The F-factor has been determined based on several catchment experiments (cf. Hultberg and Grennfelt, 1992) and resulting values were in the range of 0.4 to 0.6. An empirical mean of 0.5 was also used as well as calculated values using the following empirical formula (Brakke *et al.*, 1990):

$$F = \sin\left(\frac{[BC^*]\,90}{S}\right)$$

where $S = 400\,\mu\text{eq L}^{-1}$. The calculations involve the use of an estimate of $[SO^*_{40}]$, which is performed by two different equations (Henriksson *et al.*, 1990; Wilander, 1994). $[BC^*{}_0]$ is also estimated via iteration according to the formula given by Posch *et al.* (1993). The basic input data used in the calculations, including recent updates as time series have been extended, are given in Table 23.2.

As is obvious from Table 23.2, the results of the various approaches differ substantially and generate some concern about what input data to use. To state that any of the relationships used should be more appropriate than another is unwise and calculations using the "extremes" at least give a range of possible CL values. However, it is obvious that estimates of the pre-industrial base cation concentrations in runoff from the small catchments (G1 and G2) cannot be correct in some of the formulations, and result in negative critical loads. Introducing these data from Table 23.2 into the formula for exceedance of critical load according to Sverdrup *et al.* (1990) gives:

$$\Delta CC = [BC^*]_0 - [SO_4{}^*]_t - [ANC_{limit}] \tag{23.4}$$

**Table 23.2.** Input parameters for use in the Steady-State Water Chemistry method. [BC], Ca+Mg+K concentration; *, non-marine; $_t$, present; $_0$, pre-acidification concentrations. The F-factor is defined as: $F = \frac{\Delta(Ca^{*}+Mg^{*}+K^{*})}{\Delta SO_4{}^{*}}$ (Henriksen *et al.*, 1990)

| Parameter | Basis/formulation used for calculations | Gårdsjön | F1 long-term | F1 | G1 | G2 |
|---|---|---|---|---|---|---|
| Runoff, mm $yr^{-1}$ | | 580 | 580 | 580 | 450 | 580 |
| $[BC]_t$, $\mu$eq $L^{-1}$ | | 205 | 211 | 260 | 270 | 161 |
| $[BC^*]_t$, $\mu$eq $L^{-1}$ | | 136 | 72 | 113 | 71 | 18 |
| $[SO_4{}^*]_t$, $\mu$eq $L^{-1}$ | | 184 | 201 | 222 | 305 | 270 |
| F-factor | empirical | ............... | ............... | 0.5 | ............... | ............... |
| | Brakke *et al.*, 1990 | 0.51 | 0.28 | 0.43 | 0.28 | 0.07 |
| | Iteration, Posch *et al.*, 1993 | 0.42 | 0.22 | 0.32 | 0.17 | 0.05 |
| | Bernes, 1991 (Monitor 12) | 0.45 | 0.25 | 0.22 | −0.20 | −0.27 |
| $[SO_4{}^*]_0$, $\mu$eq $L^{-1}$ | Henriksen *et al.*, 1990 | 37 | 27 | 33 | 26 | 18 |
| | Bernes, 1991 (Monitor 12) | 14 | 13 | 14 | 13 | 11 |
| | Henriksen *et al.*, 1993 | 17 | 19 | 20 | 27 | 24 |
| | Posch *et al.*, 1993 | 28 | 21 | 25 | 21 | 16 |
| | Wilander, 1994, alternative 2 | 12 | 9 | 11 | 9 | 6 |
| | Wilander, 1994, alternative 3 | 19 | 12 | 16 | 12 | 7 |
| | AVERAGE $[SO_4{}^*]_0$ | 21 | 17 | 20 | 18 | 14 |
| $[BC^*]_0$ $\mu$eq $L^{-1}$ | empirical, using high $[SO_4{}^*]_0$ | 62 | −15 | 18 | −68 | −108 |
| | $[SO_4{}^*]_0$ Wilander, 1994, alternative 3 | 53 | −22 | 10 | −75 | −114 |
| | Brakke *et al.*, 1990 + Henriksen *et al.*, 1993 | 61 | 24 | 32 | −6 | 0 |
| | Iteration, Posch *et al.*, 1993 | 71 | 33 | 50 | 24 | 6 |
| | Based on average $[SO_4{}^*]_0$ and Brakke *et al.*, 1990 | 53 | 21 | 26 | −8 | 0 |

where $\Delta CC$ is the amount by which present $SO_4{}^*$ deposition exceeds the critical load. For Lake Gårdsjön, with $[SO_4{}^*]_t = 184\,\mu$eq $L^{-1}$ and $[ANC_{limit}] = 20\,\mu$eq $L^{-1}$, the equation gives $\Delta CC = 144\,\mu$eq $L^{-1}$ if a roughly averaged $[BC^*]_0$ of $60\,\mu$eq $L^{-1}$ is used (Table 23.2). At the CL the sulphate loading is reduced such that there is no exceedance. The critical load exceedance using the most recent data for Lake Gårdsjön then equals 0.84 keq $ha^{-1}\,yr^{-1}$ compared with the earlier estimate of 0.92 keq $ha^{-1}\,yr^{-1}$ calculated by Sverdrup *et al.* (1990).

Calculations of critical load then can be made by either of the following formulas:

$$CL = SO_{4\,deposition} - CL_{exceedance} \tag{23.5}$$

or

$$CL = BC^*_{0L} - BC^*_{D} - ANC_{L} \tag{23.6}$$

Equation (23.5) has later been modified to include the acidity by nitrogen by taking into account the leaching of nitrogen when calculating exceedance of *CL* (Henriksen and Posch, 1995). At present, there is probably no change of the terrestrial nitrogen leaching at Gårdsjön, so this extension of the formulation is ignored. Filling in data with a non-marine sulphur deposition, earlier being 1.34 keq $ha^{-1} yr^{-1}$ and most recently 0.86 keq $ha^{-1} yr^{-1}$, gives critical load values of 0.50 and 0.02 keq $ha^{-1} yr^{-1}$ respectively. For equation (23.6), data are calculated using runoff to estimate fluxes, which gives $BC^*_{0L} = 0.35$ keq $ha^{-1} yr^{-1}$, with the alkalinity limit set at 20 $\mu$eq $L^{-1}$, the $ANC_L = 0.12$ keq $ha^{-1} yr^{-1}$ and the non-marine base cation deposition given in Table 23.1 as 0.13 keq $ha^{-1} yr^{-1}$. By application of these data the alternative calculation formula gives a $CL$ value of 0.10 keq $ha^{-1} yr^{-1}$. In both cases, resultant values are higher than calculated earlier by Sverdrup *et al.* (1990) who used a dataset with larger uncertainties. If calculations according to equation (23.6) are based on the Swedish operational $ANC_L = 50\,\mu$eq $L^{-1}$, the resultant $CL$ would be negative (−0.7 keq $ha^{-1} yr^{-1}$). Similarly, trying to use small catchment runoff data would result in generally negative values even if an $ANC_L = 0\,\mu$eq $L^{-1}$ is used. Calculations according to Equation (23.5) also result in values around or below zero.

*Method 4. Critical load for acidity on water — "Diatom Model"*

This method is based on observations from the British Isles that in the stratigraphic record of sensitive lakes a definitive horizon usually characterises the onset of acidification, i.e. critical load exceedance (Battarbee *et al.*, 1993). The observations indicate a strong correlation between taxonomic changes of diatoms and the relationship between the base cation concentration of the lake water and the acid loading due to sulphur deposition at the site. The dose–response model is said to be so good that the sulphur deposition can be derived empirically for the time of the "critical load" passage. Based on a comparison of contemporary water chemistry and sulphur deposition, a ratio for Ca/S of 70 : 1, where Ca is in $\mu$eq $L^{-1}$ and S in keq $ha^{-1} yr^{-1}$, is suggested as the critical ratio, with sites having lower ratios showing exceedance. Further developments of this method and application on UK sites have followed, and Allott *et al.* (1995) suggested a ratio of 94 : 1 as the CL point. Current work is considering the CL of total acidity including nitrogen, resulting in a somewhat lower ratio, 89 : 1, as given by Allott *et al.* (1995).

The critical load can be calculated for any site given the value of $[Ca]_0$, the pre-acidification calcium concentration. This can be determined using the F-factor approach (Henriksen *et al.*, 1990), or the iterative method used above during calculations for the SSWC method. As it is not explicitly stated which $[Ca]$ should be used in calculations, values for both non marine $[Ca^*]_0$ and total $[Ca]_0$ have been estimated resulting in 33 $\mu$eq $L^{-1}$ and 52 $\mu$eq $L^{-1}$ respectively for Lake Gårdsjön. Applying these values into the suggested relationship:

$$CL\,(Sulphur) = [Ca_0]/R \qquad (23.7)$$

where $R$ equals 70 (early estimate) or 94 and 89 (recent estimate) gives $CL$ values of 0.74, 0.55 and 0.58 keq $ha^{-1} yr^{-1}$ using total calcium, while applying the non-marine calcium concentrations results in 0.47, 0.35 and 0.37 keq $ha^{-1} yr^{-1}$. In both cases, $CL$ values are substantially higher than estimates based on the SSWC method. For the moment, it is not clear whether the relationship between Ca and S is different for Swedish lakes and thus whether any adaptation is needed to Swedish conditions. This could probably be done without too much work as to date a large number of

Swedish lakes have been analysed for stratigraphic diatom records (Renberg *et al.*, 1993). In general, these lakes have been chosen with respect to available water chemistry which should facilitate the assessment of pre-acidification water concentration of calcium.

*Method 5. Critical load for nitrogen and sulphur on soil—*
*Mass Balance Method (Lökeberg formulation)*

Much of the initial work on critical loads was focused on sulphur as the important pollutant for observed acidification. The fact that sulphur in many environments is an element without intensive biochemical reactions or transformations between different pools in the soil or lakes waters facilitated the calculation of critical loads. Gradually, a similar understanding of the importance of nitrogen has developed in conjunction with increased knowledge as assessment of the different processes has proceeded. In its biogeochemical cycling, nitrogen is subject to several important transformations both within terrestrial and aquatic parts of the ecosystems. A general critical load equation similar to the original mass balance equation for calculation of critical load of "actual acidity" has been formulated (Hettelingh *et al.*, 1992):

$$CL(S) + CL(N) = BC^*_{deposition} + BC_{weathering} - BC_{uptake} + N_{denitrification} + N_{uptake} + N_{immobilisation} - ANC_{crit.\ leaching} \quad (23.8)$$

which, with respect to nitrogen, transforms to:

$$CL(N) = N_{uptake} + N_{immobilisation} + N_{denitrification} + N_{crit.\ leaching} \quad (23.9)$$

Another formulation of equation (23.9) was proposed by Hettelingh *et al.* (1992) to enhance the use of transfer functions:

$$CL(N) = N_{uptake} + N_{immobilisation} + (1/1 - f_{denitrification})\, N_{crit.\ leaching} \quad (23.10)$$

Equation (23.9) forms the basis for calculations of *CL* for nitrogen on soil as a eutrophying substance and this equation uses the same parameters, though differently denoted, as defined by Gundersen (1992).

Forests in the Gårdsjön area currently are subject to an excessive nitrogen deposition probably stressing the system and giving higher uptake rates of nitrogen and base cations (Sverdrup and de Vries, 1994). Uptake values presented in Table 23.1 are based on old estimates of base cation concentrations in trees from F1 and an assumed tree nitrogen uptake rate of 0.4 keq $ha^{-1}\,yr^{-1}$ and values are only scaled for G1 in proportion to total biomass. Other input data needed for calculations are from Table 23.1 or are discussed below.

The denitrification rate in these well-drained forest soils amounts to about 0.1 kg $ha^{-1}\,yr^{-1}$ in the form of $N_2O$ and it is suggested that the total denitrification should not be more than double this value, consequently about 0.2 kg $ha^{-1}\,yr^{-1}$ (Klemedtsson, pers. comm.) which would be 0.014 keq $ha^{-1}\,yr^{-1}$. The $f_{den.}$ is assumed to be a maximum of 0.02 for G1 and 0.04 for F1, taking into account that there are more waterlogged soils and small peat areas in F1. With respect to soil characteristics of the Lake Gårdsjön area, a denitrification fraction of 0.1 is assumed in the calculations. The critical leaching of nitrogen is assumed to approximate the long-term observed values in the area (between 0.1 and 0.2 keq $ha^{-1}$ $yr^{-1}$ and defined to be background), here 0.15 keq $ha^{-1}$ $yr^{-1}$ is used. Filling in the values to Equation (23.9) gives for catchment G1 a *CL* value for nitrogen of 1.07 keq $ha^{-1}$ $yr^{-1}$ and for F1 0.74 keq $ha^{-1}$ $yr^{-1}$. Filling in values to Equation (23.10) gives very similar values for G1 and F1, respectively 1.07 and 0.72 keq $ha^{-1}$ $yr^{-1}$. This is not surprising as they differ only in the denitrification term which is very small. These *CL* values seem to be low compared with what is empirically found for managed coniferous forests. Bobbink *et al.* (1992) and Bobbink and Roelofs (1995) give values that are approximately in the range of 1.1 to 1.4 keq $ha^{-1}$ $yr^{-1}$ for ground layer vegetation changes, but this limit may be extended downwards to 0.7 keq $ha^{-1}$ $yr^{-1}$ when tree nutrient imbalances are included as an effect.

Considering the total dominance of the tree growth uptake term in these calculations and the fact that this is an estimated value, the calculated *CL* values should be regarded as rather rough estimates. This is especially so as the value used for catchment G1, 0.84 keq $ha^{-1}$ $yr^{-1}$, may be comparatively high (cf. Sverdrup *et al*, 1990). Other processes taking part in this nitrogen balance seem to be more accurate, and inclusions of other terms as suggested in the formulation of a very general equation by de Vries and Latour (1995) would not improve the assessment at Gårdsjön.

A *CL* for sulphur can be calculated from the derived formulation (Hettelingh *et al.*, 1992):

$$CL(S) = BC^*_{deposition} + BC_{weathering} - BC_{uptake} - ANC_{crit.\ leaching} - N_{crit.\ leaching} \quad (23.11)$$

Filling in values from Table 23.1, the critical nitrogen leaching of 0.15 keq $ha^{-1}$ $yr^{-1}$ used earlier and a critical ANC leaching of 1.35 keq $ha^{-1}$ $yr^{-1}$ results in a *CL* for sulphur of 1.25 keq $ha^{-1}$ $yr^{-1}$ for catchment F1 and a value of 0.62 keq $ha^{-1}$ $yr^{-1}$ for G1. Summing up both sulphur and nitrogen *CL* values for the two catchments gives 1.69 for G1 and 1.99 for F1. The value chosen for the critical ANC leaching is crucial for the sulphur part of this estimate. This large dependence on a single term merits a more detailed evaluation.

*Method 6. Critical load for nitrogen and sulphur on water— First-order Acidity Balance (FAB) Method and determination of an ecosystem protection isoline for Lake Gårdsjön*

A general equation for calculation of critical load for sulphur and acidifying nitrogen can be formulated in a similar way to that used for forest soils. Basic formulations were given in Kämäri *et al.* (1993) and Posch *et al.* (1993) and further developed in Henriksen *et al.* (1993). Below is the formulation by Hettelingh *et al.* (1992):

$$CL(N) + CL(S) = f\,N_{uptake} + (1 - r)\,(N_{immobilisation} + N_{denitrification}) + r\,N_{retention} + r\,S_{retention} + Q\,([BC^*_0] - [ANC]_{limit}) \quad (23.12)$$

This should be combined with limiting criteria related to sensitive target species preferably typical for the actual surface waters. In calculations with the SSWC method earlier (Equations (23.4 and 23.6), the $[ANC]_{limit}$ used was 20 $\mu$eq $L^{-1}$ which is the value used in Norway (cf. Henriksen *et al.*, 1992), while the Swedish operational limit is defined as 50 $\mu$eq $L^{-1}$. Whichever is used has profound implications for the result of calculations. In addition to acidification, there is an upper limit on nitrogen deposition to avoid eutrophication. The mass balance for nitrogen gives:

$$CL(N) = f\,N_{uptake} + (1 - r)\,(N_{immobilisation} + N_{denitrification}) + r\,N_{retention} + N_{leaching} \quad (23.13)$$

where $f$ = the fraction of the catchment that is forested, $r$ = the lake to catchment area ratio, and $N_{retention}$ =in-lake retention of nitrogen.

Other parameters are defined as in earlier formulations above. For calculations according to the equation, the following values for the Lake Gårdsjön and its catchment (Table 23.1) have been used and supplemented from derivations above:

| | |
|---|---|
| $f$ | $= 0.716$ |
| $r$ | $= 0.142$ |
| $N_{ret}$ | $= 1$ keq $ha^{-1}$ $yr^{-1}$ |
| $N_{upt}$ | $= 0.5$ keq $ha^{-1}$ $yr^{-1}$ |
| $N_{imm}$ | $= 0.06$ keq $ha^{-1}$ $yr^{-1}$ |
| $N_{den}$ | $= 0.1$ keq $ha^{-1}$ $yr^{-1}$ |
| $N_{leach}$ | $= 0.15$ keq $ha^{-1}$ $yr^{-1}$ |

For the whole catchment of Lake Gårdsjön, larger areas of peat and waterlogged soils probably imply a significant increase of the total denitrification, hence the value 0.1 keq $ha^{-1}$ $yr^{-1}$. For the uptake rate a conservative value of 0.5 keq $ha^{-1}$ $yr^{-1}$ has been assumed, even though there are areas in the upper catchment where younger forest stands predominate. The in-lake retention is estimated based on data for annual mass balance figures and independent measurements of denitrification in lake sediments which both balance around 20 keq $ha^{-1}$ $yr^{-1}$ (cf. Andersson and Gahnström, 1985; Persson and Broberg, 1985; Gahnström *et al.*, 1993). Putting these figures into the equation gives a nitrogen CL value of 0.79 keq $ha^{-1}$ $yr^{-1}$.

An alternative equation for calculation is given by:

$$CL(N) = f\,a\,N_{upt} + (1 - r)\,a\,N_{imm.} + b\,N_{crit.\ leach.} \quad (23.14)$$

where $a = (1 - f_{den})/(1 - f_{den}(1 - r))$, $b = 1/(1 - \rho_N)(1 - f_{den}(1 - r))$ and $\rho_N = s_N/(Q/r + s_N)$. With $s_N = 5$ m $yr^{-1}$ (Henriksen *et al.*, 1993), $f_{den} = 0.1$ (a roughly estimated average value for the catchment based on assumptions from method 5) and $\rho_N = 0.55$, this gives $a = 0.98$ and $b = 2.43$. Inserting these values into Equation (23.14) gives 0.76 keq $ha^{-1}$ $yr^{-1}$, very similar to that from Equation (23.13).

Combining Equations (23.12 and 23.13) gives an estimate for the critical load for S of:

$$CL(S) = r\,S_{retention} + Q\,([BC^*{}_0] - [ANC]_{limit}) - N_{crit.\ leaching} \quad (23.15)$$

where $r\,S_{retention} = \rho_S\,S_{total\ deposition}$, and $\rho_s = s_S/(Q/r + s_S)$.

With $s_S = 0.5\,\text{m yr}^{-1}$ (Henriksen *et al.*, 1993) and sulphur deposition according to Table 23.1 equalling $1.58\,\text{keq ha}^{-1}\,\text{yr}^{-1}$ during the early 1980s or $1.16\,\text{keq ha}^{-1}\,\text{yr}^{-1}$ a decade later; $\rho_S$ equals 0.11 and therefore sulphur retention varies between 1.22 and the present $0.90\,\text{keq ha}^{-1}\,\text{yr}^{-1}$. These numbers transfer to a total in-lake retention of around 30 to $40\,\text{keq ha}^{-1}\,\text{yr}^{-1}$, which is approximately two to three times the actual numbers determined around 1980 (Andersson, 1985). Inserting model values into the equation gives CL values for S between 0.08 and $0.04\,\text{keq ha}^{-1}\,\text{yr}^{-1}$ using an $[ANC]_{limit}$ concentration of $50\,\mu\text{eq L}^{-1}$, which would increase to just above $0.2\,\text{keq ha}^{-1}\,\text{yr}^{-1}$ if the lower limit $20\,\mu\text{eq L}^{-1}$ is used. Actual values for sediment retention implies a lower CL value, in the first case around zero and in the latter case approximately $0.15\,\text{keq ha}^{-1}\,\text{yr}^{-1}$.

In construction of a "protection isoline" for Lake Gårdsjön, the formulations of Posch *et al.* (1993) and Henriksen *et al.* (1993) are used. Applying these equations on Lake Gårdsjön data results in Figure 23.1, which gives a picture of how different abatement strategies will affect the change from an overall substantial CL exceedance to an acceptable situation with no exceedance. Both S and N deposition reductions are required to meet the CL. The current situation for sulphur deposition implies a three to tenfold exceedance although the CL value of $0.1\,\text{keq ha}^{-1}\,\text{yr}^{-1}$, which neglects the immobilisation and export of nitrogen, seems less probable. Current exceedance for nitrogen is moderate.

*Method 7. Critical load for sulphur and nitrogen respectively, using a "sulphur fraction" approach*

The critical load of actual acidity has been mapped for Europe using the reformulated SMB (Equation 23.2). In order to discriminate between effects due

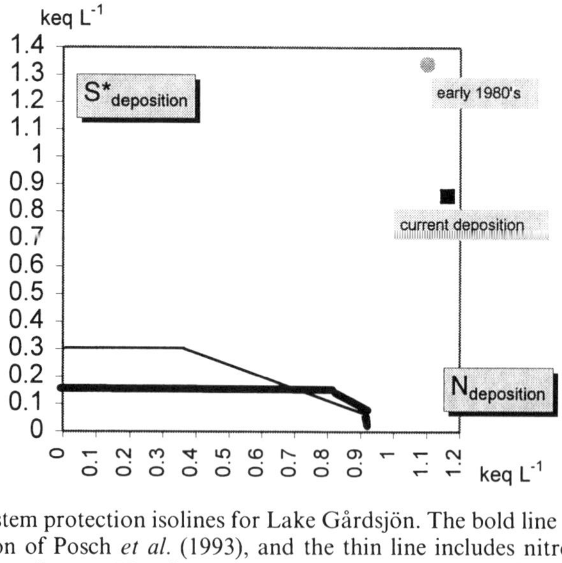


**Figure 23.1.** Ecosystem protection isolines for Lake Gårdsjön. The bold line is according to the original formulation of Posch *et al.* (1993), and the thin line includes nitrogen leaching and immobilisation according to Henriksen *et al.* (1993). Also shown is the development of deposition reduction since the early 1980s

to loading of sulphur or nitrogen, part of the critical load for acidity has been allocated to sulphur by defining a sulphur fraction $S_f$:

$$CL(S) = S_f\, CL(Ac)_{actual} \tag{23.16}$$

where:

$$Sf = \begin{cases} S_{td}/(S_{td} + N_{td} - N_u - N_i) & if \quad N_{td} > N_u + N_i \\ 1 & otherwise \end{cases}$$

Then the CL for nitrogen acidity, when acidification is caused by the fraction that is not taken up or immobilised, becomes:

$$CL(N) = N_{uptake} + N_{immobilised} + (1 - S_f)\, CL(A) \tag{23.17}$$

Using Gårdsjön values in Table 23.1, except for a standardised $N_{uptake}$ value of 0.5 keq $ha^{-1}\,yr^{-1}$, $S_f = 0.70$. Applying the $S_f$ value to Method 2, results in $CL(S)$ values of 0.34 to 0.54 keq $ha^{-1}\,yr^{-1}$ for the original formulation, and of 0.27 to 0.42 keq $ha^{-1}\,yr^{-1}$ for the current formulation. The corresponding nitrogen CL values would be 0.56 keq $ha^{-1}\,yr^{-1}$+0.3 $CL(A)$ giving a range of $CL(N)$ values of 0.68 to 0.75 keq $ha^{-1}\,yr^{-1}$ and 0.71 to 0.79 keq $ha^{-1}\,yr^{-1}$ respectively. If this sulphur fraction approach is applied to Methods 1, 3 and 4, the calculations result in $CL(S)$ values from 0.14 to 0.25 keq $ha^{-1}\,yr^{-1}$ for Method 1 to at most 0.25 to 0.52 keq $ha^{-1}\,yr^{-1}$ for the "Diatom" Model. All $CL(N)$ values fall in a much narrower range of 0.59 to 0.79 keq $ha^{-1}\,yr^{-1}$, which are close to the lower end of the estimates according to the Lökeberg formulation (Grennfelt and Thörnelöf, 1992).

*Method 8. Critical load for acidity on soil—Steady-State Mass Balance, the PROFILE variant*

The method has been applied earlier to Gårdsjön (Sverdrup and Warfvinge, 1991) which indicated a CL of potential acidity value somewhere in the range of 1.23 to 1.64 keq $ha^{-1}\,yr^{-1}$, for runoff from various small catchments and the forest soil of catchment F1 respectively. In their discussion they note that the calculations assume present nitrogen uptake/reactions to largely exceed the base cation uptake, indicating nutrient imbalance. Adjusting for this by reverting to a weathering and ANC leaching based formulation gives much lower CL values, 0.44 keq $ha^{-1}\,yr^{-1}$ for surface water runoff and 0.84 keq $ha^{-1}\,yr^{-1}$ for the forest soil.

Using the current dataset from catchment G1 for calculating CL values with PROFILE results in values in between earlier estimates. CL for runoff (lake) is 0.21 and 0.33 keq $ha^{-1}\,yr^{-1}$, variation dependent on the $[ANC]_{limit}$ used (50 or 20 $\mu$eq $L^{-1}$ respectively) for the simulation runs. For the soil of catchment G1 a CL value of 1.07 was calculated when internal cycling of base cations and nutrients was accounted for in the modelling. If the latter ecosystem-related processes are excluded, the resultant CL value is only 0.54 keq $ha^{-1}\,yr^{-1}$. Considering existing soil and vegetation differences between the G1 and F1 catchments, a probable CL value for F1 is still similar to the lower estimate made by Sverdrup and Warfvinge (1991).

# CONCLUSIONS

### Uncertainties in calculated critical loads

A summary of the results of the above calculations is presented in Table 23.3. There is clearly some variability, less for N than S, but the results generally fall into the range 0–1 keq $ha^{-1}$ $yr^{-1}$. The variability within methods derives from variability in input parameters, whereas the variability between methods arises from the incorporation of different processes. The Gårdsjön site has been studied for more than 20 years, and a large database has been accumulated, but there are still uncertainties and gaps which affect critical load calculations, particularly in estimates of the uptake and storage of base cations and N by the growing forest. On a regional scale, away from intensively studied experimental sites, these uncertainties will be even greater. More information on these processes, and how they relate to forest management, remains a priority. For the steady-state water chemistry methods, regional patterns of base cation deposition and estimates of pre-acidification conditions are important sources of uncertainty. Good estimates of weathering rates are crucial to most methods, though this is being addressed by the development and continued refinement of models such as PROFILE.

**Table 23.3.** Summary of calculated critical load values, embracing the variation caused by using catchment or lake data. For Methods 1–4, the acidity critical load has been split into S and N components using the "sulphur fraction" approach described in Method 7

| Method | | CL(A) (keq $ha^{-1}$ $yr^{-1}$) | CL(S) (keq $ha^{-1}$ $yr^{-1}$) | CL(N) (keq $ha^{-1}$ $yr^{-1}$) |
|---|---|---|---|---|
| **FOREST** | | | | |
| Level zero—rough mineral composition | 1 | 0.20–0.35 | 0.14–0.25 | 0.62–0.66 |
| SMB formulation, Sverdrup *et al.*, 1990 | 2 | 0.49–0.77 | 0.35–0.54 | 0.71–0.79 |
| SMB, with BC:Al criteria | 2 | 0.40–0.60 | 0.28–0.42 | 0.68–0.74 |
| As above but with total BC deposition | 2 | 0.39–0.64 | 0.27–0.45 | 0.68–0.75 |
| PROFILE, no circulation in the system | 8 | 0.54 | | |
| PROFILE, circulation in vegetation included | 8 | 1.07 | | |
| **LAKE** | | | | |
| SSWC, according to Sverdrup *et al.*, 1990 | 3 | 0.02–0.50 | 0.01–0.35 | 0.57–0.71 |
| SSWC Gårdsjön, alternative $SO_4{}^*{}_0$ derivation | 3 | 0.1 | 0.07 | 0.59 |
| Diatom model, old formulation | 4 | 0.47–0.74 | 0.33–0.52 | 0.70–0.78 |
| Diatom model, new formulation | 4 | 0.35–0.55 | 0.25–0.39 | 0.64–0.73 |
| Diatom model, new formulation with nitrogen | 4 | 0.37–0.58 | 0.26–0.41 | 0.64–0.73 |
| CL formulation based on Grennfelt and Thörnelöf, 1992 | 5 | | 0.64–1.25 | 0.74–1.07 |
| FAB, with $N_{imm}$ and $N_{Leach}$ set to zero | 6 | | 0.04–0.08 | 0.79 |
| FAB, according to Henriksen *et al.*, 1993 | 6 | | 0.29 | 0.92 |
| PROFILE, lake with $[ANC]_{limit}$ 50 or 20 $\mu$eq $L^{-1}$ | 8 | 0.21, 0.33 | | |

**Validation of critical loads**

Sites such as Lake Gårdsjön are important for testing and validation of critical load formulations. It should be possible to improve these methods by giving indications of how well parameters and transfer functions describe the real system. At Gårdsjön, deposition of S and N currently exceeds virtually all the calculated critical load values. The adverse ecological effects observed in Lake Gårdsjön (Andersson and Olsson, 1985) show that historical deposition at least must have exceeded the critical load. A variety of paleoecological evidence (e.g. Renberg and Wallin, 1985) showed that severe acidification of the lake started around 1950. The S deposition in 1950 is estimated to have been about 0.44 keq $ha^{-1}$ $yr^{-1}$ (Chapter 3, this volume) which provides an upper bound for the "true" S critical load. It is an upper bound because soil acidification is likely to have been occurring in the years before 1950, but not at sufficient rate to have much effect on the quality of runoff water and thus on lake organisms. Acidification modelling using MAGIC (Chapter 19, this volume) or SAFE (Chapter 20, this volume) predicts acidification during this period, from around 1900 or 1920 respectively. Indeed, diatom stratigraphy (Renberg and Wallin, 1985) shows Lake Gårdsjön to have been acidifying slowly (from pH 7 to pH 6) from its formation in about 12 500 BP until the 20th century. This indicates that natural acid deposition combined with acid production processes in the catchment probably exceeds the "level zero" critical load (criterion: no change in base saturation). It should not however exceed critical loads calculated by using absolute criteria, such as the mass balance methods, since it has to be assumed that the lake and surrounding forest were functioning without "significant adverse effects" during this long period. If it is assumed that the prehistoric rain pH was around 5.0, then prehistoric deposition would be about 0.1 keq $ha^{-1}$ $yr^{-1}$, which puts a lower bound on the true value of the acidity critical load.

It is much harder to suggest a range for nitrogen critical loads using the historical data, as there are no obvious deleterious effects. Exceedance of the critical load for N is more a prediction of adverse effects in the future.

The roof itself should provide some evidence on the validity of the calculated critical loads. Acid deposition has been reduced to 0.05 keq $ha^{-1}$ $yr^{-1}$, and N and non-marine $SO_4^-$ deposition to zero (Chapter 4, this volume). Recovery is taking place, but this does not in itself indicate that the critical load is not exceeded. ANC and BC/Al ratios in runoff are still much worse than criterion values. This illustrates one problem of validating critical loads: they are intended to indicate the deposition which is sustainable in the long term, whereas all experimental evidence is obtained in the short term, since the response time of these ecosystems is so long (decades to centuries). Acidification model predictions can be used to extend the validation period with some loss of confidence. MAGIC predictions (Chapter 19, this volume) give an ANC of $-100$ $\mu$eq $L^{-1}$ in the year 2007, rising slowly but still well short of the criterion values of 0 to $+50$ $\mu$eq $L^{-1}$. SAFE predicts values approaching $+100$ $\mu$eq $L^{-1}$ as the ultimate ANC. Thus it is not really possible to use the roof results to put bounds on the true value of the critical load of acidity and S. A virtually non-existent $NO_3^-$ leaching shows no signs of a change during the course of the experiment and it is therefore impossible to evaluate critical load for N based on the roof results.

Historical data therefore indicate that the critical load (S/acidity) for the Gårdsjön catchments ought to be in the range 0.1–0.44 keq $ha^{-1}$ $yr^{-1}$. Inspection of Table 23.3 shows that the lower limits of the calculated critical loads fall within this range, but the upper limits tend to be above it. The Gårdsjön data thus provide some justification for making precautionary assumptions about some of the default parameters used in calculating critical loads.

## ACKNOWLEDGEMENTS

This work was funded by the National Power–PowerGen Joint Environmental Programme and IVL.

## REFERENCES


Abrahamsen, G. and Tveite, B. (1983) Effects of air pollutants on forest and forest growth. In: *Ecological Effects of Acid Deposition*. SNV Report PM 1636, pp. 199–219.

Allott, T. E. H., Battarbee, R. W., Curtis, C., Kreiser, A. M., Juggins, S. and Harriman, R. (1995) An empirical model of critical acidity loads for surface waters based on palaeolimnological data. In: Hornung, M., Sutton, M. A. and Wilsson, R. B. (eds), *Mapping and Modelling of Critical Loads for Nitrogen — A Workshop Report*. ITE, Bush Estate, Penicuik, pp. 50–54.

Andersson, B. I. (1985) Properties and chemical composition of surficial sediments in the acidified Lake Gårdsjön, SW Sweden. *Ecological Bulletins (Stockholm)*, **37**, 251–263.

Andersson, F. and Olsson, B. (eds) (1985) Lake Gårdsjön — An Acid Forest Lake and its Catchment. *Ecological Bulletins (Stockholm)*, **37**, 336 pp.

Andersson, G. and Gahnström, G. (1985) An attempt to quantify sediment processes in the acidified Lake Gårdsjön, SW Sweden. *Ecological Bulletins (Stockholm)*, **37**, 327–336.

Ashmore, M. R., Bell, N. and Rutter, J. (1985). The role of ozone in forest damage in West Germany. *Ambio*, **14**, 81–87.

Ashmore, M. R. and Wilson, R. B. (eds) (1994) *Critical Levels of Air Pollutants for Europe*. Report of the Egham workshop. Air Quality Division, Department of the Environment, London.

Battarbee, R. W., Allott, T. E. H., Kreiser, A. M. and Juggins, S. (1993) Setting critical loads for UK surface waters: the diatom model. In: Hornung, M. and Skeffington, R. A. (eds), *Critical Loads: Concept and Applications*. ITE Symposium no. 28, pp. 99–102.

Bernes, C. (1991) *Acidification and Liming of Swedish Freshwaters*. Monitor 12. Swedish Environmental Protection Agency, 144 pp.

Bobbink, R. and Roelofs, J. (1995) Empirical nitrogen critical loads: update since Lökeberg (1992). In: Hornung, M., Sutton, M. A. and Wilsson, R. B. (eds), *Mapping and Modelling of Critical Loads for Nitrogen — A Workshop Report*. ITE, Bush Estate, Penicuik, pp. 9–19.

Bobbink, R., Boxman, D., Fremstad, E., Hei, G., Houdijk, A. and Roelofs, J. (1992) Critical loads for nitrogen eutrophication of terrestrial and wetland ecosystems based upon changes in vegetation and fauna. In: Grennfelt, P. and Thörnelöf, E. (eds), *Critical Loads for Nitrogen*. Report from a Workshop held at Lökeberg, Sweden, 6–10 April 1992. Nordic Council of Ministers, NORD Report 1992: 41, Copenhagen, pp. 111–159.

Brakke, D. F., Henriksen, A. and Norton, S. A. (1990) A variable F-factor to explain changes in base cation concentrations as a function of strong acid deposition. *Verhandlungen Internationale Vereinigung Für Theoretische und Angewandte Limnologie*, **24**, 146–149.

Cronan, C. S., April, R., Bartlett, R. J. *et al.* (1989) Aluminum toxicity in forests exposed to acidic deposition: The ALBIOS results. *Water, Air, and Soil Pollution*, **48**, 181–192.

deVries, W. (1988) Critical deposition levels for nitrogen and sulphur on Dutch forest ecosystems. *Water, Air, and Soil Pollution*, **42**, 221–239.

de Vries, W. and Latour, J. B. (1995) Methods to derive critical loads for nitrogen for terrestrial systems. In: Hornung, M., Sutton, M. A. and Wilson, R. B. *Mapping and Modelling of Critical Loads for Nitrogen — A Workshop Report*. ITE, Bush Estate, Penicuik, pp. 20–33.

de Vries, W., Reinds, G. J. and Posch, M. (1994) Assessment of critical loads and their exceedance on European forests using a one-layer steady-state model. *Water, Air, and Soil Pollution*, **72**, 357–394.

Dise, N. B. and Wright, R. F. (eds) (1992) *The NITREX project (Nitrogen Saturation Experiments)*. Ecosystems Research Report No. 2, Commission of the European Communities, Directorate-General for Science, Research and Development, Brussels.

Downing, R. J., Hettelingh, J.-P. and de Smet, P. A. M. (eds) (1993) *Calculation and Mapping of Critical Loads in Europe. Status Report 1993*. Coordination Centre for Effects. RIVM-Report No. 259101003, 163 pp.

Gahnström, G., Blomqvist, P. and Fleischer, S. (1993) Are key nitrogen fluxes changed in the acidified aquatic ecosystem? *Ambio*, **22**, 318–324.

Grennfelt, P. and Thörnelöf, E. (eds) (1992) *Critical Loads for Nitrogen*. Report from a Workshop held at Lökeberg, Sweden, 6–10 April 1992. Nordic Council of Ministers, NORD Report 1992: 41, Copenhagen, 428 pp.

Gundersen, P. (1992) Mass balance approaches for establishing critical loads for nitrogen in terrestrial ecosystems. In: Grennfelt, P. and Thörnelöf, E. (eds) *Critical Loads for Nitrogen*. Report from a Workshop held at Lökeberg, Sweden, 6–10 April 1992. Nordic Council of Ministers, NORD Report 1992: 41, Copenhagen, pp. 55–109.

Hedin, L. O., Granat, L., Likens, G. E., Buishand, T. A., Galloway, J. N., Butler, T. J. and Rodhe, H. (1994) Steep declines in atmospheric base cations in regions of Europe and North America. *Nature*, **367**, 351–354.

Henriksen, A. and Brakke, D. F. (1988) Increasing contributions of nitrogen to the acidity of surface waters in Norway. *Water, Air, and Soil Pollution*, **42**, 183–201.

Henriksen, A. and Posch, M. (1995) Critical loads for nitrogen: Surface waters. In: Hornung, M., Sutton, M. A. and Wilson, R. B. (eds), *Mapping and Modelling of Critical Loads for Nitrogen — A Workshop Report*. ITE, Bush Estate, Penicuik, pp. 55–62.

Henriksen, A., Kämäri, J., Posch, M., Lövblad, G., Forsius, M. and Wilander, A. (1990) *Critical Loads to Surface Waters in Fennoscandia. Intra- and Inter-grid Variability of Critical Loads and their Exceedance*. Nordic Council of Ministers, NORD Report 1990: 124, Copenhagen, 33+9 pp.

Henriksen, A., Kämäri, J., Posch, M. and Wilander, A. (1992) Critical loads of acidity: Nordic surface waters. *Ambio*, **21**, 356–363.

Henriksen, A., Forsius, M., Kämäri, J., Posch, M. and Wilander, A. (1993) *Exceedance of Critical Loads for Lakes in Finland, Norway and Sweden: Reduction Requirements for Nitrogen and Sulfur Deposition*. NIVA-REPORT 32/1993, 46 pp.

Henrikson, L. and Brodin, Y. W. (1995) *Liming of Acidified Surface Waters. A Swedish Synthesis*. Springer-Verlag, Berlin, 458 pp.

Hettelingh, J.-P., Downing, R. J. and de Smet, P. A. M. (1991) *Mapping Critical Loads for Europe*. CCE Technical Report No. 1, Co-ordination Centre for Effects. RIVM Report No. 259101001, 86 pp + Appendices.

Hettelingh, J.-P., Posch, M., de Vries, W., Bull, K. and Sverdrup, H. U. (1992) Guidelines for the computation and mapping of nitrogen critical loads and exceedances in Europe. In: Grennfelt, P. and Thörnelöf, E. (eds) *Critical Loads for Nitrogen*. Report from a Workshop held at Lökeberg, Sweden, 6–10 April 1992. Nordic Council of Ministers, NORD Report 1992: 41, Copenhagen, pp. 287–303.

Hornung, M. and Skeffington, R. A. (eds) (1993) *Critical Loads: Concept and Applications*. ITE Symposium no. 28.

Hornung, M., Sutton, M. A. and Wilson, R. B. (1995) *Mapping and Modelling of Critical Loads for Nitrogen — A Workshop Report*. ITE, Bush Estate, Penicuik, pp. 55–62.

Hultberg, H. (1985) Budgets of base cations, chloride, nitrogen and sulphur in the acid Lake Gårdsjön catchment, SW Sweden. *Ecological Bulletins (Stockholm)*, **37**, 133–158.

Hultberg, H. and Grennfelt, P. (1992) Sulphur and seasalt deposition as reflected by throughfall and runoff chemistry in forested catchments. *Environment Pollution*, **75**, 215–222.

Kämäri, J., Forsius, M. and Posch, M. (1993) Critical loads of sulfur and nitrogen for lakes II: regional extent and variability in Finland. *Water, Air, and Soil Pollution*, **66**, 77–96.

Melkerud, P.-A. (1983) *Quaternary Deposits and Bedrock Outcrops in an Area Around Gårdsjön, South-West Sweden, with Physical, Mineralogical and Geochemical Investigations*. Reports on Forest Ecology and Forest Soils No: 44, Swedish University of Agricultural Sciences, Uppsala, Sweden, 87 pp.

Moldan, F., Hultberg, H. and Andersson, I. (1995) Covered Catchment Experiment at Gårdsjön: Changes in runoff chemistry after four years of experimentally reduced acid deposition. *Water, Air, and Soil Pollution*, **85**, 1599–1604.

Nilsson, J. (1986) *Critical Loads for Nitrogen and Sulphur*. Nordic Council of Ministers, NORD Report 1986: 11, Copenhagen, 232 pp.

Nilsson, J. and Grennfelt, P. (1988) *Critical Loads for Sulphur and Nitrogen*. Report from a workshop at Skokloster, Sweden, 19–24 March, 1988. Nordic Council of Ministers, NORD Report 1988: 15, Copenhagen, 418 pp.

Odén, S. (1968) Nederbördens och luftens försurning. Dess orsaker, förlopp och verkan i olika miljöer. *Ecological Bulletins (Stockholm)*, **1**, 1–86. (In Swedish with English summary).

Olsson, B., Hallbäcken, L., Johansson, S., Melkerud, P.-A., Nilsson, S. I. and Nilsson, T. (1985) The Lake Gårdsjön area — physiographical and biological features. *Ecological Bulletins (Stockholm)*, **37**, 10–28.

Persson, G. and Broberg, O. (1985) Nutrient concentration in the acidified Lake Gårdsjön: The role of transport and retention of phosphorus, nitrogen and DOC in watershed and lake. *Ecological Bulletins (Stockholm)*, **37**, 158–176.

Posch, M., Forsius, M. and Kämäri, J. (1993) Critical loads of sulphur and nitrogen for lakes. I: Model description and estimation of uncertainty. *Water, Air, and Soil Pollution*, **66**, 173–192.

Renberg, I. and Wallin, J. E. (1985) The history of the acidification of Lake Gårdsjön as deduced from diatoms and *Sphagnum* leaves in the sediment. *Ecological Bulletins (Stockholm)*, **37**, 47–52.

Renberg, I., Korsman, T. and Anderson, N. J. (1993) A temporal perspective of lake acidification in Sweden. *Ambio*, **22**, 264–271.

Sverdrup, H. and de Vries, W. (1994) Calculating critical loads for acidity with the simple mass balance method. *Water, Air, and Soil Pollution*, **72**, 143–162.

Sverdrup, H. and Warfvinge, P. (1988a) Assessment of critical loads of acid deposition on forest soils. In: Nilsson, J. and Grennfelt, P. (eds), *Critical Loads for Sulphur and Nitrogen*. NORD Report 1988: 15, Copenhagen, pp. 81–129.

Sverdrup, H. and Warfvinge, P. (1988b) Chemical weathering of minerals in the Gårdsjön catchment in relation to a model based on laboratory rate coefficients. In: Nilsson, J. and Grennfelt, P. (eds), *Critical Loads for Sulphur and Nitrogen*. NORD Report 1988: 15, Copenhagen, pp. 131–149.

Sverdrup, H. and Warfvinge, P. (1991) Calculating critical loads of acid deposition with PROFILE — a steady-state soil chemistry model. *Water, Air, and Soil Pollution*, **63**, 119–143.

Sverdrup, H., de Vries, W. and Henriksen, A. (1990) Mapping critical loads. A guidance to the criteria, calculations, data collection and mapping of critical loads. Nordic Council of Ministers, NORD Report 1990: 98, 123 pp.

Sverdrup, H., de Vries, W., Hornung, M., Cresser, M., Langan, S., Reynolds, B., Skeffington, R. and Robertson, W. (1995) Modification of the Simple Mass Balance equation for calculation of critical loads of acidity. In: Hornung, M., Sutton, M. A. and Wilson, R. B. (eds) *Mapping and Modelling of Critical Loads for Nitrogen — A Workshop Report*. ITE, Bush Estate, Penicuik, pp. 55–62.

Ulrich, B. (1983) An ecosystem oriented hypothesis on the effect of air pollution on forest ecosystems. In: *Ecological Effects of Acid Deposition*. SNV Report PM 1636, pp. 221–231.

Wilander, A. (1994) Estimation of background sulphate concentrations in natural surface waters in Sweden. *Water, Air, and Soil Pollution*, **75**, 371–387.

Wright, R. F. and van Breemen, N. (1995). The NITREX project: an introduction. *Forest Ecology and Management*, **71**, 1–5.

# 24 Summary and Conclusions from the Gårdsjön Covered Catchment Experiment—and What Remains to be Discovered

R. A. SKEFFINGTON & H. HULTBERG

This chapter aims to provide a summary and overview of the whole Project. Unanswered questions and possibilities for further work are then enumerated.

## INTRODUCTION

It is not easy to obtain an overview of a large and complex project such as the Gårdsjön Covered Catchment Experiment. This chapter attempts to summarize the individual chapters from the point of view of the aims of the experiment, and tries to synthesize data from several chapters where they address the same questions.

## EXPERIMENTAL DESIGN

The design of the Roof Experiment is very simple in concept, but in practice some difficult choices and compromises have to be made. These arise from the fact that it is impossible within a normal research budget to isolate a whole ecosystem the size of the Roof Catchment from its surroundings and return it to the imperfectly known conditions of 250 years ago, while maintaining the integrity in space and time of natural processes which are themselves not well understood. By thinking through the compromises and the reasoning behind them, it is possible to learn a lot about ecosystems and how they function. The chapters by Moldan *et al.* (Chapter 4) and Beier *et al.* (Chapter 22) respectively contain good discussions of the reasoning behind the experimental design of the Roof Catchment, and the problems of roofed catchments in general. A few of the less obvious problems are outlined below.

The structures necessary for the experiment are one source of problems, and much effort has gone into making their effects on light, temperature, water distribution and chemical conditions as small as possible (Moldan *et al.*, Chapter 4). Having the roof under the forest canopy avoids strong heating effects and allows dry deposition

*Experimental Reversal of Acid Rain Effects: The Gårdsjön Roof Project.*
Edited by Hans Hultberg and Richard Skeffington. 

to be virtually eliminated. It has disadvantages in that the tree crowns are still exposed to the modern pollution climate, and the roof interrupts the normal cycle of return of nutrients to the forest floor in throughfall. These can be compensated for, as throughfall composition in the pre-treatment period is known, but some of the cations in pre-treatment throughfall are due to cation deposition and some to acid leaching of the foliage. Only the latter should be replaced, but there is considerable uncertainty about the proportions of deposition and crown leaching. Counter ions have to be found for these cations, which either introduces an additional mobile anion into the soil system (if $SO_4^{2-}$, $NO_3^-$ or $Cl^-$ are used) or extra alkalinity if the anion of a weak acid such as bicarbonate is the counter ion. The counter ion chosen was originally $Cl^-$ but after Year 4 some $NO_3^-$ was included since this also would be present in pre-industrial rain and be taken up by trees and micro-organisms and would thus be less mobile. It was decided to replace all the $K^+$, which cycles very tightly, but only the marine part of the $Ca^{2+}$. This is probably leading to a slower recovery than would be seen naturally, as there would have been some non-marine $Ca^{2+}$ under pre-industrial conditions, and crown leaching of $Ca^{2+}$ is losing base cations from the system (Moldan *et al.*, Chapter 4). The roof inevitably prevents dissolved organic carbon, an important source of energy for soil micro-organisms, from entering the forest floor. The composition of this is not adequately known, so it cannot be replaced artificially. Though litter is collected and re-spread, it is litter which has been exposed to the modern pollution climate. It would have been better to collect from an unpolluted site elsewhere, but the logistics of dealing with the several tonnes involved would have been very difficult. Finally, the experiment differs from classical experimental design in being unreplicated and without the roofed control which some experiments have (see Beier *et al.*, Chapter 22). This is partly because of the expense, but also because the diversity of catchments is such that genuine controls are not possible.

Notwithstanding all this, the care taken to minimize the effects of these factors, and the strength of the experimental treatments, means that the major clear-cut results can be taken as a good simulation of what would happen in the real situation. The experiment has been designed to be slightly conservative in that recovery in the real situation would be expected to be slightly quicker. There is no reason to believe, however, that the main features seen in the Gårdsjön Roof Experiment will not be seen under ambient conditions.

## ATMOSPHERIC DEPOSITION

### Major elements

Deposition of major elements at the Gårdsjön research site is described in this book by Ferm and Hultberg (Chapter 3). The proximity of Gårdsjön to the Swedish west coast gives deposition a strong marine component. The position of the site in relation to typical weather patterns means it is exposed to a considerable amount of deposition of anthropogenic pollutants derived from the industrialized areas of Europe. Moderately high rainfall, averaging 1140 mm annually since 1979, enhances wet deposition, while the physiography of the area, with hilly terrain covered in fairly mature coniferous forest, tends to enhance dry deposition. Deposition of $SO_4^{2-}$, $NO_3^-$ and $NH_4^+$, while not as high as in Central Europe, is much higher than in unpolluted regions further north in Scandinavia. This moderately high acid deposition, coupled with the sensitivity of the terrain (see below), makes this a suitable site for an acidification recovery experiment.

The historical development of acid deposition in the Gårdsjön area is of interest both for assessing the degree of acidification in relation to other areas and for evaluating the results of the current experiment. Reductions in $SO_2$ emissions in Europe have led to a pronounced decline in ambient sulphur (and acid) deposition during the course of the experiment. EMEP Model calculations (see Barrett and Seland, 1995) estimate that 24% of the S deposition on the Gårdsjön area in 1985–93 originated in Germany, with the next highest contributors the United Kingdom (13%), Sweden (11%) and Poland (11%). Political changes in Eastern Europe since 1989 have allowed a sharp reduction in $SO_2$ emissions. Emissions of $SO_2$ from Germany fell by 2.3 Mt (37%) between 1989 and 1993, and those from Poland by 1.2 Mt (30%) over the same period (Barrett and Seland, 1995). Emissions from the UK and Sweden fell by 0.5 Mt (14%) and 0.06 Mt (37%) during the same time. It is not surprising therefore that ambient S deposition to the Gårdsjön area has declined by about 40% during the course of the experiment (Ferm and Hultberg, Chapter 3). This has had the effect of turning the control catchment (F1) into another acid reduction experiment, though with a lower rate of reduction than the Roof Catchment (G1). Nitrogen oxide emissions have not declined to the same extent (10% Germany; 23% Poland; 14% UK; 1% Sweden) so deposition of oxidized N at Gårdsjön has not declined so markedly. Base cation deposition is heavily influenced by sea-salt deposition and is variable year-to-year: it has not obviously declined although there is a long-term trend towards lower deposition in the area (Hedin *et al.*, 1994). The net effect of these changes is that the pH of wet deposition rose from 4.30 in 1980/81 to 4.38 in 1991/93, and the pH of throughfall rose from 3.97 to 4.17 over the same period (Ferm and Hultberg, Chapter 3).

It is also interesting to consider what current deposition is in relation to historical deposition. Ferm and Hultberg (Chapter 3) estimate on the basis of emission data that non-marine S deposition in 1993 was similar to that around 1900 at about 25 meq $m^{-2}$ $yr^{-1}$. Barrett and Seland (1995), using a more sophisticated model for an area slightly further north, on a line from Bergen to Stockholm, estimate that current deposition is similar to that in about 1950 at 21 meq $m^{-2}$ $yr^{-1}$, and considerably higher than that at the turn of the century of about 6 meq $m^{-2}$ $yr^{-1}$. The difference between these figures reflects the uncertainties and the effect of spatial averaging over large areas compared with individual sites. All values are greater those measured under the roof, where deposition of non-marine $SO_4^{2-}$ was reduced to zero.

## Trace substances

Munthe *et al.* (Chapter 12) describe the deposition of mercury species. Though deposition of mercury and methyl mercury is small in weight terms, of the order of 10–20 g $km^{-2}$ $yr^{-1}$ and 0.2–0.6 g $km^{-2}$ $yr^{-1}$ respectively, the ecological effects of the mercury once deposited are profound because of its toxicity and tendency to accumulate in food chains. Deposition is comparable with other areas where long-range transport is the major component, and comprises both wet and dry components, and litterfall where much of the mercury is thought to have originally been dry-deposited. The study described by Brorström-Lundén and Löfgren (Chapter 13) of deposition of persistent organic pollutants (POPs) is at an early stage, but sampling of needles under and above the roof is tending to show the importance of particulate deposition of POPs to forest ecosystems.

## THE ROOF CATCHMENT

### Description

For putting the results of the Roof Project into context it is very important to assess how representative of acidified areas are the Gårdsjön catchments, and especially the Roof Catchment. The Gårdsjön area was described by Andersson and Olsson (1985), and the three research catchments G1 ROOF, G2 NITREX and F1 CONTROL are described in detail in the present book by Andersson *et al.* (Chapter 2). Only a brief synopsis is thus given here.

The Gårdsjön area is characterized by moderate rainfall of around 1150 mm, well distributed through the year. The rainfall regime is quite variable, and the experimental period has been characterized by some unusually dry summers. The climate is maritime, and thus buffered against extremes, with mean monthly temperatures ranging from 16 °C in July to −2.7 °C in February. The Gårdsjön catchments are underlain by igneous and metamorphic rocks of low weathering rate, mostly granites and gneissic granodiorite. The soils developed on these materials are thin and podzolic on slopes and peaty in the valley bottoms. They have developed from till left after the retreat of the last ice sheet, and have thus been developing for no more than 8000 years. They are acid, with low base saturations and base cation contents except in the organic surface layers. The B horizons of the soils have high concentrations of exchangeable aluminium and inorganic aluminium in the soil solution (Giesler *et al.*, Chapter 6). There is a pool of adsorbed sulphate in the B horizon, and a larger pool of organic sulphur (Torssander and Mörth, Chapter 8). The vegetation of the catchment is mature coniferous forest, with an understorey of acid-tolerant shrubs, grasses and moss species.

Andersson *et al.* (Chapter 2) compare the catchment characteristics with those of other research catchments around the world. But in terms of whole landscapes, the combination of characteristics noted in the above paragraph is very common in Scandinavia and other areas of northern and western Europe. The results from the Gårdsjön Covered Catchment project may however still be of relevance to areas where some of the catchment characteristics deviate from those above. The crucial process which has determined the rate of recovery so far is sulphate desorption from a relatively large pool of sulphur in the soil. Most acidic podzolic soils in areas exposed to S deposition could be expected to be in a similar state, and thus to behave in the way the G1 ROOF catchment has done.

## Hydrology

The importance of an understanding of hydrology in evaluating the Roof Project is recognized by the inclusion of two chapters — on runoff generation by Bishop *et al.* (Chapter 5) and on modelling by Hauhs *et al.* (Chapter 21). Water has a dual importance in acting both as a reagent in the reactions which affect recovery and as the medium which transports the products of the soil reactions out of the catchment. As the rate and timing of precipitation can be controlled by the roof and sprinkler system, the Roof Catchment can be turned into a laboratory for the study of hydrological reactions, and this has enabled a lot of information about runoff processes within the catchment to be built up.

The pattern of runoff in the Roof Catchment is in many ways typical of the hydrology of till soils which are widespread in Scandinavia (Bishop *et al.*, Chapter 5). The soil moisture was similar to natural conditions in the catchments outside, in spite of the switch to sprinkler irrigation. The catchment responds quickly to precipitation inputs, largely by the "transmissivity feedback" mechanism in which precipitation inputs saturate progressively more superficial layers of the soil which exhibit a strong increase in hydraulic conductivity towards the soil surface. Most of the runoff takes place in the top 30 cm of the soils. Flows increase when the catchment has approximately 190 mm of water in storage, the variation over a 3 year period being from 140 to 260 mm. Comparison of these amounts of storage with precipitation inputs gives an impression of the turnover time of water in the catchment — calculated residence times varied from 5 to 192 days. The topography of the catchment leads to the valley bottom transmitting much larger amounts of water than the side slopes, and this appears to have created some unusual soils there, having B horizons of high porosity and organic matter content.

The difficulties encountered in the attempts to model the hydrological responses of the catchment reflect some unsolved problems of contemporary hydrology (Hauhs *et al.*, Chapter 21). The switch from natural rainwater to deionized lakewater created a step shift in the $\delta^{18}O$ concentration of the water input which could be used as a tracer. It was 7 months before the $\delta^{18}O$ content of the runoff water matched that of the input. During this time 544 mm of water was added to the catchment, and 163 mm left in runoff, of which 60 mm was "old" water derived from before the roof was built according to a mixing model. A further 126 mm of "old" water left the catchment in transpiration. The mixing model gave a good simulation of the pattern of $^{18}O$ response, implying that under transient flow conditions the whole of the soil water is involved in the mixing process. This is not however compatible with hydrological tracer experiments done under steady-state flow conditions, which imply a large number of soil water pools with differing reaction times, as do measurements of soil solution concentrations (Giesler *et al.*, Chapter 6). Application of three more elaborate models did not derive a consistent representation of internal hydrological dynamics (Hauhs *et al.*, Chapter 21). The roof could potentially continue to be used as a tool to explore the reasons for some of these difficulties and advance the understanding of hydrology in general.

The important questions about how hydrological processes interact with soil chemistry in influencing recovery remain unexplored as yet. As it appears that there are flowpaths which transmit greater volumes of water than the bulk of the catchment soils, do the soils along these paths determine the catchment response? In that case catchments might react more rapidly than predicted by acidification models which treat the soil as a continuously stirred tank reactor. Or do the soils not on flowpaths still influence runoff concentrations? Another aspect not yet investigated is whether acid episodes due to increases in precipitation input are ameliorated more quickly than might have been predicted from the properties of the catchment soils as a whole, because of the importance of certain flowpaths during these episodes. The next phase of the Roof Project should try to address some of these questions.

## BIOGEOCHEMISTRY

Understanding the biogeochemistry of the reversal of acidification has been a major part of the Project. The output fluxes and concentrations from the G1 ROOF catchment are discussed in detail in Hultberg *et al.* (Chapter 7). Deposition of non-marine $SO_4^{2-}$ and dissolved inorganic N to the catchment was virtually completely eliminated by the roof. Over the next 5 years, water quality steadily improved, driven by reducing $SO_4^{2-}$ concentrations. The acid neutralizing capacity (ANC) of the output water grew steadily less negative, and the flux of inorganic aluminium was reduced considerably. The concentrations of base cations decreased, however, but not sufficiently to offset the improvement in aluminium concentrations from the point of view of fish toxicity. This picture is complicated by the fact that ambient sea-salt deposition has declined markedly during the experiment. Sulphur deposition to the F1 CONTROL catchment also declined, leading to a lesser degree of recovery. The concentrations and fluxes of $SO_4^{2-}$ and Al from G1 remain high, however, even after 5 years. Predictions of the time course of recovery in the future depend on understanding what is happening inside the catchment. In the next sections we attempt to synthesize the data from several chapters to address this question.

### Mass balances

On the reduction of S deposition by construction of the roof, the catchment reacted by desorbing S from its stores. Over the first 5 years of the experiment, the G1 catchment lost 327 meq m$^{-2}$ of S, generating acidity equivalent to 1000 mm of rain at pH 4.18 over the same period. The store, moreover, is much larger than this flux, comprising 11 500 meq m$^{-2}$ of total S, which includes 1410 meq m$^{-2}$ of adsorbed $SO_4^{2-}$ (Torssander and Mörth, Chapter 8). The catchment thus contains enough adsorbed $SO_4^{2-}$ to maintain the present rate of $SO_4^{2-}$ output for a further 16 years, and enough total S to maintain this rate for 217 years. Prediction of S release is thus the key to prediction of recovery.

### Sulphur

Soil solution sampling by Giesler *et al.* (Chapter 6) showed that $SO_4^{2-}$ concentrations fell substantially in the O and E horizons of the Roof Catchment soils within 2 months of roof construction, but in the $B_s$ horizon concentrations fell more gradually, being between 20–50% of pre-treatment levels after 2.5 years. Sulphate in the output water paralleled the reactions of the $B_s$ horizon rather than the O and E horizons. Torssander and Mörth (Chapter 8) conducted an innovative study of sulphur dynamics which exploited the fact that sprinkler water, in which the $SO_4^{2-}$ is derived from seawater, has a different isotopic signature ($\delta^{34}S$ value) from inorganic S in the catchment, and anthropogenic S in acid deposition. The $\delta^{34}S$ value of runoff water reflected the $\delta^{34}S$ value of the inorganic S pool in the catchment at the start of the experiment, and in 2 years it hardly increased at all towards the value characteristic of the sprinkled water. This implies that the adsorbed pool of $SO_4^{2-}$ controls output concentrations, and that incoming $SO_4^{2-}$ exchanges readily and possibly reversibly with it. The isotopic composition of the adsorbed pool indicates that it is largely derived from atmospheric deposition, whereas the composition of the much larger organic S pool indicates it is largely derived from sea salt, presumably over a much longer period. Oxygen isotope analysis shows that at most one third of the $SO_4^{2-}$ transfer occurs through the organic S pool of the A horizon. This is fortunate because, as noted above, the organic S pool is very large and if mineralization of this was an important mechanism of net S release, recovery would be delayed for centuries. Even the $SO_4^{2-}$ pool, if entirely reversibly adsorbed, could on release delay recovery for decades. Gobran *et al.* (Chapter 9), however, suggest on the basis of laboratory experiments that 40% of the total $SO_4^{2-}$ is retained irreversibly. They point out the interaction between $SO_4^{2-}$ adsorbtion/desorption and dissolved organic carbon and pH, and suggest this needs to be taken into account in predicting the fate of the catchment stores of inorganic S. Gobran *et al.* (Chapter 9) nevertheless agree with the other authors that the store of inorganic S in the podzolic "B" horizon will dominate recovery processes, and that desorption is likely to take some decades.

**Base cations**

Base cations have declined in the soil solution in parallel with a reduction in output concentration (Geisler *et al.*, Chapter 6). They attribute this to three causes: decline in the concentration of the mobile $SO_4^{2-}$ anion; decline in sea salt inputs and decline in input of recycled base cations as discussed above. Though the decline in $SO_4^{2-}$ was not large relative to the total ionic strength of the soil solution, the source of the base cations is largely the upper soil horizons, where the decline in sulphate has been greatest.

**Acidic cations**

The pH has changed more slowly than the other ions, in keeping with previous results, but there are signs after 3 years in the soil solution and 4 years in the output that it is starting to rise. Most of the reduction in acidity of the output water has been expressed as declining aluminium concentrations, both in the soil solution and in the output (Hultberg *et al.*, Chapter 7). There was no clear relationship between $H^+$ and Al in the soil solution which might indicate control by an $Al(OH)_3$ phase, an assumption of most acidification models. Likewise there was no consistent change in Ca/Al ratio, which has been used as a criterion value in critical load calculations (see

Andersson *et al.*, Chapter 23). The sharp decline in Al output has been one of the most encouraging features of the results, and contrasts with the very slight reduction observed in output from the organic soils of the RAIN Project (see Wright *et al.*, 1993, and discussion by Skeffington *et al.* in Chapter 1).

### Acidification modelling

Modelling is the only way to predict the future course of acidification over Europe, to test the effects of different scenarios, and to generalize the results of the study geographically to somewhat different ecosystems. Two of the leading acidification models, MAGIC (Moldan *et al.*, Chapter 19) and SAFE (Sverdrup *et al.*, Chapter 20) were applied to the results from the catchment. Both models could be fitted to the observed data and reproduced some major features of the catchment response. SAFE, however, does not have a routine for $SO_4^{2-}$ desorption, and treats $SO_4^{2-}$ as a conservative ion like chloride—it therefore overpredicts the rate of $SO_4^{2-}$ decline. Since $SO_4^{2-}$ desorption is the major controlling process in catchment response (see above), it is rather disconcerting that the model can be fitted to the data so well. This illustrates the difficulties of validating complex models. MAGIC gives a better fit to the $SO_4^{2-}$ data as it contains a desorption routine based on a Langmuir isotherm. However, Gobran *et al.* (Chapter 9) suggest that models need to include the possibilities of irreversible adsorption and interactions with DOC and pH to give realistic predictions. Predictions into the future differ in detail between the two models, reflecting a number of model assumptions. SAFE clearly needs to incorporate $SO_4^{2-}$ desorption as a process if it is to give realistic predictions of the reactions of podzolic soils.

### Trace metals

Lyvén *et al.* (Chapter 11) compared the concentrations of a number of trace metals in the G1 ROOF and F1 CONTROL catchments over 3.5 years. Concentrations of cobalt, manganese, vanadium and zinc declined significantly in G1 relative to F1 over this period, and arsenic, cadmium and copper also declined. Lyvén *et al.* (Chapter 11) attributed this to increasing pH rather than reduced deposition. These results indicate that reduction of deposition may have a fairly rapid effect on heavy metal availability.

Munthe *et al.* (Chapter 12) showed that mercury and methyl mercury in runoff declined relative to the reference catchment initially, but that this difference became insignificant in subsequent years. In view of the high temporal variability of mercury outputs, more monitoring will be needed before this part of the Project yields results.

## BIOTIC PROCESSES

The complexity of the biologically mediated processes has led to their relative neglect in recovery studies, but they are of course of vital importance. Indeed, the deleterious effects of acidification on living organisms are the reason for devoting resources to reducing acid deposition and studying recovery. Soil and water chemistry are investigated so we can predict the effect on living organisms. As the preceding sections show, the effects on the chemistry of the soils and waters are multifarious, and this implies that it is not possible to attribute any observed effects on the biota unequivocally to changes in one aspect of chemistry or another. Independent evidence does often suggest what the important causal factors might be, however. The most important findings at Gårdsjön of the effect of reduction of acid deposition on living organisms are summarized below.

**Composition and activity of the soil microflora**

Soil micro-organisms play a vital part in nutrient cycling, especially in the decomposition of organic matter and consequent release of immobilized nutrients, and also conversely in immobilization of nutrients in short supply. It was postulated long ago that acidification might inhibit these vital processes in forest soils, but it has turned out that the effects of acidification as such on decomposition rates of forest soils are hard to detect under realistic conditions (Cronan, 1985; Berdén *et al.*, 1987). Much more significant is the effect of inputs of nitrogen (e.g. Wright and Tietema, 1995). It has, however, been postulated that removal of acid stress may induce an increased mineralization of soil organic matter, which could result in increased leaching of organic acids, and mobilization of mercury, methyl mercury, other heavy metals transported in organic complexes, or nutrients like phosphorus. The increase in ionized organic acid concentrations observed in the RAIN Project in Norway after removal of incoming acidity (Wright *et al.*, 1993) may partly have been due to increased decomposition. In the present book, however, Hogervorst *et al.* (Chapter 14) demonstrate that there was no change in decomposition under the Roof after the first year's treatment in both the L- and F-layers. There was a significant reduction in bacterial numbers and hyphal length, however, an effect Hogervorst *et al.* (Chapter 14) attribute to the reduction in nitrogen inputs. They support this contention with reference to other studies in which changes in nitrogen inputs were the only variable, while recognizing that there could be other causes. One of these is increased activity by the soil microfauna. The diversity of soil microarthropods was shown to increase in another study in which S and N inputs were reduced by means of a roof (Boxman *et al.*, 1995). This highlights the need for a more comprehensive study of changes in populations of soil organisms under the roof, as soil fauna have also a vital role in decomposition and nutrient cycling. Another clear need is for a re-assessment of decomposition rates now a much longer period has passed without the soil of the G1 ROOF catchment being subject to acid deposition.

Mycorrhizal fungi are another vital component of forest ecosystems. One of the earliest observable effects of the roof at Gårdsjön was a striking increase in fruiting body production, almost six times the production in the control catchment in 1994 (Brandrud, Chapter 15). Even so, production is still less than that observed in unpolluted sites. The increase in fruiting body production was again attributed to reduced nitrogen deposition, as the adjacent NITREX catchment, which is subjected to added nitrogen, showed a decrease in production (Brandrud, Chapter 15). There were however no changes in the density and vitality of mycorrhizal root tips in the Roof Catchment, which remained similar to those in the Control Catchment at least by 1993. It is possible to regard fruit body production as a stress response to lack of available N, and even though it is a response characteristic of natural forests it may not be a good indicator of the health of the mycorrhizae underground. The rapid recovery in fruit body production observed at Gårdsjön was not seen in more heavily polluted sites where N deposition was decreased (e.g. Boxman *et al.*, 1995), though the trend is similar. It seems likely that the fungal flora of the Dutch and German sites, which have been subjected to larger amount of N deposition for a longer time than the Swedish sites, will take longer to re-establish normal fruit body production.

**Tree health and nutrition**

Intimately related to the health of the mycorrhizal fungi are the fine roots of the trees. Here there were only slight signs of any improvement in condition after 3 years of experimental treatment (Clemensson-Lindell and Persson, Chapter 16). There was towards the end of this period a small increase in the percentage of living fine roots in the Roof Catchment, but the large amount of temporal and spatial variability in fine-root condition makes it difficult to attribute this unequivocally to the reduction in deposition.

Standard measurements of tree health (crown density and discoloration) have been made every year in the G1 ROOF, G2 NITREX and F1 CONTROL catchments (Stuanes *et al.*, Chapter 17). Crown density has declined in parallel in all three catchments, and thus the improvement in soil conditions (Giesler *et al.*, Chapter 6) in the Roof Catchment has had no effect on crown density. Using the official classification (UNECE, 1993), the trees on the experimental catchments would be classified as moderately defoliated (Class 2) and hence "damaged". Attempts to model this response using SAFE (Sverdrup *et al.*, Chapter 20) attribute the increase in defoliation to drought (1992 and 1994 had extremely dry summers), which the observed decline in soil solution aluminium concentration was not strong enough to influence. Crown discoloration has also increased over the experimental period in all treatments (Stuanes *et al.*, Chapter 17), from slightly to moderately discoloured according to the UNECE classification. In 1994 the trees in the Roof Catchment were slightly less discoloured than those in the control and NITREX catchments, though Stuanes *et al.* (Chapter 17) attribute this to a slightly delayed watering in the Roof Catchment alleviating the subsequent summer drought. Nutrient concentrations in needles mirror the crown discoloration patterns. Radial growth of the trees has been low and unaffected by treatment. Overall there has been no obvious improvement in tree health and nutrition due to the reduction in acid deposition in the first 4 years of the experiment.

**Fish**

Bioassays with brown trout (*Salmo trutta* L.) showed that runoff from the Roof, NITREX and Control catchments was rapidly lethal (Carlsson and Kroglund, Chapter 18). There were no treatment-related differences. This was an early study (after 1 year's roof treatment) aimed at establishing a baseline and appropriate methodology. It would be worth repeating now water quality is closer to that expected to support fish.

**Conclusions**

The overall picture of recovery presented by the study of biotic responses is not one of rapid change. This is so even though the composition of the soil solution has changed quite sharply. There is little evidence here that biological recovery will be evident as soon as acid deposition is reduced, except that reductions in nitrogen deposition may increase mushroom production in moderately polluted areas. The effects of reduction of nitrogen deposition appear to be much stronger than those of a reduction in sulphur deposition. This is not too surprising in view of the importance of N in biological processes, and the fact that the elimination of N deposition under the roof has returned the catchment to its natural state as a nitrogen-deficient system. Most of the studies of biotic processes need to be repeated now the experiment is more advanced. This is not only because effects may be more pronounced, but to enable real effects to be distinguished from variability due to environmental conditions. The experiment has coincided with some unusually dry summers and warm winters.

Many important groups of organisms and biological processes remain uninvestigated. For example, soil animals are very important in nutrient cycling; root pathogens have been postulated to interact with pollution levels; aquatic biota other than brown trout are affected by acidification. While the experimental treatments continue, the Gårdsjön Roof and the adjacent NITREX and Control catchments are a resource which can be used to study these neglected interactions.

## CRITICAL LOADS AND POLICY ISSUES

The large amount of data collected about the Gårdsjön catchments, and the nature of the experiment itself, makes it a suitable place for testing the concept of a critical load which plays a large part in negotiations about pollutant reductions. Andersson *et al.* (Chapter 23) calculated critical loads for Gårdsjön using a number of official methods. The "true" critical load for acid deposition at Gårdsjön is probably between $0.1\,keq\,ha^{-1}\,yr^{-1}$ (the long-term historical deposition) and $0.44\,keq\,ha^{-1}\,yr^{-1}$ (the deposition at which acidification was first observed in Lake Gårdsjön). The lower limits of the calculated critical loads mostly fall within this range, though a few are lower, but the upper limits are often higher. This provides some justification for making precautionary assumptions about the default parameters being used to calculate critical loads. Deposition was reduced to about $0.05\,keq\,ha^{-1}\,yr^{-1}$ in the experiment, less than most of the calculated critical loads, and the catchment duly began to recover. However, recovery in itself does not demonstrate that the critical load is not exceeded: the catchment needs to reach the appropriate criterion value (e.g. an ANC of $+20\,\mu eq\,L^{-1}$) and the Roof Catchment experiment is still well short of doing so. This illustrates one of the difficulties of validating critical loads: they are intended to indicate the deposition which is sustainable in the long term, whereas all the experimental evidence is obtained in the short term.

The policy issues which flow from the experiment are addressed in the final chapter, by Hultberg *et al.* (Chapter 25). The chapter is structured around 21 policy-relevant questions which it attempts to answer using the best evidence available from the Roof and elsewhere. The general picture that emerges is one of a relatively rapid initial improvement in water quality provided emission reductions are large enough, but restoration to a pristine state will require deeper reductions and take much longer. Soil quality and forest health are unlikely to react quickly. This chapter ought to be of interest to those responsible for formulating policy on acidification, and it provides one of the few discussions of the influence of time scales of recovery on policy.

## CONCLUSIONS

It is worth while at this point considering the original aims of the Project and whether they have been fulfilled. They are listed below (from Skeffington *et al.*, Chapter 1) and brief answers are given:

*Aim:* to assess the rate and direction of response of a forested catchment to a reduction in sulphate, nitrate and mercury deposition.

*Specific objectives:* to answer the following major questions:

1. Does reducing acid input to this forest lead to an improvement in the quality of runoff water? Will the output of acidity and toxic ions such as aluminium be reduced, and if so by how much? Will there be an initial deterioration in water quality, as predicted by some models?

Reducing acid inputs has led to an improvement in runoff water quality. The output of acidity and toxic ions (aluminium and some heavy metals) has been reduced significantly. There was no initial deterioration in water quality.

2. Will the runoff water be suitable for fish if acid input from the atmosphere is reduced? Or will the forest itself still produce toxic water?

The runoff is probably not yet suitable for fish, and the forest is still producing toxic water. However, water quality is improving and on a landscape scale the results imply that fish status would have improved.

3. How *fast* does any improvement in water quality take place? What implications does this have for the rate at which FGD and $NO_x$ reduction have to be fitted to power stations?

Some improvement takes place immediately, but 5 years of more than 95% reduction in acid deposition has left the catchment far from complete recovery. Predictions are that this will take decades at least. The rate at which emission control has to be fitted is a political question in which the benefits of faster action have to be balanced against the costs. The results of the Project need to be generalized onto a wider scale to allow policymakers to answer this question.

4. Do existing models of soil and water acidification and deacidification fit the response of the catchment? If not, how do they need to be modified?

Two leading acidification models, MAGIC and SAFE, have been fitted to the data. SAFE clearly requires a modification to allow it to model sulphate desorption. MAGIC apparently gives a good simulation, though only time will tell how well it predicts the next few years' data.

5. Does a reduction in atmospheric mercury input lead to a reduction of mercury outputs, and can this be quantified?

This question cannot so far be answered, as a clear signal has not yet appeared over a considerable amount of noise.

## WHAT REMAINS TO BE DISCOVERED?

The conclusions above show that it is possible to give partial answers to most of the questions posed at the start of the Project. However, more work is clearly needed on many questions which are listed below.

1. Do the patterns and trends observed so far persist if the long-term biogeochemical measurements are continued? These include: inputs and outputs of major ions and mercury; tree health in terms of crown condition and discoloration; trace metals; and persistent organic pollutants.
2. Do soils along hydrological flowpaths, rather than the bulk soil, determine catchment response?
3. Do acid episodes become less severe as treatment continues?
4. Can hydrological models of the catchment response be reconciled with the data?
5. Does recovery continue if sea-salt concentrations increase above their presently low levels?
6. What is the effect of meteorological fluctuations in temperature, rainfall, snow, frost, etc.?
7. Will the present rate of release of $SO_4^{2-}$ from the soils continue?
8. What is the best way to model $SO_4^{2-}$ release: do factors like DOC and pH need to be taken into account?
9. Does the mineralization of the organic S pool contribute significantly to $SO_4^{2-}$ release under deacidifying conditions?
10. Will the biogeochemical results obtained in future years fit the model predictions which already exist (adjusting for environmental fluctuations if necessary)?
11. Are there now any differences in decomposition rate between the experimental and control catchments and have there been any developments over time?
12. Are other aspects of the soil flora and fauna affected by experimental treatment?
13. Have any differences in fine-root health and distribution developed?
14. Has mycorrhizal species composition, infection, or fruiting body production changed?
15. Is the catchment output water now less toxic to fish?
16. Is the catchment output water toxic to other aquatic organisms?

The Gårdsjön Covered Catchment Project presents an opportunity to study many aspects of the biogeochemistry and ecology of recovering ecosystems. It is to be hoped that this opportunity can be exploited while treatments continue.

## REFERENCES

Andersson, F. and Olsson, B. (eds) (1985) *Lake Gårdsjön, an Acid Forest Lake and its Catchment. Ecological Bulletins (Stockholm)*, **37**.

Barrett, K. and Seland, Ø. (1995) *European Transboundary Acidifying Air Pollution. Ten Years Calculated Fields and Budgets to the End of the First Sulphur Protocol.* EMEP/MSC-W Report 1/95, Norwegian Meteorological Institute, Oslo, Norway.

Berdén, M., Nilsson, I., Rosen, K. and Tyler, G. (1987) *Soil Acidification—Extent, Causes and Consequences.* SNV Report 3292, Swedish Environmental Protection Board, Solna, Sweden.

Boxman, A. W., Van Dam, D., Van Dijk, H. F. G., Hogervorst, R. F. and Koopmans, C. J. (1995) Ecosystem responses to reduced nitrogen and sulphur inputs into two coniferous forest stands in The Netherlands. *Forest Ecology and Management*, **71**, 7–29.

Cronan, C. S. (1985) Comparative effects of precipitation acidity on three forest soils: carbon cycling responses. *Plant and Soil*, **88**, 101–112.

Hedin, L. O., Granat, L., Likens, G. E., Buishand, T. A., Galloway, J. N., Butler, T. J. and Rodhe, H. (1994) Steep declines in atmospheric base cations in regions of Europe and North America. *Nature*, **367**, 351–354.

UNECE (1993) *Forest Condition in Europe: 1993 Report.* United Nations Economic Commission for Europe, Geneva, and Commission of the European Communities, Brussels.

Wright, R. F. and Tietema, A. (eds) (1995) NITREX Special Issue. *Forest Ecology and Management*, **71**.

Wright, R. F., Lotse, E. and Semb, A. (1993) RAIN Project: results after 8 years of experimentally reduced acid deposition to a whole catchment. *Canadian Journal of Fisheries and Aquatic Sciences*, **50**, 258–268.

# 25 Reversal of Acidification: Policy Implications Derived from the Roof Project

H. HULTBERG, Å. IVERFELDT, B. I. ANDERSSON, P. GRENNFELT, & R. A. SKEFFINGTON

This concluding chapter attempts to draw out the implications of the Gårdsjön Roof Project for emission control policy. The results of a scientific experiment like the Gårdsjön Project are not enough on their own to determine policy, and we have tried in this chapter to answer the question "What does the Project tell us about the extent and rate of recovery of waters, soils and forests when acid deposition is reduced?". The chapter is structured around 21 policy-relevant questions which it attempts to answer using the best evidence available from the Roof Project and elsewhere. The general picture that emerges is one of a relatively rapid initial improvement in water quality provided emission reductions are large enough, but restoration to a pristine state will require deeper reductions and take much longer. Soil quality and forest health are unlikely to react quickly. Some uncertainties and major outstanding questions are also discussed.

## INTRODUCTION

In this chapter, we consider the implications for control of acidifying emissions which can be deduced from the results of the Gårdsjön Roof Project so far. The Roof Project is a scientific experiment, and scientific knowledge is not sufficient on its own to *determine* what an emission control policy should be. Policymakers have many other aspects to consider—essentially they have to weigh the benefits of emission reduction against the costs. The position at which the balance is struck depends on the many factors which influence the perception of the benefits and the ability to pay the costs. The factors influencing the perception of the benefits include information on which aspects of the environment would be protected or restored by the policy, how quickly this is likely to happen, and the cultural and economic value which is placed on the natural resources in question. These values may go beyond a purely economic calculus of the value of natural resources and may include such wider environmental benefits as a desire for unpolluted ecosystems, increased biodiversity, and other environmental desires as epitomized by the Rio Declarations. This book is concerned with acidification of the natural environment, but there may

*Experimental Reversal of Acid Rain Effects: The Gårdsjön Roof Project.*
Edited by Hans Hultberg and Richard Skeffington. 

of course be other benefits to reducing acidifying emissions: improvements in human health; reduced weathering of cultural artefacts; lessened direct impacts of the gases, for instance. Factors affecting the ability to pay the costs include other spending priorities; the ability of the overall economy to sustain the expenditure (e.g. the governments of western Europe are more likely to be able to afford large expenditures than those from the East); interactions with macroeconomic policy (e.g. there was resistance to expenditure on S control from British power stations because it would have increased the UK Government's borrowing requirement when the industry was publicly owned); concerns about international competitiveness; pressure from powerful economic interests; and institutional resistances.

The transboundary nature of acidifying pollution adds complexity to this picture: the mix of pressures is different for different countries, and, even more problematic, the distribution of costs and benefits falls unevenly so that expenditure in one country may largely give rise to benefits in others. A detailed discussion of how these pressures led to the negotiation in 1988 of the European Community's Large Combustion Plant Directive can be found in Boehmer-Christiansen and Skea (1991). More recently, European countries have agreed to reduce S emissions further by signing the second Sulphur Protocol to the UNECE Convention on Long-range Transboundary Air Pollution in Oslo in 1994. Once again, the various ecosystem sensitivities and deposition loads across Europe as well as the differing political pressures on countries led to substantial differences in the degree and rate of emission reductions which were agreed in Oslo.

In this chapter, therefore, we will be concerned with the first two aspects of the policy balance: what the results from the Roof Project tell us about the restoration of acidified ecosystems; and how quickly restoration will occur. Strictly the results apply only to the Roof Catchment itself, but we can use our knowledge of ecosystem function and the results from similar experiments to extrapolate more widely. These extrapolations naturally become less certain the further removed the system conditions are from those of this particular area of southern Sweden. Knowledge of reversibility of acidification from other studies is reviewed in other chapters of this book, particularly in Chapters 1 and 24, and these should be consulted for comprehensive references to the literature.

The chapter is structured around a set of questions which might be asked by policymakers. Each question is answered succinctly, and then the scientific reasoning which supports the answer is given in outline, together in most cases with some amplification and qualification.

## POLICY-RELATED QUESTIONS AND ANSWERS: RECOVERY OF LAKES

### 1. Will sulphur emission reductions lead to recovery of acidified lakes to a status where sustainable natural fish reproduction can occur?

*Answer:* Yes . . .

Provided:

1. Emission reductions are sufficiently large.

2. Enough time is allowed for recovery.
3. The rate of weathering of catchment soils is sufficient to sustain an adequate water quality.
4. Catchments are managed sustainably.

These statements are not in themselves very helpful as they need to be quantified, and the reasoning behind them and how they derive from the results of the Roof Project is not explicit. Some more specific questions are needed to elucidate this.

**2. Do the Roof Project results so far show that the quality of runoff water improves when acid deposition is eliminated?**

*Answer:* Yes.

After 5 years, the flux of toxic Al ions from the Gårdsjön Roof Catchment is less than half that in the two pre-treatment years. The pH of runoff may have increased very slightly, and export of acid neutralizing capacity has also increased. On the debit side, the flux and concentration of base cations have decreased, but not sufficiently to prevent the overall changes in water quality from being favourable for fish survival and reproduction. These improvements in water quality are reflected to a lesser extent in output from the Control Catchment, where ambient sulphur deposition has declined to some extent due to emission control programmes already implemented.

**3. Is the output water from the Roof Catchment likely to be suitable for fish, now or in the near future?**

*Answer:* No.

Although it has not yet been tested experimentally, knowledge of the minimum water quality necessary for salmonid fish indicates that runoff water from the Roof Catchment is after 5 years still too acidic and contains too much aluminium for fish survival and reproduction. Predictions by the acidification models SAFE and MAGIC show that this will continue to be the case until at least the years 2000 or 2007 respectively, and probably for well beyond that. However, the models are not so well developed that these predictions can be relied on absolutely. Species of fish other than salmonids ought also to be considered.

**4. Does this imply that there will be no improvement in fish status in areas such as southern Sweden for a very long time, even if emissions are reduced drastically?**

*Answer:* No.

Typically the suitability of water for fish increases as it passes from headwater catchments, such as the Roof Catchment, into larger streams and lakes. This is for two main reasons: a higher proportion of the water in these systems is drawn from deeper groundwaters which typically have a larger buffering capacity than the water in the shallow runoff from headwater catchments; and because some processes in lakes, especially sulphate and nitrate reduction, have the capacity to destroy acidity. Crudely, water in larger streams and lakes can be looked on as a mixture of waters from these sources. Therefore, if the output of acidity from headwater catchments is reduced, the water quality further downstream will improve. At some point, water which is now not suitable for fish will cross the threshold and become suitable again. What we cannot do at present is predict how extensive this improvement would be on a geographical basis. What is also unclear is whether the headwaters would still produce acidic episodes, which may determine fish distribution in the lower reaches of streams even if average water quality is suitable for fish.

**5. The Roof Project reduced acid deposition by 95%. Will there be positive effects with smaller reductions?**

*Answer:* It depends: there will be positive effects if further soil acidification is prevented.

Any reduction in S deposition is likely to produce positive effects in that the output flux and concentration of aluminium and acidity will tend to be reduced as shown by the Roof Project. If, however, the remaining acid deposition is high enough for soil acidification to continue, this tendency will be offset, completely or partially. For instance, although sulphur emissions in Europe were reduced by 18% from 1980 to 1986, very few lakes in Sweden and Norway showed signs of recovery, and some continued to acidify. A major reason for this was the continuing long-term depletion of base cations in soils by acid deposition (another was the increasing influence of N deposition). After 1990, however, with further reductions, there have been some signs of recovery (e.g. Skjelkvåle and Henriksen, 1995). Thus, depending on the amount of S reduction, it is possible to envisage several scenarios:

1. If the amount of S reduction is enough to prevent further soil acidification, water quality will improve permanently.
2. If the amount of S reduction is not enough to prevent further soil acidification, but is enough for the positive effects of a reduced flux of S into surface waters to outweigh this, then water quality will improve. Eventually, however, it will start to deteriorate again.
3. If the amount of S reduction is enough only to slow the rate of soil acidification to the point where the positive effects of a reduced flux of S into surface waters are offset, then water quality will not improve initially, and will eventually deteriorate.
4. If the amount of S reduction is less than in (3) above, then water quality will deteriorate from the beginning.

**6. So what factors, apart from the rate of S reduction, determine which of these recovery scenarios apply?**

*Answer:* Mainly those which affect soil acidification.

These are:

1. the absolute rate of acid (S and N) deposition;
2. the rate of mineral weathering;
3. the rate of base cation deposition from the atmosphere;
4. land management practices which can affect soil acidification (e.g. forestry).

This is the reason that weathering rate and catchment management appear as qualifiers in the answer to Question 1. It is possible that for some catchments the rates of weathering and atmospheric deposition are not enough to sustain an adequate base cation concentration in surface waters, and these will therefore never recover. This is only likely to apply in very low weathering areas with high rainfall, however. Current deposition and weathering rates calculated for the Roof Catchment, for instance, will maintain a calcium concentration of $1.2\,mg\,L^{-1}$, which is low, but adequate for salmonid fish. Any land use which causes a drain on the soil base cation reserves will tend to reduce the base cation concentration in surface waters, and with it the degree of recovery from acidification.

**7. Does the value of the critical load predict whether surface waters in an area will recover from acidification?**

*Answer:* It depends on which critical load is selected.

Chapter 23 of this book applies at least eight different "official" methods of calculating the critical load for acid deposition to the Roof Catchment. It is clear that the definition of recovery (implicit or explicit) which is used in calculation of the critical load must be considered very carefully. Of the methods designed for water, most (i.e. the Steady-State Water Chemistry Method; the Mass Balance Methods; the First-order Acid Balance Method and the PROFILE Method) adopt some criterion for recovery such as acid neutralizing capacity (ANC) of $20\,\mu eq\,L^{-1}$. This may not represent the historical water chemistry very well: it will be too high in slow-weathering areas and too low elsewhere. These methods are thus useful for calculating relative risk, but do not calculate recovery to pristine conditions. The "Diatom Method" (Battarbee *et al.*, 1996) should be free from this limitation, though it suffers from being empirically based, so far on UK data only. The *soil* critical loads proposed at the Skokloster Workshop (Nilsson and Grennfelt, 1988) should, if accurate, give a good estimate of whether the waters of an area will recover, according to the logic in Question 5 above. The other methods for calculating soil critical loads are not suitable for predicting recovery of lakes, as they normally allow the leaching of large quantities of acidity into surface waters. None of the critical load calculations give any information about the *time-scale* of recovery: only the final equilibrium state.

**8. Can the results from the Roof Project be used to validate any of the critical load methodologies?**

*Answer:* Yes, within limits.

The "true" critical load for acid deposition at Gårdsjön is probably between $0.1\,keq\,ha^{-1}\,yr^{-1}$ (the long-term historical deposition) and $0.44\,keq\,ha^{-1}\,yr^{-1}$ (the deposition at which acidification was first observed in Lake Gårdsjön). The lower limits of the calculated critical loads mostly fall within this range, though a few are lower, but the upper limits are often higher. This provides some justification for making precautionary assumptions about the default parameters being used to calculate critical loads. Deposition was reduced to about 0.05 keq $ha^{-1}$ $yr^{-1}$, less than most of the calculated critical loads, and the catchment duly began to recover. However, recovery in itself does not demonstrate that the critical load is not exceeded: the catchment needs to reach the appropriate criterion value (e.g. an ANC of $+20\,\mu$eq $L^{-1}$) and the Roof Catchment experiment is still well short of doing so. This illustrates one of the difficulties of validating critical loads: they are intended to indicate the deposition which is sustainable in the long term, whereas all the experimental evidence is obtained in the short term. Chapter 23 contains a more extensive discussion of these points.

**9. Were there any negative effects of the reduced sulphur deposition?**

*Answer:* Very few.

The long-term acidification of soils has resulted in unnatural changes such as extensive losses of base cations, reduced mineralization of organic matter and increased accumulation of carbon, sulphur, nitrogen, inorganic and methyl mercury. The recovery may therefore induce an increased mineralization of soil organic matter, which could result in increased leaching of organic acids, as in the RAIN Project in Norway. This would buffer pH at a lower level than it would otherwise be, and increase the colour of the water. Studies of fish mortality in runoff from the RAIN Project showed a lower fish survival in water from the catchment with reduced sulphur input, presumably as a result of decreased Ca concentration in combination with low pH and a high Al concentration. Also as a result of increased soil microbial activity, leaching of inorganic mercury, methyl mercury and other metals co-transported with the organic matter in runoff may increase, and these toxics will accumulate in biota in lakes and streams.

Fortunately, organic acid concentrations have not increased in the runoff from the Roof Catchment at Gårdsjön during the first 5 years, and the potential deleterious effects associated with this have thus not occurred. The only deleterious effect has been a reduction in base cation concentrations, and as noted above this has not been enough to offset the improvement in aluminium and hydrogen ion concentrations.

**10. Will the improvements in chemical conditions in runoff water observed during the Roof Project lead automatically to biological improvements?**

*Answer:* Not necessarily.

Recovery in fish and invertebrate reproduction, and population density is often remarkably fast if the change to a better pH situation shows a steady positive trend. However, species that have become locally extinct do not generally recover, or recover only over a long time. Introduction by man may not always be successful.

Some undesirable changes may occur as a result of water quality improvements. Organic matter, which has accumulated in lake sediments during acid conditions, may be decomposed at a higher rate when pH increases, resulting in excessive amounts of micronutrients like P and $CO_2$. During the recovery of acidic lakes, blooms of filamentous algae are therefore likely to occur at pH values of 5.0 to 5.5, when $CO_2$ still is the main carbon source. Under specific situations other aquatic plants may contribute to this "weed" problem in lakes, with heavy growth of submerged *Juncus bulbosus* var. *fluitans* on organic bottoms at depths down to 2 m. These problems may be temporary, but they will need monitoring and management.

**11. What does the Roof Project tell us about the speed of recovery from acidification of surface waters?**

*Answer:* There is likely to be an immediate partial response, but complete recovery may take decades or centuries.

Recovery in the form of reduced sulphate concentrations in runoff water was visible within a few months in response to the virtually complete elimination of acid deposition by the roof. Within 2 years (the delay being partly due to masking by changes in sea-salt inputs) this was translated into reduced inorganic aluminium concentrations and increased acid neutralizing capacity. All these parameters progressively improved over the 5 years of treatment, but over the last 2 years of the experiment, sulphate output concentrations may perhaps have stabilized at around 55% of their pre-roof values. This means that the catchment is releasing some of the acidity which it had stored during the period it was subject to acidic deposition. There is enough adsorbed sulphate stored in the catchment to maintain this rate of release for about 16 years. However, there is a much larger pool of organic sulphur—if this were all to be mineralized to sulphate as well, the catchment could maintain this output rate for 217 years. Clearly, not all this sulphur is in fact available for release, and the controls on sulphate production are one major uncertainty which continuation of the Project could be expected to resolve. These figures, however, do illustrate the time-scales which might be involved. The results from Gårdsjön contrast sharply with those from the Norwegian RAIN Project, in which sulphur inputs and outputs were close to balance after 6 years. This is due to the different characteristics of the respective catchments, especially the presence of mineral soils at Gårdsjön and in particular the existence of sulphate-absorbing minerals.

Although sulphate release has turned out to be the major factor limiting the rate of recovery, the underlying rate of soil acidification or recovery from acidification will ultimately limit recovery towards historical conditions. There is no evidence for soil recovery at present (see below) and theoretical considerations suggest that when recovery occurs it will also take decades.

**12. Is the *rate* of recovery dependent on the time-scale of reduction of sulphur deposition?**

*Answer:* Yes.

Recovery to a pristine state will be prolonged while sulphur deposition continues to acidify the soil, and while the soil continues to accumulate reversibly absorbed sulphur. Quantification of these effects (e.g. how many more years will be required to attain a certain state given various sulphur deposition scenarios) requires long-term modelling, and has not yet been attempted.

**13. Is the *amount* of recovery dependent on the time-scale of reduction of sulphur deposition?**

*Answer:* Possibly in places.

This is a more difficult question to answer than Question 12, as it requires the demonstration that acid deposition is causing some *irreversible* effects in catchment soils or waters. Soil acidification in the sense of loss of exchangeable base cations is in principle reversible. Sulphate absorbtion is either reversible or if not reversible is not of ecological significance. It has been suggested that some clay minerals have been irreversibly changed by continued acid deposition, though not in soils related to those at Gårdsjön. Likewise it has been suggested that prolonged acidification will limit phosphorus availability, which in turn will prevent a full recovery of soils and waters. Where the soil mineral weathering rate is very low, natural processes may be unable to restore an acidified soil. These mechanisms all imply that the amount of recovery may be reduced, but there is no evidence from the results of the Roof Project that they are occurring at present.

**14. Are parallel reductions in European nitrogen emissions also necessary before recovery of lakes can begin?**

*Answer:* Yes and no, depending on location.

In some areas there is evidence that lake chemistry has not responded to reductions in sulphur deposition because of the increasing importance of nitrogen. Nitrate concentrations in high elevation lakes in southern Norway increased substantially between 1975 and 1986, counteracting a decrease in sulphate concentrations. Both pH and aluminium concentrations remained the same, and fish status declined markedly. From 1986 onwards, however, nitrate concentrations have not apparently increased in these waters, and some recovery has occurred (Skjelkvåle and Henriksen, 1995) illustrating the complexity of the situation. In most parts of Sweden, however, forest soils, lakes and streams generally have low nitrate concentrations. High nitrate concentrations with low pH and high Al concentrations indicating nitrogen saturation are only found in south Sweden in areas with large local ammonia emission sources. For these Swedish ecosystems, sulphur reduction is most important. This is illustrated by the behaviour of the Roof Catchment, where a leakage of nitrate which was already low was reduced to negligible amounts by the Roof. The adjacent NITREX catchment has shown that addition of extra nitrogen results in a small increase in nitrate leakage, though the amount of this increased over time (Moldan *et al.*, 1995). This, and the other work in the NITREX Project (Wright and Tietema, 1995) has led to the conclusion that some ecosystems are nitrogen saturated and will require reductions in nitrate and ammonia deposition if full recovery is to happen; others are on their way to nitrogen saturation and will require reductions in N emissions eventually, and others are likely to be able to continue to retain nitrogen without ill effects.

**15. Will a reduction in sulphur and mercury emissions in Europe lead to acceptable methyl mercury levels in freshwater fish from a human health perspective?**

*Answer:* Yes and not yet clear.

Atmospheric deposition of mercury to ecosystems in Sweden contains about 1% methyl mercury and about 99% inorganic mercury. More than 95% of the mercury in fish is in the methyl mercury form, which also is the most dangerous form in a human health perspective. Methylation occurs in both the aquatic and the terrestrial ecosystems, so the pathways leading to fish are complex. Given that an improved surface water quality due to reduced sulphur deposition results in a recovery of fish populations and a larger biomass, methyl mercury concentrations in each individual fish should be reduced. The effect of a reduction of mercury deposition on lake methyl mercury concentrations is less easy to predict. In the Roof Catchment, both mercury and methyl mercury output have been reduced, but there are large inter-annual variations and parallel reductions of output in the Control Catchment, so it is not yet clear whether the roof has had any effect. More time is thus needed to answer this question.

## POLICY-RELATED QUESTIONS AND ANSWERS: RECOVERY OF FORESTS AND SOILS

**1. Will a decrease in European sulphur emissions lead to an improvement in the long-term nutrient status of forest soils?**

*Answer:* Yes, provided the decrease is deep enough, but any improvement will be very slow.

The forest soils in acid-sensitive areas of northern Europe have been subject to a long-term depletion of base cations in the upper soil horizons. A recovery can only occur if reductions in sulphur emissions reach levels which allow the alkali generation by atmospheric deposition and weathering to exceed the loss by leaching and biomass uptake. Spruce forests have a shallow root system which restricts redistribution of base cations from weathering in deeper soil horizons. A shift to tree species which have deeper root systems might enhance the recovery rate via more basic throughfall and litterfall. The rate of recovery will, however, be slow even with these species changes in the terrestrial ecosystem. This is because the size of the pool of base cation exchange sites is very large compared with the probable inputs of base cations from weathering and atmospheric deposition. It follows from this that soils with a larger pool of base cation exchange sites, while they are more resistant to acidification, will tend to recover more slowly. Base cation deposition is decreasing substantially: for instance in Sweden, France, Germany and parts of the USA, it has recently been shown that Ca deposition by precipitation has decreased by some 50% over the last 20 years, most likely as a result of emission control of particle emissions from coal burning and industrial processes (Hedin *et al.*, 1994). Mitigation of forest soils by adding base cations may therefore be necessary to maintain a sustainable forest production along with reductions of sulphur emissions. This must be performed carefully and in appropriate places only if it is not to disrupt the ecosystem. The soil sulphur pool may continue to release sulphur and acidity into soil and surface water, even after a large reduction of the deposition to the forest. This may in turn cause a continued loss of base cations from the soils, which will prolong or even prevent recovery altogether. Areas with large stores of nitrogen available for mineralization may have similar problems.

The results from the Roof Project at Gårdsjön illustrate this account well. There has been no measurable change in exchangeable cations. Acidity is being released from the stores of soil sulphur, and the catchment soils continue to lose some base cations in spite of the 95% reduction in acid deposition. Any improvement in nutrient status due to natural processes is calculated to take a long time (e.g. the SAFE Model predicts 60 years for an increase in base saturation in the uppermost soil layer of 10%). Over the past century or so the Roof Catchment soils have accumulated a large acidification debt in the form of stored sulphur. This debt will have to be largely repaid before there can be much progress in improving their nutrient status.

**2. Will there be any shorter-term improvements in soil quality as a result of reduced sulphur and nitrogen deposition?**

*Answer:* Yes.

Although the properties of bulk soil at the Roof Catchment have not changed, the reduction in anion flux has reduced ionic concentrations in the soil solution. This has led in turn to reduced toxic aluminium concentrations and an improvement in the calcium/aluminium ratio. These changes took place quickly, being readily visible within 2 years. There were some indications that soil biology was responding to these chemical changes. There was a marked increase in fungal fruiting body production (i.e. more mushrooms), though this could be regarded as a stress response to reduced nitrogen. Numbers of soil bacteria and fungi declined. After 3 years there were signs that the percentage of living fine roots of the trees had increased. The conclusion appears to be that the organisms and processes that respond to changes in the soil solution may alter quickly, while alterations in those properties which depend on bulk soil (such as nutrient status, decomposition rate) will be slow.

**3. Will decreased European sulphur emissions lead to improved forest vitality and production?**

*Answer:* Not immediately.

Forest declines in Europe are complex phenomena with multiple causes. Crown thinning and crown discoloration are among the main symptoms of Norway spruce decline, and are assessed in annual damage surveys. The Roof results throw some light on the probable effects of reduction of sulphur inputs on these parameters, although no one experiment can be definitive in such a complex situation. Crown thinning has increased slightly in all the Gårdsjön catchments over 5 years. Virtual elimination of sulphur deposition has thus had no effect in this time-scale in spite of the improvements in calcium/aluminium ratios in the soil solution. Crown discoloration was slightly less pronounced under the Roof after 3 years' treatment, though this may perhaps be due to a slightly more favourable water regime in that year. There was no effect on growth. It seems therefore that there will not be a rapid effect on forest vitality and growth from reduced sulphur deposition. In the long term, however, improvements in soil nutrient status should lead to improved vitality, though this may take some decades (see Question 1, above).

**4. Will decreased European nitrogen emissions lead to improved forest vitality and production?**

*Answer:* In some places but not others.

Excessive nitrogen deposition can cause problems in forest nutrition in various ways. High concentrations of $NH_4^+$ in the soil solution may interfere with base cation uptake, leading to low cation/N ratios in foliage. Nitrogen may increase tree growth, putting too great a strain on soil base cation supply rates. If the system is nitrogen-saturated, nitrate leaching may contribute to acidification and cation depletion. There are also various indirect effects, such as promotion of pathogen attack. Of these mechanisms, the reduction of nitrogen inputs will have a rapid effect only on the first. This has been demonstrated in Roof experiments in the Netherlands within the NITREX Project. Otherwise, reduction of nitrogen inputs will cause only a slow response, as it requires changes in large soil pools such as those of exchangeable base cations or organic matter. In areas where nitrogen inputs are still limiting to forest growth, a reduction in nitrogen input will of course cause a reduction in productivity, though not necessarily vitality. This will apply largely to forests in the peripheral areas of Europe. Where forest decline is due to some other cause, such as high ozone concentrations or water stress, a decline in nitrogen deposition will not necessarily have an effect.

## METHODOLOGICAL ISSUES

**1. Are there any outstanding issues which the Roof Experiment might be expected to resolve in the future?**

*Answer:* Yes.

The long response time of many of the processes underlying recovery from acidification implies that the experiment needs to be continued for a few further years in order to have any chance of detecting them. Into this category come changes in soil base status and many of the predicted biotic responses such as changes in tree vitality, nutrient status and health of fine roots. Mercury and methyl mercury outputs are also expected to respond slowly. Many other aspects of forest ecosystem function have not been investigated, although they could be while treatments continue. However, the crucial question which continuation of the experiment should answer is: *how long, and at what concentration, is sulphate going to continue to leach from the catchment*? This is the major process delaying recovery, and potentially it could do so for a very long time in view of the large stores of sulphur in the catchment. On the other hand, it may be that the readily mineralizable sources of sulphate are close to being used up, and the output concentration of sulphate will soon start to fall. Or it may be that only the hydrologically active portion of the catchment needs to be considered, reducing the effective stores of sulphur considerably. Continued monitoring of sulphate output, and associated process studies in the soil, should resolve these issues. The validation of predictive models which allow the extrapolation of results in time and space is also a vital reason to continue the experiment.

**2. Do we have adequate tools to assess the consequences of different sulphur emission reduction scenarios without further experiments?**

*Answer:* No, but they are improving.

The present generation of acidification models cannot yet be used with confidence as prediction instruments for recovery processes without a great degree of uncertainty. The extensive database on trees, soils, soil water, groundwater and surface water from the Forest Roof Project at Gårdsjön has been used by several research groups for model testing and validation. Though the major models could be calibrated to the Gårdsjön data, they predicted rather different futures for the catchment under reduced deposition. Process studies have clarified the reasons for some of these differences. For instance, the SAFE model tends to predict a faster response because it does not contain a mechanism for sulphate release which is the major process delaying recovery at Gårdsjön. The MAGIC model, on the other hand, assumes this process to be reversible, which may be an oversimplification. Though these models are being used as predictive tools, as they are the best available, it is important to understand their limitations and that experimental tests may lead to their modification. In this respect, the Gårdsjön Roof Project should play a leading role.

## CONCLUSIONS

The Gårdsjön Roof Project has over the last few years enabled us to answer the question "What would happen to an acidified forest catchment if acid deposition were to be eliminated?". The results have confirmed some pre-existing views and refuted others, while various uncertainties remain. Recovery starts quickly, but its rate soon becomes limited by some rather slow ecosystem processes. It is these very processes which have protected ecosystems from rapid acidification under the influence of acid deposition, and this "acidification debt" must now be repaid. Quite how long this will take is still uncertain.

How these results translate into the wider landscape of Europe is not very clear at present. Soils which can absorb and desorb sulphate (like the soil at Gårdsjön) are very common in acid-affected areas, and might be expected to react in a rather similar way. Some illustrative scenario modelling might clarify the picture for policymakers. But improvements will not be immediate and dramatic as far as water quality is concerned, and still less so for forest health. This contrasts with mechanisms where pollutants affect organisms directly—the effect goes away quickly once the pollutant is removed. The unmanaged soils of Europe have had to cope with increasing amounts of deposition from the atmosphere since the Industrial Revolution began in around 1780. It is likely to take natural processes a similar amount of time to restore the pre-industrial *status quo*—if that is what we want to do. If a faster response is required, direct intervention to restore ecosystems will be necessary. It will be essential to do this with sensitivity.

## REFERENCES

Battarbee, R. W., Allott, T. E. H., Juggins, S., Kreiser, A. M., Curtis, C. and Harriman, R. (1996) Critical loads of acidity to surface waters: an empirical diatom-based palaeolimnological model. *Ambio*, **25**, 366–369.

Boehmer-Christiansen, S. and Skea, J. (1991). *Acid Politics: Environmental and Energy Policies in Britain and Germany*. Bellhaven Press, London, 296 pp.

Hedin, L. O., Granat, L., Likens, G. E., Buishand, T. A., Galloway, J. N., Butler, T. J. and Rodhe, H. (1994). Steep declines in atmospheric base cations in regions of Europe and North America. *Nature*, **367**, 351–354.

Moldan, F., Hultberg, H., Nyström, U. and Wright, R. F. (1995). Nitrogen saturation at Gårdsjön, southwest Sweden, induced by experimental addition of ammonium nitrate. *Forest Ecology and Management*, **71**, 89–98.

Nilsson, J. and Grennfelt, P. (eds) (1988). *Critical Loads for Sulphur and Nitrogen*. Nordic Council Report 1988:15. Nordic Council of Ministers, Copenhagen.

Skjelkvåle, B. L. and Henriksen, A. (1995). Acidification in Norway—status and trends. *Water, Air, and Soil Pollution*, **85**, 629–634.

Wright, R. F. and Tietema, A. (eds) (1995). NITREX Special Issue. *Forest Ecology and Management*, **71**.

# Index

Page numbers in **bold** type refer to figures and page numbers in *italics* refer to tables.